Walter Roth

Integral Representation

De Gruyter Expositions in Mathematics

Volume 74

Walter Roth

Integral Representation

Choquet Theory for Linear Operators on Function Spaces

DE GRUYTER

Mathematics Subject Classification 2020
46A03, 46A20, 46A32, 46A55, 46E40, 46G10

Author
Dr. Walter Roth
House 705, Purok Diamond 2
Tacip Mirafuentes, Tagum City 1800
Davao del Norte
Philippines
walter.roth47@gmail.com

ISBN 978-3-11-131450-1
e-ISBN (PDF) 978-3-11-131547-8
e-ISBN (EPUB) 978-3-11-131558-4
ISSN 0938-6572

Library of Congress Control Number: 2023940890

Bibliographic information published by the Deutsche Nationalbibliothek
The Deutsche Nationalbibliothek lists this publication in the Deutsche Nationalbibliografie; detailed bibliographic data are available on the Internet at http://dnb.dnb.de.

Typesetting: VTeX UAB, Lithuania
Printing and binding: CPI books GmbH, Leck

www.degruyter.com

Introduction

In a groundbreaking paper in 1957 [15], G. Choquet established an integral representation theorem for positive linear functionals on continuous affine functions on a compact convex metric space. Any such functional, so it states, can be represented by a regular measure that is supported by the set of extreme points. This work generated considerable interest and was later generalized to nonmetric compact spaces by E. Bishop and K. de Leeuw in [11] with substantial contributions from P. A. Meyer [17], G. Mokobozki [37] and others. The result was regarded to be profound, and the technology employed to be difficult. It broadly works as follows. Given a subspace G of the Banach space $\mathcal{C}(X)$ of all continuous functions on a compact set X, an order relation is established on the dual of $\mathcal{C}(X)$, that is, the space of all regular Borel measures on X. This order is defined using a suitable subcone of $\mathcal{C}(X)$, and the choice of this subcone guarantees that the set of functionals that dominate a given one is weak* compact in the dual. This assures the existence of maximal elements, which deliver integral representations for positive linear functionals on the subspace G with particular properties. The support of any such maximal representation measure is localized on the Choquet boundary, a subset of X characterized by G. In the case of a compact convex set X and the space of all continuous affine functions for G, the Choquet boundary stands for the set of all extreme points of X. The details and a comprehensive presentation of the theory can, for example, be found in the excellent books by E. M. Alfsen [1] and by R. R. Phelps [41].

There have been several advances to generalize these by now classical concepts (see, e. g., [28] and [59]). The approach to integral representation in this text leans on ideas in earlier publications by the author [47] and a generalization of the techniques therein by Batty [9]. These address Choquet theory for complex- and Banach space-valued functions on a compact Hausdorff space, respectively. A more recent paper [51] contains a classification of spaces of vector-valued functions on a locally compact space and establishes an integral representation for continuous linear functionals. Our take on Choquet theory can be perceived as a generally nonconstructive optimization procedure for linear operators on spaces of vector-valued functions. Given such an operator, an optimized version is found, which preserves certain properties of the original operator and optimizes others. The details of this process depend on a highly flexible set of parameters. We deal with locally compact rather than compact domains X, spaces of vector-valued rather than real- or complex-valued functions and linear operators into a second vector space rather than linear functionals. The transfer of the classical techniques to our situation is far from trivial, and often not possible in a straightforward way.

Since our operators are defined on function spaces, as in the classical version of Choquet theory, their optimized counterparts can be partially characterized by the localization of their representation measures. In order to achieve this objective, we shall introduce a suitable integration theory for vector-valued functions with respect to operator-valued measures. Vector integration was introduced in the early twentieth century, and comprehensive presentations of the subject can, for example, be found in the works of

https://doi.org/10.1515/9783111315478-201

N. Dunford and J. Schwartz [26] and [25], J. Diestel and J. J. Uhl [21], N. Dinculeanu [23] or W. H. Graves [29]. In a paper in 1955, R. G. Bartle [5] formulated the basic setup for the integration of vector-valued functions with respect to vector-valued measures. It requires the utilization of a continuous bilinear form on the concerned vector spaces. Equivalently, it was observed that one might use vector-valued functions and measures whose values are continuous linear operators from the range of these functions into a second vector space. This is the formulation that best suits our intentions, but available models do not provide sufficient generality for our purpose.

The first part of this text is therefore dedicated to the development of a compatible and thoroughly broad approach to integration theory. There are several substantial differences to existing concepts. We use a σ-ring of subsets, which does not necessarily contain the whole space, rather than a σ-field for the domain of our measures. This is crucial since in the vector- or operator-valued case, infinity-type elements are not available. Integrals are first defined over sets in this σ-ring, and then extended via limits to an associated σ-field. We consider functions whose values are elements or, more generally, convex subsets of a vector space E and, given a second vector space F, measures whose values are linear operators from a cone of convex subsets of E into the cone of sublinear functionals on the dual of F. These sublinear functionals can at instances be interpreted to be elements of F, its completion, its second dual or indeed convex subsets of either of these spaces. The same selections are available for the values of integrals with respect to such measures of functions that are vector-valued or convex set-valued. For our approach, we employ some elementary aspects of the theory of locally convex cones, a generalization of locally convex topological vector spaces, which was introduced by the author together with K. Keimel in [33] and [48]. Some of its basic concepts will be presented in Section 1.4. Integrals of set-valued functions have been considered before, notably in an influential paper by R. J. Aumann [3], but they arise naturally in our approach. We also offer an understanding of the modulus or semivariation of a measure as its canonical extension to the cone of convex subsets of E. Our main result in this part establishes an integral representation for continuous linear operators on spaces of vector-valued functions. The properties of the representation measures are closely related to the properties of the respective operators. This representation is then extensively used in the second part of this text, which deals with our expansion of Choquet theory.

This manuscript is organized in the following way. Chapter 1 contains a variety of preparatory subjects. Most importantly, we introduce function space neighborhoods, define and provide examples for function spaces, that is, spaces of continuous vector-valued functions endowed with particular topologies. We outline aspects of the theory of locally convex cones in preparation for the development of measure and integration theory in Chapter 2. There we define operator-valued measures, measurability and integrability of vector- and set-valued functions. The main result is an integral representation theorem for linear operators on function spaces. In Chapter 3, we develop a

Choquet-type representation theory for linear operators on function spaces utilizing the results of the earlier parts.

Reading this text requires familiarity with the general theory of locally convex topological vector spaces, and for the second part, with classical Choquet theory. We shall recall some of the basic facts in Section 1.1 and in the sequel frequently refer to several of the more advanced results, mostly in the well-known books by H. H. Schäfer [56], N. Dunford and J. Schwartz [26], W. Rudin [54], A. Wilansky [62] and A. P. and W. Robertson [45] for the former, and by E. M. Alfsen [1] and R. R. Phelps [41] for the latter.

Contents

1 Function spaces

The first section of this chapter compiles some general prerequisites and frequently used terminology and notation. We do not provide proofs for the cited facts and results, but refer to standard textbooks instead. In Section 1.2, we consider vector-valued functions and introduce our concept of function space neighborhoods, which will be used throughout this text. For our purpose, we adapt a similar notion in [51] and [48] from locally convex cones to locally convex topological vector spaces. This leads to the definition of function spaces, that is, suitable spaces of continuous vector-valued functions on a locally compact space that are endowed with particular topologies. In Section 1.3, we review aspects of continuous linear operators defined on locally convex topological vector spaces, different operator space topologies and notions of compactness in such spaces. Section 1.4 briefly introduces locally convex cones in preparation for the development of operator-valued measures in Chapter 2.

1.1 Prerequisites and notation

Throughout this text, we shall use the following basic facts and notation. We broadly refer to standard textbooks on topology such as [24, 81] and [63], on topological vector spaces such as [45, 56], and [62] and on functional analysis such as [26, 53, 54, 87] and [61], for details, proofs and further explanations.

Locally convex topological vector spaces

A locally convex topological vector space $(E, \mathcal{V})$ is a real or complex vector space E endowed with a basis $\mathcal{V}$ of balanced convex neighborhoods of the origin. We assume that the neighborhood system $\mathcal{V}$ is closed for multiplication by positive scalars and downward directed with respect to set inclusion. We do not require that E is Hausdorff, that is, that $\bigcap_{V \in \mathcal{V}} = \{0\}$. A subset A of E is *convex* if $\alpha a + \beta b \in A$ whenever $a, b \in A$ and $\alpha, \beta \geqslant 0$ such that $\alpha + \beta = 1$, and *balanced* if $\alpha a \in A$ whenever $a \in A$ and $|\alpha| \leqslant 1$. A *subcone* of E is a nonempty convex subset that is closed for multiplication by nonnegative reals. The addition and multiplication by scalars of subsets of E are defined as

$$A + B = \{a + b \mid a \in A,\ b \in B\} \quad \text{and} \quad \alpha A = \{\alpha a \mid a \in A,\ \alpha \geqslant 0\}$$

for $A, B \subset E$ and elements α in the scalar field of E. Sums and scalar multiples of convex or of balanced sets are again convex or balanced. For a subset of E, there are a smallest convex set, a smallest balanced set, a smallest balanced convex set and a smallest subcone of E containing it. These are called its *convex hull, balanced hull, balanced convex hull* and *conic hull*, respectively. The convex and the balanced hull of a subset A of E are given by

https://doi.org/10.1515/9783111315478-001

$$\operatorname{conv}(A) = \left\{ \sum_{i=1}^{n} \alpha_i a_i \;\middle|\; a_i \in A,\ \alpha_i \geqslant 0,\ \sum_{i=1}^{n} \alpha_i = 1 \right\}$$

and $\operatorname{bal}(A) = \bigcup_{|\alpha|\leqslant 1} \alpha A$. Whereas the convex hull of a balanced set is balanced, easy examples show that the balanced hull of a convex set is not necessarily convex. The balanced convex hull $\operatorname{bconv}(A)$ of a set $A \subset E$ is therefore the convex hull of $\operatorname{bal}(A)$. The conic hull $\operatorname{cone}(A)$ of a set $A \subset E$ is the convex hull of $\bigcup_{\alpha\geqslant 0} \alpha(A \cup \{0\})$.

The *interior* $\mathring{A}$ of a subset A of E is the largest open subset of A. The *closure* $\overline{A}$ of A is the smallest closed subset of E containing A, and is given by

$$\overline{A} = \bigcap_{V \in \mathcal{V}} (A + V).$$

The interior and the closure of a convex or a balanced set are again convex or balanced. The closure of a subcone is again a subcone of E. If $A \subset E$ is a convex neighborhood of $0 \in E$, that is, if $V \subset A$ for some $V \in \mathcal{V}$, then its closure is given by $\overline{A} = \bigcap_{\gamma>1} \gamma A$.

For subsets A, B of E, we say that A *absorbs* B if there is $\alpha \geqslant 0$ such that $B \subset \beta A$ for all $|\beta| \geqslant \alpha$. A set $A \subset E$ is *absorbing* if it absorbs every finite subset of E. All neighborhoods $V \in \mathcal{V}$ are absorbing. A set $A \subset E$ is *bounded* if it is absorbed by all neighborhoods in $\mathcal{V}$, that is, if for every $V \in \mathcal{V}$ there is $\lambda \geqslant 0$ such that $A \subset \lambda V$.

Linear operators

If both $(E, \mathcal{V})$ and $(F, \mathcal{W})$ are locally convex topological vector spaces over the same scalar field $\mathbb{R}$ or $\mathbb{C}$, a *linear operator* from E into F is a mapping $T : E \to F$ such that

$$T(a + b) = T(a) + T(b) \quad \text{and} \quad T(\alpha a) = \alpha T(a)$$

for all $a, b \in E$ and $\alpha \in \mathbb{R}$ or $\alpha \in \mathbb{C}$. Such an operator is continuous if for every neighborhood $W \in \mathcal{W}$ there is $V \in \mathcal{V}$ such that $T(V) \subset W$. The linear operators from E into F form a vector space over the same scalar field containing the continuous linear operators as a subspace.

The dual space

The *algebraic dual* $E^{\circledast}$ of E is the space of all linear operators from E into its scalar field, that is, the space of all continuous real- or complex-valued *linear functionals* on E. The *topological dual* E^* is the subspace of $E^{\circledast}$ consisting of all continuous linear functional on E. In the complex case, both $E_{\mathbb{R}}^{\circledast}$ and $E_{\mathbb{R}}^*$ denote the corresponding spaces of all real-valued real-linear, and all continuous real-valued real-linear functionals on E, that is, the real parts of the functionals in $E^{\circledast}$ and in E^*, respectively. The *polar* A°

of a subset A of E is usually taken in E^* and consists of all linear functionals $\mu \in E^*$ such that $\mathfrak{Re}\,\mu(a) \leqslant 1$ for all $a \in A$. The *weak topology* $\sigma(E, E^*)$ on E is generated by the neighborhoods

$$V_{\{\mu_1,\dots,\mu_n\}} = \{a \in E \mid \mathfrak{Re}\,\mu_i(a) \leqslant 1 \text{ for all } i = 1,\dots,n\}$$

corresponding to finitely elements $\mu_1,\dots,\mu_n \in F^*$. The weak topology is coarser that the given one, but creates the same dual space, the same closure of convex sets and the same notion of boundedness for subsets of E. It is Hausdorff if and only if E is Hausdorff. The weak* topology on E^* is the topology $\sigma(E^*, E)$. It is generated by the neighborhoods

$$U_{\{a_1,\dots,a_n\}} = \{\mu \in E^* \mid \mathfrak{Re}\,\mu(a_i) \leqslant 1 \text{ for all } i = 1,\dots,n\}$$

corresponding to finitely elements $a_1,\dots,a_n \in E$ and is always Hausdorff. The dual of E^* in the weak topology is the quotient space E/E_0, where $E_0 = \bigcap_{V \in \mathcal{V}} V$. In case that E is Hausdorff, we have $E_0 = \{0\}$, hence $E/E_0 = E$. In our settings, we will assume that the vector spaces involved are either all real or all complex. Correspondingly, all linear operators are real- or complex-linear, and all linear functionals are $\mathbb{R}$- or $\mathbb{C}$-valued.

Polar topologies

A *polar topology* is induced by a dual pair (E, F) of vector spaces together with a real- or complex-valued bilinear form $\langle\, ,\, \rangle$ on $E \times F$. The bilinear form is required to be *nondegenerate*, that is,

$$\langle a, c \rangle = 0 \quad \text{for } a \in E \text{ and all } c \in F$$

implies that $a = 0$, and likewise,

$$\langle a, c \rangle = 0 \quad \text{for } c \in F \text{ and all } a \subset E$$

implies that $c = 0$. Every element of F defines a linear functional on E, and in this way F corresponds to a subspace of $E^{\circledast}$. The *polar* in F of a subset A of E is the convex set

$$A^\circ = \{c \in F \mid \mathfrak{Re}\langle a, c \rangle \leqslant 1 \text{ for all } a \in A\}.$$

The polar in E of a subset of F is similarly defined. The polar of a set and the polar of its convex hull coincide. A subset A of E is *weakly bounded* if $\sup\{\mathfrak{Re}\langle a, c \rangle \mid a \in A\} < +\infty$ for all $c \in F$.

A *polar topology* on E or on F is a locally convex topology on one of the spaces, whose neighborhood system is rendered by the polars of a family $\mathfrak{S}$ of weakly bounded subsets of the opposite space with the following properties: (i) $\alpha A \in \mathfrak{S}$ whenever $A \in \mathfrak{S}$ and $\alpha > 0$, (ii) for $A, B \in \mathfrak{S}$, there is $C \in \mathfrak{S}$ such that $A \cup B \subset C$, and (iii) the union of all sets in $\mathfrak{S}$ spans the whole space. The coarsest polar topology on E is the weak topology $\sigma(E,F)$ generated by the family $\mathfrak{S}$ of all finite subsets of F, the finest is the strong topology $\beta(E,F)$ generated by the family $\mathfrak{S}$ of all weakly bounded subsets of F. A polar topology is *consistent with the duality* if it generates the opposite space as its topological dual. The weak topology is consistent, and the finest consistent polar topology on E is the *Mackey topology* $\tau(E,F)$ generated by the family $\mathfrak{S}$ of all weakly compact convex subsets of F (see Section IV.3 in [56]). Every Banach space (more generally, every barreled space) E carries the strong topology $\beta(E,E^*)$. A Hausdorff locally convex topological vector space E is called *semireflexive* if the strong topology $\beta(E^*,E)$ on E^* is consistent with the duality, that is, if the dual of E^* in the strong topology coincides with E. It is *reflexive* if it is semireflexive and carries the strong topology $\beta(E,E^*)$. Closures of convex subsets of E or F and the concept of boundedness coincide in all topologies consistent with the duality (E,F). The second polar (or bipolar) $A^{\circ\circ}$ in E of a subset A of E is the polar of A° in E and coincides with the closed convex hull of $A \cup \{0\}$ (Theorem IV.1.5 in [56]).

Every Hausdorff locally convex topology on a vector space E is a polar topology of the dual pair (E,E^*), endowed with the nondegenerate bilinear form $(a,\mu) \mapsto \mu(a)$. In general, since the elements of E^* vanish on the subspace $E_0 = \bigcap_{V\in\mathcal{V}} V$ and separate elements of E whose difference is not in E_0, the transfer of the bilinear form to $E/E_0 \times E^*$ is nondegenerate, and the preceding results apply to the dual pair $(E/E_0,E^*)$. The projection of E onto E/E_0 is continuous, open and closed. Further details and clarifications can, for example, be found in Chapter IV of [56] or in Chapter 13 of [61]. Dual pairs with degenerate bilinear forms are considered in [60].

σ-rings

Let X be a set. For the sake of generality, we shall formulate our take on integration theory without the imposition of a topological structure on X. Measures suitable for our purposes can generally not be defined on a σ-field containing the whole space X, since this would require such a measure to take infinite values, which are not available in the vector- or operator-valued case. We shall therefore use a *σ-ring* in X, that is, a nonempty family $\mathfrak{R}$ of subsets of X, which contains finite unions, set differences and countable intersections of its elements. These properties imply that $\emptyset \in \mathfrak{R}$, and if the sets $A_i \in \mathfrak{R}$ for $i \in \mathbb{N}$ all are subsets of a set $A \in \mathfrak{R}$, that their union $\bigcup_{i\in\mathbb{N}} A_i$ is also contained in $\mathfrak{R}$. For the latter, one easily verifies that $\bigcup_{i\in\mathbb{N}} A_i = A \setminus \bigcap_{i\in\mathbb{N}}(A \setminus A_i)$. Any σ-field (or σ-algebra) is of course a σ-ring in this sense, and a σ-ring is a σ-field if and only if it contains X. We may, however, associate with $\mathfrak{R}$ in a canonical way the σ-field

$$\mathfrak{A} = \{B \subset X \mid A \cap B \in \mathfrak{R} \text{ for all } A \in \mathfrak{R}\}.$$

We shall also refer to the topological case, that is, the case that X is a locally compact Hausdorff space and that $\mathfrak{R}$ consists of all relatively compact Borel subsets of X. Then the σ-field $\mathfrak{A}$ contains all Borel subsets of X and exactly those if X is countably compact. By $\mathfrak{O}$ we shall denote the subfamily of all open sets in $\mathfrak{R}$. For every compact subset K of X, there exists $O \in \mathfrak{O}$ such that $K \subset O$ (Theorem 2.7 in [55]). For a subset Y of a topological space, the sets $\overline{Y}$ and $\mathring{Y}$ denote its topological closure, and interior, respectively, whereas $\partial Y = \overline{Y} \setminus \mathring{Y}$ stands for its topological boundary. A subset of a topological space is *relatively compact* if its closure in this space is compact.

Vector-valued functions

Let $\mathcal{F}(X, E)$ be the space of all E-valued functions on X, endowed with the pointwise algebraic operations. In case that $E = \mathbb{R}$, we write $\mathcal{F}(X)$ for short. A function $f \in \mathcal{F}(X, E)$ is *supported* by a subset Y of X if $f(x) = 0$ for all $x \in X \setminus Y$. For a scalar-valued function φ and $f \in \mathcal{F}(X, E)$, the function $\varphi \cdot f \in \mathcal{F}(X, E)$ is the mapping

$$x \mapsto \varphi(x)f(x) : X \to E.$$

If f is the constant function $x \mapsto a$ for some $a \in E$, we write $\varphi \cdot a$ for the function $x \mapsto \varphi(x)a$. The *characteristic function* $\chi_Y \in \mathcal{F}(X)$ of a subset Y of X takes the value 1 on Y and 0 else. Following the usual convention, a real-valued function $\varphi \in \mathcal{F}(X)$ is said to be *nonnegative* if $\varphi(x) \geqslant 0$ and *positive* if $\varphi(x) > 0$ for all $x \in X$. In the topological case, that is, if X is a locally compact Hausdorff space, we denote by $\mathcal{C}(X, E)$ the subspace of all continuous functions in $\mathcal{F}(X, E)$. The smallest closed stet in X that supports a function $f \in \mathcal{F}(X, E)$ is called the *support* of f and referred to by $\operatorname{supp}(f)$. The subspace of all functions in $\mathcal{C}(X, E)$ with compact support is denoted by $\mathcal{C}_{\mathcal{K}}(X, E)$. In case that $E = \mathbb{R}$, we write $\mathcal{C}(X)$ and $\mathcal{C}_{\mathcal{K}}(X)$ for short. If $\varphi \in \mathcal{C}(X)$ and $f \in \mathcal{C}(X, E)$, then $\varphi \cdot f \in \mathcal{C}(X, E)$.

Given a compact subset K of X and open sets $O_1 \dots, O_n \subset X$ whose union contains K, a *partition of the unity* (see Proposition 9.4.16 in [53] or Theorem 2.13 in [55]) yields a collection of nonnegative functions $\varphi_1, \dots, \varphi_n \in \mathcal{C}_{\mathcal{K}}(X)$ such that $\operatorname{supp}(\varphi_i) \subset O_i$ and $\sum_{i=1}^{n} \varphi_i(x) = 1$ for all $x \in K$.

Extended number systems

The set $\overline{\mathbb{R}} = \mathbb{R} \cup \{+\infty\}$ denotes the extended real number system with the usual order and algebraic operations. In particular, we designate $a + \infty = +\infty$ for all $a \in \overline{\mathbb{R}}$, $a \cdot (+\infty) = +\infty$ for all $0 < a \leqslant +\infty$ and $0 \cdot (+\infty) = 0$. The multiplication of $+\infty$ with negative numbers is not defined. We shall similarly use $\underline{\mathbb{R}} = \mathbb{R} \cup \{-\infty\}$ and $\overline{\underline{\mathbb{R}}} = \mathbb{R} \cup \{+\infty, -\infty\}$. In $\overline{\underline{\mathbb{R}}}$, the sum of $+\infty$ and $-\infty$ is however not defined. $\overline{\mathbb{R}}_+$ denotes the subcone of all nonnegative numbers in $\overline{\mathbb{R}}$, that is, $\overline{\mathbb{R}}_+ = [0, +\infty]$.

Semicontinuous functions

An $\overline{\mathbb{R}}$-valued function φ on a topological space X is *lower semicontinuous* if for every $\alpha \in \mathbb{R}$ the set $\{x \in X \mid \varphi(x) > \alpha\}$ is open in X, or equivalently, if for every $x \in X$ and $\alpha < \varphi(x)$ there is a neighborhood U of x such that $\varphi(y) > \alpha$ for all $y \in U$. Sums, positive multiples and pointwise suprema of families of lower semicontinuous functions are again lower semicontinuous. An $\underline{\mathbb{R}}$-valued function is *upper semicontinuous* if its negative is lower semicontinuous. A real-valued function is continuous if and only if it is both lower and upper semicontinuous.

Convex sets

Let $\operatorname{Conv}(E)$ be the collection of all nonempty convex subsets of E. If endowed with the addition of sets and the multiplication by nonnegative scalars, the usual associative and distributive properties hold. The singleton set $\{0\} \in \operatorname{Conv}(E)$ serves as the neutral element of the addition. Hence, while not being a subcone of some vector space, $\operatorname{Conv}(E)$ forms a cone in the sense of Section 1.4 below, where we shall also introduce suitable topologies for cones. $\operatorname{Conv}(E)$ is canonically ordered by set inclusion, that is, we write

$$A \leqslant B \quad \text{if} \quad A \subset B$$

for $A, B \in \operatorname{Conv}(E)$. For a nonnegative real-valued function φ on X and $A \in \operatorname{Conv}(E)$, we denote $\varphi \cdot A$ for the function $x \mapsto \varphi(x)A : X \to \operatorname{Conv}(E)$. If $A \geqslant \{0\}$, that is, if $0 \in A$, we allow φ to take the value $+\infty$, where $(+\infty)A$ is meant to be the conic hull of A, that is, $(+\infty)A = \bigcup_{\rho \geqslant 0} \rho A$. Thus, if A is absorbing, then $(+\infty)A = E$. Sums, multiples by nonnegative scalars and the order relation for $\operatorname{Conv}(E)$-valued functions are defined pointwise on X.

There is also the multiplication by all elements of the scalar field of E in $\operatorname{Conv}(E)$, but it does no longer satisfy the distributive law. We have

$$(\alpha + \beta)A \subset \alpha A + \beta A$$

instead. In particular, $(-1)A$ is not necessarily an additive inverse of A. We write $-A$ for $(-1)A$ and $A - B$ for $A + (-B)$. Moreover, $\alpha A \leqslant \alpha B$ holds whenever $A \leqslant B$ for $A, B \in \operatorname{Conv}(E)$ and α in $\mathbb{R}$ or $\mathbb{C}$.

1.2 Spaces of vector-valued functions that vanish at infinity

We shall use the previously introduced basic facts and notation. Let $(E, \mathcal{V})$ be a locally convex topological vector space. The *extended neighborhood system* $\mathcal{V}_c$ of E is the subcone of $\operatorname{Conv}(E)$ generated by $\mathcal{V} \cup \{E\}$. That is,

$$\mathcal{V}_c = \left\{ \sum_{i=1}^{n} V_i \,\middle|\, V_i \in \mathcal{V} \right\} \bigcup \{E, \{0\}\}.$$

Neighborhood functions

Let X be a set. A function $n : X \to \mathcal{V}_c$ is called a *neighborhood function.* Clearly, sums and positive multiples of neighborhood functions are again neighborhood functions.

Function space neighborhoods

A *function space neighborhood* $\mathfrak{v}$ is defined by a convex family $\mathcal{N}_\mathfrak{v}$ of neighborhood functions provided that for every $A \in \mathfrak{R}$ there is $V \in \mathcal{V}$ and $n \in \mathcal{N}_\mathfrak{v}$ such that $V \subset n(x)$ for all $x \in A$. The neighborhood $\mathfrak{v}$ then determines a balanced convex set $[\mathfrak{v}]$ of functions in $\mathcal{F}(X,E)$ by

$$[\mathfrak{v}] = \{f \in \mathcal{F}(X,E) \mid \text{there is } n \in \mathcal{N}_\mathfrak{v} \text{ such that } f(x) \in n(x) \text{ for all } x \in X\}.$$

We shall write $f \leqslant \mathfrak{v}$ if $f \in [\mathfrak{v}]$ for functions $f \in \mathcal{F}(X,E)$. The reference to an order relation in this notation will be justified later on when we shall extend the use of function space neighborhoods to $\mathrm{Conv}(E)$-valued functions. Likewise, will be this less than straightforward definition altogether. Due to the absence of negatives of set-valued functions, function space neighborhoods can no longer be identified with convex subsets of functions in the more general set-valued case (see Section 1.4 below).

We observe the following basic properties.

1.2.1 Lemma. *Let $\mathfrak{v}$ be a function space neighborhood.*
(a) *If $f \leqslant \mathfrak{v}$, then $\varphi \cdot f \leqslant \mathfrak{v}$ for all real- or complex-valued functions φ such that $|\varphi(x)| \leqslant 1$ for all $x \in X$. In particular, $\chi_A \cdot f \leqslant \mathfrak{v}$ for all $A \in \mathfrak{R}$.*
(b) *If $\mathcal{N}_\mathfrak{v}$ is upward directed, then $\varphi \cdot f + (1-\varphi) \cdot g \leqslant \mathfrak{v}$ whenever $f, g \leqslant \mathfrak{v}$ and φ is a real-valued function such that $0 \leqslant \varphi(x) \leqslant 1$ for all $x \in X$.*

Proof. Part (a) is obvious from the definition of a function space neighborhood, since all values of neighborhood functions are convex and balanced subsets of E. For (b), if $\mathcal{N}_v$ is upward directed, then for $f, g \leqslant \mathfrak{v}$ there is $n \in \mathcal{N}_v$ such that both $f(x), g(x) \in n(x)$ for all $x \in X$. □

Sums and positive multiples of function space neighborhoods are defined via the generating sets of neighborhood functions. That is, we set

$$\mathcal{N}_{\mathfrak{v}+\mathfrak{u}} = \{n + m \mid n \in \mathcal{N}_\mathfrak{v},\ m \in \mathcal{N}_\mathfrak{u}\} \quad \text{and} \quad \mathcal{N}_{a\mathfrak{v}} = \{an \mid n \in \mathcal{N}_\mathfrak{v}\}$$

for function space neighborhoods $\mathfrak{v}$ and $\mathfrak{u}$ and $a > 0$. The associative, commutative and distributive rules for these operations transfer from the corresponding ones for set-valued functions. We shall also establish an order relation among function space neighborhoods setting $\mathfrak{v} \leqslant \mathfrak{u}$ if for every $n \in \mathcal{N}_{\mathfrak{v}}$ there is $m \in \mathcal{N}_{\mathfrak{u}}$ such that $n \leqslant m$. Recall that the latter means $n(x) \subset m(x)$ for all $x \in X$. The intersection $\mathfrak{v} \cap \mathfrak{u}$ of $\mathfrak{v}$ and $\mathfrak{u}$ is generated by the neighborhood functions $n \cap m$ for all $n \in \mathcal{N}_{\mathfrak{v}}$ and $m \in \mathcal{N}_{\mathfrak{u}}$, whereby $(n \cap m)(x) = n(x) \cap m(x)$ for all $x \in X$.

Function space neighborhood systems

A family $\mathfrak{V}$ of function space neighborhoods is called a *function space neighborhood system* if it is closed for multiplication by positive scalars and downward directed. That is, for $\mathfrak{v}, \mathfrak{u} \in \mathfrak{V}$ there is $\mathfrak{w} \in \mathfrak{V}$ such that both $\mathfrak{w} \leqslant \mathfrak{v}$ and $\mathfrak{w} \leqslant \mathfrak{u}$. The collection $\mathfrak{V}_*$ of all function space neighborhoods satisfies this condition. Indeed, if $\mathfrak{u}$ and $\mathfrak{v}$ are function space neighborhoods, then for every $A \in \mathfrak{R}$ there are $n \in \mathcal{N}_{\mathfrak{u}}$, $m \in \mathcal{N}_{\mathfrak{v}}$ and $V_A \in \mathcal{V}$ such that both $V_A \subset n(x)$ and $V_A \subset m(x)$ for all $x \in A$. Then the convex hull $\mathcal{N}_{\mathfrak{w}}$ of the collection of the neighborhood functions $\chi_A \cdot V_A$, for all $A \in \mathfrak{R}$, generates a function space neighborhood $\mathfrak{w}$ such that both $\mathfrak{w} \leqslant \mathfrak{u}$ and $\mathfrak{w} \leqslant \mathfrak{v}$.

If both $\mathfrak{V}$ and $\mathfrak{U}$ are function space neighborhood systems, we shall say that $\mathfrak{V}$ is *finer* than $\mathfrak{U}$, or equivalently, that $\mathfrak{U}$ is *coarser* than $\mathfrak{V}$ if for every $\mathfrak{u} \in \mathfrak{U}$ there is $\mathfrak{v} \in \mathfrak{V}$ such that $\mathfrak{v} \leqslant \mathfrak{u}$. $\mathfrak{V}$ and $\mathfrak{U}$ are *equivalent* if $\mathfrak{V}$ is both finer and coarser than $\mathfrak{U}$.

Functions that vanish at infinity

A function $f \in \mathcal{F}(X,E)$ *vanishes at infinity* relative to a function space neighborhood system $\mathfrak{V}$ if for every $\mathfrak{v} \in \mathfrak{V}$ there is $A \in \mathfrak{R}$ such that $\chi_{(X \setminus A)} \cdot f \leqslant \mathfrak{v}$. Clearly, sums and multiples of functions that vanish at infinity share the same property.

For the remainder of this section, we shall consider the topological case, that is, the case that X is a locally compact Hausdorff space and that $\mathfrak{R}$ consists of all relatively compact Borel subsets of X. For continuous E-valued functions on X, we observe the following as an immediate consequence of our buildup.

1.2.2 Lemma. *Let $\mathfrak{v}$ be a function space neighborhood. For every $A \in \mathfrak{R}$ and $f \in C(X,E)$, there is $\lambda \geqslant 0$ such that $\chi_A \cdot f \leqslant \lambda \mathfrak{v}$.*

Proof. For $A \in \mathfrak{R}$ and $\mathfrak{v} \in \mathfrak{V}$, there is $V \in \mathcal{V}$ and $n \in \mathcal{N}_{\mathfrak{v}}$ such that $V \subset n(x)$ for all $x \in A$. If f is continuous, then $f(A)$ is relatively compact, hence bounded in E. Thus there is $\lambda \geqslant 0$ such that $f(A) \subset \lambda V$, that is, $\chi_A \cdot f(x) \in \lambda n(x)$ for all $x \in X$. □

Weakly lower continuous neighborhood functions

A neighborhood function n is said to be *weakly lower continuous* if for every $x \in X$, every $a \in n(x)$ and every $V \in \mathcal{V}$ there is a neighborhood U of x in X such that $a \in n(y) + V$ for all $y \in U$. The suitability of this definition will become apparent in the later stages of this text. Clearly, sums or multiples by positive reals of weakly lower continuous neighborhood functions are again weakly lower continuous.

Similarly, a function space neighborhood $\mathfrak{v}$ is said to be *weakly lower continuous* if all of its defining neighborhood functions $n \in \mathcal{N}_{\mathfrak{v}}$ are weakly lower continuous. A function space neighborhood system $\mathfrak{V}$ is called *weakly lower continuous* if it consists of weakly lower continuous function space neighborhoods.

1.2.3 Lemma. *If φ is a lower semicontinuous $\overline{\mathbb{R}}_+$-valued function and if $V \in \mathcal{V}_c$, then $n = \varphi \cdot V$ is a weakly lower continuous neighborhood function.*

Proof. Let $V \in \mathcal{V}_c$ and φ be as stated and let $n = \varphi \cdot V$. Let $x \in X$, let $a \in n(x)$ and let $W \in \mathcal{V}$. There is $\lambda > 0$ such that $a \in \lambda W$. If either $\varphi(x) = 0$ or $V = \{0\}$, then $a = 0$ and the requirement for weak lower continuity is trivially satisfied. If $V \neq \{0\}$ and $\varphi(x) = +\infty$, then $\varphi(x)V = E$, and we find a neighborhood U of x such that $\varphi(y) \geqslant \lambda$, hence $a \in \varphi(y)V = n(y)$ for all $y \in U$. If $0 < \varphi(x) < +\infty$, then there is a neighborhood U of x such that $(\lambda/(1+\lambda))\varphi(x) \leqslant \varphi(y)$ for all $y \in U$. The latter yields $(\lambda/(1+\lambda))a \in n(y)$, and

$$a = \frac{\lambda}{1+\lambda}a + \frac{1}{1+\lambda}a \in n(y) + \frac{\lambda}{1+\lambda}W \subset n(y) + W.$$

The requirement for the weak lower continuity of n is therefore met. □

In particular, for every open subset O of X and every $V \in \mathcal{V}_c$ the neighborhood function $\chi_O \cdot V$ is weakly lower continuous. So is the neighborhood function $\chi_B \cdot V + \chi_{X \setminus B} \cdot E$ for every closed subset B of X. For the following, recall that $\mathfrak{O}$ denotes the collection of all open sets in $\mathfrak{R}$.

1.2.4 Lemma. *Let $\mathfrak{v}$ be a weakly lower continuous function space neighborhood.*

(a) *If $\chi_A \cdot f \leqslant \mathfrak{v}$ for $A \in \mathfrak{R}$ and $f \in C(X,E)$, then for every $\varepsilon > 0$ there is an open set $O \in \mathfrak{O}$ such that $A \subset O$ and $\chi_O \cdot f \leqslant (1+\varepsilon)\mathfrak{v}$.*

(b) *If $a \in n(x)$ for some $n \in \mathcal{N}_{\mathfrak{v}}$ and $x \in X$, then for every $\varepsilon > 0$ there is a function $f \in C_K(X,E)$ such that $f(x) = a$ and $f \leqslant (1+\varepsilon)\mathfrak{v}$.*

Proof. Suppose that $\mathfrak{v}$ is a weakly lower continuous function space neighborhood. For Part (a), suppose that $\chi_A \cdot f \leqslant \mathfrak{v}$ for $A \in \mathfrak{R}$ and $f \in C(X,E)$. Let $U \in \mathfrak{O}$ be a relatively compact open set containing A. There is a neighborhood function $n \in \mathcal{N}_{\mathfrak{v}}$ and $V \in \mathcal{V}$ such that $V \subset n(x)$ for all $x \in U$, and there is $m \in \mathcal{N}_{\mathfrak{v}}$ such that $(\chi_A \cdot f)(x) \in m(x)$ for all $x \in X$, that is, $f(x) \in m(x)$ for all $x \in A$. Given $\varepsilon > 0$, for every point $x \in A$ there is an open neighborhood $U_x \subset U$ of x such that both $f(y) - f(x) \in \varepsilon V$ and $f(x) \in m(y) + \varepsilon V$ holds for all $y \in U_x$. The latter is due to the weak lower continuity of the neighborhood function m. This renders

$$f(y) \in m(y) + 2\varepsilon V \subset m(y) + 2\varepsilon n(y).$$

As a convex combination of m and n, the neighborhood function

$$s = \frac{1}{1+2\varepsilon} m + \frac{2\varepsilon}{1+2\varepsilon} n$$

is contained in $\mathcal{N}_{\mathfrak{v}}$, and we have

$$f(y) \in (1+2\varepsilon)\, s(y)$$

for all $y \in U_x$. Setting $O = \bigcup_{x \in A} U_x$ we have $f(y) \in (1+2\varepsilon)s(y)$ for all $y \in O$, hence $\chi_O \cdot f \in (1+2\varepsilon)\mathfrak{v}$, and our claim in (a) follows. For Part (b), suppose that $a \in n(x)$ for some neighborhood function $n \in \mathcal{N}_{\mathfrak{v}}$ and $x \in X$. We shall use Part (a) with the constant function $x \mapsto a$ for f and the singleton set $\{x\}$ for $A \in \mathfrak{R}$. Then $\chi_A \cdot a \leqslant \mathfrak{v}$, and for every $\varepsilon > 0$ there is $O \in \mathfrak{O}$ such that $x \in O$ and $\chi_O \cdot a \leqslant (1+\varepsilon)\mathfrak{v}$. Using Urysohn's lemma (see Proposition 15.6 in [63], also Proposition III.4.15 in [53]) we find a nonnegative function $\varphi \in C_{\mathcal{K}}(X)$ such that $\varphi(x) = 1$ and $\varphi \leqslant \chi_O$. Then $f = \varphi \cdot a \in C_{\mathcal{K}}(X, E)$ and $f = \varphi \cdot (\chi_O \cdot a) \leqslant (1+\varepsilon)\mathfrak{v}$ by Lemma 1.2.1(a); hence our claim in (b). □

Lemma 1.2.2 implies that every function in $f \in C(X, E)$, which vanishes at infinity is absorbed by the neighborhoods in $\mathfrak{V}$. Indeed, given $\mathfrak{v} \in \mathfrak{V}$ there is $A \in \mathfrak{R}$ and $\lambda \geqslant 0$ such that $\chi_{(X \setminus A)} \cdot f \leqslant \mathfrak{v}$ and $\chi_A \cdot f \leqslant \lambda\mathfrak{v}$. Thus $f \leqslant (\lambda+1)\mathfrak{v}$.

The function spaces $\mathcal{C}_{\mathfrak{V}}(X, E)$

Our upcoming investigations will focus on spaces of continuous E-valued functions that vanish at infinity relative to a function space neighborhood system $\mathfrak{V}$. This subspace of $C(X, E)$ will be denoted by $\mathcal{C}_{\mathfrak{V}}(X, E)$. The balanced convex and absorbing subsets

$$\langle \mathfrak{v} \rangle = \{f \in \mathcal{C}_{\mathfrak{V}}(X, E) \mid f \leqslant \mathfrak{v}\}$$

of $\mathcal{C}_{\mathfrak{V}}(X, E)$, for all $\mathfrak{v} \in \mathfrak{V}$, form a basis $\langle \mathfrak{V} \rangle$ for a locally convex topology on $\mathcal{C}_{\mathfrak{V}}(X, E)$. We note that $\langle \mathfrak{v} \rangle + \langle \mathfrak{u} \rangle \subset \langle \mathfrak{v} + \mathfrak{u} \rangle$ and $\langle \alpha\mathfrak{v} \rangle = \alpha\langle \mathfrak{v} \rangle$ holds for all $\mathfrak{v}, \mathfrak{u} \in \mathfrak{V}$ and $\alpha > 0$. For the sake of keeping our notation manageable and since the meaning will be clear from the context, we shall avoid the angled brackets and use the same symbols $\mathfrak{v}$ for function space neighborhoods and $\mathfrak{V}$ for the neighborhood system also for the balanced convex neighborhoods and the neighborhood system, which they induce on $\mathcal{C}_{\mathfrak{V}}(X, E)$. In particular, the notions $f \leqslant \mathfrak{v}$ and $f \in \mathfrak{v}$ will be synonymous for functions $f \in \mathcal{C}_{\mathfrak{V}}(X, E)$. In this way, $(\mathcal{C}_{\mathfrak{V}}(X, E), \mathfrak{V})$ is a locally convex topological vector space. We shall abbreviate $\mathcal{C}_{\mathfrak{V}}(X)$ if $E = \mathbb{R}$. Note that for any choice of the neighborhood system $\mathfrak{V}$ all functions with compact support vanish at infinity, hence $C_{\mathcal{K}}(X, E)$ is a subspace of $\mathcal{C}_{\mathfrak{V}}(X, E)$. Since for every $x \in X$, there is $\varphi \in C_{\mathcal{K}}(X)$ such that $\varphi(x) = 1$ and because $\varphi \cdot a \in C_{\mathcal{K}}(X, E)$ for all $a \in E$, the latter implies in particular that $\{f(x) \mid f \in \mathcal{C}_{\mathfrak{V}}(X, E)\} = E$. Moreover, if $f \in \mathcal{C}_{\mathfrak{V}}(X, E)$ and if φ is a bounded continuous real-or complex valued (depending on the scalar field of E) function on X, then $\varphi \cdot f \in \mathcal{C}_{\mathfrak{V}}(X, E)$ according to Lemma 1.2.1(a). We shall make use of the following observations.

1.2.5 Proposition. *Every function $f \in \mathcal{C}_{\mathfrak{V}}(X,E)$ can be approximated with respect to the neighborhoods in $\mathfrak{V}$ by:*

(i) *elementary functions $e = \varphi_1 \cdot a_1 + \cdots + \varphi_n \cdot a_n \in \mathcal{C}_{\mathcal{K}}(X,E)$ for $\varphi_i \in \mathcal{C}_{\mathcal{K}}(X)$ and $a_i \in f(X)$.*

(ii) *step functions $h = \chi_{A_1} \cdot a_1 + \cdots + \chi_{A_n} \cdot a_n \in \mathcal{F}(X,E)$ for $A_i \in \mathfrak{R}$ and $a_i \in f(A_i)$.*

Proof. Let $f \in \mathcal{C}_{\mathfrak{V}}(X,E)$ and $\mathfrak{v} \in \mathfrak{V}$. There is a compact subset K of X such that $\chi_{X\setminus K} \cdot f \leqslant \mathfrak{v}$. We choose an open set $O \in \mathfrak{O}$ containing K. According to the definition of a function space neighborhood, there is $V \in \mathcal{V}$ and $n \in \mathcal{N}_{\mathfrak{v}}$ such that $V \subset n(x)$ for all $x \in O$. There are open sets $O_1, \ldots, O_n \subset O$, which cover K and such that $f(x) - f(y) \in V$ whenever $x, y \in O_i$. We choose $a_i \in f(O_i)$. Now for (i) we select a corresponding partition of the unity (see Section 1.1), that is, a collection of nonnegative functions $\varphi_1, \ldots, \varphi_n \in \mathcal{C}_{\mathcal{K}}(X)$ such that $\operatorname{supp}(\varphi_i) \subset O_i$ and $\varphi(x) = 1$ for all $x \in K$, where $\varphi = \sum_{i=1}^n \varphi_i$. We may also assume that $0 \leqslant \varphi(x) \leqslant 1$ holds for all $x \in X$, because otherwise we can divide each of the functions φ_i by the continuous function $x \mapsto \max\{\varphi(x), 1\}$. Let $e = \sum_{i=1}^n \varphi_i \cdot a_i$. We have $\varphi_i(x)(f(x) - a_i) \leqslant \varphi_i(x)V$ for all $x \in X$ and $i = 1, \ldots, n$, hence

$$\varphi(x)f(x) - e(x) \leqslant \varphi(x)V.$$

This renders $f(x) - e(x) \in V$ for all $x \in K$, hence $\chi_K \cdot (f - e) \leqslant \mathfrak{v}$ by our choice of V. Moreover, since O contains the support of e and since $X \setminus O \subset X \setminus K$, we have

$$\chi_{X\setminus O} \cdot (f - e) = \chi_{X\setminus O} \cdot f \leqslant \mathfrak{v}.$$

For all $x \in O \setminus K$, on the other hand we have

$$\varphi(x)f(x) - e(x) \leqslant \varphi(x)V \subset V$$

by the above, hence $\chi_{O\setminus K} \cdot (\varphi \cdot f - e) \leqslant \mathfrak{v}$ by our choice of the neighborhood $V \in \mathcal{V}$. Furthermore, since $0 \leqslant 1 - \varphi(x) \leqslant 1$ for all $x \in X$ and since $\chi_{O\setminus K} f \leqslant \mathfrak{v}$, Lemma 1.2.1(a) yields that $\chi_{O\setminus K} \cdot (1 - \varphi) \cdot f \leqslant \mathfrak{v}$. Combining the last two observations, we obtain

$$\begin{aligned} \chi_{O\setminus K} \cdot (f - e) &= \chi_{O\setminus K} \cdot (\varphi \cdot f - e) + \chi_{O\setminus K} \cdot (1 - \varphi) \cdot f \\ &\leqslant \mathfrak{v} + \mathfrak{v} = 2\mathfrak{v}. \end{aligned}$$

Finally, summarizing all of the above,

$$\begin{aligned} f - e &= \chi_{X\setminus O} \cdot (f - e) + \chi_{O\setminus K} \cdot (f - e) + \chi_K \cdot (f - e) \\ &\leqslant \mathfrak{v} + 2\mathfrak{v} + \mathfrak{v} = 4\mathfrak{v}. \end{aligned}$$

This proves our claim for the approximation by elementary functions. The argument for step functions is more straightforward. With the same choice for K, V, O, O_i from above, we set $A_1 = O_1$ and $A_i = O_i \setminus \bigcup_{k=1}^{i-1} A_i$ for $i = 2, \ldots, n$. The sets A_i are pairwise disjoint, their union A is a subset of O and contains X. We choose $a_i \in f(A_i)$ and set $h = \sum_{i=1}^{n} \chi_{A_i} \cdot a_i$. Then $f(x) - h(x) \in V$ for all $x \in A$, hence $\chi_A \cdot (f - h) \leqslant \mathfrak{v}$. Since $X \setminus A \subset X \setminus K$, we also have

$$\chi_{X\setminus A} \cdot (f - h) = \chi_{X\setminus A} \cdot f \leqslant \mathfrak{v}.$$

Thus $f - h \in 2\mathfrak{v}$. Our claim follows. □

1.2.6 Proposition.

(a) *If $\mathcal{C}_{\mathfrak{V}}(X,E)$ is Hausdorff, then E is Hausdorff and for every nonempty set $O \in \mathfrak{O}$ and $0 \neq a \in E$ there is $\mathfrak{v} \in \mathfrak{V}$ such that $\chi_O \cdot a \nleqslant \mathfrak{v}$.*

(b) *If for every $x \in X$ and $0 \neq a \in E$, there is $\mathfrak{v} \in \mathfrak{V}$ such that $\chi_{\{x\}} \cdot a \nleqslant \mathfrak{v}$, then $\mathcal{C}_{\mathfrak{V}}(X,E)$ is Hausdorff.*

Proof. For Part (a), suppose that $\mathcal{C}_{\mathfrak{V}}(X,E)$ is Hausdorff. Let $a \in \bigcap_{V\in\mathcal{V}} V$ and $0 \neq \varphi \in \mathcal{C}_{\mathcal{K}}(X)$. For every $\mathfrak{v} \in \mathfrak{V}$, there is $V \in \mathcal{V}$ and $n \in \mathcal{N}_{\mathfrak{v}}$ such that $V \subset n(x)$ for all $x \in \operatorname{supp}(\varphi)$. Hence $\varphi \cdot a \leqslant \mathfrak{v}$. The latter holds true for all $\mathfrak{v} \in \mathfrak{V}$ and implies that $\varphi \cdot a = 0$. Thus $a = 0$, and we infer that E is Hausdorff. For the second part of our claim in (a), suppose to the contrary that there are $\emptyset \neq O \in \mathfrak{O}$ and $0 \neq a \in E$ such that $\chi_O \cdot a \leqslant \mathfrak{v}$ for all $\mathfrak{v} \in \mathfrak{V}$. We choose $0 \neq \varphi \in \mathcal{C}_{\mathcal{K}}(X)$ such that $|\varphi| \leqslant \chi_O$. Then $0 \neq \varphi \cdot a \in \mathcal{C}_{\mathfrak{V}}(X,E)$ and $\varphi \cdot a = \varphi \cdot (\chi_O \cdot a) \leqslant \mathfrak{v}$ for all $\mathfrak{v} \in \mathfrak{V}$ by Lemma 1.2.1(a). Hence $\mathcal{C}_{\mathfrak{V}}(X,E)$ is not Hausdorff, contradicting the assumption of Part (a). For Part (b), let $0 \neq f \in \mathcal{C}_{\mathfrak{V}}(X,E)$. There is $x \in X$ such that $a = f(x) \neq 0$. Under the assumptions of Part (b), there is $\mathfrak{v} \in \mathfrak{V}$ such that $\chi_{\{x\}} \cdot a \nleqslant \mathfrak{v}$. Then $f \nleqslant \mathfrak{v}$, because otherwise we had $\chi_{\{x\}} \cdot f = \chi_{\{x\}} \cdot a \leqslant \mathfrak{v}$ by Lemma 1.2.1(a). We infer that $\mathcal{C}_{\mathfrak{V}}(X,E)$ is Hausdorff. □

We shall list a few examples of function spaces. Our very general approach allows for a wide variety of settings. More details for some of these can be found in [51], Examples 2.2.

1.2.7 Examples. Examples (a) to (d) below deal with cases where the function space neighborhoods are generated by singleton sets $\mathcal{N}_{\mathfrak{v}} = \{n_{\mathfrak{v}}\}$, that is, by a single weakly lower continuous neighborhood function.

(a) The topology of uniform convergence on X is generated by the function space neighborhoods

$$\mathfrak{v}_V = \{f \in \mathcal{F}(X,E) \mid f(x) \in V \text{ for all } x \in X\},$$

rendered by the constant neighborhood function $x \mapsto V$, where $V \in \mathcal{V}$. They give rise to the neighborhood system

$$\mathfrak{V} = \{\mathfrak{v}_V \mid V \in \mathcal{V}\}.$$

The function space $\mathcal{C}_{\mathfrak{V}}(X,E)$ consists of all functions $f \in \mathcal{C}(X,E)$ such that for every $V \in \mathcal{V}$ there is a compact subset K of X and $f(x) \in V$ holds for all $x \in X \setminus K$.

(b) The topology of compact convergence is generated by the function space neighborhoods

$$\mathfrak{v}_{(K,V)} = \{f \in \mathcal{F}(X,E) \mid f(x) \in V \text{ for all } x \in K\},$$

where $V \in \mathcal{V}$ and K is a compact subset of X, rendered by the weakly lower continuous neighborhood function $x \mapsto V$ for $x \in K$ and $x \mapsto E$ else. They give rise to the neighborhood system

$$\mathfrak{V} = \{\mathfrak{v}_{(K,V)} \mid V \in \mathcal{V},\ K \subset X \text{ compact}\}.$$

We have $\mathcal{C}_{\mathfrak{V}}(X,E) = \mathcal{C}(X,E)$ in this case.

(c) The topology of pointwise convergence is generated by the function space neighborhoods

$$\mathfrak{v}_{(Y,V)} = \{f \in \mathcal{F}(X,E) \mid f(x) \in V \text{ for all } x \in Y\},$$

where $V \in \mathcal{V}$ and Y is a finite subset of X, rendered by the weakly lower continuous neighborhood function $x \mapsto V$ for $x \in Y$ and $x \mapsto E$ else. They give rise to the neighborhood system

$$\mathfrak{V} = \{\mathfrak{v}_{(Y,V)} \mid V \in \mathcal{V},\ Y \subset X \text{ finite}\}.$$

We have $\mathcal{C}_{\mathfrak{V}}(X,E) = \mathcal{C}(X,E)$.

(d) Examples (a), (b) and (c) render weakly lower continuous function space neighborhood systems and are special cases for *weighted space topologies*, which were introduced for real-valued functions by Nachbin and Prolla (see [39] and [42]). A family Ω of nonnegative real-valued upper semicontinuous functions on X is called a *family of weights* if for all $\omega_1, \omega_2 \in \Omega$ there are $\omega_3 \in \Omega$ and $\rho > 0$ such that $\omega_1 \leqslant \rho\,\omega_3$ and $\omega_2 \leqslant \rho\,\omega_3$. The functions $\rho_\omega = 1/\omega$ then are $(0,+\infty]$-valued and lower semicontinuous, and for $V \in \mathcal{V}$ and $\omega \in \Omega$ the function space neighborhood

$$\mathfrak{v}_{(\rho_\omega,V)} = \{f \in \mathcal{F}(X,E) \mid f(x) \in \rho_\Omega V \text{ for all } x \in X\},$$

where $V \in \mathcal{V}$, is rendered by the neighborhood function $\rho_\omega \cdot V$. (Following our convention from Section 1.1, we set $(+\infty)\,V = E$.) According to Lemma 1.2.3, this neighborhood function is weakly lower continuous. The neighborhoods of this type give rise to the function space neighborhood system

$$\mathfrak{V} = \{\mathfrak{v}_{(\rho_\omega,V)} \mid V \in \mathcal{V},\ \omega \in \Omega\}$$

and a corresponding function space $\mathcal{C}_{\mathfrak{V}}(X,E)$.

(e) We can modify the preceding examples replacing the singleton sets $\mathcal{N}_v = \{n_{\mathfrak{v}}\}$ by the upward directed sets $\mathcal{N}_{\mathfrak{v}}$ consisting of all convex combinations of the weakly lower continuous neighborhood functions $\chi_O \cdot n_{\mathfrak{v}}$, where $O \in \mathfrak{O}$. Since all involved neighborhood functions have compact support, any function in $\mathcal{C}(X,E)$ that vanishes at infinity must itself have a compact support. Thus $\mathcal{C}_{\mathfrak{V}}(X,E) = \mathcal{C}_{\mathcal{K}}(X,E)$ in this setting.

(f) The finest function space topology for $\mathcal{F}(X,E)$ is created by the neighborhood system $\mathfrak{V}_*$ consisting of all function space neighborhoods. It is straightforward to construct an easily accessible basis for this topology: We consider collections of neighborhood functions $\chi_A \cdot V$, where $A \in \mathfrak{R}$ and $V \in \mathcal{V}$, with the following property: For every $A \in \mathfrak{R}$, there is $B \in \mathfrak{R}$ and $V \in \mathcal{V}$ such that $A \subset B$ and $\chi_B \cdot V$ is contained in the collection. The convex hull $\mathcal{N}$ of such a collection defines a function space neighborhood, and the system $\mathfrak{V}_\blacksquare$ of all function space neighborhoods of this type is closed for multiplication by positive scalars, downward directed and easily seen to be equivalent to $\mathfrak{V}_*$. If in the topological case one considers only collections of neighborhood functions $\chi_O \cdot V$ for open sets $O \in \mathfrak{O}$, these functions are weakly lower continuous and define weakly lower continuous function space neighborhoods. The system $\mathfrak{V}_\tau$ of these neighborhoods is therefore weakly lower continuous, clearly coarser than $\mathfrak{V}_\blacksquare$, but also finer since every set in $\mathfrak{R}$ is contained in some set in $\mathfrak{O}$. Consequently, both systems $\mathfrak{V}_\blacksquare$ and $\mathfrak{V}_\tau$ are equivalent to $\mathfrak{V}_*$. In summary, in the topological case the finest function space topology admits a weakly lower continuous basis. Furthermore, all neighborhood functions involved in the systems $\mathfrak{V}_\blacksquare$ and $\mathfrak{V}_\tau$ are supported by sets in $\mathfrak{R}$. Thus, in the topological case, any function in $\mathcal{C}(X,E)$ that vanishes at infinity relative to $\mathfrak{V}_*$ has a compact support. We infer that $\mathcal{C}_{\mathfrak{V}_\star}(X,E) = \mathcal{C}_{\mathcal{K}}(X,E)$. The finest function space topology is indeed the usual strict inductive limit topology on $\mathcal{C}_{\mathcal{K}}(X,E)$, generated by the family of subspaces, each consisting of those functions with support in a fixed compact subset of X and endowed with the topology of uniform convergence on this subset (see II.6 in [56]).

(g) Let μ be a regular real-valued Borel measure on X, let $1 \leqslant p \leqslant +\infty$ and consider the function space neighborhood $\mathfrak{v}_p$ generated by the collection of all weakly lower continuous neighborhood functions n_ρ, that is, functions $x \mapsto \rho(x)\mathbb{B}$, where $\mathbb{B} = [-1,+1]$ is the unit ball in $E = \mathbb{R}$ and ρ is a nonnegative real-valued lower semicontinuous function such that $\|\rho\|_p \leqslant 1$. Then $\mathfrak{V} = \{\lambda \mathfrak{v}_p \mid \lambda > 0\}$ forms a function space neighborhood system, and $\mathcal{C}_{\mathfrak{V}}(X)$ consists of all continuous functions in $L^p(\mu)$ that vanish at infinity with respect to $\mathfrak{v}_p$. For $p < +\infty$, the latter is guaranteed if either X is σ-compact or if $\mu(X)$ is finite. The neighborhood $\mathfrak{v}_p$ in $\mathcal{C}_{\mathfrak{V}}(X)$ is the unit ball with respect to the L^p-norm.

(h) The case $X = \mathbb{N}$ with the discrete topology covers a variety of sequence spaces.

All function space systems in the preceding Examples 1.2.7 are weakly lower continuous. In Parts (a) to (e), the families $\mathcal{N}_\mathfrak{v}$ of neighborhood functions generating the respective function space neighborhoods are upward directed. Thus Lemma 1.2.1(b) applies. If E is Hausdorff, then according to Proposition 1.2.6(b), examples (a) to (f) render Hausdorff function space topologies for $\mathcal{C}_{\mathfrak{V}}(X,E)$, provided that in (d) for every $x \in X$ there is $\omega \in \Omega$ such that $\omega(x) > 0$.

Several special cases appear worth mentioning.

An *ordered topological vector space* E over $\mathbb{R}$ or $\mathbb{C}$ is a locally convex topological vector space that carries an order relation $\leqslant$ defined by the subcone

$$E_+ = \{a \in E \mid a \geqslant 0\}$$

of its positive elements. That is, $a \leqslant b$ for $a, b \in E$ if and only if $b - a \in E_+$. Note that $0 \in E_+$, hence the element $0 \in E$ is called positive, other than in the conventional notation for real numbers. The positive cone E_+ is supposed to be closed in E (see [56] V.4). We do however not require that $E_+ \cap (-E_+) = \{0\}$. The latter holds true if and only if the order is antisymmetric and implies that the set $\{0\}$ is closed in E and, therefore, that the topology of E is Hausdorff. The positive cone in the dual E^* of E is

$$E^*_+ = \{\mu \in E \mid \mathfrak{Re}\, \mu(a) \geqslant 0 \text{ for all } a \in E_+\},$$

that is, the polar of $-E_+$ in E^*. We have $a \leqslant b$ for $a, b \in E$ if and only if $\mathfrak{Re}\, \mu(a) \leqslant \mathfrak{Re}\, \mu(b)$ for all $\mu \in E^*_+$.

The order of E transfers pointwise to the functions in $\mathcal{C}_{\mathfrak{V}}(X, E)$. If convergence in $\mathcal{C}_{\mathfrak{V}}(X, E)$ implies pointwise convergence, that is, if $\mathfrak{V}$ is finer than the function space topology of pointwise convergence from Example 1.2.7(c), then the subcone of all E_+-valued functions is closed. Otherwise, we use its closure for the positive cone $\mathcal{C}_{\mathfrak{V}}(X, E)_+$ of $\mathcal{C}_{\mathfrak{V}}(X, E)$.

A *topological vector lattice* E is an orderd topological vector space over $\mathbb{R}$ with an antisymmetric order and lattice operations $\vee$ and $\wedge$ satisfying the usual rules (see [99] or [95]). The operator $a \mapsto |a| : E \to E$ is supposed to be continuous. This implies that the lattice operations $a \mapsto a_+ = a \vee 0$, $a \mapsto a_- = (-a) \vee 0$ are continuous from E to E and both $(a, b) \mapsto a \vee b$ and $(a, b) \mapsto a \wedge b$ are continuous from $E \times E$ to E. Antisymmetry of the order implies that E is Hausdorff. The lattice operations in E transfer pointwise to the functions in $\mathcal{F}(X, E)$.

A *topological algebra* E over $\mathbb{R}$ or $\mathbb{C}$ is a locally convex topological vector space with a multiplication satisfying the rules of an algebra, and such that for every fixed element a the operators $b \mapsto ab$ and $b \mapsto ba$ are continuous from E to E. The multiplication transfers pointwise to the functions in $\mathcal{F}(X, E)$. An *involution* is a continuous real-linear operator $a \mapsto a^*$ on E such that $(\alpha a)^* = \bar{\alpha} a^*$, $(ab)^* = b^* a^*$ and $(a^*)^* = a$ for $a, b \in E$ and α in $\mathbb{R}$ or $\mathbb{C}$. Again, this operation transfers pointwise to the functions in $\mathcal{F}(X, E)$.

1.2.8 Proposition. *Let $(\mathcal{C}_{\mathfrak{V}}(X, E), \mathfrak{V})$ be a function space.*

(a) *If E is an ordered topological vector space, then $\mathcal{C}_{\mathfrak{V}}(X, E)$ is also an ordered topological vector space. Its positive cone $\mathcal{C}_{\mathfrak{V}}(X, E)_+$ is the closure of the cone of all E_+-valued functions.*

(b) *If E is a topological vector lattice over $\mathbb{R}$, if the cone of all E_+-valued functions is closed in $\mathcal{C}_{\mathfrak{V}}(X, E)$ and if for every $\mathfrak{v} \in \mathfrak{V}$ there is $\mathfrak{u} \in \mathfrak{V}$ such that $f \leqslant \mathfrak{v}$ whenever $|f| \leqslant |g|$ for some $g \leqslant \mathfrak{u}$, for $f, g \in \mathcal{C}_{\mathfrak{V}}(X, E)$, then $\mathcal{C}_{\mathfrak{V}}(X, E)$ is also a topological vector lattice.*

(c) *If E is a topological algebra over $\mathbb{R}$ or $\mathbb{C}$ and if for every $\mathfrak{v} \in \mathfrak{V}$, there is $\mathfrak{u} \in \mathfrak{V}$ such that $fg \leqslant \mathfrak{v}$ whenever $f, g \leqslant \mathfrak{u}$, for $f, g \in \mathcal{C}_{\mathfrak{V}}(X, E)$, then $\mathcal{C}_{\mathfrak{V}}(X, E)$ is also a topological algebra. If E contains an involution and if for every $\mathfrak{v} \in \mathfrak{V}$ there is $\mathfrak{u} \in \mathfrak{V}$ such that $f^* \leqslant \mathfrak{v}$ whenever $f \in \mathfrak{u}$, then $f \mapsto f^*$ is an involution on $\mathcal{C}_{\mathfrak{V}}(X, E)$.*

Proof. Part (a) was argued before. For Part (b), that is, the case of a topological vector lattice E, we first observe that, due to our assumption that the cone of all E_+-valued functions is closed in $\mathcal{C}_{\mathfrak{V}}(X,E)$, it coincides with the positive $\mathcal{C}_{\mathfrak{V}}(X,E)_+$. Hence the order of $\mathcal{C}_{\mathfrak{V}}(X,E)$ is antisymmetric. Indeed, if both $f \leqslant 0$ and $f \geqslant 0$ holds for a function $f \in \mathcal{C}_{\mathfrak{V}}(X,E)$, then $f(x) \leqslant 0$ and $f(x) \geqslant 0$ for all $x \in X$. Since the order of E is antisymmetric, we infer that $f = 0$. Because of the continuity of the lattice operations in E, the functions $f \vee g$, $f \wedge g$ and $|f|$ are continuous whenever f and g are continuous. Moreover, if $f \in \mathcal{F}(X,E)$ vanishes at infinity relative to $\mathfrak{V}$, then for every $\mathfrak{v} \in \mathfrak{V}$ there is $\mathfrak{u} \in \mathfrak{V}$, as in the condition of (b). Hence there is $A \in \mathfrak{R}$ such that $\chi_{X\setminus A} \cdot |f| \leqslant \mathfrak{v}$, as $\chi_{X\setminus A} \cdot |f| \leqslant \mathfrak{u}$. The function $|f|$ therefore also vanishes at infinity. If both $f,g \in \mathcal{F}(X,E)$ vanish at infinity, since $|f \vee g| \leqslant |f| + |g|$ and $|f \wedge g| \leqslant |f| + |g|$, the same argument shows that both functions $f \vee g$ and $f \wedge g$ also vanish at infinity. Thus $\mathcal{C}_{\mathfrak{V}}(X,E)$ is indeed a lattice, and the condition of (b) implies that the operation $f \mapsto |f|$ is continuous. Likewise, for Part (c) the continuity of the multiplication in E guarantees that the product of two continuous functions is again continuous, and the condition in (c) implies that the product of functions that vanish at infinity also vanishes at infinity. Hence $\mathcal{C}_{\mathfrak{V}}(X,E)$ is an algebra, and the continuity of its multiplication follows in a similar way. If E contains an involution satisfying the condition in (c), then the continuity of the involution operator guarantees that the function f^* is continuous, whenever f is continuous. Given $\mathfrak{v} \in \mathfrak{V}$ and $A \in \mathfrak{R}$, there is $\mathfrak{u} \in \mathfrak{V}$ such that $\chi_{X\setminus A} \cdot f^* \leqslant \mathfrak{v}$ whenever $\chi_{X\setminus A} \cdot f \in \mathfrak{u}$. Thus if $f \in \mathcal{F}(X,E)$ vanishes at infinity relative to $\mathfrak{V}$, so does f^*. The involution is therefore a real-linear operator on $\mathcal{C}_{\mathfrak{V}}(X,E)$ with the required properties. Its continuity follows directly from the condition in (c). □

Suppose that E is indeed a function space $\mathcal{C}_{\mathfrak{U}}(Y,F)$, where Y is another locally compact Hausdorff space, $(F,\mathcal{W})$ is a locally convex topological vector space and $\mathfrak{U}$ is a function space neighborhood system for F-valued functions on Y. With a function $f \in \mathcal{F}(X,\mathcal{F}(Y,F))$, we associate in a canonical way a function $\tilde{f} \in \mathcal{F}(X \times Y,F)$ setting $\tilde{f}(x,y) = f(x)(y)$. This correspondence is one-to one and onto. If we assume that $\mathfrak{U}$ is the function space neighborhood system of compact convergence on Y (see Example 1.2.7(b)), then $\mathcal{C}_{\mathfrak{U}}(Y,F) = \mathcal{C}(Y,F)$, and we claim that $\tilde{f} \in \mathcal{C}(X \times Y,F)$ if and only if $f \in \mathcal{C}(X,\mathcal{C}(Y,F))$. Indeed, suppose that the function $f \in \mathcal{F}(X,\mathcal{C}(Y,F))$ is continuous. Let $(x,y) \in X \times Y$ and let $W \in \mathcal{W}$ be a neighborhood in F. We find a compact neighborhood L of y in Y such that

$$f(x)(y') - f(x)(y) = \tilde{f}(x,y') - \tilde{f}(x,y) \in (1/2)W$$

for all $y' \in L$. Since convergence in $\mathcal{C}_{\mathfrak{U}}(Y,F)$ means compact convergence, there is $\mathfrak{u} \in \mathfrak{U}$ such that $g \leqslant \mathfrak{u}$ for $g \in \mathcal{C}_{\mathfrak{U}}(Y,F)$ implies that $g(y') \in (1/2)W$ holds for all $y' \in L$. In turn, there is a neighborhood U of x in X such that $f(x') - f(x) \leqslant \mathfrak{u}$ for all $x' \in U$. Hence

$$\tilde{f}(x',y') - \tilde{f}(x,y') \in (1/2)W$$

holds for all $x' \in U$ and $y' \in L$. Combining the above, we obtain that

$$\tilde{f}(x',y') - \tilde{f}(x,y) = (\tilde{f}(x',y') - \tilde{f}(x,y')) + (\tilde{f}(x,y') - \tilde{f}(x,y)) \in W$$

holds for all $(x',y') \in U \times L$. Hence the function $\tilde{f}$ is indeed continuous. For the converse, suppose that the function $\tilde{f} \in \mathcal{F}(X \times Y, F)$ is continuous and for every $x \in X$ define the function $f(x) \in \mathcal{F}(Y,F)$ by $f(x)(y) = \tilde{f}(x,y)$. The continuity of $\tilde{f}$ implies that $f(x) \in \mathcal{C}(Y,F)$. We proceed to establish that the function $f \in \mathcal{F}(X, \mathcal{C}(Y,F))$, that is, the mapping $x \mapsto f(x)$ is continuous. Let $x \in X$ and $\mathfrak{u} \in \mathfrak{U}$. There is a compact subset K of Y and $W \in \mathcal{W}$ such that $g(y) \in W$ for all $y \in K$ implies that $g \leqslant \mathfrak{u}$ for a function $g \in \mathcal{F}(Y,F)$. By the continuity of the function $\tilde{f}$, for every $y \in K$, there are open neighborhoods U_y of x in X and L_y of $y \in Y$ such that

$$f(x',y') - f(x,y) \in (1/2)W$$

for all $(x',y') \in U_y \times L_y$. Hence there are $y_1, \ldots, y_n \in K$ such that the open sets L_{y_i} cover K. Then $U = \bigcap_{i=1}^n U_{y_i}$ is a neighborhood of x. Every element $y \in K$ is contained in one of the sets $y \in L_{y_i}$ and, therefore,

$$(x',y) \in U \times L_{y_i} \subset U_{y_i} \times L_{y_i}$$

holds for all $x' \in U$. Thus

$$\tilde{f}(x',y) - \tilde{f}(x,y) = (\tilde{f}(x',y) - \tilde{f}(x,y_i)) + (\tilde{f}(x,y_i) - \tilde{f}(x,y)) \in W$$

holds for all $(x',y) \in U \times K$. This shows that $f(x')(y) - f(x)(y) \in W$, and consequently, that $f(x') - f(x) \leqslant \mathfrak{u}$ for all $x' \in U$. Our claim follows.

Going forward with this special case, we further narrow our assumptions on the function space $\mathcal{C}_{\mathfrak{U}}(Y,F)$ and assume that F is a normed space, that Y is compact and that $\mathfrak{U}$ renders the topology of uniform convergence on Y (see Example 1.2.7(a)). Let $\mathbb{B}$ denote the unit ball of F and $\mathfrak{u}$ the unit ball of $\mathcal{C}_{\mathfrak{U}}(Y,F)$, that is, the function space neighborhood defined by the constant neighborhood function $y \mapsto \mathbb{B}$. Then $\mathfrak{U}$ consists of all positive multiples of $\mathfrak{u}$ and induces the neighborhood system $\langle \mathfrak{U} \rangle$ on $\mathcal{C}_{\mathfrak{U}}(Y,F)$. We have

$$\langle \mathfrak{U} \rangle_c = \{\lambda \langle \mathfrak{u} \rangle \mid \lambda \in \overline{\mathbb{R}}_+\},$$

where

$$\langle \mathfrak{u} \rangle = \{g \in \mathcal{C}_{\mathfrak{U}}(Y,F) \mid \|g(y)\| \leqslant 1 \text{ for all } y \in Y\}$$

and $(+\infty)\langle \mathfrak{u} \rangle = \mathcal{C}_{\mathfrak{U}}(Y,F)$ (see Section 1.1). We continue to outline a natural correspondence between $\mathcal{C}_{\mathfrak{V}}(X, \mathcal{C}_{\mathfrak{U}}(Y,F))$ and a function space $\mathcal{C}_{\widetilde{\mathfrak{V}}}(X \times Y, F)$ where $\widetilde{\mathfrak{V}}$ is a suitable function space neighborhood system for F-valued functions on $X \times Y$. We note that the neighborhood functions on X that define the function space neighborhood system

$\mathfrak{V}$ and are $\langle\mathfrak{U}\rangle_c$-valued, while neighborhood functions on $X\times Y$ are $\mathcal{W}_c$-valued, where $\mathcal{W}_c = \{\lambda\mathbb{B} \mid \lambda \in \overline{\mathbb{R}}_+\}$. Every neighborhood function n on X is defined by an $\overline{\mathbb{R}}_+$-valued function φ_n on X such that $n(x) = \varphi_n(x)\langle\mathfrak{u}\rangle$ for all $x \in X$. In a canonical way, we associate with n a neighborhood function $\widetilde{n}$ on $X \times Y$ defined as

$$\widetilde{n}(x,y) = \varphi_n(x)\mathbb{B}$$

for all $(x,y) \in X \times Y$. Using this correspondence, we can now define the function space neighborhood system $\widetilde{\mathfrak{V}}$ for $X \times Y$ as follows. With every $\mathfrak{v} \in \mathfrak{V}$, we associate a neighborhood $\widetilde{\mathfrak{v}} \in \widetilde{\mathfrak{V}}$ defined by the convex family of neighborhood functions

$$\mathcal{N}_{\widetilde{\mathfrak{v}}} = \{\widetilde{n} \mid n \in \mathcal{N}_{\mathfrak{v}}\}.$$

The requirements for a function space neighborhood are satisfied for $\widetilde{\mathfrak{v}}$. Indeed, let K be a compact subset of X. There is $n \in \mathcal{N}_{\mathfrak{v}}$ and $\lambda > 0$ such that $\lambda\langle\mathfrak{u}\rangle \subset n(x)$ for all $x \in K$. Hence $\lambda\mathbb{B} \subset \widetilde{n}(x,y)$ for all $(x,y) \in K \times Y$. Since every relatively compact Borel subset of $X\times Y$ is contained in some set of this type, our claim follows. The family $\widetilde{\mathfrak{V}} = \{\widetilde{\mathfrak{v}} \mid \mathfrak{v} \in \mathfrak{V}\}$ then forms a function space neighborhood system for $\mathcal{F}(X \times Y, F)$. It is now obvious that $f \leqslant \mathfrak{v}$ holds if and only if $\widetilde{f} \in \widetilde{\mathfrak{v}}$ for all $f \in \mathcal{F}(X, \mathcal{F}(Y,F))$. Combining all of the above, we realize that the mapping $f \mapsto \widetilde{f}$ is indeed a linear bijection between

$$\mathcal{C}_{\mathfrak{V}}(X, \mathcal{C}_{\mathfrak{U}}(Y,F)) \quad \text{and} \quad \mathcal{C}_{\widetilde{\mathfrak{V}}}(X \times Y, F),$$

which preserves the respective function space neighborhoods. Moreover, if the function space neighborhood system $\mathfrak{V}$ is weakly lower continuous, so is $\widetilde{\mathfrak{V}}$. Indeed, let $\widetilde{\mathfrak{v}} \in \widetilde{\mathfrak{V}}$ and let $\widetilde{n} \in \mathcal{N}_{\widetilde{\mathfrak{v}}}$. For weak lower continuity of the neighborhood function $\widetilde{n}$, let $(x,y) \in X \times Y$, let $c \in \widetilde{n}(x,y)$ and let $W = \varepsilon\mathbb{B} \in \mathcal{W}$ for $\varepsilon > 0$. We set $\lambda = \varphi_n(x)$, that is, $n(x) = \lambda\langle\mathfrak{u}\rangle$. Hence $\|c\| \leqslant \lambda$. There is a function $g \in \mathcal{C}_{\mathfrak{U}}(Y,F)$ such that $g \leqslant \lambda\mathfrak{u}$ and $g(y) = c$. We have $g \in n(x)$, and by the weak lower continuity of the neighborhood function n there is a neighborhood U of x such that $g \in n(x') + \varepsilon\langle\mathfrak{u}\rangle$ for all $x' \in U$. The latter means that $\|g(y')\| \leqslant \varphi_n(x') + \varepsilon$ holds for all $(x'y') \in U \times Y$. Hence

$$c = g(y) \in \widetilde{n}(x',y') + \varepsilon\mathbb{B}.$$

Since $U \times Y$ is a neighborhood of (x,y), our claim follows. We summarize.

1.2.9 Proposition. *If $\mathcal{C}_{\mathfrak{U}}(Y,F)$ is a function space, where F is a normed space, Y is compact and $\mathfrak{U}$ renders the topology of uniform convergence on Y, then $\mathcal{C}_{\mathfrak{V}}(X, \mathcal{C}_{\mathfrak{U}}(Y,F))$ is isomorphic to a function space $\mathcal{C}_{\widetilde{\mathfrak{V}}}(X \times Y, F)$. If the function space neighborhood system $\mathfrak{V}$ is weakly lower continuous, so is $\widetilde{\mathfrak{V}}$.*

Function subspaces

We conclude this section with the following observation. If G is a subspace of E, then the sets $V_G = V \cap G$ for $V \in \mathcal{V}$ form a neighborhood basis for G. Thus if n is a neighborhood function for $\mathcal{F}(X,E)$, then the intersections of its values with G define a neighborhood function for $\mathcal{F}(X,G)$. In this way, a functions-space neighborhood system $\mathfrak{V}$ induces a functions-space neighborhood system for $\mathcal{F}(X,G)$ in a canonical way. Accordingly, $(\mathcal{C}_{\mathfrak{V}}(X,G),\mathfrak{V})$ is again a function space and indeed a subspace of $(\mathcal{C}_{\mathfrak{V}}(X,E),\mathfrak{V})$.

1.3 Compactness in spaces of linear operators

Let $(E,\mathcal{V})$ and $(F,\mathcal{W})$ be real or complex locally convex topological vector spaces such that F is Hausdorff. Let F^* denote the topological dual of F. The second dual (or bidual) F^{**} of F is the dual of F^* if the latter is endowed with its strong topology, that is, the topology generated by the polars of balanced bounded subsets of F. We consider the following standard topologies on the space $\mathfrak{L}(E,F)$ of all continuous real- or complex- (depending on the common scalar field) linear operators $T : E \to F$:

(i) the *uniform operator topology* generated by the neighborhoods

$$\mathcal{U}_A^W = \{T \in \mathfrak{L}(E,F) \mid T(A) \subset W\}$$

for bounded subsets A of E and $W \in \mathcal{W}$;

(ii) the *strong operator topology* generated by the neighborhoods

$$\mathcal{U}_A^W = \{T \in \mathfrak{L}(E,F) \mid T(A) \subset W\}$$

for finite subsets A of E and $W \in \mathcal{W}$;

(iii) the *weak operator topology* generated by the neighborhoods

$$\mathcal{U}_A^Y = \{T \in \mathfrak{L}(E,F) \mid |\nu(T(a))| \leqslant 1 \text{ for all } a \in A \text{ and } \mu \in Y\}$$

for finite subsets A of E and Y of F^*. If $F = G^*$ is the dual of a locally convex ordered topological vector space G, then we also refer to

(iv) the *weak* operator topology* generated by the neighborhoods

$$\mathcal{U}_A^Y = \{T \in \mathfrak{L}(E,F) \mid |\nu(T(a))| \leqslant 1 \text{ for all } a \in A \text{ and } \mu \in Y\}$$

for finite subsets A of E and Y of G. Using any of the above topologies $\mathfrak{L}(E,F)$ is again a locally convex topological vector space. It is Hausdorff since F is supposed to be Hausdorff.

For later use, we recall the well-known fact that the operators in $\mathfrak{L}(E,F)$ are also *weakly continuous*, that is, continuous if we consider the spaces E and F in their respective weak topologies $\sigma(E,E^*)$ and $\sigma(F,F^*)$ (see IV.7.4 in [56] or II.6.13 in [45]).

This follows with a simple argument. Let $T \in \mathfrak{L}(E,F)$. A neighborhood W in the $\sigma(F,F^*)$-topology of F is defined by finitely many elements $\mu_1,\dots,\mu_n \in F^*$, that is,

$$W = \{c \in F \mid \mathfrak{Re}\, \mu_i(c) \leqslant 1, \text{ for } i = 1,\dots,n\}.$$

The linear functionals $\mu_i \circ T$ on E, that is, $a \mapsto \mu_i(T(a))$, are continuous, hence elements of E^*. Thus

$$V = \{a \in F \mid \mathfrak{Re}(\mu_i \circ T)(a) \leqslant 1, \text{ for } i = 1,\dots,n\}$$

defines a neighborhood in the $\sigma(E,E^*)$-topology of E. Our claim follows, because $T(V) \subset W$.

Bounded operators

A linear operator $T : E \to F$ is called *bounded* if it maps bounded subsets of E into bounded subsets of F. Clearly, all continuous operators are bounded. The converse holds true if E is a normed space, but not in general. The families of bounded subsets of E are identical for all topologies consistent with the dual pair (E,E^*). The same holds true for F and (F,F^*) (see IV.3.2, Corollary 2 in [56] and the remark below). We shall use the same set of topologies (i) to (iv) from above for the space $\mathfrak{L}_b(E,F)$ of all bounded linear operators from E to F.

While we always assume that the locally convex space F is Hausdorff, we do not impose the same requirement on E. The set $E_0 = \bigcap_{V\in\mathcal{V}} V$ forms a closed subspace of E, and we have $E_0 = \{a \in E \mid \mu(a) = 0 \text{ for all } \mu \in E^*\}$ by the Hahn–Banach theorem. Thus $A = A + E_0$ for every subset A of E that is either open or closed. We abbreviate $\widetilde{E}$ for the quotient space

$$E/E_0 = \{a + E_0 \mid a \in E\},$$

endowed with the quotient space topology (see I.2 in [56]). The space $\widetilde{E}$ is Hausdorff, and the canonical projection from E to $\widetilde{E}$ is linear, onto, continuous and both open and closed. The space E itself is Hausdorff if and only if $E_0 = \{0\}$, that is, if and only if $E = \widetilde{E}$. For an element $a \in E$, we write $\tilde{a}$ for $a + E_0 \in \widetilde{E}$, that is the image of a under the canonical projection of E onto $\widetilde{E}$, and

$$\widetilde{A} = \{\tilde{a} \mid a \in A\} \subset \widetilde{E}$$

for a subset A of E. Clearly, $\widetilde{A} = \widetilde{A + E_0}$. If A is open, closed, convex or balanced in E, then $\widetilde{A}$ is open, closed, convex or balanced in $\widetilde{E}$. Conversely, if $\widetilde{A}$ is open, closed, convex or balanced in $\widetilde{E}$, then $A + E_0$ is open, closed, convex or balanced in E. A subset

A is bounded, compact or relatively compact in E if and only if $\widetilde{A}$ is bounded, compact or relatively compact in $\widetilde{E}$.

Moreover, the subspace E_0 is contained in the kernel of every operator $T \in \mathfrak{L}_b(E,F)$. Indeed, E_0 is a bounded subset of E, hence $T(E_0)$ is bounded in F and, therefore, $T(E_0) = \{0\}$, since F is supposed to be Hausdorff. We infer that T takes the same value at all elements of an equivalence class $\widetilde{a} = a + E_0$ in $\widetilde{E}$ and corresponds to an operator $\widetilde{T} \in \mathfrak{L}_b(\widetilde{E},F)$, where

$$\widetilde{T}(\widetilde{a}) = T(a)$$

for all $\widetilde{a} \in \widetilde{E}$. Conversely, for every operator $\widetilde{T} \in \mathfrak{L}_b(\widetilde{E},F)$ the formula $T(a) = \widetilde{T}(\widetilde{a})$, for all $a \in E$, defines an operator in $\mathfrak{L}_b(E,F)$. Thus there is a bijective linear correspondence between $\mathfrak{L}_b(\widetilde{E},F)$ and $\mathfrak{L}_b(E,F)$, which also preserves the respective operator space topologies. The same observations apply to the spaces of continuous linear operators $\mathfrak{L}(\widetilde{E},F)$ and $\mathfrak{L}(E,F.)$ In particular, E and $\widetilde{E}$ share the same dual, that is, $\widetilde{E}^* = E^*$.

Every operator $\mathcal{T} \in \mathfrak{L}_b(E,F)$ renders a bilinear form on $\widetilde{E} \times F^*$, that is, $(\widetilde{a},\mu) \mapsto \mu(\mathcal{T}(a))$, where a is any element of the equivalence class $\widetilde{a} \in \widetilde{E}$. Thus the elements of the tensor product $\widetilde{E} \otimes F^*$ are linear functionals on $\mathfrak{L}_b(E,F)$, and

$$(\mathfrak{L}_b(E,F), \widetilde{E} \otimes F^*)$$

constitutes a dual pair with the bilinear form

$$\left\langle \mathcal{T}, \sum_{i=1}^{n} \widetilde{a}_i \otimes \mu_i \right\rangle = \sum_{i=1}^{n} \mu_i(\mathcal{T}(a_i)),$$

where $a_i \in \widetilde{a}_i$. According to the entry about polar topologies in Section 1.1 (see also IV.1 in [56]), we need to establish that the above bilinear form is nondegenerate, that is, (i) $\langle \mathcal{T}, \omega\rangle = 0$ for all $\omega \in \widetilde{E} \otimes F^*$ implies hat $\mathcal{T} = 0$, and (ii) $\langle \mathcal{T}, \omega\rangle = 0$ for all $\mathcal{T} \in \mathfrak{L}_b(E,F)$ implies that $\omega = 0$. While (i) is obvious, (ii) requires a brief argument. Let

$$\omega = \sum_{i=1}^{n} \widetilde{a}_i \otimes \mu_i \in E \otimes F^*.$$

We may assume that both sets $\{\widetilde{a}_i\}$ and $\{\mu_i\}$ are linearly independent in $\widetilde{E}$ and F^*, respectively. If $\omega \neq 0$, then $n \geqslant 1$, and there is $\nu \in \widetilde{E}^* = E^*$ such that $\nu(\widetilde{a}_1) = 1$ and $\nu(\widetilde{a}_i) = 0$ for all $i > 1$ (see IV.1.1 in [56]). Since F is Hausdorff, there is $c \in F$ such that $\mu_1(c) = 1$. We define $\mathcal{T}(a) = \nu(a)c \in F$ for all $a \in E$ and observe that $\mathcal{T} \in \mathfrak{L}(E,F)$. Since $\langle \mathcal{T}, \omega\rangle = 1$, our claim follows.

For the following, recall our brief review of polar topologies in Section 1.1.

1.3.1 Proposition. *The weak, strong and uniform operator topologies of* $\mathfrak{L}_b(E,F)$ *are polar topologies of the dual pair* $(\mathfrak{L}_b(E,F), \widetilde{E} \otimes F^*)$. *The weak and the strong operator* *topologies are both consistent with this duality, and in either of these* $\mathfrak{L}(E,F)$ *is a dense subspace of* $\mathfrak{L}_b(E,F)$, *hence features the same dual* $\widetilde{E}\otimes F^*$.

Proof. For our first statement, we observe that the weak, strong and uniform operator topologies of $\mathfrak{L}_b(E,F)$ are all generated by families of polars of subsets

$$\Omega_A^{\Upsilon} = \{\tilde{a}\otimes\mu \mid \tilde{a}\in\widetilde{A},\ \mu\in\Upsilon\}$$

of $\widetilde{E}\otimes F^*$. For the weak operator topology, we allow all finite subsets of E for A and all finite subsets of F^* for Υ. For the strong operator topology, we choose all finite subsets for A and the polars of neighborhoods $W\in\mathcal{W}$ for Υ. We obtain the uniform operator topology if we choose all bounded subsets of E for A and again all polars of neighborhoods in $\mathcal{W}$ for Υ. In all of these instances, Υ is compact in the weak* topology of F^*. We shall first verify that all the sets Ω_A^{Υ} involved are weakly bounded in $\widetilde{E}\otimes F^*$, that is, families of their polars in $\mathfrak{L}_b(E,F)$ create polar topologies of the dual pair $(\mathfrak{L}_b(E,F),\widetilde{E}\otimes F^*)$. Indeed, for any of the above choices for A and Υ, the set A is bounded in E and $\Upsilon\subset W^\circ$ for some $W\in\mathcal{W}$. Thus for every $\mathcal{T}\in\mathfrak{L}_b(E,F)$ there is $\lambda\geqslant 0$ such that $\mathcal{T}(A)\subset\lambda W$, hence $\langle\mathcal{T},\tilde{a}\otimes\mu\rangle = \mu(\mathcal{T}(a))\leqslant\lambda$ for all $\tilde{a}\otimes\mu\in\Omega_A^W$. This yields our claim. For our second statement, we shall prove that the sets Ω_A^{Υ} that generate the strong or the weak operator topology of $\mathfrak{L}_b(E,F)$, respectively, are weakly compact in $\widetilde{E}\otimes F^*$. That is, the family of their polars in $\mathfrak{L}_b(E,F)$ creates a topology consistent with the dual pair $(\mathfrak{L}_b(E,F),\widetilde{E}\otimes F^*)$. For this, let $(\tilde{a}_i\otimes\mu_i)_{i\in\mathcal{I}}$ be a net in Ω_A^{Υ}. Since A is finite and Υ is weak* compact, there is a subnet $(\tilde{a}\otimes\mu_j)_{j\in\mathcal{J}}$ such that $a\in A$ and μ_j converges to $\mu\in\Upsilon$. Thus $\tilde{a}\otimes\mu\in\Omega_A^{\Upsilon}$ and for every $\mathcal{T}\in\mathfrak{L}_b(E,F)$ we have

$$\lim_{j\in\mathcal{J}}\langle\mathcal{T},\tilde{a}\otimes\mu_j\rangle = \lim_{j\in\mathcal{J}}\mu_j(\mathcal{T}(a)) = \mu(\mathcal{T}(a)) = \langle\mathcal{T},\tilde{a}\otimes\mu\rangle.$$

We conclude that the subnet $(\tilde{a}\otimes\mu_j)_{j\in\mathcal{J}}$ converges to $\tilde{a}\otimes\mu\in\Omega_A^{\Upsilon}$ in the weak topology of this duality, thus completing our argument. Our final assertion that $\mathfrak{L}(E,F)$ is a dense subspace of $\mathfrak{L}_b(E,F)$ in the strong topology and, therefore, also in the weak operator topology follows directly from the observation (ii) from above: Only the zero element in $\widetilde{E}\otimes F^*$, that is, the dual of $\mathfrak{L}_b(E,F)$ in the strong operator topology vanishes on all operators in $\mathfrak{L}(E,F)$. The Hahn–Banach theorem then renders our claim. □

Proposition 1.3.1 shows in particular that the weak operator topology on $\mathfrak{L}_b(E,F)$ is indeed the weak topology of the duality $(\mathfrak{L}_b(E,F),\widetilde{E}\otimes F^*)$.

1.3.2 Corollary. *If* $\mathfrak{L}_b(E,F)$ *is endowed with the weak or the strong operator topology, then for every linear functional* $\omega\in\mathfrak{L}_b(E,F)^*$ *there are* $a_1,\dots,a_n\in E$ *and* $\mu_1,\dots,\mu_n\in F^*$ *such that*

$$\omega(\mathcal{T}) = \sum_{i=1}^{n}\mu_i(\mathcal{T}(a_i))$$

for all $\mathcal{T} \in \mathfrak{L}_b(E, F)$.

Proof. We have $\mathfrak{L}_b(E,F)^* = \widetilde{E} \otimes F^*$ by Proposition 1.3.1. Thus $\omega = \sum_{i=1}^n \tilde{a}_i \otimes \mu_i$ for $\tilde{a}_i \in \widetilde{E}$ and $\mu_i \in F^*$. If we choose $a_i \in \tilde{a}_i$, then

$$\omega(\mathcal{T}) = \left\langle \mathcal{T}, \sum_{i=1}^n \tilde{a}_i \otimes \mu_i \right\rangle = \sum_{i=1}^n \mu_i(\mathcal{T}(a_i))$$

holds for all $\mathcal{T} \in \mathfrak{L}_b(E,F)$, our claim. □

Compact and weakly compact operators

An operator in $\mathfrak{L}_b(E,F)$ is called *compact* (or *weakly compact*) if it maps bounded subsets of E into relatively compact (or relatively weakly compact) subsets of F, that is, the closure in the topology of F (or in the weak topology of F) of the image of a bounded set is compact. Because the balanced convex hull of a bounded set in E is again bounded, it suffices to verify these conditions for balanced convex subsets of E. Consequently, since the closures of a convex set in the given and in the weak topologies of F coincide and since compactness implies weak compactness, this reflection shows in particular that every compact operator is also weakly compact. The compact and the weakly compact operators form respective subspaces of $\mathfrak{L}_b(E,F)$. We observe the following.

1.3.3 Proposition. *Let τ be a polar topology of the dual pair (F, F^*) on F.*
(a) *Every weakly compact operator in $\mathfrak{L}_b(E,F)$ is also bounded if F is endowed with τ.*
(b) *If every subset of F that is bounded with respect to τ is relatively weakly compact in the given topology, then every operator $\mathcal{T} \in \mathfrak{L}_b(E,F)$ that is also bounded if F is endowed with τ, is weakly compact.*

Proof. (a) Suppose that $\mathcal{T} \in \mathfrak{L}_b(E,F)$ is weakly compact and let A be a bounded subset of E. Then $\mathcal{T}(A)$ is relatively compact in F with respect to its weak topology. Thus the polar of $T(A)$ in F^* is a neighborhood consistent with the duality (F, F^*). Let U be a closed convex neighborhood in F for the polar topology τ. Then its polar U° is bounded in F^*. Thus there is $\lambda > 0$ such that $U^\circ \subset \lambda T(A)^\circ$. This yields $T(A) \subset \lambda U$, and our claim follows. (b) Suppose that τ satisfies our assumption and that $\mathcal{T} \in \mathfrak{L}_b(E,F)$ is also bounded if F is endowed with τ. Let A be a bounded subset of E. Then $\mathcal{T}(A)$ is bounded in F with respect to τ, hence relatively weakly compact in the given topology of F. □

If F is semireflexive, that is, if the canonical embedding of F into F^{**} is onto (see IV.5 in [56]), then the condition on τ in 1.3.3(b) holds with τ for the given topology of F (see IV.5.5 in [56]). Consequently, all operators in $\mathfrak{L}_b(E,F)$ are weakly compact in this case.

In general, no straightforward description appears to be available for the dual of $\mathfrak{L}_b(E,F)$ under the uniform operator topology. However, we list the following useful observation concerning its second dual, that is, the dual of $\mathfrak{L}_b(E,F)^*$ endowed with the strong topology from its duality with $\mathfrak{L}_b(E,F)$.

1.3.4 Proposition. *If $\mathfrak{L}_b(E,F)$ is endowed with the uniform operator topology, then for every functional $\Omega \in \mathfrak{L}_b(E,F)^{**}$ there is an operator $\mathcal{T}_\Omega \in \mathfrak{L}_b(E,F^{**})$, where F^{**} is endowed with its weak* topology, such that*

$$\mathcal{T}_\Omega(a)(\mu) = \Omega(\tilde{a} \otimes \mu)$$

for all $a \in E$ and $\mu \in F^*$, that is, $\tilde{a} \otimes \mu \in \tilde{E} \otimes F^* \subset \mathfrak{L}_b(E,F)^*$.

Proof. Following Proposition 1.3.1, the dual $\mathfrak{L}_b(E,F)^*$ of $\mathfrak{L}_b(E,F)$ with respect to the uniform operator topology contains $\tilde{E} \otimes F^*$ as a subspace. Let $\Omega \in \mathfrak{L}_b(E,F)^{**}$. There is a neighborhood $\mathfrak{B}^\circ$ in $\mathfrak{L}_b(E,F)^*$, the polar of a balanced bounded subset $\mathfrak{B}$ in $\mathfrak{L}_b(E,F)$ such that $|\Omega(\Theta)| \leqslant 1$ for all $\Theta \in \mathfrak{B}^\circ$, that is, for all $\Theta \in \mathfrak{L}_b(E,F)^*$ such that $|\Theta(T)| \leqslant 1$ for all $T \in \mathfrak{B}$. If A is a bounded subset of E, we note that

$$\mathfrak{B}_A = \{T(a) \mid a \in A \text{ and } T \in \mathfrak{B}\}$$

is a balanced bounded subset of F, since for every $W \in \mathcal{W}$ the uniform neighborhood $\mathcal{U}_A^W$ in $\mathfrak{L}_b(E,F)$ absorbs $\mathfrak{B}$. Its polar $\mathfrak{B}_A^\circ$ is therefore a neighborhood in the strong topology of F^*, and for every $\mu \in \mathfrak{B}_A^\circ$ and $a \in A$ we have $|(\tilde{a} \otimes \mu)(T)| = |\mu(T(a))| \leqslant 1$ for all $T \in \mathfrak{B}$. Thus $\tilde{a} \otimes \mu \in \mathfrak{B}^\circ$ and $|\Omega(\tilde{a} \otimes \mu)| \leqslant 1$ by the above. Next, for every fixed element $a \in E$ we associate with Ω the linear functional Ω_a on F^* setting $\Omega_a(\mu) = \Omega(\tilde{a} \otimes \mu)$ for all $\mu \in F^*$. Then Ω_a is an element of F^{**}. Surely, if μ is an element of the neighborhood $\mathfrak{B}_{\{a\}}^\circ$, then $|\Omega_a(\mu)| = |\Omega(\tilde{a} \otimes \mu)| \leqslant 1$ by the above. The mapping $\mathcal{T}_\Omega$ such that $\mathcal{T}_\Omega(a) = \Omega_a$ is therefore a linear operator from E into F^{**}. Moreover, this operator is bounded and, therefore, an element of $\mathfrak{L}_b(E,F^{**})$, provided that F^{**} is endowed with its weak* topology. Indeed, let A be a bounded subset of E and let $\mu \in F^*$, that is, the dual of F^{**}. The polar $\mathfrak{B}_A^\circ$ of $\mathfrak{B}_A$ is a neighborhood in F^* and, therefore, absorbs μ, that is, $\mu \in \rho\mathfrak{B}_A^\circ$ for some $\rho \geqslant 0$. Thus $|\mu(\Omega_a)| = |\Omega(\tilde{a} \otimes \mu)| \leqslant \rho$ for all $a \in A$ by the above. Consequently, the set $\{\mathcal{T}_\Omega(a) \mid a \in A\}$ is bounded in F^{**} as claimed. □

Proposition 1.3.4 is of particular interest in the case that F is reflexive, that is, if the canonical embedding of F into F^{**} is both onto and an isomorphism (see IV.5 in [56]). If combined with the inverse of the embedding of F into F^{**}, the operator $\mathcal{T}_\Omega$ becomes F-valued and bounded with respect to the weak, hence also the given topology of F. That is, we have $\mathcal{T}_\Omega \in \mathfrak{L}_b(E,F)$ in this case.

We go forward to investigate spaces of continuous linear operators and for our purposes are notably interested in identifying compact subsets of $\mathfrak{L}_b(E,F)$. We consider both the strong and weak operator topologies. We note that any subset of $\mathfrak{L}(E,F)$ that is relatively compact in $\mathfrak{L}(E,F)$ is also relatively compact in $\mathfrak{L}_b(E,F)$. The reverse conclusion does however not hold true in general.

1.3.5 Proposition. *If $\mathfrak{H} \subset \mathfrak{L}_b(E,F)$ is relatively compact in the weak operator topology, then it is bounded in the uniform operator topology.*

Proof. Suppose that $\mathfrak{H} \subset \mathfrak{L}_b(E,F)$ is relatively compact in the weak operator topology and let U_A^W be a neighborhood in the uniform operator topology. The polar $\mathfrak{H}^\circ$ of $\mathfrak{H}$ in $\tilde{E}\otimes F^*$ is a neighborhood consistent with the duality $(\mathfrak{L}_b(E,F), \tilde{E}\otimes F^*)$, whereas $(U_A^W)^\circ$ is bounded in $\tilde{E}\otimes F^*$ since U_A^W is a neighborhood in a polar topology of this duality (see Proposition 1.3.1). Thus $(U_A^W)^\circ \subset \lambda\mathfrak{H}^\circ$ for some $\lambda > 0$ and, therefore, $\mathfrak{H} \subset \lambda U_A^W$. □

For compactness in the weak operator topology, a complete characterization is available in case that F is reflexive using Proposition 1.3.4.

1.3.6 Proposition. *Suppose that F is reflexive. A subset of $\mathfrak{L}_b(E,F)$ is relatively compact in the weak operator topology if and only if it is bounded with respect to the uniform operator topology.*

Proof. In order to verify that a uniformly bounded subset $\mathfrak{B}$ of $\mathfrak{L}_b(E,F)$ is relatively compact in the weak operator topology, it suffices to verify that its polar in $\tilde{E} \otimes F^*$ is consistent with the duality $(\mathfrak{L}_b(E,F), \tilde{E} \otimes F^*)$. That is, we shall verify that every linear functional on $\tilde{E} \otimes F^*$ that is bounded on $\mathfrak{B}^\circ$ is indeed an element of $\mathfrak{L}_b(E,F)$. Let Ω be such a functional. Since $\mathfrak{B}$ is bounded with respect to the uniform operator topology, its polar is a neighborhood in the strong topology of $\mathfrak{L}_b(E,F)^*$, the dual of $\mathfrak{L}_b(E,F)$ with respect to the uniform topology, and we have $\Omega \in \mathfrak{L}_b(E,F)^{**}$. According to Proposition 1.3.4, there is $\mathcal{T}_\Omega \in \mathfrak{L}_b(E,F^{**})$, where F^{**} is endowed with its weak* topology, such that $\mathcal{T}_\Omega(a)(\mu) = \Omega(\tilde{a} \otimes \mu)$ for all $a \in E$ and $\mu \in F^*$. Since F is reflexive, we have $\mathcal{T}_\Omega \in \mathfrak{L}_b(E,F)$ by the observation following Proposition 1.3.4, and our claim follows. □

The general case is however less accessible. We shall in the sequel restrict ourselves to the space $\mathfrak{L}(E,F)$ of continuous linear operators rather than bounded ones. The following proposition establishes a useful sufficient criterion for compactness in the strong and weak operator topologies. A subset $\mathfrak{H}$ of $\mathfrak{L}(E,F)$ is *equicontinuous* if for every $W \in \mathcal{W}$ there is $V \in \mathcal{V}$ such that $T(V) \subset W$ for all $T \in \mathfrak{H}$.

1.3.7 Proposition. *If $\mathfrak{H} \subset \mathfrak{L}(E,F)$ is equicontinuous, closed in the strong (or weak) operator topology of $\mathfrak{L}(E,F)$ and if for every $a \in E$ the set $\{T(a) \mid T \in \mathfrak{H}\}$ is relatively compact (or relatively weakly compact) in F, then $\mathfrak{H}$ is compact in the strong (or weak) operator topology of $\mathfrak{L}(E,F)$.*

Proof. Let $(T_i)_{i\in\mathcal{I}}$ be an ultranet in $\mathfrak{H}$. For every $a \in E$, the ultranet $(T_i(a))_{i\in\mathcal{I}}$ is contained in the relatively compact (or relatively weakly compact) subset $\{T(a) \mid T \in \mathfrak{H}\}$ of F and, therefore, converges toward an element $T(a) \in F$. The mapping $a \mapsto T(a) : E \to F$ is obviously linear. Because $\mathfrak{H}$ is equicontinuous, given $W \in \mathcal{W}$ there is $V \in \mathcal{V}$ such that $T_i(V) \subset W$ for all $i \in \mathcal{I}$. Thus $T(a) \in W$ whenever $a \in V$, hence

$T(V) \subset W$ and, therefore, of $T \in \mathfrak{L}(E,F)$. Furthermore, our construction implies that $(T_i)_{i \in \mathcal{I}}$ converges to T in the strong (or weak) operator topology. Thus $T \in \mathfrak{H}$, which was supposed to be closed. Hence every ultranet in $\mathfrak{H}$ is convergent, and our claim follows. □

A locally convex topological vector space E is *barreled* (see II.7 in [56]) if every *barrel*, that is, every closed absorbing balanced convex subset of E, is a neighborhood of the origin. This property holds if and only if E carries the strong topology $\beta(E, E^*)$ of the dual pair (E, E^*). Every Banach space is barreled.

1.3.8 Corollary. *If E is barreled, then $\mathfrak{H} \subset \mathfrak{L}(E,F)$ is compact in the strong (or weak) operator topology of $\mathfrak{L}(E,F)$ if and only if it is closed in the strong (or weak) operator topology of $\mathfrak{L}(E,F)$ and for every $a \in E$ the set $\{T(a) \mid T \in \mathfrak{H}\}$ is relatively compact (or relatively weakly compact) in F.*

Proof. The necessity of the conditions for the compactness of $\mathfrak{H}$ is obvious. On the other hand, if E is barreled, $\mathfrak{H}$ is closed and for all $a \in E$ the sets $\{T(a) \mid T \in \mathfrak{H}\}$ are relatively compact, then for every $W \in \mathcal{W}$ the set $U = \bigcap_{T \in \mathfrak{H}} T^{-1}(\overline{W})$ is closed, convex and absorbing, hence a barrel in E. The latter follows since for every $a \in E$ there is $\lambda \geqslant 0$ such that $\{T(a) \mid T \in \mathfrak{H}\} \subset \lambda W$, hence $a \in \lambda U$. Since E is barreled, there is $V \in \mathcal{V}$ such $V \subset U$. Thus

$$T((1/2)V) \subset (1/2)\overline{W} \subset W$$

for all $T \in \mathfrak{H}$, and $\mathfrak{H}$ is understood to be equicontinuous. Our claim follows with Proposition 1.3.7. □

For a subset U of E and a convex subset S of F, we denote by $\mathfrak{L}_U^S(E,F)$ the subset of operators $T \in \mathfrak{L}(E,F)$ such that $T(U) \subset S$. Clearly, $\mathfrak{L}_U^S(E,F)$ is convex, and closed in the weak operator topology of $\mathfrak{L}(E,F)$ provided that S is closed in F.

The balanced convex hull of a subset A of a locally convex topological vector space was defined in Section 1.1 and is given by

$$\operatorname{bconv}(A) = \left\{ \sum_{i=1}^{n} \alpha_i a_i \,\middle|\, a_i \in A,\ \sum_{i=1}^{n} |\alpha_i| \leqslant 1 \right\}.$$

It is the smallest balanced convex set containing A and coincides with the convex hull of its balanced hull, that is, the convex hull of the set $\operatorname{bal}(A) = \bigcup_{|\gamma| \leqslant 1} \gamma A$. If A is convex and relatively compact, so is $\operatorname{bconv}(A)$. For the latter, we observe that $\operatorname{bconv}(A)$ is a subset of $(A - A)$ in the real, and of $(A - A) + i(A - A)$ in the complex case, hence is also relatively compact, since multiples and sums of relatively compact sets are known to be relatively compact (Lemma III.7 in [45]). There is also a largest balanced subset of A, called its *balanced core* and given by

$$\operatorname{bcore}(A) = \begin{cases} \bigcap_{|\gamma|\geq 1} \gamma A, & \text{if } 0 \in A, \\ \varnothing, & \text{if } 0 \notin A. \end{cases}$$

If A is convex, compact or relatively compact, so is $\operatorname{bcore}(A)$.

1.3.9 Corollary. *Let U be a subset of E and let S be a closed convex subset of F. If either U is a neighborhood of $0 \in E$ and $\operatorname{bcore}(S)$ is compact (or weakly compact), or $\operatorname{bconv}(U)$ is a neighborhood of $0 \in E$ and S is compact (or weakly compact), then the set*

$$\mathfrak{L}_U^S(E,F) = \{T \in \mathfrak{L}(E,F) \mid T(U) \subset S\}$$

is convex and compact in the strong (or weak) operator topology of $\mathfrak{L}(E,F)$. The operators in $\mathfrak{L}_U^S(E,F)$ are themselves compact (or weakly compact).

Proof. Since S is supposed to be closed in F, the set $\mathfrak{L}_U^S(E,F)$ is closed in $\mathfrak{L}(E,F)$ in both the strong and weak operator topology. If U is a neighborhood of $0 \in E$, then it contains a balanced neighborhood $V \in \mathcal{V}$. Thus $T(V) \in \operatorname{bcore}(S)$ for every operator $T \in \mathfrak{L}_U^S(E,F)$ and, therefore,

$$\mathfrak{L}_U^S(E,F) \subset \mathfrak{L}_V^{\operatorname{bcore}(S)}(E,F).$$

We shall verify the criterion of Proposition 1.3.7 for $\mathfrak{H} = \mathfrak{L}_V^{\operatorname{bcore}(S)}(E,F)$ provided that $\operatorname{bcore}(S)$ is compact (or weakly compact). Indeed, $\mathfrak{H}$ is closed in $\mathfrak{L}(E,F)$ and given $W \in \mathcal{W}$ there is $\alpha > 0$ such that $\operatorname{bcore}(S) \subset \alpha W$. Hence $T((1/\alpha)V) \subset W$ for all $T \in \mathfrak{H}$, and $\mathfrak{H}$ is equicontinuous. Furthermore, for every $a \in E$ we have $a \in \alpha V$ for some $\alpha > 0$, hence $T(a) \in \alpha \operatorname{bcore}(S)$ for all $T \in \mathfrak{H}$. Consequently, the set $\{T(a) \mid T \in \mathfrak{H}\}$ is relatively compact (or relatively weakly compact) in F as required. As a closed subset of $\mathfrak{H}$, the set $\mathfrak{L}_U^S(E,F)$ is therefore also compact. Finally, for every bounded subset A of E there is $\alpha > 0$ such that $A \subset \alpha V$. Thus $T(A) \subset \alpha \operatorname{bcore}(S)$ for all $T \in \mathfrak{L}_U^S(E,F)$. Since $\alpha \operatorname{bcore}(S)$ is relatively compact (or relatively weakly compact), all operators in $\mathfrak{L}_U^S(E,F)$ are for that reason compact (or weakly compact). For the second case, that is, if $\operatorname{bconv}(U)$ is a neighborhood of $0 \in E$ and S is compact, we observe that

$$\mathfrak{L}_U^S(E,F) \subset \mathfrak{L}_{\operatorname{bconv}(U)}^{\operatorname{bconv}(S)}(E,F).$$

Since $\operatorname{bconv}(S)$ is relatively compact, we may use the first case with $\operatorname{bconv}(U)$ in place of U and the closure of $\operatorname{bconv}(S)$ in place of S. □

If $S \subset F$ is the polar in F of a subset Π of F^* (this will be the case in many of our applications below), then an easy argument using IV.1.3 in [56] shows that

$$\operatorname{bcore}(S) = \bigcap_{|\gamma|\geq 1} \gamma\Pi^\circ = \bigcap_{|\gamma|\leq 1} (\gamma\Pi)^\circ = \Big(\bigcup_{|\gamma|\leq 1} \gamma\Pi\Big)^\circ = \operatorname{bconv}(\Pi)^\circ.$$

The last of the above equalities uses the aforementioned fact that the balanced convex hull of Π coincides with the convex hull of $\mathrm{bal}(\Pi) = \bigcup_{|\gamma|\leqslant 1} \gamma\Pi$, and then the fact that the polar of a set equals the polar of its convex hull. The requirement from Corollary 1.3.9 that $\mathrm{bcore}(S)$ is weakly compact therefore implies that its polar in F^*, that is, the weak* closure of $\mathrm{bconv}(\Pi)$ (see Theorem IV.1.5 in [56]), is absorbing in F^*.

1.3.10 Proposition. *If $\mathfrak{H} \subset \mathfrak{L}(E,F)$ is compact in the strong (or weak) operator topology, then for any choice of $T_i \in \mathfrak{H}$ and $\lambda_i \geqslant 0$ such that $\sum_{i=1}^{\infty} \lambda_i = 1$ the series $\sum_{i=1}^{\infty} \lambda_i T_i$ converges in the strong (or weak) operator topology. Its sum is an element of $\mathfrak{H}$.*

Proof. First, assume that $0 \in \mathfrak{H}$. For $n \in \mathbb{N}$, let $\sigma_n = \sum_{i=1}^{n} \lambda_i$ and

$$S_n = \sum_{i=1}^{n} \lambda_i T_i = \sum_{i=1}^{n} \lambda_i T_i + (1-\sigma_n)\, 0 \in \mathfrak{H},$$

that is, the partial sums of the given operator series. Given a neighborhood $\mathcal{U}$ in the strong (or weak) operator topology of $\mathfrak{L}(E,F)$, there is $\rho > 0$ such that $\mathfrak{H} \subset \rho\,\mathcal{U}$ and $n_0 \in \mathbb{N}$ such that $1-\sigma_n \leqslant 1/\rho$. Then for all $n_0 \leqslant n < m$ we have

$$S_m - S_m = \sum_{i=n+1}^{m} \lambda_i T_i \in (1-\sigma_n)\mathfrak{H} \subset \mathcal{U}.$$

Hence $(S_n)_{n\in\mathbb{N}}$ is a Cauchy sequence and, therefore, convergent in $\mathfrak{H}$. Now for the general case choose $T_0 \in \mathfrak{H}$ and set $\mathfrak{H}_0 = \mathfrak{H} - T_0$, and $S_i = T_i - T_0$. The series $\sum_{i=1}^{\infty} \lambda_i S_i$ then converges by the preceding to $\sum_{i=1}^{\infty} \lambda_i T_i - T_0 \in \mathfrak{H}_0$. Hence $\sum_{i=1}^{\infty} \lambda_i T_i \in \mathfrak{H}$ as claimed. □

We shall illustrate these results in a few examples.

1.3.11 Examples. (a) If $F = \mathbb{R}$ or $\mathbb{C}$, that is, the scalar field of E, then $\mathfrak{L}(E,F) = E^*$ with both the strong and weak operator topologies coinciding with the weak* topology of E^*, that is, the topology $\sigma(E^*,E)$. We have $\tilde{E} \otimes F = \tilde{E}$ in this case, and Proposition 1.3.1 confirms that the dual of E^* under its weak* topology is $\tilde{E}$. Recall that $\tilde{E} = E$, provided that E is Hausdorff. The uniform operator topology on $\mathfrak{L}(E,F)$, on the other hand, is the strong topology $\beta(E^*,E)$ on E^*. The statements of Propositions 1.3.5 and 1.3.6 are of some interest in this case. While every continuous linear functional in E^* is bounded, that is, it maps bounded subsets of E into bounded subsets of $\mathbb{R}$ or $\mathbb{C}$, the reverse conclusion does not hold in general. It does hold true however, if E is a normed space or, more generally, if E is *bornological*, that is, if every balanced convex subset of E that absorbs every bounded subset of E is a neighborhood (see II.8 in [56]). In general, we have $E^* \subset E_b^*$, whereby the latter denotes the space of all bounded linear functionals on E. Proposition 1.3.6 says that a subset of E_b^* is $\sigma(E_b^*,E)$-relatively compact if and only if it $\beta(E_b^*,E)$-bounded, a statement that does not hold true in general, if one replaces E_b^* by E^*.

If U is a convex neighborhood of $0 \in E$ and $S = \{z \in F \mid \mathfrak{Re}(z) \leqslant 1\}$, then $\mathfrak{L}_U^S(E,F) = U^\circ$. Since bcore$(S)$ is the compact unit ball of $\mathbb{R}$ or $\mathbb{C}$, Corollary 1.3.9 recovers the familiar result that U° is weak* compact.

(b) If F is semireflexive, that is, if the canonical embedding of F into F^{**} is onto, then every closed convex bounded subset S of F is weakly compact. Thus for any neighborhood U of $0 \in E$ the set $\mathfrak{L}_U^S(E,F)$ is compact in the weak operator topology. In case that both E and F are normed spaces with the respective unit balls $\mathbb{B}_E$ and $\mathbb{B}_F$, then the unit ball $\mathfrak{L}_{\mathbb{B}_E}^{\mathbb{B}_F}(E,F)$ is compact in the weak operator topology. In case that F is indeed reflexive, the last observation is transcended by the result of Proposition 1.3.6 since $\mathfrak{L}_b(E,F)$ and $\mathfrak{L}(E,F)$ coincide in this case.

(c) Suppose that F is the dual of a normed space G and endowed with the weak topology of the dual pair (F,G). Then the strong topology $\beta(F,G)$ on F, that is, the topology of the dual norm, satisfies the assumption of Proposition 1.3.3 by the Banach–Alaoglu theorem. Hence any operator $T \in \mathfrak{L}_b(E,F)$ that is bounded with respect to the strong topology of F is weakly compact. If E is also a normed space, then all of these operators are continuous with respect to the respective norm topologies of E and F. For a concrete example, consider an integration space $L^1(\mu)$ for G and $L^\infty(\mu)$ for F.

(d) Suppose that both E and F are ordered topological vector spaces with positive cones E_+ and F_+, respectively. If U is a neighborhood of $0 \in E$, which contains E_+, and if S is a closed convex subset S of F such that bcore(S) is weakly compact, then according to the first alternative in Corollary 1.3.9 the set $\mathfrak{L}_U^S(E,F)$ is compact in the weak operator topology. Its elements are weakly compact operators, and positive, provided that $\bigcap_{\lambda>0} \lambda S \subset F_+$. Alternatively, if U is a convex subset of E_+ such that bconv(U) is a neighborhood of $0 \in E$ and if S is a weakly compact convex subset of F_+, then according to the second alternative in Corollary 1.3.9 the set $\mathfrak{L}_U^S(E,F)$ is compact in the weak operator topology. Its elements are weakly compact operators, and positive, provided that U absorbs all elements of E_+. In an equivalent setup, we may replace the positive cones E_+ and F_+ by their negative counterparts $E_- = -E_+$ and $F_- = -F_+$ in all of the above settings.

1.4 Vector space extensions

Our upcoming integration theory involving operator-valued measures will require to consider extensions of locally convex topological vector spaces into locally convex cones, a concept developed in [33] and [48]. We shall provide the outlines of this theory as far as it is essential for our progress and globally refer to the above mentioned texts for further details. Locally convex cones are generalizations of locally convex topological vector spaces.

Locally convex cones

In this vein, a *cone* is a set $\mathcal{P}$ endowed with an addition $(a,b) \mapsto a+b$ and a scalar multiplication $(\alpha,a) \mapsto \alpha a$ for $a,b \in \mathcal{P}$ and real numbers $\alpha \geqslant 0$. The addition is supposed to be associative and commutative, and there is a neutral element $0 \in \mathcal{P}$. For the scalar multiplication, the usual associative and distributive properties hold, that is, $\alpha(\beta a) = (\alpha\beta)a$, $(\alpha+\beta)a = \alpha a + \beta a$, $\alpha(a+b) = \alpha a + \alpha b$, $1a = a$ and $0a = 0$ for all $a,b \in \mathcal{P}$ and $\alpha,\beta \geqslant 0$. The *cancellation law*, stating that $a+c = b+c$ implies $a = b$, however, is not required in general. It holds if and only if the cone $\mathcal{P}$ can be embedded into a real vector space. An *ordered cone* $\mathcal{P}$ carries a reflexive transitive relation $\leqslant$ such that $a \leqslant b$ implies $a+c \leqslant b+c$ and $\alpha a \leqslant \alpha b$ for all $a,b,c \in \mathcal{P}$ and $\alpha \geqslant 0$. Antisymmetry is not required. Equality on $\mathcal{P}$ is obviously such an order. The extended real number system $\overline{\mathbb{R}}$ and the collection $\mathrm{Conv}(E)$ of nonempty subsets of a vector space E are examples of ordered cones. Both do not satisfy the cancellation law.

A *full locally convex cone* $(\mathcal{P},\mathcal{V})$ is an ordered cone $\mathcal{P}$ that contains an *abstract neighborhood system* $\mathcal{V}$, that is, a subset of positive (that is $\geqslant 0$) elements that is directed downward and closed for multiplication by positive scalars. The elements V of $\mathcal{V}$ define *upper, lower* and *symmetric neighborhoods* for the elements $a \in \mathcal{P}$ by

$$V(a) = \{b \in \mathcal{P} \mid b \leqslant a + V\}, \quad (a)V = \{b \in \mathcal{P} \mid a \leqslant b + V\}$$

and

$$V^s(a) = V(a) \cap (a)V,$$

which create the *upper, lower* and *symmetric topologies* on $\mathcal{P}$, respectively. All elements of $\mathcal{P}$ are required to be *bounded below*, that is, for every $a \in \mathcal{P}$ and $V \in \mathcal{V}$ we have $0 \leqslant a + \lambda V$ for some $\lambda \geqslant 0$. They need however not be bounded above. An element $a \in \mathcal{P}$ is called *bounded (above)* if for every $V \in \mathcal{V}$ there is $\lambda \geqslant 0$ such that $a \leqslant \lambda V$. The presence of unbounded elements constitutes the main difference between locally convex cones and subcones of locally convex topological vector spaces and accounts for much of the richness and subtlety of this setting. The upper, lower and symmetric topologies are each compatible with the algebraic operations, that is, the mappings $(a,b) \mapsto a+b : \mathcal{P}^2 \to \mathcal{P}$ and $a \mapsto \alpha a : \mathcal{P} \to \mathcal{P}$ for a fixed $\alpha \geqslant 0$ are continuous (see Proposition I.1.1 in [48]). Hence the usual limit rules apply for nets in $\mathcal{P}$ with sums and multiples by positive scalars. The mapping $\alpha \mapsto \alpha a : \mathbb{R}_+ \to \mathcal{P}$ is however assured to be continuous only if the element $a \in \mathcal{P}$ is bounded.

Finally, a *locally convex cone* $(\mathcal{P},\mathcal{V})$ is a subcone of a full locally convex cone not necessarily containing the abstract neighborhood system $\mathcal{V}$. Every locally convex ordered topological vector space over $\mathbb{R}$ is a locally convex cone in this sense, as it can be canonically embedded into a full locally convex cone (Example I.2.7 in [33]). We shall further elaborate on the embedding of vector spaces into locally convex cones below. First examples also include the cone $\mathcal{P} = \overline{\mathbb{R}}$ with the neighborhood system $\mathcal{V} = \{\varepsilon > 0\}$, cones of convex subsets of a locally convex topological vector space and cones of $\overline{\mathbb{R}}$-valued functions with suitable neighborhood systems. These are indeed the types of locally convex cones that we shall employ for our purposes. We note that the symmetric topology on $\overline{\mathbb{R}}$ is finer than the usual one-point compactification topology on this set. Instead, it is the usual topology of $\mathbb{R}$ with $+\infty$ adjoined as an isolated point. Thus convergence toward $+\infty$ requires that this value is taken cofinally. For a wide range of examples of locally convex cones, we refer to Examples I.1.4 in [50]. Subcones of a locally convex cone, if endowed with the inherited neighborhood system are again locally convex cones. A subset A of $\mathcal{P}$ is called *bounded below (or above)* if for every $V \in \mathcal{V}$ there is $\lambda \geqslant 0$ such that $0 \leqslant a + \lambda V$ (or $a \leqslant \lambda V$) for all $a \in A$, and *bounded* if it satisfies both.

We remark that other than in the case of topological vector spaces the mapping $(\alpha, a) \mapsto \alpha a : [0, +\infty) \times \mathcal{P} \to \mathcal{P}$ is not guaranteed to be continuous in any of the cone topologies at points (α, a) where the element $a \in \mathcal{P}$ is unbounded (see Proposition I.1.1 in [48]). This shortcoming is at least partially remedied by the use of the slightly coarser *relative topologies*, which we shall briefly introduce and employ in Section 2.4 below.

If $(\mathcal{P}, \mathcal{V})$ and $(\mathcal{Q}, \mathcal{W})$ are locally convex cones, an operator $T : \mathcal{P} \to \mathcal{Q}$ is called *linear* if $T(a + b) = T(a) + T(b)$ and $T(\alpha a) = \alpha T(a)$ holds for all $a, b \in \mathcal{P}$ and $\alpha \geqslant 0$. T is *continuous* if for every $W \in \mathcal{W}$ there is $V \in \mathcal{V}$ such that $T(a) \leqslant T(b) + W$ whenever $a \leqslant b + V$ for $a, b \in \mathcal{P}$. By $\mathfrak{L}(\mathcal{P}, \mathcal{Q})$, we denote the cone of all continuous linear operators from $\mathcal{P}$ to $\mathcal{Q}$. The operators in $\mathfrak{L}(\mathcal{P}, \mathcal{Q})$ map bounded subsets of $\mathcal{P}$ into bounded subsets of $\mathcal{Q}$ and are monotone in the sense that $a \leq b + V$ for all $V \in \mathcal{V}$ implies that $T(a) \leq T(b) + W$ for all $W \in \mathcal{W}$.

The dual cone

The *dual cone* $\mathcal{P}^*$ of a locally convex cone $\mathcal{P}$ is the cone $\mathfrak{L}(\mathcal{P}, \overline{\mathbb{R}})$ of all continuous $\overline{\mathbb{R}}$-valued linear functionals on $\mathcal{P}$. The *polar* V° of a neighborhood $V \in \mathcal{V}$ consists of all $\mu \in \mathcal{P}^*$ such that $\mu(a) \leqslant \mu(b) + 1$ whenever $a \leqslant b + V$ for $a, b \in \mathcal{P}$. The polars of neighborhoods are known to be compact in the weak* topology of $\mathcal{P}^*$ (Proposition II.2.4 in [33]), that is, the topology of pointwise convergence of the elements of $\mathcal{P}^*$ considered as $\overline{\mathbb{R}}$-valued functions on $\mathcal{P}$. In this context, $\overline{\mathbb{R}}$ is meant to carry its usual one-point compactification topology. We have

$$\mathcal{P}^* = \bigcup_{V \in \mathcal{V}} V^\circ.$$

If $\mathcal{P}$ is indeed a locally convex topological vector space E, then $\mathcal{P}^*$ coincides with its real dual $E^*_{\mathbb{R}}$. If E is ordered and if this order is implemented in its embedding into a full cone (see I.2.7 in [33]), then $\mathcal{P}^*$ consists of the real parts of the functionals in E^*_+, that is, $\mathcal{P}^*$ is the cone of all real-valued monotone real-linear functionals on E.

Locally convex cones yield a rich duality theory, including powerful Hahn–Banach-type extension and separation theorems (see [33, 48] and [50]), some of which we shall use. We recall some facts about sublinear and superlinear functionals on locally convex cones.

Sub and superlinear functionals

A *sublinear functional* on a locally convex cone $\mathcal{P}$ is a mapping $p : \mathcal{P} \to \overline{\mathbb{R}}$ such that

$$p(\lambda a) = \lambda p(a) \quad \text{and} \quad p(a+b) \leqslant p(a) + p(b)$$

for all $a, b \in \mathcal{P}$ and $\lambda \geqslant 0$. A *superlinear functional* on $\mathcal{P}$ is a mapping $q : \mathcal{P} \to \underline{\mathbb{R}}$ such that $-q$ is sublinear. An *extended superlinear functional* is allowed to take both values $+\infty$ and $-\infty$. The inequality $q(a+b) \geqslant q(a) + q(b)$ is expected to hold if the sum on the right-hand side is defined in $\overline{\underline{\mathbb{R}}}$. The Sandwich theorem for locally convex cones (Theorem 3.1 in [50]) states the following.

1.4.1 Theorem. *For a sublinear functional $p : \mathcal{P} \to \overline{\mathbb{R}}$ and an extended superlinear functional $q : \mathcal{P} \to \underline{\mathbb{R}}$, there exists a linear functional $\mu \in \mathcal{P}^*$ such that $q \leqslant \mu \leqslant p$ if and only if there is $V \in \mathcal{V}$ such that*

$$q(a) \leqslant p(a) + 1 \quad \textit{whenever} \quad a \leqslant b + V$$

for $a, b \in \mathcal{P}$.

The relation $q \leqslant \mu \leqslant p$ is meant pointwise on $\mathcal{P}$. With suitable insertions for p and q, this result guarantees in particular that every continuous linear functional on a subcone of $\mathcal{P}$ can be extended to a continuous linear functional on $\mathcal{P}$. We shall also make use of the powerful Range theorem (Theorem 5.1 in [50]).

1.4.2 Theorem. *Let p and q be sublinear and extended superlinear functionals on $\mathcal{P}$ such that there is at least one linear functional $\mu \in \mathcal{P}^*$ satisfying $q \leqslant \mu \leqslant p$. Then for all $a \in \mathcal{P}$,*

$$\sup_{\substack{\mu \in \mathcal{P}^* \\ q \leqslant \mu \leqslant p}} \mu(a) = \sup_{V \in \mathcal{V}} \inf\{p(b) - q(c) \mid b, c \in \mathcal{P},\ q(c) \in \mathbb{R},\ a + c \leqslant b + V\},$$

and for all $a \in \mathcal{P}$ such that $\mu(a)$ is finite for at least one $\mu \in \mathcal{P}^*$ satisfying $q \leqslant \mu \leqslant p$,

$$\inf_{\substack{\mu \in \mathcal{P}^* \\ q \leqslant \mu \leqslant p}} \mu(a) = \inf_{V \in \mathcal{V}} \sup\{q(c) - p(b) \mid b, c \in \mathcal{P},\ p(b) \in \mathbb{R},\ c \leqslant a + b + V\}.$$

These results apply of course also to a locally convex topological vector space $(E, \mathcal{V})$, where $a \leqslant b + V$ for $a, b \in E$ and $V \in \mathcal{V}$ reads as $a \in b + V$, that is, $a - b \in V$. The continuous linear functionals in consideration then are real-valued and real-linear on E, that is, elements of $E^*_{\mathbb{R}}$. The classical Hahn–Banach theorem, although purely algebraic, is an immediate consequence of Theorem 1.4.1. Indeed, if p is a real-, that is, finite-valued sublinear functional on a real or complex vector space E, then the set

$$V = \{a \in E \mid p(\gamma a) \leqslant 1 \text{ for all } |\gamma| = 1\}$$

is balanced, convex and absorbing in E. The first two properties are obvious. For the latter, given $a \in E$, let $\rho = \max\{p(a), p(-a)\}$ in the real and $\rho = \max\{p(a), p(-a), p(ia), p(-ia)\}$ in the complex case. Then for any γ in the scalar field such that $|\gamma| = 1$, one easily calculates that $p(\gamma a) \leqslant 2\rho$. Thus $a \in 2\rho V$, our claim. Hence the neighborhood system $\mathcal{V} = \{\rho V \mid \rho > 0\}$ defines a locally convex topology on E. We use Theorem 1.4.1 with the given sublinear functional p, any $\underline{\mathbb{R}}$-valued superlinear functional q such that $q \leqslant p$, and the neighborhood $V \in \mathcal{V}$. Surely, for $a - b \in V$ we have

$$q(a) \leqslant p(a) \leqslant p(b) + p(a - b) \leqslant p(b) + 1.$$

Thus by 1.4.1 there is a linear functional $\mu \in E_{\mathbb{R}}^{\circledast}$ such that $q \leqslant \mu \leqslant p$. In particular, given a subspace F of E and a linear functional $\nu \in F_{\mathbb{R}}^{\circledast}$ that is dominated by p on F, we set $q(a) = \nu(a)$ for $a \in F$, and $q(a) = -\infty$ else. Then $q \leqslant \mu$ means that μ coincides with ν on F, hence is an extension of ν. This is the version of the Hahn–Banach theorem most popular in standard texts (see, e. g., Theorem II.3.10 in [26] or Theorem II.3.2 in [56]).

We note that these particular assertions hold no longer true in general if p is allowed to take the value $+\infty$ (see Example 1.4.8 below). In fact, Theorem 1.4.1 implies that an $\overline{\mathbb{R}}$-valued sublinear functional p on a vector space E dominates a linear functional in $E_{\mathbb{R}}^{\circledast}$ if and only if it is bounded below on some balanced, convex and absorbing subset of E (see Theorem 1.8 in [2]). Indeed, if p dominates a linear functional, the latter is bounded below on some balanced, convex and absorbing subset of E, and so is p. Conversely, if $p(a) \geqslant -1$ for all elements a of such a subset V of E, this subset generates a locally convex topology on E, and we use 1.4.1 with p and the superlinear functional q defined as $q(0) = 0$ and $q(a) = -\infty$ for all $a \neq 0$. Then for $a - b \in V$ we have either $q(a) = -\infty$ or $a = 0$ and $p(b) \geqslant -1$, hence $q(0) = 0 \leqslant p(b) + 1$, and Theorem 1.4.1 yields our claim. In case that p is finite-valued, since $p(a) \geqslant -p(-a)$ holds for all $a \in E$, this condition is satisfied with the set $V = \{a \in E \mid p(\gamma a) \leqslant 1 \text{ for all } |\gamma| = 1\}$ from above.

Convex sets and sublinear functionals

Let (E, F) be a dual pair of real or complex vector spaces endowed with a nondegenerate real- or complex-valued bilinear form $\langle\, , \rangle$, and let $\mathcal{V}$ and $\mathcal{W}$ be neighborhood systems for polar topologies on E and F, respectively, both consistent with this duality. There is a canonical correspondence between nonempty convex subsets of E and lower semicontinuous sublinear functionals on F, and vice versa. For $A \in \mathrm{Conv}(E)$, we set

$$p_A(c) = \sup\{\mathfrak{Re}\langle a, c\rangle \mid a \in A\}$$

for all $c \in F$. Clearly, p_A is sublinear and lower semicontinuous on F, and we have $p_{\alpha A} = \alpha p_A$ and $p_{A+B} = p_A + p_B$ for $A, B \in \mathrm{Conv}(E)$ and $\alpha \geqslant 0$. For $A \in \mathrm{Conv}(E)$ and scalars α in $\mathbb{R}$ or in $\mathbb{C}$ in general, we have

$$p_{\alpha A}(c) = \sup\{\mathfrak{Re}\langle \alpha a, c\rangle \mid a \in A\} = \sup\{\mathfrak{Re}\langle a, \alpha c\rangle \mid a \in A\} = p_A(\alpha c)$$

for all $c \in F$. Conversely, given a lower semicontinuous sublinear functional p on F, we set

$$A_p = \{a \in E \mid \mathfrak{Re}\langle a, c\rangle \leqslant p(c) \text{ for all } c \in F\}.$$

It follows from Theorem 1.4.1 that A_p is not empty, hence an element of $\mathrm{Conv}(E)$. Indeed, we use F in place of $\mathcal{P}$, the given sublinear functional p on F, and set $q(0) = 0$ and $q(c) = -\infty$ for $0 \neq c \in F$. Due to the lower semicontinuity of p at $0 \in F$, there is $W \in \mathcal{W}$ such that $p(c) \geqslant -1$ for all $c \in W$. Then $c - d \in W$ for $c, d \in F = E^*$ implies that $q(c) \leqslant p(d) + 1$. This is obvious for $c \neq 0$, and for $c = 0$ it follows from the above. Hence there is a linear functional $a \in E = F^*$ such that $\mathfrak{Re}\langle a, c\rangle \leqslant p(c)$ for all $c \in F$. That is $a \in A_p$. We shall confirm that $p_{A_p} = p$. Surely, the inequality $p_{A_p} \leqslant p$, pointwise on F, is evident. The reverse inequality is however less trivial and uses the first statement in Theorem 1.4.2 with the same insertions for p and q from above. We have $a \in A_p$ if and only if $q(d) \leqslant \mathfrak{Re}\langle a, d\rangle \leqslant p(d)$ for all $d \in F$. That is to say,

$$p_{A_p}(c) = \sup_{a \in A_p} \mathfrak{Re}\langle a, c\rangle = \sup_{W \in \mathcal{W}} \inf\{p(d) \mid d \in F,\ c - d \in W\}$$

holds by 1.4.2 for every $c \in F$. As p is supposed to be lower semicontinuous on F, given $c \in F$ and any $\alpha < p(c)$ there is a neighborhood $W \in \mathcal{W}$ such that $p(d) > \alpha$ for all $d \in c + W$, that is, $c - d \in -W = W$. Hence

$$\inf\{p(d) \mid d \in F,\ c - d \in W\} \geqslant \alpha.$$

This proves $p_{A_p}(c) \geqslant p(c)$, and our claim follows. Furthermore, according to the Hahn–Banach separation theorem (see Theorem II.9.2 in [56]), for every $A \in \mathrm{Conv}(E)$ and $b \notin \overline{A}$, the topological closure of A in E, there is $c \in F$ such that $\mathfrak{Re}\langle b, c\rangle > \sup\{\mathfrak{Re}\langle a, c\rangle \mid a \in A\}$. Therefore, we have $b \in \overline{A}$ if and only if

$$\mathfrak{Re}\langle b, c\rangle \leqslant \sup\{\mathfrak{Re}\langle a, c\rangle \mid a \in A\} = p_A(c)$$

holds for all $c \in F$. We infer that $p_A \leqslant p_B$ holds for A, B in $\mathrm{Conv}(E)$ if and only if $A \subset \overline{B}$, and that $p_A = p_B$ if and only if $\overline{A} = \overline{B}$.

1.4.3 Proposition. *Let (E, F) be a dual pair, and let $\mathcal{V}$ and $\mathcal{W}$ be neighborhood systems for polar topologies on E and F, respectively, both consistent with the duality.*

(a) *The embedding* $A \mapsto p_A$ *from* $\mathrm{Conv}(E)$ *into the cone of all lower semicontinuous sublinear functionals on* F *is linear, onto and* $p_A \leqslant p_B$ *holds for* A, B *in* $\mathrm{Conv}(E)$ *if and only if* $A \subset \overline{B}$.

(b) *For* $A \in \mathrm{Conv}(E)$, *the sublinear functional* p_A *is real-valued on* F *if and only if* A *is bounded.*

(c) *A sublinear functional on* F *is continuous if and only if it is upper semicontinuous at* $0 \in F$, *that is, if and only if it is bounded above on some neighborhood* $W \in \mathcal{W}$.

(d) *For* $A \in \mathrm{Conv}(E)$, *the sublinear functional* p_A *is continuous in the Mackey topology* $\tau(F, E)$ *of* F *if and only if* A *is relatively compact in the weak topology* $\sigma(E, F)$ *of* E.

Proof. Part (a) was argued before. Part (b) is obvious since the topology of E is supposed to be consistent with the duality (E, F). Therefore, boundedness of sets in $\mathrm{Conv}(E)$ in the given and in the weak topology $\sigma(E, F)$ of E coincides. For (c), let p be a sublinear functional on F. Clearly, continuity of p implies upper semicontinuity at $0 \in F$. Hence there is $W \in \mathcal{W}$ such that p is bounded above on W. Conversely, suppose that there is $W \in \mathcal{W}$ such that $p(c) \leqslant 1$ for all $c \in W$. Let $c \in F$ and $\varepsilon > 0$. For every $d \in c + \varepsilon W$, that is, $d - c \in \varepsilon W$, we have both

$$p(d) \leqslant p(c) + p(d - c) \leqslant p(c) + \varepsilon$$

and

$$p(c) \leqslant p(d) + p(c - d) \leqslant p(d) + \varepsilon$$

since W is supposed to be balanced. Thus p is continuous at c. For Part (d), suppose that $A \in \mathrm{Conv}(E)$ is relatively compact in the weak topology $\sigma(E, F)$ of E. Then the polar A° of A is a neighborhood in the Mackey topology $\tau(F, E)$ of F and

$$\sup\{p_A(c) \mid c \in A^\circ\} = \sup\{\mathfrak{Re}\langle a, c\rangle \mid a \in A,\ c \in A^\circ\} \leqslant 1.$$

Thus p_A is bounded above on A° and, therefore, continuous in the Mackey topology $\tau(F, E)$ of F by Part (c). Conversely, if p_A is $\tau(F, E)$-continuous on F, then there is a balanced $\sigma(E, F)$-compact set $B \in \mathrm{Conv}(E)$ such that $p_A(c) \leqslant 1$ for all $c \in B^\circ$. Thus $B^\circ \subset A^\circ$ and $A \subset B^{\circ\circ} = B$. Hence A is relatively $\sigma(E, F)$-compact. □

Proposition 1.4.3(c) implies in particular that a sublinear functional, if dominated by a continuous one, is itself continuous.

1.4.4 Corollary. *Let* E *be a Hausdorff locally convex topological vector space. A sublinear functional on* E *is lower semicontinuous in the given topology of* E *if and only if it is lower semicontinuous in the weak topology* $\sigma(E, E^*)$.

Proof. Clearly, lower semicontinuity in the weak topology implies lower semicontinuity in the given topology. For the converse, we apply Proposition 1.4.3(a) with the dual pair

(E^*, E), the given topology on E and the weak* topology, that is, $\sigma(E^*, E)$, on E^*. If a sublinear functional p on E is lower semicontinuous in the given topology, then according to Proposition 1.4.3(a) there is a convex subset Ω of E^* such that

$$p(a) = \sup\{\mathfrak{Re}\,\mu(a) \mid \mu \in \Omega\}$$

for all $a \in E$. As the pointwise supremum of $\sigma(E, E^*)$-continuous functions on E, the functional p itself is therefore $\sigma(E, E^*)$-lower semicontinuous. □

We remark that the ordered cone $\mathrm{Conv}(E)$ contains the supremum of any family $\{A_i\}_{i\in\mathcal{I}}$ of its elements. Indeed we observe that

$$\bigvee_{i\in\mathcal{I}} A_i = \mathrm{conv}\Big(\bigcup_{i\in\mathcal{I}} A_i\Big),$$

that is, the convex hull of the union of the sets A_i. We have

$$p_{\bigvee_{i\in\mathcal{I}} A_i} = \bigvee_{i\in\mathcal{I}} p_{A_i},$$

that is, the pointwise supremum on F of the sublinear functionals p_{A_i}. The distributive law is easily verified, that is to say,

$$\bigvee_{i\in\mathcal{I}} (A_i + B) = \bigvee_{i\in\mathcal{I}} A_i + B$$

holds for all $B \in \mathrm{Conv}(E)$. Similar statements are generally not available for infima in $\mathrm{Conv}(E)$, even if the family $\{A_i\}_{i\in\mathcal{I}}$ is directed downward and its intersection is not empty.

The locally convex cone $\mathcal{P}_E$

We shall utilize these concepts to develop our approach to integration theory. The following extensions of a vector space are of our particular interest. Given a locally convex topological vector space $(E, \mathcal{V})$, the family $\mathrm{Conv}(E)$ of all nonempty convex subsets of E endowed with its usual addition and multiplication by nonnegative scalars forms a cone for which the cancellation does not hold in general. It is canonically ordered by set inclusion. A subcone $\mathcal{P}_E$ of $\mathrm{Conv}(E)$ that contains all singleton sets $\{a\}$ for $a \in E$ is called a *cone extension of* E. It is a *full cone extension* provided that $\mathcal{V} \subset \mathcal{P}_E$. With the set inclusion as its canonical order, we use the vector space neighborhoods in $\mathcal{V}$ to generate a corresponding neighborhood system for $\mathcal{P}_E$. That is, we set

$$A \in V(B) \quad \text{if} \quad A \leqslant B + V$$

for $A, B \in \mathcal{P}_E$ and $V \in \mathcal{V}$. Clearly, for every $A \in \mathrm{Conv}(E)$ and $V \in \mathcal{V}$ there is $\lambda \geqslant 0$ such that $\{0\} \leqslant A + \lambda V$. In this way, $(\mathcal{P}_E, \mathcal{V})$ becomes a locally convex cone as previously introduced. It is a full locally convex cone provided that $\mathcal{V} \subset \mathcal{P}_E$. The embedding

$$a \mapsto \{a\} : E \to \mathcal{P}_E$$

is one-to-one and preserves the algebraic operations of E and $\mathcal{P}_E$. The upper, lower and symmetric topologies of $\mathcal{P}_E$ all coincide with the given topology of E on the image of this embedding. In this manner, $(E, \mathcal{V})$ becomes a real-linear subspace of $\mathcal{P}_E$. We shall identify the elements a of E with the singleton sets $\{a\} \in \mathcal{P}_E$. There is also a multiplication by negative numbers in $\mathrm{Conv}(E)$, which does however not result in additive inverses (see Section 1.1).

For every linear functional $\nu \in E^*$, the mapping

$$C \mapsto \sup\{\mathfrak{Re}\, \mu(a) \mid a \in C\} : \mathcal{P}_E \to \overline{\mathbb{R}}$$

defines an element of the dual cone $\mathcal{P}_E^*$ of $\mathcal{P}_E$.

The locally convex cone $\mathcal{Q}_F$

We shall also make use of a second cone extension of a vector space, called its *lattice cone extension*. For this, let $(F, \mathcal{W})$ be a Hausdorff locally convex topological vector space and set

$$\mathcal{Q}_F = \left\{ \varphi : F^* \to \overline{\mathbb{R}} \;\middle|\; \begin{array}{l} \varphi \text{ is sublinear and bounded below} \\ \text{on the polars of all neighborhoods in } \mathcal{W} \end{array} \right\}.$$

Endowed with the pointwise addition and multiplication by nonnegative scalars for functions on F^*, together with the pointwise order on F^*, $\mathcal{Q}_F$ forms an ordered cone. With every neighborhood $W \in \mathcal{W}$, we associate an $\overline{\mathbb{R}}_+$-valued sublinear functional $\sigma_W \in \mathcal{Q}_F$ defined as

$$\begin{aligned} \sigma_W(\mu) &= \sup\{\mathfrak{Re}\, \mu(c) \mid c \in W\} \\ &= \inf\{\lambda \geqslant 0 \mid \mu \in \lambda W^\circ\}, \end{aligned}$$

for $\mu \in F^*$. (As usual, we set $\inf \emptyset = +\infty$.) The second equality in this expression is easily verified. For this, we abbreviate $\alpha = \sup\{\mathfrak{Re}\, \mu(c) \mid c \in W\}$ and $\beta = \inf\{\lambda \geqslant 0 \mid \mu \in \lambda W^\circ\}$. We have $\mathfrak{Re}\, \mu(c) \leqslant \lambda$ whenever $c \in \mathcal{W}$ and $\mu \in \lambda W^\circ$. Thus $\alpha \leqslant \beta$. The inverse inequality is obvious if $\alpha = +\infty$. Otherwise, we observe that $\mu \in \rho W^\circ$ holds for every $\rho > \alpha$. Hence $\beta \leqslant \rho$ and, therefore, $\beta \leqslant \alpha$, our claim. We also notice that $\sigma_W(\mu) > 0$ at all nonzero elements $\mu \in F^*$, which follows from the fact that W is absorbing in F. The elements of $W \in \mathcal{W}$ then define a corresponding neighborhood system on $\mathcal{Q}_F$, that is to say, we set

$$\varphi \in W(\psi) \quad \text{if} \quad \varphi \leqslant \psi + \sigma_W$$

for $\varphi, \psi \in \mathcal{Q}_F$ and $W \in \mathcal{W}$. In this way, $(\mathcal{Q}_F, \mathcal{W})$ becomes a full locally convex cone. Indeed, every $\varphi \in \mathcal{Q}_F$ is bounded below on the polar of every neighborhood $W \in \mathcal{W}$. Hence there is $\lambda \geqslant 0$ such that $\varphi(\mu) \geqslant -\lambda$ for all $\mu \in W^\circ$. We claim that $0 \leqslant \varphi(\mu) + \lambda\sigma_W(\mu)$ holds for all $\nu \in F^*$. There is nothing to prove if $\sigma_W(\mu) = +\infty$. Otherwise, we have $\mu \in \rho W^\circ$ for all $\rho > \sigma_W(\mu)$. Thus $\varphi(\mu) \geqslant -\lambda\rho$, that is, $0 \leqslant \varphi(\mu) + \lambda\rho$. Our claim follows, and we infer that $0 \leqslant \varphi + \lambda\sigma_W$. Hence $\varphi \in \mathcal{Q}_F$ is bounded below as is required for the elements of a locally convex cone. We further note that $\varphi \leqslant \psi$ holds for $\varphi, \psi \in \mathcal{Q}_F$ if and only if $\varphi \leqslant \psi + \sigma_W$ for all $W \in \mathcal{W}$, and that $\varphi \leqslant \psi + \sigma_W$ holds if and only if $\varphi(\mu) \leqslant \psi(\mu) + 1$ for all $\mu \in W^\circ$. Indeed, the former obviously implies the latter, and if the latter holds and if $\mu \in \lambda W^\circ$ for $\lambda > 0$, then $(1/\lambda)\mu \in W^\circ$, hence $\varphi((1/\lambda)\mu) \leqslant \psi((1/\lambda)\mu) + 1$, that is, $\varphi(\mu) \leqslant \psi(\mu) + \lambda$.

The symmetric topology of $\mathcal{Q}_F$ is the topology of uniform convergence on the polars in F^* of all neighborhoods $W \in \mathcal{W}$. In other words, convergence of a net $(\varphi_i)_{i \in \mathcal{I}}$ in $\mathcal{Q}_F$ toward an element $\varphi \in \mathcal{Q}_F$ in the symmetric topology requires that for every $W \in \mathcal{W}$ there is $i_0 \in \mathcal{I}$ such that for every $\mu \in W^\circ$ we have either $|\varphi_i(\mu) - \varphi(\mu)| \leqslant 1$ or $\varphi_i(\mu) = \varphi(\mu) = +\infty$ for all $i \geqslant i_0$. Any Cauchy net $(\varphi_i)_{i \in \mathcal{I}}$ in $\mathcal{Q}_F$ with respect to the symmetric topology is therefore a pointwise Cauchy net on F^* with values in $\overline{\mathbb{R}}$ endowed with its symmetric topology. Accordingly, it is pointwise convergent to an $\overline{\mathbb{R}}$-valued function φ on F^*, which is easily seen to be sublinear and bounded below on the polars of all neighborhoods in $\mathcal{W}$, hence an element of $\mathcal{Q}_F$. The convergence of $(\varphi_i)_{i \in \mathcal{I}}$ to φ is indeed uniform on the polars of all neighborhoods in $\mathcal{W}$ and, therefore, refers to the symmetric topology of $\mathcal{Q}_F$. We summarize.

1.4.5 Proposition. *The locally convex cone $\mathcal{Q}_F$ is complete in its symmetric topology.*

If F is a normed space with unit ball $\mathbb{B}$, then the dual unit ball $\mathbb{B}^\circ$ in F^*, which determines the topology of $\mathcal{Q}_F$ also defines the strong topology $\beta(F^*, F)$, that is, the dual norm topology, of F^*. A sublinear functional φ on F^* is an element of $\mathcal{Q}_F$ if and only if it is bounded below on $\mathbb{B}^\circ$, that is, if and only if it is $\beta(F^*, F)$-lower semicontinuous at $0 \in F^*$. Because the topology $\beta(F^*, F)$ is consistent with the duality (F^*, F^{**}), according to Corollary 1.4.4 a sublinear functional φ on F^* is $\beta(F^*, F)$-lower semicontinuous on all of F^* if and only if it is lower semicontinuous in the weak topology $\sigma(F^*, F^{**})$. According to Proposition 1.4.3(c), a sublinear functional φ on F^* is $\beta(F^*, F)$-continuous on F^* if and only if it is bounded above on $\mathbb{B}^\circ$.

1.4.6 Proposition. *Let F be a normed space.*

(a) *The subcone of all $\beta(F^*, F)$-continuous functionals on F^* is closed in the lower topology of $\mathcal{Q}_F$.*
(b) *The subcone of all $\beta(F^*, F)$-lower semicontinuous functionals on F^* is closed in the symmetric topology of $\mathcal{Q}_F$.*

Proof. Suppose that F is a normed space with unit ball $\mathbb{B}$. Both sets $\mathcal{Q}_c$ of $\beta(F^*, F)$-continuous and $\mathcal{Q}_l$ of $\beta(F^*, F)$-lower semicontinuous sublinear functionals on F^*

form subcones of $\mathcal{Q}_F$. First, let $\varphi \in \mathcal{Q}_F$ be an element of the lower closure of $\mathcal{Q}_c$. There is $\psi \in \mathcal{Q}_c$ such that $\psi \in (\varphi)\mathbb{B}$, that is, $\varphi \leqslant \psi + \sigma_{\mathbb{B}}$, and since ψ is $\beta(F^*,F)$-continuous, there is $\lambda \geqslant 0$ such that $\psi(\mu) \leqslant \lambda$ for all $\mu \in \mathbb{B}^\circ$. Thus

$$\varphi(\mu) \leqslant \psi(\mu) + \sigma_{\mathbb{B}}(\mu) \leqslant \lambda + 1$$

for all $\mu \in \mathbb{B}^\circ$. Hence φ is also $\beta(F^*,F)$-continuous by 1.4.3(c), our claim. Now let $\varphi \in \mathcal{Q}_F$ be an element of the symmetric closure of $\mathcal{Q}_l$. We shall verify that φ is also lower semicontinuous on F^* with respect to $\beta(F^*,F)$. For this, let $\mu \in F^*$ and $\alpha < \varphi(\mu)$. Let $\lambda > 0$ such that $\mu \in \lambda\mathbb{B}^\circ$ and choose $\varepsilon > 0$ such that $\beta = \alpha + 3\varepsilon\lambda < \varphi(\mu)$. There is $\psi \in \mathcal{Q}_l$ such that $\psi \in (\varepsilon\mathbb{B})^s(\varphi)$, that is, both

$$\varphi \leqslant \psi + \varepsilon\sigma_{\mathbb{B}} \quad \text{and} \quad \psi \leqslant \varphi + \varepsilon\sigma_{\mathbb{B}}.$$

We infer that $\beta < \varphi(\mu) \leqslant \psi(\mu) + \varepsilon\sigma_{\mathbb{B}}(\mu) \leqslant \psi(\mu) + \varepsilon\lambda$, hence $\beta - \varepsilon\lambda < \psi(\mu)$. Because ψ is $\beta(F^*,F)$-lower semicontinuous at μ, we find $0 < \delta \leqslant \lambda$ such that

$$\beta - \varepsilon\lambda < \psi(\nu) \leqslant \varphi(\nu) + \varepsilon\sigma_{\mathbb{B}}(\nu)$$

for all $\nu \in \mu + \delta\mathbb{B}^\circ$. Since $\mu + \delta\mathbb{B}^\circ \subset 2\lambda\mathbb{B}^\circ$, we have $\sigma_{\mathbb{B}}(\nu) \leqslant 2\lambda$ and, therefore,

$$\alpha = \beta - 3\varepsilon\lambda \leqslant \varphi(\nu)$$

for all $\nu \in \mu + \delta\mathbb{B}^\circ$. This demonstrates that φ is lower semicontinuous at μ, hence on all of F^*, in the strong topology $\beta(F^*,F)$. Our claim follows. □

We shall however mostly use the coarser topology of pointwise convergence for the $\overline{\mathbb{R}}$-valued sublinear functionals in $\mathcal{Q}_F$. That is, convergence of a net $(\varphi_i)_{i\in\mathcal{I}}$ in $\mathcal{Q}_F$ toward an element $\varphi \in \mathcal{Q}_F$ in this topology requires that $(\varphi_i(\mu))_{i\in\mathcal{I}}$ converges to $\varphi(\mu)$ for every $\mu \in F^*$ in the usual one-point compactification topology of $\overline{\mathbb{R}}$. The topology of pointwise convergence for $\mathcal{Q}_F$ is of course still Hausdorff. The following observations are apparent from the definition of $\mathcal{Q}_F$. We recall that a subset of Φ of $\mathcal{Q}_F$ is bounded below if for every $W \in \mathcal{W}$ there is $\lambda \geqslant 0$ such that $\varphi \in (0)(\lambda W)$, that is, $0 \leqslant \varphi + \lambda\sigma_W$, for all $\varphi \in \Phi$.

The following observation is evident.

1.4.7 Proposition.

(a) *Every nonempty subset* $\Phi \subset \mathcal{Q}_F$ *has a supremum* $\sup\Phi \in \mathcal{Q}_F$. *We have* $\sup(\Phi + \varphi) = \sup\Phi + \varphi$ *for all* $\varphi \in \mathcal{Q}_F$.

(b) *The pointwise limit of a bounded below net in* $\mathcal{Q}_F$ *is contained in* $\mathcal{Q}_F$.

The intersection of $\mathcal{Q}_F$ and $-\mathcal{Q}_F$ consists of all real-valued linear functionals on F^* that are bounded on the polars of all neighborhoods in $\mathcal{W}$. For this reason, we shall also consider the space $F^{*\blacksquare}$ of all real- or complex-valued linear functionals on F^* that are bounded on the polars of all neighborhoods in $\mathcal{W}$, endowed with the topology of uniform convergence on these polars. That is, the topology generated by the second polars $W^{\circ\circ}$ in $F^{*\blacksquare}$ of the neighborhoods $W \in \mathcal{W}$. We have

$$F \subset F^{**} \subset F^{*\blacksquare} \subset F^{*\circledast},$$

where F^{**} is the second dual of F, that is, the dual of F^* under its strong topology, and $F^{*\circledast}$ is the algebraic dual of F^*. Indeed, for every element $\omega \in F^{**}$ there is a bounded subset B of F such that ω is bounded on the polar B° of B in F^*. For every $W \in \mathcal{W}$, we have $B \subset \lambda W$ for some $\lambda \geqslant 0$, hence $W^\circ \subset \lambda B^\circ$, and ω is seen to be bounded on W°, hence an element of $F^{*\blacksquare}$. If F is indeed a normed space, that is, if F admits a bounded neighborhood, then $F^{*\blacksquare} = F^{**}$. The topology of $F^{*\blacksquare}$ induces the given topology on F and the *natural* topology on F^{**} (see IV.5.3 in [56]). We note that the natural topology is generally weaker than the strong topology of F^{**}, which is generated by the polars of all (strongly) bounded subsets of F^*. Via the embedding

$$\omega \mapsto \varphi_\omega : F^{*\blacksquare} \to \mathcal{Q}_F,$$

where $\varphi_\omega(\mu) = \mathfrak{Re}\, \omega(\mu)$ for all $\mu \in F^*$, the spaces F, F^{**} and $F^{*\blacksquare}$ may be considered to be real-linear subspaces of $\mathcal{Q}_F$. The elements φ_ω are real-linear functionals on F^*, bounded on the polars of the neighborhoods in $\mathcal{W}$. The functional φ_ω is continuous with respect to the strong topology of F^* provided that $\omega \in F^{**}$, and continuous with respect to the weak* topology provided that $\omega \in F$. The element $\omega \in F^{*\blacksquare}$ can be reconstructed from φ_ω by $\omega(\mu) = \varphi_\omega(\mu)$ in the real and $\omega(\mu) = \varphi_\omega(\mu) - i\varphi_\omega(i\mu)$ in the complex case. The embedding $\omega \mapsto \varphi_\omega$ of F, F^{**} and $F^{*\blacksquare}$ into $\mathcal{Q}_F$ is real-linear and one-to-one. The image of $F^{*\blacksquare}$ is the intersection of $\mathcal{Q}_F$ and $-\mathcal{Q}_F$. The upper, lower and symmetric topologies of $\mathcal{Q}_F$ all coincide with the given topology of F and the *natural* topology of F^{**} and $F^{*\blacksquare}$ on this embedding. The completion $\widehat{F}$ of F is generally not a subspace of F^{**}, but of $F^{*\blacksquare}$ since it carries the same neighborhood system and the same dual F^* as F. Thus we may use the same insertion $c \mapsto \varphi_c : \widehat{F} \to \mathcal{Q}_F$ as for $F^{*\blacksquare}$. The topology of pointwise convergence on $\mathcal{Q}_F$ induces the weak topology on F and $\widehat{F}$ and the weak* topology on F^{**}.

Similarly, using the embedding of F, F^{**} and $F^{*\blacksquare}$ into $\mathcal{Q}_F$, the spaces $\mathfrak{L}(E,F)$, $\mathfrak{L}(E,F^{**})$ and $\mathfrak{L}(E,F^{*\blacksquare})$ may be considered to be real-linear subspaces of $\mathfrak{L}(E,\mathcal{Q}_F)$. Indeed, with an operator $T \in \mathfrak{L}(E,F^{*\blacksquare})$ we associate the real-linear operator $a \mapsto \varphi_{T(a)} : E \to \mathcal{Q}_F$. Again, $F^{*\blacksquare}$ is supposed to carry the topology of uniform convergence on the polars of the neighborhoods in $\mathcal{W}$. The linearity of $T \in \mathfrak{L}(E,F^{*\blacksquare})$ returns that

$$\varphi_{T(\alpha a)}(\mu) = \varphi_{\alpha T(a)}(\mu) = \varphi_{T(a)}(\alpha\mu)$$

holds for all $a \in E$, all α in the common scalar field of E and F and all $\mu \in F^*$. If $F = \mathbb{C}$, for example, then $F^* = \mathbb{C}$ and $\mathcal{Q}_F$ consists of all $\overline{\mathbb{R}}$-valued functions on $\mathbb{C}$ that are sublinear and bounded below on the unit ball of $\mathbb{C}$. The elements $z \in \mathbb{C}$ are embedded into $\mathcal{Q}_F$ via $z \mapsto \varphi_z \in \mathcal{Q}_F$, where $\varphi_z(w) = \mathfrak{Re}\, zw$ for all $w \in \mathbb{C}$. The case that $F = \mathbb{R}$ is similar.

As was established before, the embedding $c \mapsto \varphi_c : F \to \mathcal{Q}_F$ can be extended canonically to an embedding $C \mapsto \varphi_C$ of $\mathrm{Conv}(F)$ into $\mathcal{Q}_F$ setting $\varphi_C = \sup\{\varphi_c \mid c \in C\} \in \mathcal{Q}_F$, that is,

$$\varphi_C(\mu) = \sup\{\mathfrak{Re}\,\mu(c) \mid c \in C\}$$

for all $C \in \mathrm{Conv}(F)$ and $\mu \in F^*$. The sublinear functional $\varphi_C \in \mathcal{Q}_F$ is lower semicontinuous on F^* with respect to its weak* topology. Proposition 1.4.3(a) if applied to the dual pair (F, F^*) with the given topology of F and the weak* topology for F^* yields that this embedding is linear, and that its range consists of all sublinear functionals on F^* that are lower semicontinuous in a topology consistent with the duality (F^*, F) (see Corollary 1.4.4). The operator

$$C \mapsto \varphi_C : \mathrm{Conv}(F) \to \mathcal{Q}_F$$

is continuous if both $\mathrm{Conv}(F)$ and $\mathcal{Q}_F$ are endowed with their respective locally convex cone topologies. Indeed, given $W \in \mathcal{W}$, we notice that $\varphi_W = \sigma_W$. Then $C \in W(D)$, that is, $C \leqslant D + W$ for $C, D \in \mathrm{Conv}(F)$ implies by 1.4.3(a) that

$$\varphi_C \leqslant \varphi_{D+W} = \varphi_D + \varphi_W = \varphi_D + \sigma_W,$$

that is, $\varphi_C \in W(\varphi_D)$. A similar embedding can be established for $\mathrm{Conv}(F^{**})$ if we use Proposition 1.4.3 for the dual pair (F^{**}, F^*) with the weak* topology of F^{**} and the strong topology $\beta(F^*, F)$ for F^*. The elements of $\mathrm{Conv}(F^{**})$ correspond to the sublinear functionals in $\mathcal{Q}_F$ that are lower semicontinuous on F^* in a topology consistent with the duality (F^*, F^{**}), and the operator

$$C \mapsto \varphi_C : \mathrm{Conv}(F^{**}) \to \mathcal{Q}_F$$

is continuous if $\mathrm{Conv}(F^{**})$ is endowed with the locally convex cone topology generated by the second polars $W^{\circ\circ}$ in F^{**} of the neighborhoods in $\mathcal{W}$, in other words, by the natural topology of F^{**}. Indeed, given $W \in \mathcal{W}$, again we notice that $\varphi_{W^{\circ\circ}} = \sigma_W$ since

$$\varphi_{W^{\circ\circ}}(\mu) = \sup\{\mathfrak{Re}\,\omega(\mu) \mid \omega \in W^{\circ\circ}\} \leqslant 1$$

holds for all $\mu \in W^\circ$. Then $C \in W^{\circ\circ}(D)$, that is, $C \leqslant D + W^{\circ\circ}$ for $C, D \in \mathrm{Conv}(F^{**})$ implies that

$$\varphi_C \leqslant \varphi_{D+W^{\circ\circ}} = \varphi_D + \varphi_{W^{\circ\circ}} = \varphi_D + \sigma_W.$$

For $C \in \mathrm{Conv}(F)$ and an element α in the scalar field of F, we have $\varphi_{\alpha C}(\mu) = \varphi_C(\alpha\mu)$ for all $\mu \in F^*$.

Elements of $\mathcal{Q}_F$ representing convex sets

Let F^{*0} be a subspace of $F^{*\blacksquare}$ containing F. In the light of the preceding considerations, we shall say that an element φ of $\mathcal{Q}_F$ *represents a set in* $\mathrm{Conv}(F^{*0})$ if $\varphi = \varphi_C$ for some $C \in \mathrm{Conv}(F^{*0})$, that is, if the sublinear functional φ is lower semicontinuous on F^* in a topology that is consistent with the duality (F^*, F^{*0}). We are most interested in the cases that

$$F^{*0} = F, \quad F^{*0} = F^{**}, \quad F^{*0} = \widehat{F} \quad \text{or} \quad F^{*0} = F^{*\blacksquare}.$$

Obviously, if F^{*1} is a subspace of $F^{*\blacksquare}$ containing F^{*0}, then lower semicontinuity in $\sigma(F^*, F^{*0})$ implies lower semicontinuity in $\sigma(F^*, F^{*1})$. Thus, if an element $\varphi \in \mathcal{Q}_F$ represents a set C_0 in $\mathrm{Conv}(F^{*0})$, then it also represents a set C_1 in $\mathrm{Conv}(F^{*1})$. According to Proposition 1.4.3, however, the closures of C_0 and C_1 as subsets of F^{*1} coincide in any topology that is consistent with the duality (F^{*1}, F^*).

The following example will demonstrate that in general not every element of $\mathcal{Q}_F$ represents a convex subset of some subspace of $F^{*\blacksquare}$.

1.4.8 Example. Let $(F, \mathcal{W})$ be the space of all sequences $(\alpha_i)_{i\in\mathbb{N}}$ in $\mathbb{R}$ endowed with the topology of pointwise convergence (see Example 1.2.7(c)). The neighborhoods in $\mathcal{W}$ are defined by finite subsets Y of $\mathbb{N}$ and $\varepsilon > 0$ as

$$W_{(Y,\varepsilon)} = \{(\alpha_i)_{i\in\mathbb{N}} \in F \mid |\alpha_i| \leqslant \varepsilon \text{ for all } i \in Y\}.$$

If $\alpha_i = 0$ for all $i \in Y$ for a sequence $(\alpha_i)_{i\in\mathbb{N}} \in F$, then every linear functional μ in the polar $W^\circ_{(Y,\varepsilon)}$ of $W_{(Y,\varepsilon)}$ takes the value 0 on this sequence. Hence μ concurs on any two sequences in F whose terms with indexes in Y coincide. If e_n denotes the n-th unit sequence in F, then $\mu(e_i) = 0$ for all $i \notin Y$ and

$$\mu((\alpha_i)_{i\in\mathbb{N}}) = \mu\Big(\sum_{i\in Y} \alpha_i e_i\Big) = \sum_{i\in Y} \alpha_i \mu(e_i)$$

holds for all $(\alpha_i)_{i\in\mathbb{N}} \in F$. Thus the dual F^* of F corresponds to the space of all sequences $(\gamma_i)_{i\in\mathbb{N}}$ in $\mathbb{R}$ with only finitely many nonzero elements, and the polar $W^\circ_{(Y,\varepsilon)}$ of $W_{(Y,\varepsilon)}$ consists of all sequences $(\gamma_i)_{i\in\mathbb{N}}$ in F^* such that $\gamma_i = 0$ for all $i \notin Y$ and $\sum_{i\in Y} |\gamma_i| \leqslant 1/\varepsilon$. For every $c = (\gamma_i)_{i\in\mathbb{N}} \in F^*$, we set

$$\varphi(c) = \begin{cases} +\infty, & \text{if at least one } \gamma_i > 0, \\ n(c) \sum_{i\in\mathbb{N}} \gamma_i, & \text{if all } \gamma_i \leqslant 0. \end{cases}$$

whereby $n(c) = \max\{i \in \mathbb{N} \mid \gamma_i \neq 0\}$ if $c \neq 0$ and $n(0) = 0$. It is easily verified that φ defines a sublinear functional on F^*, and indeed an element of $\mathcal{Q}_F$ as it is bounded below on the polar of every neighborhood in F. Surely, for every $W_{(Y,\varepsilon)} \in \mathcal{W}$ we have $n(c) \leqslant \max\{i \in Y\}$ for all $c \in W^{\circ}_{(Y,\varepsilon)}$, hence

$$\varphi(c) \geqslant -(1/\varepsilon)\max\{i \in Y\}.$$

There is however no linear functional $\omega \in F^{*\circledast}$ on F^* such that $\omega \leqslant \varphi$, because assuming otherwise, we set $\varepsilon_n = \omega(c_n)$, where c_n is the n-th unit sequence in F^* and obtain a contradiction as follows. We choose $n \in \mathbb{N}$ and $\lambda \geqslant 0$ such that $n \geqslant \varepsilon_1 + 1$ and $\lambda > \varepsilon_n - n$ and set $c = -(\lambda c_1 + c_n) \in F^*$. Then $\omega(c) = -(\lambda\varepsilon_1 + \varepsilon_n)$, and as $\omega(c) \leqslant \varphi(c) = -n(\lambda+1)$, we conclude that $\lambda\varepsilon_1 + \varepsilon_n \geqslant n(\lambda+1)$, that is, $\varepsilon_n - n \geqslant \lambda(n-\varepsilon_1) \geqslant \lambda$, contradicting the above.

As a consequence of Propositions 1.4.5 and 1.4.6, we obtain the following.

1.4.9 Proposition. *If E is a reflexive normed space, then the locally convex cone* $\operatorname{Conv}(E)$ *is complete in its symmetric topology.*

Proof. We shall use Proposition 1.4.6 with E in place of F. Let E be a reflexive normed space with unit ball $\mathbb{B}$ and let $(A_i)_{i\in\mathcal{I}}$ be a symmetric Cauchy net in the locally convex cone $\mathcal{P}_E = \operatorname{Conv}(E)$. For every $\varepsilon > 0$, there is $i_0 \in \mathcal{I}$ such that $A_i \in (\varepsilon\mathbb{B})^s(A_k)$, that is,

$$A_i \leqslant A_k + \varepsilon\mathbb{B} \quad \text{and} \quad A_k \leqslant A_i + \varepsilon\mathbb{B}$$

for $i, k \geqslant i_0$. The corresponding sublinear functionals $\varphi_{A_i} \in \mathcal{Q}_E$ are lower semicontinuous in $\sigma(E^*, E)$, hence also in the finer topology $\beta(E^*, E)$. The net $(\varphi_{A_i})_{i\in\mathcal{I}}$ is Cauchy in the symmetric topology of $\mathcal{Q}_E$ and, therefore, convergent by Proposition 1.4.5. Its limit φ is also $\beta(E^*, E)$-lower semicontinuous by Proposition 1.4.6(b). Because E is supposed to be reflexive, the strong topology $\beta(E^*, E)$ is consistent with the duality (E^*, E). Thus following Proposition 1.4.3(a) there is a convex set $A \in \operatorname{Conv}(E)$ such that $\varphi = \varphi_A$. It is now straightforward to verify that $\lim_{i\in\mathcal{I}} A_i = A$ holds in the symmetric topology of $\operatorname{Conv}(E)$. □

We proceed with a few observations about continuous linear operators from the locally convex cone $\mathcal{P}_E$ into the locally convex cone $\mathcal{Q}_F$.

1.4.10 Lemma. *Every operator $T \in \mathfrak{L}(\mathcal{P}_E, \mathcal{Q}_F)$ is monotone in the sense that $A \subset \overline{B}$ for $A, B \in \mathcal{P}_E$ implies that $T(A) \leqslant T(B)$.*

Proof. Let $T \in \mathfrak{L}(\mathcal{P}_E, \mathcal{Q}_F)$ and $A \subset \overline{B}$ for $A, B \in \mathcal{P}_E$. For $\mu \in F^*$ and $\varepsilon > 0$, there is $W \in \mathcal{W}$ such that $\mu \in \varepsilon W^{\circ}$. Because T is continuous, there is $V \in \mathcal{V}$ such that $T(C) \in W(T(D))$, that is, $T(C) \leqslant T(D) + \sigma_W$, whenever $C \in V(D)$ for $C, D \in \mathcal{P}_E$. But $A \subset \overline{B}$ implies that $A \subset B + V$, and we infer that

$$T(A)(\mu) \leqslant T(B)(\mu) + \sigma_W(\mu) \leqslant T(B)(\mu) + \varepsilon$$

holds for all $\varepsilon > 0$. Our claim follows. □

We note that every operator $T \in \mathfrak{L}(\mathcal{P}_E, \mathcal{Q}_F)$ is real-linear on the subspace E of $\mathcal{P}_E$. Indeed, for $a \in E$ we have $\{a\} + \{-a\} = \{0\}$. Hence $T(-a) = -T(a)$ and $T(\alpha a) = (-\alpha)T(-a) = \alpha T(a)$ for $\alpha < 0$. Thus $T(\alpha a) = \alpha T(a)$ holds for all $\alpha \in \mathbb{R}$. The image $T(a)$ of an element $a \in E$ is therefore contained in the embedding of F^{*}■ as a subspace of $\mathcal{Q}_F$.

1.4.11 Theorem. *For an operator $T \in \mathfrak{L}(\mathcal{P}_E, \mathcal{Q}_F)$ and $A \in \operatorname{Conv}(E)$, the formula*

$$\overline{T}(A) = \sup\{T(a) \mid a \in A\} \in \mathcal{Q}_F$$

defines an operator in $\mathfrak{L}(\operatorname{Conv}(E), \mathcal{Q}_F)$ such that $\overline{T}(A) \leqslant T(A)$ for all $A \in \mathcal{P}_E$. We have $\overline{T}(A) \leqslant S(A)$ for all $A \in \operatorname{Conv}(E)$ for every other operator $S \in \mathfrak{L}(\operatorname{Conv}(E), \mathcal{Q}_F)$ with this property.

Proof. Let $T \in \mathfrak{L}(\mathcal{P}_E, \mathcal{Q}_F)$. Let us first verify that the mapping

$$A \mapsto \overline{T}(A) : \operatorname{Conv}(E) \to \mathcal{Q}_F$$

as defined in our statement is linear: For $A, B \in \operatorname{Conv}(E)$, for $\lambda \geqslant 0$ and every $\mu \in F^*$, we have

$$\begin{aligned}\overline{T}(A+B)(\mu) &= \sup\{T(a+b)(\mu) \mid a \in A,\ b \in B\} \\ &= \sup\{T(a)(\mu) \mid a \in A\} + \sup\{T(b)(\mu) \mid b \in B\} \\ &= \overline{T}(A)(\mu) + \overline{T}(B)(\mu)\end{aligned}$$

and

$$\begin{aligned}\overline{T}(\lambda A)(\mu) &= \sup\{T(a)(\mu) \mid a \in \lambda A\} \\ &= \lambda \sup\{T(a)(\mu) \mid a \in A\} \\ &= \lambda \overline{T}(A)(\mu).\end{aligned}$$

For continuity, let $W \in \mathcal{W}$ and choose $V \in \mathcal{V}$ such that $T(C) \in W(T(D))$ whenever $C \in V(D)$ for $C, D \in \mathcal{P}_E$. Let $A, B \in \operatorname{Conv}(E)$ such that $A \in V(B)$, that is, $A \leqslant B + V$. We shall verify that $\overline{T}(A) \in W(\overline{T}(B))$. Indeed, for every $a \in A$ we have $a \in b + V$, that is, $a \in V(b)$, for some $b \in B$. Since $a, b \in \mathcal{P}_E$, this yields $T(a) \in W(T(b))$, that is, $T(a) \leqslant T(b) + \sigma_W$. Now using Proposition 1.4.7(a), we infer that

$$T(a) \leqslant \sup\{T(b) + \sigma_W \mid b \in B\} = \sup\{T(b) \mid b \in B\} + \sigma_W = \overline{T}(B) + \sigma_W$$

holds for all $a \in A$ and, therefore, $\overline{T}(A) \leqslant \overline{T}(B) + \sigma_W$, our claim. We conclude that $\overline{T} \in \mathfrak{L}(\operatorname{Conv}(E), \mathcal{Q}_F)$. Moreover, for every $A \in \mathcal{P}_E$ we have $\overline{T}(A) = \sup\{T(a) \mid a \in A\} \leqslant T(A)$ by Lemma 1.4.10. Every other operator $S \in \mathfrak{L}(\operatorname{Conv}(E), \mathcal{Q}_F)$ with this property coincides with T on the elements of E, hence

$$\overline{T}(A) = \sup\{T(a) \mid a \in A\} = \sup\{S(a) \mid a \in A\} \leqslant S(A)$$

holds for all $A \in \mathrm{Conv}(E)$. □

The restriction of an operator $T \in \mathfrak{L}(\mathcal{P}_E, \mathcal{Q}_F)$ to the subspace E of $\mathcal{P}_E$ is real-linear, that is, $0 = T(a) + T(-a)$ implies that $T(a) \in F^{*\blacksquare}$ and $T(-a) = -T(a)$ holds for all $a \in E$. In other words, we have $T(-a)(\mu) = T(a)(-\mu)$ for all $\mu \in F^*$. If, in the complex case, the restriction of T to E is indeed complex-linear, that is, an element of $\mathfrak{L}(E, F^{*\blacksquare})$, then we have

$$T(\alpha a)(\mu) = T(a)(\alpha\mu)$$

for all $a \in E$, all $\alpha \in \mathbb{C}$ and all $\mu \in F^*$.

The cone $\mathcal{F}(X, \mathrm{Conv}(E))$

Let X be a set. By $\mathcal{F}(X, \mathrm{Conv}(E))$, we denote the cone of all $\mathrm{Conv}(E)$-valued functions on X, endowed with the pointwise operations and order, that is, $f \leqslant g$ for $f, g \in \mathcal{F}(X, \mathrm{Conv}(E))$ if $f(x) \leqslant g(x)$ in $\mathrm{Conv}(E)$ for all $x \in X$. On the subspace $\mathcal{F}(X, E)$ of $\mathcal{F}(X, \mathrm{Conv}(E))$, this order is of course the equality. A function $f \in \mathcal{F}(X, \mathrm{Conv}(E))$ is said to be *positive* if $f(x) \geqslant 0$ for all $x \in X$. Unfortunately, this notion conflicts with the conventional notion of positivity for real-valued functions, where the value 0 is not permitted. We continue to use the latter for functions in $\mathcal{F}(X)$. The definition of function space neighborhoods in Section 1.2 extends in a canonical way to the set-valued functions in $\mathcal{F}(X, \mathrm{Conv}(E))$. In this vein, for a function space neighborhood $\mathfrak{v}$ defined by a convex family $\mathcal{N}_{\mathfrak{v}}$ of neighborhood functions and for functions $f, g \in \mathcal{F}(X, \mathrm{Conv}(E))$ we write $f \leqslant g + \mathfrak{v}$ if $f \leqslant g + n$, that is, $f(x) \leqslant g(x) + n(x)$ for all $x \in X$, for some neighborhood function $n \in \mathcal{N}_{\mathfrak{v}}$. On the subspace $\mathcal{F}(X, E)$ of $\mathcal{F}(X, \mathrm{Conv}(E))$, this notion is consistent with our definition in Section 1.2, that is, $f \leqslant g + \mathfrak{v}$ for $f, g \in \mathcal{F}(X, E)$ means that both $f - g \leqslant \mathfrak{v}$ and $g - f \leqslant \mathfrak{v}$. We note that other than for E-valued functions the action of a function space neighborhood $\mathfrak{v}$ on $\mathcal{F}(X, \mathrm{Conv}(E))$ is not fully determined by the convex subset $\{f \in \mathcal{F}(X, \mathrm{Conv}(E)) \mid f \leqslant \mathfrak{v}\}$ of $\mathcal{F}(X, \mathrm{Conv}(E))$. One would have to employ the subset $\{(f, g) \mid f \leqslant g + \mathfrak{v}\}$ of $\mathcal{F}(X, \mathrm{Conv}(E))^2$ instead. We also write $\mathcal{A} \leqslant \mathcal{B} + \mathfrak{v}$ for nonempty subsets $\mathcal{A}, \mathcal{B}$ of $\mathcal{F}(X, \mathrm{Conv}(E))$ if for every $f \in \mathcal{A}$ there is $g \in \mathcal{B}$ such that $f \leqslant g + \mathfrak{v}$.

Using this notation for $\mathrm{Conv}(E)$-valued functions, the condition on function space neighborhoods from Section 1.2 now reads that for every $A \in \mathfrak{R}$ there is $V \in \mathcal{V}$ such that $\chi_A \cdot V \leqslant \mathfrak{v}$. Statements (a) and (b) of Lemma 1.2.1 transfer as follows to $\mathrm{Conv}(E)$-valued functions. (a): if $f \leqslant g + \mathfrak{v}$, then $\varphi \cdot f \leqslant \varphi \cdot g + \mathfrak{v}$ for all φ such that $|\varphi| \leqslant 1$, and (b): if $\mathcal{N}_{\mathfrak{v}}$ is upward directed, then $\varphi \cdot f + (1 - \varphi) \cdot g \leqslant \varphi \cdot f' + (1 - \varphi) \cdot g' + \mathfrak{v}$ whenever $f \leqslant f' + \mathfrak{v}$, $g \leqslant g' + \mathfrak{v}$ and $0 \leqslant \varphi \leqslant 1$. We shall take advantage of this simplified notation in the sequel.

A function $f \in \mathcal{F}(X, \mathrm{Conv}(E))$ is said to be *bounded above* (or *bounded below*) if for every function space neighborhood $\mathfrak{v}$ there is $\lambda \geqslant 0$ such that $f \leqslant \lambda\mathfrak{v}$ (or $0 \leqslant f + \lambda\mathfrak{v}$). A function is *bounded* if it is both bounded above and below. These concepts of boundedness are also used for subsets of $\mathcal{F}(X, \mathrm{Conv}(E))$ provided that the above conditions hold with the same $\lambda \geqslant 0$ for all functions in this subset. Since all function space neighborhoods are balanced, boundedness above and below coincide for E-valued functions.

Given a subset B of X, we shall also use the symbols $=_B$, $\leqslant_B$ or $\geqslant_B$ in order to signify that the relation $=$, $\leqslant$ or $\geqslant$ holds for $\mathrm{Conv}(E)$-valued functions pointwise on the set B.

For a function $f \in \mathcal{F}(X, \mathrm{Conv}(E))$ and an element α of the scalar field of E, we shall denote the function $x \mapsto \alpha f(x)$ by $\alpha f \in \mathcal{F}(X, \mathrm{Conv}(E))$. For an E-valued function, $-f$ is of course the additive inverse of f, but in general we only have $0 \leqslant f + (-f)$ and $-(-f) = f$. Moreover, $\alpha f \leqslant \alpha g$ holds whenever $f \leqslant g$, and $\alpha f \leqslant \alpha g + |\alpha|\mathfrak{v}$ holds whenever $f \leqslant g + \mathfrak{v}$ for $f, g \in \mathcal{F}(X, \mathrm{Conv}(E))$, a scalar α in $\mathbb{R}$ or $\mathbb{C}$ and a function space neighborhood $\mathfrak{v}$.

2 Integration

Using the notation, facts and methodology from the preceding chapter, we shall proceed to investigate continuous linear operators from a function space $(\mathcal{C}_{\mathfrak{V}}(X,E),\mathfrak{V})$ into a locally convex Hausdorff space $(F,\mathcal{W})$. The space of these operators will be denoted by $\mathfrak{L}(\mathcal{C}_{\mathfrak{V}}(X,E),F)$. Section 2.6 below will be concerned with integral representations for this type of operators, based on the classical result by Riesz [44], which states that every positive linear functional on the space of all continuous real-valued functions with compact support on a locally compact Hausdorff space X can be represented by the integral with respect to some regular positive Borel measure on X (see Theorem 13.23 in [53]). There are a number of generalizations of this result, most of them are concerned with operators on spaces of continuous real-valued functions on a compact space and representations by vector-valued measures. One of the better-known versions is the representation theorem by Bartle–Dunford–Schwartz [7] (see Theorem 1 in Chapter VI.2 in [21]). It states that a weakly compact continuous linear operator from a space of continuous real- or complex-valued functions on a compact space, endowed with the supremum norm, into some Banach space F can be represented by the integral with respect to a regular F-valued Borel measure. In case that the operator T is not weakly compact, the representing measure has to be allowed to take values in the second dual of F. These results will be recovered as special cases in the later sections of this chapter. Our approach to integral representations follows the method used for locally convex cones in [48]. In the special case of topological vector spaces, results can be transferred using vector space terminology, but at some expense of refinement and simplicity. Our more specialized setting, however, allows at times for more direct arguments and at instances more advanced results. Ideally, one would aim for a representation of an operator in $\mathfrak{L}(\mathcal{C}_{\mathfrak{V}}(X,E),F)$ by an integral using an $\mathfrak{L}(E,F)$-valued measure on X. This is, however, not possible in general as has been established by earlier approaches, even in the special case when X is compact, $E=\mathbb{R}$ and $\mathcal{C}(X)$ is endowed with the topology of uniform convergence.

We define and investigate operator-valued measures in Section 2.1. In Section 2.2 we establish a rather narrow measure independent notion of measurability of functions, which is used in Section 2.3 for the definition of integrals of vector- and set-valued functions. The general convergence theorems are treated separately in Section 2.4. In Section 2.5, we investigate the linear operators on function spaces defined by the integral. Section 2.6 then contains an integral representation theorem for linear operators on function spaces together with some of its consequences, one of our main results.

We broadly refer to the sources in [6, 10, 19] and [31] for real-valued, and to [5, 20, 21, 29, 43] and [58] for vector-valued integration theory. Set-valued functions and their integrals are considered in [3, 4, 27] and [57]. Results about integral representation can be found in [7, 8, 12, 21, 35] and [44].

https://doi.org/10.1515/9783111315478-002

2.1 Operator-valued measures

For the sake of generality, we shall formulate our take on integration theory in this and the following sections for a set X without the imposition of a topological structure. Measures utilized for integral representation of linear operators on function spaces can generally not be defined on a σ-field containing the whole space X. We shall therefore employ a *σ-ring* $\mathfrak{R}$ in X instead. This notion was introduced and developed in Sections 1.1 and 1.2. In the topological case, that is, if X is a locally compact Hausdorff space, we assume that $\mathfrak{R}$ consists of all relatively compact Borel subsets of X. As remarked earlier, the σ-field $\mathfrak{A}$ associated with $\mathfrak{R}$ contains all Borel subsets of X in this case, and exactly those if X is countably compact.

Definition of an operator-valued measure

The measures θ on X in our approach will be defined on the sets in $\mathfrak{R}$ and will be $\mathfrak{L}(\mathcal{P}_E, \mathcal{Q}_F)$-valued, where $\mathcal{P}_E$ is a cone extensions of E and $\mathcal{Q}_F$ is the lattice cone extension of F. Both were introduced in Section 1.4. More precisely, an *$\mathfrak{L}(\mathcal{P}_E, \mathcal{Q}_F)$-valued measure* θ on $\mathfrak{R}$ is a set function

$$A \mapsto \theta_A : \mathfrak{R} \to \mathfrak{L}(\mathcal{P}_E, \mathcal{Q}_F)$$

such that $\theta_\emptyset = 0$ and that θ is *countably additive*, that is,

$$\theta_{\bigcup_{i\in\mathbb{N}} A_i} = \sum_{i=1}^{\infty} \theta_{A_i}$$

holds whenever the sets $A_i \in \mathfrak{R}$ are disjoint and $\bigcup_{i\in\mathbb{N}} A_i \in \mathfrak{R}$. Convergence for the series on the right-hand side is meant pointwise for the elements of $\mathcal{P}_E$ and F^*, that is, for every element $C \in \mathcal{P}_E$ the series $\sum_{i=1}^{\infty} \theta_{A_i}(C)$ converges in the topology of pointwise convergence of $\mathcal{Q}_F$ to $\theta_{\bigcup_{i\in\mathbb{N}} A_i}(C)$.

$\mathfrak{L}(E, F^{*\blacksquare})$-valued measures

If $\mathcal{P}_E = E$, then for every $A \in \mathfrak{R}$ and $a \in E$ we have $0 = \theta_A(a) + \theta(-a)$, hence $\theta_A(a) = -\theta_A(a) \in \mathcal{Q}_F$ and, therefore, $\theta_A(a)$ is an element of the embedding of $F^{*\blacksquare}$ into $\mathcal{Q}_F$. Thus $\theta_A(a)$ is a real-linear functional on F^*, and θ_A defines a real-linear operator from E into $F^{*\blacksquare}$. We have

$$\theta_A(-a)(\mu) = -\theta_A(a)(\mu) = \theta_A(a)(-\mu)$$

for all $a \in E$ and $\mu \in F^*$. However, if both E and F are vector spaces over $\mathbb{C}$, not every $\mathfrak{L}(E, \mathcal{Q}_F)$-valued measure θ defines a family of complex-linear operators θ_A from E into $F^{*\blacksquare}$. For θ to be $\mathfrak{L}(E, F^{*\blacksquare})$-valued, we require in addition that

$$\theta_A(\alpha a)(\mu) = \theta_A(a)(\alpha\mu)$$

holds for all $A \in \mathfrak{R}$, all $a \in E$, $\alpha \in \mathbb{C}$ and $\mu \in F^*$. This condition then guarantees that $\theta_A(\alpha a) = \alpha\theta_A(a)$ if interpreted as elements of $F^{*\blacksquare}$. Considering our prior observations, it needs to be verified only in the complex case and then only for $\alpha = i$. The same considerations apply to $\mathfrak{L}(E,F)$- and $\mathfrak{L}(E,F^{**})$-valued measures.

If the measure θ is $\mathfrak{L}(E,F^{**})$-valued, pointwise convergence on F^* of its evaluations on sets in $\mathfrak{R}$ means convergence of the operators θ_A in the weak* operator topology of $\mathfrak{L}(E,F^{**})$. In case that θ is $\mathfrak{L}(E,F)$-valued, this means convergence in the weak operator topology of $\mathfrak{L}(E,F)$. The last statement may however be strengthened using the classical Orlicz–Pettis theorem about conditional convergence of series in a topological vector space.

2.1.1 Proposition. *If $\mathcal{P}_E = E$ and θ is an $\mathfrak{L}(E,F)$-valued measure, then for disjoint sets $A_i \in \mathfrak{R}$ such that $\bigcup_{i\in\mathbb{N}} A_i \in \mathfrak{R}$ the series $\theta_{\bigcup_{i\in\mathbb{N}} A_i} = \sum_{i=1}^{\infty}\theta_{A_i}$ converges in the strong operator topology of $\mathfrak{L}(E,F)$.*

Proof. We recall from Proposition 1.3.1 that the dual of $\mathfrak{L}(E,F)$ under both its weak and strong operator topologies is the tensor product $\tilde{E}\otimes F^*$, where $\tilde{E}$ denotes the quotient space E/E_0 and $E_0 = \bigcap_{V\in\mathcal{V}} V$. The weak and strong operator topologies on $\mathfrak{L}(E,F)$ are both Hausdorff. Suppose that θ is an $\mathfrak{L}(E,F)$-valued measure. Let A_i be disjoint sets in $\mathfrak{R}$ such that $\bigcup_{i\in\mathbb{N}} A_i \in \mathfrak{R}$. According to the above definition of a measure, the series in $\sum_{i=1}^{\infty}\theta_{A_i}$ in $\mathfrak{L}(E,F)$ converges in the weak topology of the duality $(\mathfrak{L}(E,F), \tilde{E}\otimes F^*)$ to the operator $\theta_{\bigcup_{i\in\mathbb{N}} A_i} \in \mathfrak{L}(E,F)$. The strong operator topology of $\mathfrak{L}(E,F)$ is consistent with this duality. Thus we may employ the Orlicz–Pettis theorem from [40] (also see [36] or [46]), which states that a series in a Hausdorff locally convex topological vector space that is weakly subseries convergent is convergent in the given topology. In this context, a series is weakly subseries convergent if the series formed by any subsequence of its terms is weakly convergent. This is clearly the case for the series $\sum_{i=1}^{\infty}\theta_{A_i}$ in $\mathfrak{L}(E,F)$ from above which is therefore seen to converge in the strong operator topology. □

The strong additivity of an $\mathfrak{L}(E,F)$-valued measure θ as established in Proposition 2.1.1 does not extend to $\mathfrak{L}(\mathcal{P}_E,\mathcal{Q}_F)$-valued measures in general. In this context, we shall say that an $\mathfrak{L}(\mathcal{P}_E,\mathcal{Q}_F)$-valued measure θ is *strongly additive* at an element $C \in \mathcal{P}_E$ if for all $A_i \in \mathfrak{R}$ such that $\bigcup_{i\in\mathbb{N}} A_i \in \mathfrak{R}$ the series $\sum_{i=1}^{\infty}\theta_{A_i}(C)$ in $\mathcal{Q}_F$ converges in the symmetric topology of $\mathcal{Q}_F$ (see Section 1.4) to $\theta_{\bigcup_{i\in\mathbb{N}} A_i}(C)$. That is, this series of $\overline{\mathbb{R}}$-valued functionals on F^* converges uniformly on the polars of all neighborhoods in $\mathcal{W}$, whereby convergence toward $+\infty \in \overline{\mathbb{R}}$ at a point in F^* requires that this value is taken cofinally.

Unless mentioned otherwise, the following results refer to pointwise convergence in $\mathcal{Q}_F$ for the evaluation of a measure at an element of $\mathcal{P}_E$.

2.1.2 Lemma. *Let $A \in \mathfrak{R}$.*

(a) *If $A_i \in \mathfrak{R}$ such that $A_i \subset A_{i+1}$ for all $i \in \mathbb{N}$ and $A = \bigcup_{i=1}^{\infty} A_i$, then $\theta_E = \lim_{i\to\infty}\theta_{A_i}$.*

(b) *If* $A_i \in \mathfrak{R}$ *such that* $A \supset A_i \supset A_{i+1}$ *for all* $i \in \mathbb{N}$, *and* $\bigcap_{i=1}^{\infty} A_i = \emptyset$, *then* $\lim_{i\to\infty} \theta_{A_i}(C)(\mu) = 0$ *for all* $C \in \mathcal{P}_E$ *and* $\mu \in F^*$ *such that* $\theta_A(C)(\mu) < +\infty$.

Proof. For Part (a), let $B_1 = A_1$ and $B_i = A_i \setminus B_{i-1}$ for $i > 1$. The sets B_i are disjoint, $A_n = \bigcup_{i=1}^{n} B_i$ and $A = \bigcup_{i=1}^{\infty} B_i$. From the countable additivity of the measure θ, we infer that $\theta_{A_n} = \sum_{i=1}^{n} \theta_{B_i}$ and $\theta_A = \sum_{i=1}^{\infty} \theta_{B_i}$, hence our claim. For Part (b), let $B_i = A \setminus A_i$ for $i \in \mathbb{N}$. Thus $B_i \subset B_{i+1}$ and $\bigcup_{i=1}^{\infty} B_i = A$. This shows $\theta_A = \lim_{i\to\infty} \theta_{B_i}$ by Part (a) and $\theta_A = \theta_{B_i} + \theta_{A_i}$. If $\theta_A(C)(\mu) < +\infty$ for $C \in \mathcal{P}_E$ and $\mu \in F^*$, our claim follows. □

If $\mathcal{P}_E = E$ and θ is $\mathfrak{L}(E,F)$-valued, then according to Proposition 2.1.1 the convergence statements of Lemma 2.1.2 hold indeed with respect to the strong operator topology of $\mathfrak{L}(E,F)$. More generally, if θ is $\mathfrak{L}(\mathcal{P}_E, \mathcal{Q}_F)$-valued and strongly additive at an element $C \in \mathcal{P}_E$, then the convergence statements of Lemma 2.1.2, with the terms evaluated at C, hold with respect to the symmetric topology of $\mathcal{Q}_F$.

Modulus of a measure

Other than in the classical approach, we proceed to define the semivariation of a measure as its extension from the elements of E to the cone of all nonempty convex subsets of E. Given an $\mathfrak{L}(\mathcal{P}_E, \mathcal{Q}_F)$-valued measure θ, we define its *modulus* or *semivariation* $|\theta|$ as follows: For $A \in \mathfrak{R}$ and a set $C \in \mathrm{Conv}(E)$, we set (see also I.3.2 in [48])

$$|\theta|_A(C) = \sup\left\{ \sum_{i=1}^{n} \theta_{A_i}(c_i) \;\middle|\; c_i \in C,\ A_i \in \mathfrak{R} \text{ disjoint},\ \bigcup_{i=1}^{n} A_i = A \right\},$$

where the elements $c_i \in E$ are identified with the singleton sets $\{c_i\} \in \mathcal{P}_E$. The supremum on the right-hand side is taken in $\mathcal{Q}_F$, that is, pointwise on F^*. Hence $|\theta|_A(C)$ is an element of $\mathcal{Q}_F$. Clearly, $|\theta|_A(a) = \theta_A(a)$ holds for all $a \in E$, and we shall argue in 2.1.3 below that $|\theta|_A(C) \leqslant \theta_A(C)$ holds for all $C \in \mathcal{P}_E$. We have $|\theta|_A(C) \leqslant |\theta|_A(D)$ whenever $C \subset D$ for $C, D \in \mathrm{Conv}(E)$. If $0 \in C$, then $0 \leqslant |\theta|_A(C) \leqslant |\theta|_B(C)$ provided that $A \subset B$ for $A, B \in \mathfrak{R}$, and in the definition of $|\theta|_A(C)$ we only require that the disjoint sets $A_i \in \mathfrak{R}$ are contained in A, not that their union is all of A, since the set $A_{n+1} = A \setminus \bigcup_{i=1}^{n} A_i$ with $c_{n+1} = 0$ may be added. We recognize that the modulus of an $\mathfrak{L}(\mathcal{P}_E, \mathcal{Q}_F)$-valued measure is determined by its values on E alone. If $\mathcal{P}_E = E$ and θ is an $\mathfrak{L}(E, F^{*\blacksquare})$-valued measure, our definitions of the modulus and the multiplication of the elements of $\mathrm{Conv}(E)$ with general scalars yield immediately that

$$|\theta|_A(\alpha C)(\mu) = |\theta|_A(C)(\alpha\mu)$$

holds for all $A \in \mathfrak{R}$, all $C \in \mathrm{Conv}(E)$, α in $\mathbb{R}$ or $\mathbb{C}$ and $\mu \in F^*$. If θ is $\mathfrak{L}(E, F^{**})$-valued, then for every $A \in \mathfrak{R}$ and $C \in \mathrm{Conv}(E)$ the element $\theta_A(C) \in \mathcal{Q}_F$ is a $\sigma(F^*, F^{**})$-lower semicontinuous sublinear functional on F^*, hence in the light of Proposition 1.4.3(a) represents a convex subset of F^{**}. If θ is $\mathfrak{L}(E,F)$-valued, then a similar argument shows that $\theta_A(C)$ represents a convex subset of F.

Bounded measures

An $\mathfrak{L}(\mathcal{P}_E, \mathcal{Q}_F)$-valued measure θ is *bounded* or of *bounded semivarion* if for every $A \in \mathfrak{R}$ and $W \in \mathcal{W}$ there is a neighborhood $V \in \mathcal{V}$ such that $|\theta|_A(V) \leqslant \sigma_W$, that is, $|\theta|_A(V)(\mu) \leqslant 1$ for all $\mu \in W^\circ$. Since every element $\mu \in F^*$ is contained in the polar of some neighborhood in $\mathcal{W}$, boundedness of a measure θ implies in particular that for every $A \in \mathfrak{R}$ and $\mu \in F^*$ there is a neighborhood $V \in \mathcal{V}$ such that $|\theta|_A(V)(\mu) < +\infty$.

2.1.3 Theorem. *The modulus $|\theta|$ of a bounded $\mathfrak{L}(\mathcal{P}_E, \mathcal{Q}_F)$-valued measure θ is itself a bounded $\mathfrak{L}(\mathrm{Conv}(E), \mathcal{Q}_F)$-valued measure. The modulus of $|\theta|$ is $|\theta|$. We have $|\theta|_A(C) \leqslant \theta_A(C)$ for all $A \in \mathfrak{R}$ and $C \in \mathcal{P}_E$, and for any other $\mathfrak{L}(\mathrm{Conv}(E), \mathcal{Q}_F)$-valued measure ϑ with this property we have $|\theta|_A(C) \leqslant \vartheta_A(C)$ for all $A \in \mathfrak{R}$ and $C \in \mathrm{Conv}(E)$.*

Proof. First, we shall argue that for a fixed set $A \in \mathfrak{R}$ the mapping $C \mapsto |\theta|_A(C) : \mathrm{Conv}(E) \to \mathcal{Q}_F$ is linear. Indeed, let $C, D \in \mathrm{Conv}(E)$, let $A_i, B_k \in \mathfrak{R}$ such that $\bigcup_{i=1}^n A_i = \bigcup_{k=1}^m B_k = A$. For any choice of elements $c_i \in C$ and $d_k \in D$, we calculate that $\theta_{A_i}(c_i) = \sum_{k=1}^m \theta_{A_i \cap B_k}(c_i)$ for all $i = 1, \ldots, n$ and $\theta_{B_k}(d_k) = \sum_{i=1}^n \theta_{A_i \cap B_k}(d_k)$ for all $k = 1, \ldots, m$. Thus

$$\sum_{i=1}^n \theta_{A_i}(c_i) + \sum_{k=1}^m \theta_{B_k}(d_k) = \sum_{i=1}^n \sum_{k=1}^m \theta_{A_i \cap B_k}(c_i + d_k)$$

The sets $A_i \cap B_k$ are disjoint and we have $\bigcup_{i=1}^n \bigcup_{k=1}^m A_i \cap B_k = A$. The above therefore yields that

$$|\theta|_A(C) + |\theta|_A(D) \leqslant |\theta|_A(C + D).$$

For the reverse inequality, let $A_i \in \mathfrak{R}$ such that $\bigcup_{i=1}^n A_i = A$ and let $c_i + d_i \in C + D$. Then

$$\sum_{i=1}^n \theta_{A_i}(c_i + d_i) = \sum_{i=1}^n \theta_{A_i}(c_i) + \sum_{i=1}^n \theta_{A_i}(d_i).$$

We infer that $|\theta|_A(C + D) \leqslant |\theta|_A(C) + |\theta|_A(D)$. The equality $|\theta|_A(\lambda C) = \lambda|\theta|_A(C)$ for $\lambda \geqslant 0$ is evident from the definition of the modulus. Next, we shall verify that $|\theta|_A \in \mathfrak{L}(\mathrm{Conv}(E), \mathcal{Q}_F)$, that is, that the linear operator $|\theta|_A$ is continuous. Clearly, $|\theta|_A$ is monotone, that is, $|\theta|_A(C) \leqslant |\theta|_A(D)$ holds whenever $C \leqslant D$ for $C, D \in \mathrm{Conv}(E)$. Now let $W \in \mathcal{W}$. Since θ is supposed to be bounded, there is $V \in \mathcal{V}$ such that $|\theta|_A(V) \leqslant \sigma_W$. Accordingly, $C \leqslant D + V$ for $C, D \in \mathrm{Conv}(E)$ implies that

$$|\theta|_A(C) \leqslant |\theta|_A(D + V) = |\theta|_A(V) + |\theta|_A(D) \leqslant |\theta|_A(D) + \sigma_W$$

as required. Next, we shall demonstrate that the set function

$$A \mapsto |\theta|_A : \mathfrak{R} \to \mathfrak{L}(\mathrm{Conv}(E), \mathcal{Q}_F)$$

is indeed a measure. Clearly, $|\theta|_\varnothing = 0$. For countable additivity, suppose that $\bigcup_{i=1}^\infty A_i = A \in \mathfrak{R}$ for disjoint sets $A_i \in \mathfrak{R}$ and let $C \in \mathrm{Conv}(E)$. In a first step, let us assume that $0 \in C$. The series $\sum_{i=1}^\infty |\theta|_{A_i}(C)$ then has positive terms and converges (pointwise on F^*) in $\mathcal{Q}_F$. Let $B_1, \dots, B_n \in \mathfrak{R}$ be disjoint sets such that $\bigcup_{k=1}^n B_k = A$ and let $c_k \in C$. Then

$$\theta_{B_k}(c_k) = \sum_{i=1}^\infty \theta_{B_k \cap A_i}(c_k)$$

for each $k = 1, \dots, n$ by the countable additivity of θ, hence

$$\begin{aligned}\sum_{k=1}^n \theta_{B_k}(c_k) &= \sum_{k=1}^n \left(\sum_{i=1}^\infty \theta_{B_k \cap A_i}(c_k) \right) \\ &= \sum_{i=1}^\infty \left(\sum_{k=1}^n \theta_{B_k \cap A_i}(c_k) \right) \leqslant \sum_{i=1}^\infty |\theta|_{A_i}(C).\end{aligned}$$

This renders $|\theta|_A(C) \leqslant \sum_{i=1}^\infty |\theta|_{A_i}(C)$. For the reverse inequality, let $n \in \mathbb{N}$ and for each $i = 1, \dots, n$, let $B_1^i, \dots, B_{n_i}^i \in \mathfrak{R}$ be disjoint sets such that $\bigcup_{k=1}^{n_i} B_k^i = A_i$ and let $c_1^i, \dots, c_{n_i}^i \in C$. Then

$$\sum_{i=1}^n \left(\sum_{k=1}^{n_i} \theta_{B_k^i}(c_k^i) \right) \leqslant |\theta|_{\bigcup_{i=1}^n A_i}(C) \leqslant |\theta|_A(C),$$

as the sets B_k^i and are pairwise disjoint and their union is $\bigcup_{i=1}^n A_i$. Now taking the supremum over all such choices of sets B_i^k and elements $c_k^i \in C$ yields

$$\sum_{i=1}^n |\theta|_{A_i}(C) \leqslant |\theta|_A(C).$$

This holds for all $n \in \mathbb{N}$ and yields that $\sum_{i=1}^\infty |\theta|_{A_i}(C) \leqslant |\theta|_A(C)$. Our claim follows. Now in a second step, for any $C \in \mathrm{Conv}(E)$ we choose $c \in C$ and set $D = C - c$. Then $0 \in D$ and our first step applies. That is to say, we have $\sum_{i=1}^\infty |\theta|_{A_i}(D) = |\theta|_A(D)$. Moreover, for the element $c \in E$ we have $|\theta|_B(c) = \theta_B(c)$ for all $B \in \mathfrak{R}$, hence also $\sum_{i=1}^\infty |\theta|_{A_i}(c) = |\theta|_A(c)$ by the countable additivity of the measure θ. Summarizing, and using the previously established linearity of the operators $|\theta|_A, |\theta_{A_i}|$, this renders

$$\begin{aligned}\sum_{i=1}^\infty |\theta|_{A_i}(C) &= \sum_{i=1}^\infty |\theta|_{A_i}(D + c) \\ &= \sum_{i=1}^\infty |\theta|_{A_i}(D) + \sum_{i=1}^\infty |\theta|_{A_i}(c) \\ &= |\theta|_A(D) + |\theta|_A(c) = |\theta|_A(C).\end{aligned}$$

Thus $|\theta|$ is indeed an $\mathfrak{L}(\mathrm{Conv}(E), \mathcal{Q}_F)$-valued measure. The remaining statements of the theorem are easily verified. First, for every $C \in \mathcal{P}_E$ and every element $c \in C$ we have $\theta_A(c) \leqslant \theta_A(C)$ for all $A \in \mathfrak{R}$ by Lemma 1.4.10. Thus

$$\sum_{i=1}^{n} \theta_{A_i}(c_i) \leqslant \sum_{i=1}^{n} \theta_{A_i}(C) = \theta_A(C)$$

whenever $c_i \in C$ and $A_i \in \mathfrak{R}$ are disjoint sets such that $\bigcup_{i=1}^{n} A_i = A$. This yields $|\theta|_A(C) \leqslant \theta_A(C)$, our claim. Next, let ϑ be any $\mathfrak{L}(\mathrm{Conv}(E), \mathcal{Q}_F)$-valued measure with this property, let $A \in \mathfrak{R}$ and $C \in \mathrm{Conv}(E)$. Again, we argue that for every $c \in C$ we have $\theta_A(c) = \vartheta_A(c) \leqslant \vartheta_A(C)$ and $\sum_{i=1}^{n} \theta_{A_i}(c_i) \leqslant \vartheta_A(C)$ whenever $c_i \in C$ and $A_i \in \mathfrak{R}$ are disjoint sets such that $\bigcup_{i=1}^{n} A_i = A$. Thus indeed $|\theta|_A(C) \leqslant \vartheta_A(C)$. Now consider the modulus $\|\theta\|$ of $|\theta|$ for ϑ in this last argument. We have $\|\theta\|_A(C) \leqslant |\theta|_A(C)$ for all $A \in \mathfrak{R}$ and $C \in \mathrm{Conv}(E)$ by the above, hence $|\theta|_A(C) \leqslant \|\theta\|_A(C)$, also by the above. This shows $\|\theta\| = |\theta|$. Finally, for the boundedness of $|\theta|$ let $A \in \mathfrak{R}$ and $W \in \mathcal{W}$. Because θ itself is bounded, we find $V \in \mathcal{V}$ such that $\|\theta\|_A(V) = |\theta|_A(V) \leqslant \sigma_W$. This completes our arguments. □

We shall subsequently require that all $\mathfrak{L}(\mathcal{P}_E, \mathcal{Q}_F)$-valued measures are bounded without explicitly listing this condition. It is guaranteed to hold if $\mathcal{P}_E$ is a full cone extension of E. Indeed, let $A \in \mathfrak{R}$ and $W \in \mathcal{W}$. Because $\theta_A \in \mathfrak{L}(\mathcal{P}_E, \mathcal{Q}_F)$, there is $V \in \mathcal{V}$ such that $\theta_A(C) \leqslant \theta_A(D) + \sigma_W$ whenever $C \leqslant D + V$ for $C, D \in \mathcal{P}_E$. Since $V \in \mathcal{P}_E$, this yields $|\theta|_A(V) \leqslant \theta_A(V) \leqslant \sigma_W$. In particular, the modulus $|\theta|$ of a bounded $\mathfrak{L}(\mathcal{P}_E, \mathcal{Q}_F)$-valued measure is a bounded $\mathfrak{L}(\mathrm{Conv}(E), \mathcal{Q}_F)$-valued measure.

The strong additivity of an $\mathfrak{L}(E, F)$-valued measure as established in Proposition 2.1.1 does not necessarily extend to is modulus. But we observe the following.

2.1.4 Lemma. *If $|\theta|$ is strongly additive at every neighborhood $V \in \mathcal{V}$, then it is strongly additive at every bounded element of* $\mathrm{Conv}(E)$.

Proof. Suppose that $|\theta|$ is strongly additive at every $V \in \mathcal{V}$, let $C \in \mathrm{Conv}(E)$ be bounded and let $A_i \in \mathfrak{R}$ be disjoint sets such that $A = \bigcup_{i \in \mathbb{N}} A_i \in \mathfrak{R}$. Given $W \in \mathcal{W}$, there is $V \in \mathcal{V}$ such that $|\theta|_A(V) \leqslant \sigma_W$ and $\lambda \geqslant 0$ such that both $0 \leqslant C + \lambda V$ and $C \leqslant \lambda V$. We set $B_n = \bigcup_{i=n}^{\infty} A_i \in \mathfrak{R}$. Given $\varepsilon > 0$, by the strong additivity of $|\theta|$ at V there is $n_0 \in \mathbb{N}$ such that $|\theta|_{B_n}(V) \leqslant \varepsilon\sigma_W$ for all $n \geqslant n_0$. The latter implies that both $0 \leqslant |\theta|_{B_n}(C) + \varepsilon\sigma_W$ and $|\theta|_{B_n}(C) \leqslant \varepsilon\sigma_W$. Hence

$$\left| |\theta|_A(C)(\mu) - |\theta|_{\bigcup_{i=1}^{n} A_i}(C)(\mu) \right| \leqslant \varepsilon$$

holds for all $\mu \in W^\circ$ and all $n \geqslant n_0$. □

Variation of a measure

For a bounded $\mathfrak{L}(\mathcal{P}_E, \mathcal{Q}_F)$-valued measure θ on $\mathfrak{R}$, we define its *variation relative to a function space neighborhood* $\mathfrak{v}$ *and a nonempty subset* Π *of* F^* on a set $A \in \mathfrak{R}$ as (see III.1.4 in [26] or I.1 in [21]) by

$$\|\theta\|_A^{(\mathfrak{v},\Pi)} = \sup\left\{\sum_{i=1}^{n} \theta_{A_i}(a_i)(\mu_i) \;\middle|\; \begin{array}{l} A_i \in \mathfrak{R} \text{ disjoint, } A_i \subset A, \\ a_i \in E,\, \chi_{A_i} \cdot a_i \leqslant \mathfrak{v} \text{ and } \mu_i \in \Pi \end{array}\right\}.$$

Since $0 \leqslant \mathfrak{v}$, the variation of a measure is $\overline{\mathbb{R}}_+$-valued and related to its earlier defined modulus or semivariation in the following way.

2.1.5 Lemma. *Let θ be an $\mathfrak{L}(\mathcal{P}_E, \mathcal{Q}_F)$-valued measure, let $\mathfrak{v}$ be a function space neighborhood and let $\Pi \subset F^*$.*

(a) *If $\chi_A \cdot C \leqslant \mathfrak{v}$ for $C \in \mathrm{Conv}(E)$ and $A \in \mathfrak{R}$, then $|\theta|_A(C)(\mu) \leqslant \|\theta\|_A^{(\mathfrak{v},\Pi)}$ holds for all $\mu \in \Pi$.*

(b) *If $\mathfrak{v}$ is defined by the single constant neighborhood function $\chi_X \cdot V$ for $V \in \mathcal{V}$ and if $\Pi = \{\mu\} \subset F^*$ is a singleton set, then $|\theta|_A(V)(\mu) = \|\theta\|_A^{(\mathfrak{v},\{\mu\})}$ holds for all $A \in \mathfrak{R}$.*

Proof. Our claim in Part (a) is immediate from the respective definitions of the modulus and the variation of a measure; so is the inequality $|\theta|_A(V)(\mu) \leqslant \|\theta\|_A^{(\mathfrak{v},\{\mu\})}$ in Part (b). For the reverse relation, let $A_i \in \mathfrak{R}$ be disjoint subsets of A and let $a_i \in E$ such that $\chi_{A_i} \cdot a_i \leqslant \mathfrak{v}$, that is, $a_i \in V$. Then

$$\sum_{i=1}^{n} \theta_{A_i}(a_i)(\mu) \leqslant |\theta|_A(V)(\mu)$$

and, therefore, $\|\theta\|_A^{(\mathfrak{v},\{\mu\})} \leqslant |\theta|_A(V)(\mu)$. □

2.1.6 Proposition. *Let θ be an $\mathfrak{L}(\mathcal{P}_E, \mathcal{Q}_F)$-valued measure, let $\mathfrak{v}$ be a function space neighborhood and let $\Pi \subset F^*$. The variation $\|\theta\|^{(\mathfrak{v},\Pi)}$ of θ defines a countably additive $\overline{\mathbb{R}}_+$-valued measure on $\mathfrak{R}$.*

Proof. We observed before that the values of $\|\theta\|^{(\mathfrak{v},\Pi)}$ are nonnegative, and $\|\theta\|_\varnothing^{(\mathfrak{v},\Pi)} = 0$ is apparent. For disjoint sets $A, B \in \mathfrak{R}$, our definition returns immediately that

$$\|\theta\|_A^{(\mathfrak{v},\Pi)} + \|\theta\|_B^{(\mathfrak{v},\Pi)} \leqslant \|\theta\|_{A\cup B}^{(\mathfrak{v},\Pi)}.$$

In particular, we have

$$\|\theta\|_A^{(\mathfrak{v},\Pi)} \leqslant \|\theta\|_A^{(\mathfrak{v},\Pi)} + \|\theta\|_{B\setminus A}^{(\mathfrak{v},\Pi)} \leqslant \|\theta\|_B^{(\mathfrak{v},\Pi)}$$

whenever $A \subset B$. Now let $\bigcup_{i=1}^{\infty} A_i = A \in \mathfrak{R}$ for disjoint sets $A_i \in \mathfrak{R}$. Then

$$\sum_{i=1}^{n} \|\theta\|_{A_i}^{(\mathfrak{v},\Pi)} \leqslant \|\theta\|_{(\bigcup_{i=1}^{n} A_i)}^{(\mathfrak{v},\Pi)} \leqslant \|\theta\|_A^{(\mathfrak{v},\Pi)}$$

holds for all $n \in \mathbb{N}$ by the above. This yields $\sum_{i=1}^{\infty} \|\theta\|_{A_i}^{(\mathfrak{v},\Pi)} \leqslant \|\theta\|_A^{(\mathfrak{v},\Pi)}$. For the reverse inequality, let $B_k \in \mathfrak{R}$ be disjoint subsets of A, let $\chi_{B_k} \cdot b_k \leqslant \mathfrak{v}$ and $\mu_k \in \Pi$ for $k = 1, \ldots, n$. For every k, we have $B_k = \bigcup_{i=1}^{\infty}(B_k \cap A_i)$ and, therefore, $\theta_{B_k}(b_k)(\mu_k) = \sum_{i=1}^{\infty} \theta_{(B_k \cap A_i)}(b_k)(\mu_k)$. Since the sets $B_k \cap A_i \subset A$ are disjoint and since $\chi_{(B_k \cap A_i)} \cdot b_k \leqslant \mathfrak{v}$ for all k and i, we conclude that

$$\begin{aligned}\sum_{k=1}^{n}\theta_{B_k}(b_k)(\mu_k) &= \sum_{k=1}^{n}\sum_{i=1}^{\infty}\theta_{(B_k\cap A_i)}(b_k)(\mu_k)\\ &= \sum_{i=1}^{\infty}\sum_{k=1}^{n}\theta_{(B_k\cap A_i)}(b_k)(\mu_k)\\ &\leqslant \sum_{i=1}^{\infty}\|\theta\|_{A_i}^{(\mathfrak{v},\Pi)}.\end{aligned}$$

Hence $\|\theta\|_A^{(\mathfrak{v},\Pi)} \leqslant \sum_{k=1}^{\infty}\|\theta\|_{A_i}^{(\mathfrak{v},\Pi)}$, and $\|\theta\|^{(\mathfrak{v},\Pi)}$ is understood to be countably additive as claimed. □

2.1.7 Proposition. *Let θ be an $\mathfrak{L}(\mathcal{P}_E,\mathcal{Q}_F)$-valued measure, which coincides with its modulus on $\mathcal{P}_E$. If for every $A\in\mathfrak{R}$ and $W\in\mathcal{W}$, there is a function space neighborhood $\mathfrak{v}$ such that $\|\theta\|_A^{(\mathfrak{v},W^\circ)}<+\infty$, then θ is strongly additive at every bounded element of $\mathcal{P}_E$.*

Proof. Let $A=\bigcup_{i\in\mathbb{N}}A_i\in\mathfrak{R}$ for disjoint sets $A_i\in\mathfrak{R}$ and let $W\in\mathcal{W}$. There is a function space neighborhood $\mathfrak{v}$ such that $\|\theta\|_A^{(\mathfrak{v},W^\circ)}<+\infty$. Let C be a bounded element of $\mathcal{P}_E$. There is $V\in\mathcal{V}$ such that $\chi_A\cdot V\leqslant\mathfrak{v}$. Hence

$$|\theta|_B(V)(\mu)\leqslant\|\theta\|_B^{(\mathfrak{v},W^\circ)}<+\infty$$

for all $\mu\in W^\circ$ and all subsets $B\in\mathfrak{R}$ of A by Lemma 2.1.5(a). There is $\lambda>0$ such that both $0\leqslant C+\lambda V$ and $C\leqslant\lambda V$. Because θ and $|\theta|$ are supposed to coincide at C, we calculate that

$$\theta_B(C)(\mu)=|\theta|_B(C)(\mu)\leqslant\lambda|\theta|_B(V)\leqslant\lambda\|\theta\|_B^{(\mathfrak{v},W^\circ)}$$

and

$$\begin{aligned}0\leqslant|\theta|_B(C+\lambda V)(\mu)&=|\theta|_B(C)(\mu)+\lambda|\theta|_B(V)(\mu)\\ &\leqslant\theta_B(C)(\mu)+\lambda\|\theta\|_B^{(\mathfrak{v},W^\circ)}\end{aligned}$$

holds for all $\mu\in W^\circ$ and all subsets $B\in\mathfrak{R}$ of A. Hence $|\theta_B(C)|(\mu)\leqslant\lambda\|\theta\|_B^{(\mathfrak{v},W^\circ)}$. Set $B_n=\bigcup_{i=n+1}^{\infty}A_i$, that is, $\theta_{B_n}=\theta_A-\theta_{(\bigcup_{i=1}^{n}A_i)}$. According to Proposition 2.1.6 for every $\varepsilon>0$ there is $n_0\in\mathbb{N}$ such that $\|\theta\|_{B_n}^{(\mathfrak{v},W^\circ)}\leqslant\varepsilon/\lambda$ for all $n\geqslant n_0$. Thus

$$\sup\{\theta_{B_n}(C)(\mu)\mid\mu\in W^\circ\}\leqslant\varepsilon,$$

that is, $\theta_{B_n}(C)\leqslant\sigma_W$ and $\theta_{B_n}(C)\in W^s(0)$. Our claim follows. □

In a straightforward way $\|\theta\|^{(\mathfrak{v},\Pi)}$ can be extended to an $\overline{\mathbb{R}}_+$-valued measure on the σ-field $\mathfrak{A}$ that is associated with $\mathfrak{R}$ if we set

$$\|\theta\|_B^{(\mathfrak{v},\Pi)}=\sup_{A\in\mathfrak{R}}\|\theta\|_{A\cap B}^{(\mathfrak{v},\Pi)}$$

for all $B \in \mathfrak{A}$. Countable additivity on $\mathfrak{A}$ is immediate since for disjoint sets $B_i \in \mathfrak{A}$ we calculate

$$\begin{aligned}\|\theta\|^{(\mathfrak{v},\Pi)}_{\bigcup_{i\in\mathbb{N}} B_i} &= \sup_{A\in\mathfrak{R}} \|\theta\|^{(\mathfrak{v},\Pi)}_{(A\cap\bigcup_{i\in\mathbb{N}} B_i)} = \sup_{A\in\mathfrak{R}} \sum_{i\in\mathbb{N}} \|\theta\|^{(\mathfrak{v},\Pi)}_{A\cap B_i} \\ &= \sup_{A\in\mathfrak{R}} \sup_{i\in\mathbb{N}} \|\theta\|^{(\mathfrak{v},\Pi)}_{A\cap B_i} = \sup_{i\in\mathbb{N}} \sup_{A\in\mathfrak{R}} \|\theta\|^{(\mathfrak{v},\Pi)}_{A\cap B_i} \\ &= \sup_{i\in\mathbb{N}} \|\theta\|^{(\mathfrak{v},\Pi)}_{B_i} \\ &= \sum_{i\in\mathbb{N}} \|\theta\|^{(\mathfrak{v},\Pi)}_{B_i}.\end{aligned}$$

Measure of bounded variation

An $\mathfrak{L}(\mathcal{P}_E, \mathcal{Q}_F)$-valued measure θ is of *bounded variation* relative to a function space neighborhood $\mathfrak{v}$ and a nonempty subset Π of F^* if

$$\sup_{A\in\mathfrak{R}} \|\theta\|^{(\mathfrak{v},\Pi)}_A < +\infty.$$

2.1.8 Examples and Remarks. (a) Boundedness of a measure θ implies that for every $A \in \mathfrak{R}$ and every bounded set $C \in \mathrm{Conv}(E)$ the functional $|\theta|_A(C)$ is a bounded element of the locally convex cone $\mathcal{Q}_F$. Indeed, for $W \in \mathcal{W}$ there is $V \in \mathcal{V}$ such that $|\theta|_A(V) \leqslant \sigma_W$. Since $C \subset \lambda V$ for some $\lambda \geqslant 0$, this yields $|\theta|_A(C) \leqslant \lambda|\theta|_A(V) < \lambda\sigma_W$. Because for every $\mu \in F^*$ there is $W \in \mathcal{W}$ such that $\mu \in W^\circ$, this implies in particular that $|\theta|_A(C)$ is real-valued on F^*. Consequently, if θ is $\mathfrak{L}(\mathcal{P}_E, F)$-valued, then the set

$$\left\{ \sum_{i=1}^{n} \theta_{A_i}(c_i) \;\middle|\; c_i \in C,\ A_i \in \mathfrak{R} \text{ disjoint}, \bigcup_{i=1}^{n} A_i = A \right\}$$

is bounded in the given topology of F. Similarly, if θ is $\mathfrak{L}(\mathcal{P}_E, F^{**})$-valued, then this set is bounded in the natural topology of F^{**}.

(b) If X is compact, if both E and F are normed spaces with unit balls $\mathbb{B}_E$ and $\mathbb{B}_F$, respectively, and if $\mathcal{C}_{\mathfrak{V}}(X, E)$ carries the topology of uniform convergence, then the variation of an $\mathfrak{L}(E, F)$-valued measure θ relative to the unit ball $\mathbb{B}_C$ of $\mathcal{C}_{\mathfrak{V}}(X, E)$ and the polar of $\mathbb{B}_F$ for $\Pi \subset F^*$ is given by

$$\|\theta\|^{(\mathbb{B}_C, \mathbb{B}_F^\circ)}_A = \sup\left\{ \sum_{i=1}^{n} \|\theta_{A_i}\| \;\middle|\; A_i \in \mathfrak{R} \text{ disjoint}, A_i \subset A \right\}$$

for $A \in \mathfrak{A}$, where $\|\theta_A\|$ is the operator norm of the operator $\theta_A \in \mathfrak{L}(E, F)$. This is of course the usual definition of the variation of a vector-valued measure (see III.1.4 in [26]).

An example of a bounded measure, that is, a measure with bounded semivariation, but unbounded variation can be found in [21] I.1.6.

(c) If an $\mathfrak{L}(\mathcal{P}_E, \mathcal{Q}_F)$-valued measure θ is indeed $\mathfrak{L}(\mathcal{P}_E, F)$-valued (or $\mathfrak{L}(\mathcal{P}_E, F^{**})$-valued), then its values on sets $A \in \mathfrak{R}$ are linear operators from $\mathcal{P}_E$ into the subspace of $\mathcal{Q}_F$, which consists of all real-valued linear functionals on F^* that are continuous with respect to the weak* topology (or, equivalently, to the strong topology $\beta(F^*, F)$) of F^*. The values of its modulus $|\theta|$ then are linear operators from $\mathrm{Conv}(E)$ into the subcone of $\mathcal{Q}_F$ consisting of all $\overline{\mathbb{R}}$-valued sublinear functionals on F^* that are lower semicontinuous with respect to the weak* topology (or the strong topology) of F^*. This follows immediately from the above definition of $|\theta|$. In the light of the embedding of $\mathrm{Conv}(F)$ (or of $\mathrm{Conv}(F^{**})$) into $\mathcal{Q}_F$, which was elaborated on in Section 1.4, the values of $|\theta|$ may therefore be considered to be closed, convex subsets of F (or of F^{**}). Moreover, as for all $A \in \mathfrak{R}$ the elements $\theta_A(C)$ for all $C \in \mathcal{P}_E$ are real-valued linear functionals on F^* in this case, $C \subset D$ for $C, D \in \mathcal{P}_E$ implies that

$$|\theta|_A(D) = \theta_A(D) = \theta_A(C) = |\theta|_A(C).$$

Indeed, we have $|\theta|_A(C) \leqslant \theta_A(C)$ by Theorem 2.1.3. Since $|\theta|_A(C)$ is sublinear and $\theta_A(C)$ is linear, hence finite on F^*, we deduce that $|\theta|_A(C) = \theta_A(C)$ and likewise, $|\theta|_A(D) = \theta_A(D)$. The relation $\theta_A(C) \leqslant \theta_A(D)$ follows from the monotonicity of the operator $\theta_A \in \mathfrak{L}(\mathcal{P}_E, \mathcal{Q}_F)$ (Lemma 1.4.10). Hence $\theta_A(C) = \theta_A(D)$ since both are linear functionals on F^*. For elements of D, that is for singleton subsets of D, we conclude in particular that

$$\theta_A(d) = \theta_A(D) = |\theta|_A(D)$$

holds for all $d \in D$ in this case.

(d) If $F = \mathbb{C}$, then $F^* = \mathbb{C}$ and $\mathcal{Q}_F$ consists of all $\overline{\mathbb{R}}$-valued sublinear functionals on $\mathbb{C}$ that are bounded below on the unit ball of $\mathbb{C}$. If an $\mathfrak{L}(\mathcal{P}_E, \mathcal{Q}_F)$-valued measure θ is indeed $\mathfrak{L}(\mathcal{P}_E, F)$-valued, then its values on sets $A \in \mathfrak{R}$ and $C \in \mathcal{P}_E$ are complex numbers, operating as real-valued linear functionals on $F^* = \mathbb{C}$ as $z \mapsto \mathfrak{Re}(\theta_A(C)\, z) : \mathbb{C} \to \mathbb{R}$. The values of its modulus $|\theta|$ are computed for every $A \in \mathfrak{R}$ and $C \in \mathrm{Conv}(E)$ as an $\overline{\mathbb{R}}$-valued sublinear functional on $\mathbb{C}$ by

$$|\theta|_A(C)(z) = \sup\left\{ \sum_{i=1}^{n} \mathfrak{Re}(z\theta_{A_i}(c_i)) \;\middle|\; c_i \in C,\ A_i \in \mathfrak{R} \text{ disjoint}, \bigcup_{i=1}^{n} A_i = A \right\},$$

for $z \in \mathbb{C}$. For balanced sets $C \in \mathrm{Conv}(E)$, this expression simplifies to

$$|\theta|_A(C)(z) = \sup\left\{ \sum_{i=1}^{n} |\theta_{A_i}(c_i)| \;\middle|\; c_i \in C,\ A_i \in \mathfrak{R} \text{ disjoint}, A_i \subset A \right\} |z|.$$

The latter applies in particular to the neighborhoods in $\mathcal{V}$ and reveals that our requirement of boundedness for the measure θ corresponds to Dieudonné's notion of *p-domination* in [22] and to Prolla's of *finite p-semivariation* in Chapter 5.5 of [42] for measures with values in the dual of a locally convex topological vector space. The case that $F = \mathbb{R}$ is similar.

(e) If $E = \mathbb{C}$ and F is a locally convex topological vector space over $\mathbb{C}$, then every $F^{*\blacksquare}$-valued measure θ that is countably additive pointwise on the elements of F^* may be considered to be an operator-valued measure in our sense since every element $\omega \in F^{*\blacksquare}$ constitutes a continuous linear operator from $\mathbb{C}$ into $F^{*\blacksquare}$ via $z \mapsto z\omega$. Hence θ is $\mathfrak{L}(\mathbb{C}, \mathcal{Q}_F)$-valued, and we have $\theta_A(z)(\mu) = \mathfrak{Re}\,\mu(z\theta_A)$ for $A \in \mathfrak{R}$, $z \in \mathbb{C}$ and $\mu \in F^*$. For $C \in \mathrm{Conv}(\mathbb{C})$, the modulus of θ computes as

$$|\theta|_A(C)(\mu) = \sup\left\{\sum_{i=1}^{n} \mathfrak{Re}\,\mu(z_i\theta_{A_i}) \,\middle|\, z_i \in C,\ A_i \in \mathfrak{R} \text{ disjoint}, \bigcup_{i=1}^{n} A_i = A\right\}$$

for all $\mu \in F^*$. For the unit ball $\mathbb{B}$ of $\mathbb{C}$, in particular, we have

$$|\theta|_A(\mathbb{B})(\mu) = \sup\left\{\sum_{i=1}^{n} |\mu(\theta_{A_i})| \,\middle|\, A_i \in \mathfrak{R} \text{ disjoint}, A_i \subset A\right\},$$

and in case that θ is F-valued and that F is a normed space,

$$\sup_{\|\mu\|\leqslant 1} |\theta|_A(\mathbb{B})(\mu) = \sup\left\{\left\|\sum_{i=1}^{n} z_i\theta_{A_i}\right\| \,\middle|\, |z_i| \leqslant 1,\ A_i \in \mathfrak{R} \text{ disjoint}, A_i \subset A\right\}.$$

The latter is the notion of the semivariation of a vector-valued measure for a normed space as given for example in Chapter IV.10.3 of [26]. The assumptions of Proposition 2.1.1 apply in this case, hence any such measure is strongly additive. (See Theorem 1 in IV.10.3 in [26].) The assumptions of Corollary 2.1.4 for its modulus can however not be guaranteed. The case that $E = \mathbb{R}$ and that F is a vector space over $\mathbb{R}$ is similar.

(f) If $E = F$ and $\mathcal{P}_E = E$, then every real- or complex-valued measure θ may be considered to be an $\mathfrak{L}(E,E)$-valued, hence $\mathfrak{L}(E, \mathcal{Q}_E)$-valued measure in our sense since every z in $\mathbb{R}$ or $\mathbb{C}$ constitutes a continuous linear operator from E to E via $a \mapsto za$. We have $\theta_A(a)(\mu) = \mathfrak{Re}\,\mu(\theta_A a)$ for $A \in \mathfrak{R}$, $z \in \mathbb{C}$ and $\mu \in F^*$. For $C \in \mathrm{Conv}(E)$, we calculate

$$|\theta|_A(C)(\mu) = \sup\left\{\sum_{i=1}^{n} \mathfrak{Re}\,\mu(\theta_{A_i}c_i) \,\middle|\, c_i \in C,\ A_i \in \mathfrak{R} \text{ disjoint}, \bigcup_{i=1}^{n} A_i = A\right\}$$

for all $\mu \in E^*$. If C is balanced, this expression simplifies to

$$|\theta|_A(C)(\mu) = \sup\left\{\sum_{i=1}^{n} |\theta_{A_i}| \,\middle|\, A_i \in \mathfrak{R} \text{ disjoint}, A_i \subset A\right\} p_C(\mu),$$

where

$$p_C(\mu) = \sup\{|\mu(c)| \mid c \in C\} \in \overline{\mathbb{R}}_+.$$

According to Proposition 2.1.1, the measure θ is strongly additive at all elements of E, and the above implies that $|\theta|$ is strongly additive at all neighborhoods in $\mathcal{V}$. Following Lemma 2.1.4, θ is therefore strongly additive at all bounded subsets of E.

(g) If both $E = F = \mathbb{C}$, then $\mathfrak{L}(E,F) = \mathbb{C}$, $F^* = \mathbb{C}$ and $\operatorname{Conv}(E)$ consists of all nonempty convex subsets of $\mathbb{C}$. Thus an $\mathfrak{L}(E,F)$-valued measure is indeed complex-valued, and the expressions from (f) read as

$$|\theta|_A(C)(z) = \sup\left\{\sum_{i=1}^{n} \mathfrak{Re}(\theta_{A_i} c_i z) \,\middle|\, c_i \in C,\ A_i \in \mathfrak{R} \text{ disjoint}, \bigcup_{i=1}^{n} A_i = A\right\}$$

for $C \in \operatorname{Conv}(\mathbb{C})$, and for the unit ball $\mathbb{B}$ of $\mathbb{C}$ in particular

$$|\theta|_A(\mathbb{B})(z) = \sup\left\{\sum_{i=1}^{n} |\theta_{A_i}| \,\middle|\, A_i \in \mathfrak{R} \text{ disjoint}, A_i \subset A\right\}|z|$$

for all $z \in \mathbb{C}$. The positive real-valued measure $[|\theta|(\mathbb{B})(1)]$, that is, the mapping

$$A \mapsto |\theta|_A(\mathbb{B})(1) : \mathfrak{R} \to \mathbb{R}_+$$

is the usual *absolute value of a complex-valued measure* (see, e. g., III.1.4 in [26]). The case that $E = F = \mathbb{R}$ is similar.

(h) For a counterexample to strong additivity of a measure consider the following. Let $X = [0,1]$ and let $\mathfrak{R}$ be the σ-field of all Borel subsets of X. Let $E = \mathbb{R}$ and $F = C([0,1])$ endowed with the supremum norm. By the Riesz representation theorem, F^* consists of all regular Borel measures on X, and $\mathfrak{L}(E, F^{*\blacksquare})$ may be identified with $F^{*\blacksquare}$ (see Part (e)). We define an $F^{*\blacksquare}$-valued measure θ by $\theta_A(\mu) = \mu(A)$, that is, the value of the Borel measure μ at A, for all $\mu \in F^*$ and $A \in \mathfrak{R}$. The countable additivity of the measures $\mu \in F^*$ transfers and yields the countable additivity of θ, pointwise on F^*. Moreover, since

$$|\theta|_A(\mathbb{B})(\mu) = \sup\left\{\sum_{i=1}^{n} |\mu(A_i)| \,\middle|\, A_i \in \mathfrak{R} \text{ disjoint}, A_i \subset A\right\} \leqslant \|\mu\|,$$

where $\mathbb{B} = [-1,+1]$ is the unit ball of $\mathbb{R}$ and $\|\mu\|$ is the dual norm of $\mu \in F^*$, θ is seen to be bounded. It is however not countably additive with respect to the symmetric topology of $\mathcal{Q}_F$, that is, the natural topology of $F^{*\blacksquare}$. Indeed, consider the sets $A_n = (0, 1/n)$. Then $\lim_{n\to\infty} \theta_{A_n}(\mu) = 0$ for all $\mu \in F^*$, but this convergence is not uniform on the dual unit ball in F^*.

Compact and weakly compact measures

An $\mathfrak{L}(\mathcal{P}_E, \mathcal{Q}_F)$-valued measure θ is called *compact* (or *weakly compact*) if it is $\mathfrak{L}(\mathcal{P}_E, F)$-valued and for every $A \in \mathfrak{R}$ and every bounded subset $C \in \operatorname{Conv}(E)$ the set

$$\left\{ \sum_{i=1}^{n} \theta_{A_i}(c_i) \;\middle|\; c_i \in C,\ A_i \in \mathfrak{R} \text{ disjoint, } \bigcup_{i=1}^{n} A_i = A \right\}$$

is relatively compact (or relatively weakly compact) in F. This set was of course used in our definition of the modulus of θ and was shown to be bounded in F in Example 2.1.8(a). If F is semireflexive, then every bounded subset of F is relatively weakly compact (see IV.5.5 in [56]). Hence every bounded $\mathfrak{L}(\mathcal{P}_E, F)$-valued measure is weakly compact in this case.

Because the balanced convex hull of a bounded set in $\operatorname{Conv}(E)$ is again bounded, it suffices to verify the condition for compactness or weak compactness of a measure for balanced convex sets $C \in \operatorname{Conv}(E)$. Moreover, since the closures of a convex set in the given and in the weak topologies of F coincide, the latter implies that every compact measure is also weakly compact.

Restriction of a measure

The *restriction* $\theta|_B$ of a measure θ to a set $B \in \mathfrak{A}$ is the measure

$$A \mapsto \theta_{A \cap B} : \mathfrak{R} \to \mathfrak{L}(\mathcal{P}_E, \mathcal{Q}_F).$$

The following is immediate from the respective definitions.

2.1.9 Proposition. *The restriction $\theta|_B$ of a (bounded) $\mathfrak{L}(\mathcal{P}_E, \mathcal{Q}_F)$-valued measure θ to a set $B \in \mathfrak{A}$ is again bounded and its modulus is the restriction of $|\theta|$ to B. If θ is compact (or weakly compact), then $\theta|_B$ is compact (or weakly compact).*

Sets of measure zero

A set $Z \in \mathfrak{A}$ is *of measure zero (with respect to an $\mathfrak{L}(\mathcal{P}_E, \mathcal{Q}_F)$-valued measure θ)* if $\theta_A = 0$ for all subsets $A \in \mathfrak{R}$ of Z.

2.1.10 Proposition. *Let θ be an $\mathfrak{L}(\mathcal{P}_E, \mathcal{Q}_F)$-valued measure.*

(a) *A set $B \in \mathfrak{A}$ is of measure zero if and only if all subsets $A \in \mathfrak{R}$ of B are of measure zero.*

(b) *If $A \in \mathfrak{R}$ is of measure zero, then both $|\theta_A| = 0$ and $\|\theta\|_A^{(\mathfrak{v},\Pi)} = 0$ for any choice of $\mathfrak{v}$ and Π.*

(c) *If the sets $B_i \in \mathfrak{A}$, for $i \in \mathbb{N}$, are of measure zero, then their union is of measure zero.*

Proof. Parts (a) and (b) are clear from the definition. For (c), suppose that the sets $B_i \in \mathfrak{A}$ for $i \in \mathbb{N}$ are of measure zero and let $A \in \mathfrak{R}$ such that $A \subset \bigcup_{i\in\mathbb{N}} B_i$. All the sets $A_i = A \cap B_i$ are of measure zero by (a) and there union is A. We set $C_1 = A_1$ and $C_{n+1} = A_{n+1} \setminus \bigcup_{i=1}^{n} C_i$. Again, all the sets C_i are of measure zero, they are disjoint and their union is A. Thus $\theta_A = \sum_{i=1}^{\infty} \theta_{C_i} = 0$. □

For the following result, we say that, given a measure θ, a subset Π of F^* is *θ-dense in F^** if for all $A \in \mathfrak{R}$ and $a \in E$ we have $\theta_A(a) = 0$ whenever $\theta_A(a)(\mu) = 0$ for all $\mu \in \Pi$; that is to say, if the linear span of Π is dense in F^* in the weak topology of the dual pair (F^*, F^θ), where F^θ is the linear span in $F^{*\blacksquare}$ of the set $\{\theta_A(a) \mid A \in \mathfrak{R},\ a \in E\}$.

2.1.11 Proposition. *Let θ be an $\mathfrak{L}(\mathcal{P}_E, \mathcal{Q}_F)$-valued measure, which coincides with its modulus on $\mathcal{P}_E$ and let $A \in \mathfrak{R}$.*

(a) *If there is $V \in \mathcal{V}$ such that $|\theta|_A(V) = 0$, then A is of measure zero.*
(b) *If there is a function space neighborhood $\mathfrak{v}$ and a θ-dense subset Π of F^* such that $\|\theta\|_A^{(\mathfrak{v},\Pi)} = 0$, then A is of measure zero.*

Proof. Let $A \in \mathfrak{R}$. For Part (a), suppose that θ satisfies our assumptions and that there is $V \in \mathcal{V}$ such that $|\theta|_A(V) = 0$. For every $a \in E$, there is $\varepsilon > 0$ such that $\gamma a \in V$ for all $|\gamma| \leqslant \varepsilon$. Hence $\theta_B(a) = 0$ for every subset $B \in \mathfrak{R}$ of A. Inspecting the definition of the modulus, we conclude that $\theta_B(C) = |\theta|_B(C) = 0$ holds for all $C \in \mathcal{P}_E$, that is, $\theta_B = 0$, for all such sets B. Hence our claim follows. For Part (b), suppose that $\|\theta\|_A^{(\mathfrak{v},\Pi)} = 0$ for $A \in \mathfrak{R}$ and that Π is θ-dense. According to our definition of a function space neighborhood, there is $V \in \mathcal{V}$ such that $\chi_A \cdot V \leqslant \mathfrak{v}$. Then $|\theta|_A(V)(\mu) = 0$ holds for all $\mu \in \Pi$ by Lemma 2.1.5 and, therefore, $\theta_B(a) = 0$ for all subsets $B \in \mathfrak{R}$ of A and all $a \in E$ by our assumption on Π. As in our argument for Part (a), we conclude that $\theta_B = 0$ for all subsets $B \in \mathfrak{R}$ of A. □

Properties that hold almost everywhere

If $Z \in \mathfrak{A}$ is of measure zero, the measure θ is said to be *supported* by its complement $Y = X \setminus Z$, since $\theta_A = \theta_{A\cap Y}$ holds for all $A \in \mathfrak{R}$. For a subset B of X, we shall say that a pointwise defined property of functions on X holds *θ-almost everywhere on B* if it holds on $B \setminus Z$ with some set $Z \in \mathfrak{A}$ of measure zero. In particular, we shall use the symbols $\underset{a.e.B}{\leqslant}$ or $\underset{a.e.B}{=}$ if the relations $\leqslant$ or $=$ hold θ-almost everywhere on the set B, respectively. Likewise, $f \underset{a.e.B}{=} g$, $f \underset{a.e.B}{\leqslant} g$ or $f \underset{a.e.B}{\leqslant} g + \mathfrak{v}$ for functions $f, g \in \mathcal{F}(X, \mathrm{Conv}(E))$ means that $\chi_{(B\setminus Z)} \cdot f = \chi_{(F\setminus Z)} \cdot g$, $\chi_{(B\setminus Z)} \cdot f \leqslant \chi_{(F\setminus Z)} \cdot g$ or $\chi_{(B\setminus Z)} \cdot f \leqslant \chi_{(F\setminus Z)} \cdot g + \mathfrak{v}$ holds with some $Z \in \mathfrak{A}$ of measure zero. These relations are of course transitive and compatible with the algebraic operations.

Two measures are said to be *mutually singular* if they are supported by disjoint subsets of X.

Point evaluation measures

A *point evaluation measure* is a nonzero measure supported by a singleton set, that is, a measure δ_x^T, for $x \in X$ and $0 \neq T \in \mathfrak{L}(\mathcal{P}_E, \mathcal{Q}_F)$ such that for every $A \in \mathfrak{R}$ it evaluates as $(\delta_x^T)_A = T$ if $x \in A$, and $(\delta_x^T)_A = 0$, else.

Regularity of measures

We continue with some particular consideration in the topological case, that is, the case that X is a locally compact Hausdorff space and that $\mathfrak{R}$ consists of all relatively compact Borel subsets of X. We shall assume this setting for the remainder of this section. The concept of regularity of a measure then arises naturally, and we shall approach it in the following way: An $\mathfrak{L}(\mathcal{P}_E, \mathcal{Q}_F)$-valued measure θ is said to be *outer regular* (or *inner regular*) for a set $A \in \mathfrak{R}$ if for every $\mu \in F^*$ there is $V \in \mathcal{V}$ such that $|\theta|_A(V)(\mu) < +\infty$ and

$$|\theta|_A(V)(\mu) = \lim_{O \supset A} |\theta|_O(V)(\mu) \quad \Big(\text{or } |\theta|_A(V)(\mu) = \lim_{K \subset A} |\theta|_K(V)(\mu)\Big).$$

The limits are taken over the downward directed family of all open sets $O \in \mathfrak{O}$ containing A for outer regularity, and over the upward directed family of all compact subsets K of A for inner regularity. We realize that the above conditions can be rewritten as

$$\lim_{O \supset A} |\theta|_{O \setminus A}(V)(\mu) = 0 \quad \text{and} \quad \lim_{K \subset A} |\theta|_{A \setminus K}(V)(\mu) = 0,$$

respectively. This follows from the additivity of the modulus, that is,

$$|\theta|_{O \setminus A}(V)(\mu) = |\theta|_O(V)(\mu) - |\theta|_A(V)(\mu)$$

and

$$|\theta|_{A \setminus K}(V)(\mu) = |\theta|_A(V)(\mu) - |\theta|_K(V)(\mu),$$

provided that $|\theta|_A(V)(\mu) < +\infty$. Consequently, the above limits are valid with all neighborhoods $U \in \mathcal{V}$ in place of V whenever $U \subset V$.

2.1.12 Lemma. *If θ is outer regular (or inner regular) for $A \in \mathfrak{R}$, then*

$$|\theta|_A(C) = \lim_{O \supset A} |\theta|_O(C) \quad \Big(\text{or } |\theta|_A(C) = \lim_{K \subset A} |\theta|_K(C)\Big)$$

holds pointwise on F^* for all bounded sets $C \in \mathrm{Conv}(E)$.

Proof. Let $C \in \mathrm{Conv}(E)$ be a bounded subset of E and let $\mu \in F^*$. Recall from Example 2.1.8(a) that $|\theta|_A(C)(\mu) < +\infty$ for all $A \in \mathfrak{R}$. There is $V \in \mathcal{V}$ as in the definition of outer regularity. Since $C \subset \lambda V$ for some $\lambda > 0$, for every $O \in \mathfrak{O}$ containing A we have

$$\big||\theta|_O(C)(\mu) - |\theta|_A(C)(\mu)\big| = \big||\theta|_{O\setminus A}(C)(\mu)\big| \leqslant \lambda|\theta|_{O\setminus A}(V)(\mu).$$

This yields $\lim_{O\supset A} |\theta|_O(a)(\mu) = |\theta|_A(a)(\mu)$ as claimed. The argument for the bracketed case is similar. □

Recall from Theorem 2.1.3 that θ and $|\theta|$ coincide on the elements of E. If the measure θ is indeed $\mathfrak{L}(E,F)$-valued (or $\mathfrak{L}(E,F^{**})$-valued), then the statement of Lemma 2.1.12 implies convergence of the operators θ_O and θ_K to θ_A in the weak operator topology of $\mathfrak{L}(E,F)$ (or in the weak* operator topology of $\mathfrak{L}(E,F^{**})$).

A measure that is both outer and inner regular for all sets in $\mathfrak{R}$ is called *regular*.

2.1.13 Proposition. *If θ is either inner regular for all sets in $\mathfrak{R}$ or outer regular for all sets in $\mathfrak{R}$, then θ is regular.*

Proof. First, suppose that θ is inner regular for all sets in $\mathfrak{R}$. We shall demonstrate outer regularity for all sets in $\mathfrak{R}$. For this, let $A \in \mathfrak{R}$ and let $U \in \mathfrak{O}$ be an open set containing A. Since θ is inner regular for $U \setminus A$, given $\mu \in F^*$ there is $V \in \mathcal{V}$ such that both $|\theta|_U(V)(\mu) \leqslant +\infty$ and for every $\varepsilon > 0$ there exists a compact subset K of $U \setminus A$ such that $|\theta|_{(U\setminus A)\setminus K}(V)(\mu) \leqslant \varepsilon$. Let $O = U \setminus K$. Obviously, O is open and contains A. Indeed, if $x \in A$, then $x \in U$ and $x \notin K \subset U \setminus A$. Because

$$(U \setminus A) \setminus K = (U \setminus K) \setminus A = O \setminus A$$

we compute that

$$|\theta|_{O\setminus A}(V)(\mu) = |\theta|_{(U\setminus A)\setminus K}(V)(\mu) \leqslant \varepsilon.$$

Hence θ is outer regular for A. Now suppose that θ is outer regular for all sets in $\mathfrak{R}$. We shall demonstrate inner regularity for all sets in $\mathfrak{R}$. For this, let $A \in \mathfrak{R}$ and let $U \in \mathfrak{O}$ be an open set containing $\overline{A}$, the closure of A. Since θ is outer regular for $U \setminus A$, given $\mu \in F^*$ there is $V \in \mathcal{V}$ such that both $|\theta|_U(V)(\mu) < +\infty$ and for every $\varepsilon > 0$ there exists an open set $O \in \mathfrak{O}$ containing $U \setminus A$ such that $|\theta|_{O\setminus(U\setminus A)}(V)(\mu) \leqslant \varepsilon$. We may assume that $O \subset U$. Let $K = A \setminus O$. We recognize that K is a compact subset of A as $A \setminus O = \overline{A} \setminus O$. Indeed, if $x \in \overline{A} \setminus O$, then $x \in U$, but $x \notin O$, hence $x \notin U \setminus A$. That is to say, $x \in A$. Next, we observe that

$$O \setminus (U \setminus A) = O \cap A = A \setminus (A \setminus O) = A \setminus K.$$

This shows

$$|\theta|_{A\setminus K}(V)(\mu) = |\theta|_{O\setminus(U\setminus A)}(V)(\mu) \leqslant \varepsilon,$$

and θ is seen to be inner regular for A. Our claim follows. □

Some of the earlier introduced concepts can be augmented for regular measures.

2.1.14 Proposition. *Let $A \in \mathfrak{R}$. If θ is regular, then*

$$|\theta|_A(C) = \sup\left\{\sum_{i=1}^{n} \theta_{K_i}(c_i) \;\middle|\; c_i \in C,\ K_i \text{ disjoint compact},\ K_i \subset A\right\}$$

for $C \in \mathrm{Conv}(E)$ such that $0 \in C$, and

$$\|\theta\|_A^{(\mathfrak{v},\Pi)} = \sup\left\{\sum_{i=1}^{n} \theta_{K_i}(a_i)(\mu_i) \;\middle|\; \begin{array}{l} K_i \text{ disjoint compact},\ K_i \subset A, \\ a_i \in E,\ \chi_{K_i}\cdot a_i \leqslant \mathfrak{v} \text{ and } \mu_i \in \Pi \end{array}\right\}$$

for a function space neighborhood $\mathfrak{v}$ and $\Pi \subset F^*$.

Proof. For the first statement concerning the modulus of θ, since $0 \in C$, the supremum on the right-hand side of the above expression is less or equal to $|\theta|_A(C)$. For the reverse inequality, let $A_i \in \mathfrak{R}$ be disjoint subsets of A, and let $c_i \in C$ for $i = 1, \ldots, n$. Because θ is regular, we have $\theta_{A_i}(c_i) = \lim_{K_i \subset A_i} \theta_{A_i}(c_i)$, where K_i are compact subsets of A_i, for all $k = 1, \ldots, n$. Hence

$$\sum_{i=1}^{n} \theta_{A_i}(c_i) = \lim_{K_i \subset A_i} \sum_{i=1}^{n} \theta_{K_i}(c_i).$$

Thus

$$\sum_{i=1}^{n} \theta_{A_i}(c_i) \leqslant \sup\left\{\sum_{i=1}^{n} \theta_{K_i}(c_i) \;\middle|\; c_i \in C,\ K_i \text{ disjoint compact},\ K_i \subset A\right\},$$

our claim. The argument for the variation of θ is similar. □

2.1.15 Proposition. *If θ is outer (or inner) regular for a set $A \in \mathfrak{R}$, then $|\theta|$ is also outer (or inner) regular for A. If θ is regular and $C \in \mathrm{Conv}(E)$, then $|\theta|_A(C) = \lim_{K \subset A} |\theta|_K(C)$ holds for all $A \in \mathfrak{R}$. If in addition $|\theta|_A(C)(\mu) < +\infty$ holds for $\mu \in F^{**}$ and all $A \in \mathfrak{R}$, then $|\theta|_A(C)(\mu) = \lim_{O \supset A} |\theta|_O(C)(\mu)$.*

Proof. The first statement is evident from the definition of regularity together with the finding from Theorem 2.1.3 that the modulus of the modulus $|\theta|$ of a measure θ is again $|\theta|$. For the second statement, suppose that θ is regular and let $A \in \mathfrak{R}$. In a first step, we consider a set $C \in \mathrm{Conv}(E)$ such that $C \geqslant 0$, that is, $0 \in C$. The inequality

$$\lim_{K \subset A} |\theta|_K(C) = \sup_{K \subset A} |\theta|_K(C) \leqslant |\theta|_A(C)$$

is obvious in this case. For the reverse inequality, for any $\alpha < |\theta|_A(C)$, by Proposition 2.1.14 there are disjoint compact subsets K_i of A and $c_i \in C$ such that

$$\alpha \leqslant \sum_{i=1}^{n} \theta_{K_i}(c_i) \leqslant |\theta|_K(C),$$

whereby $K = \bigcup_{i=1}^{n} K_i \subset A$. Hence $|\theta|_A(C) \leqslant \sup_{K\subset A} |\theta|_K(C)$. Now for any $C \in \mathrm{Conv}(E)$ we choose an element $c_0 \in C$ and set $C_0 = C - c_0 \in \mathrm{Conv}(E)$. Then $C_0 \geqslant 0$ and we have

$$\begin{aligned}\lim_{K\subset A} |\theta|_K(C) &= \lim_{K\subset A} |\theta|_K(C_0) + \lim_{K\subset A} |\theta|_K(c_0)\\ &= |\theta|_A(C_0) + |\theta|_A(c_0)\\ &= |\theta|_A(C)\end{aligned}$$

by our first step together with Lemma 2.1.12. If $|\theta|_A(C)(\mu) < +\infty$ for $\mu \in F^{**}$ and all $A \in \mathfrak{R}$, then the set function

$$A \mapsto |\theta|_A(C)(\mu) : \mathfrak{R} \to \mathbb{R}$$

is countably additive by Theorem 2.1.3 and defines a positive real-valued measure $[|\theta|_A(C)(\mu)]$ on $\mathfrak{R}$. This measure is inner regular for all sets in $\mathfrak{R}$ and, therefore, regular by Proposition 2.1.13. Our claim follows. □

The transfer of regularity of a measure to its $\overline{\mathbb{R}}_+$-valued variation is however less obvious.

2.1.16 Proposition. *Let $\mathfrak{v}$ be a function space neighborhood and $\Pi \subset F^*$. If θ is regular and if $\|\theta\|_O^{(\mathfrak{v},\Pi)} < +\infty$ for $O \in \mathfrak{O}$, then the $\overline{\mathbb{R}}_+$-valued measure $\|\theta\|^{(\mathfrak{v},\Pi)}$ is both outer and inner regular for all subsets $A \in \mathfrak{R}$ of O.*

Proof. Suppose that θ is regular. Using Proposition 2.1.14, we realize that in the calculation of $\|\theta\|_A^{(\mathfrak{v},\Pi)}$ it suffices to consider only disjoint compact subsets K_i of A. Let $O_0 \in \mathfrak{O}$ such that $\|\theta\|_{O_0}^{(\mathfrak{u},\Pi)} < +\infty$ and assume that for some subset $A \in \mathfrak{R}$ of O_0 we have $\|\theta\|_A^{(\mathfrak{u},\Pi)} < \inf_{O\supset A} \|\theta\|_O^{(\mathfrak{u},\Pi)}$. Hence there is $\varepsilon > 0$ such that $\|\theta\|_{O\setminus A}^{(\mathfrak{u},\Pi)} > \varepsilon$ holds for all $O \in \mathfrak{O}$ such that $O \supset A$. Choose disjoint compact subsets K_i of $O_0 \setminus A$, $\chi_{K_i} \cdot a_i \leqslant \mathfrak{u}$ and $\mu_i \in \Pi$ for $i = 1, \ldots, n$ such that $\sum_{i=1}^{n} \theta_{K_i}(a_i)(\mu_i) \geqslant \varepsilon$. Now we set $O_1 = O_0 \setminus \bigcup_{i=1}^{n} K_i \in \mathfrak{O}$. Then $O_0 \supset O_1 \supset A$ and $\|\theta\|_{O_0\setminus O_1}^{(\mathfrak{u},\Pi)} \geqslant \varepsilon$. Repeating the preceding step with O_1 in place of O_0, we find $O_2 \in \mathfrak{O}$ such that $O_1 \supset O_2 \supset A$ and $\|\theta\|_{O_1\setminus O_2}^{(\mathfrak{u},\Pi)} \geqslant \varepsilon$, that is, $\|\theta\|_{O_0\setminus O_2}^{(\mathfrak{u},\Pi)} \geqslant 2\varepsilon$. After n such steps, we have $O_n \in \mathfrak{O}$ such that $O_{n-1} \supset O_n \supset A$ and $\|\theta\|_{O_0\setminus O_n}^{(\mathfrak{u},\Pi)} \geqslant n\varepsilon$. Since $O_0 \setminus O_n \subset O_0 \setminus A$, this renders $\|\theta\|_{O_0\setminus A}^{(\mathfrak{u},\Pi)} = +\infty$, thus contradicting our assumption that $\|\theta\|_{O_0}^{(\mathfrak{u},\Pi)} < +\infty$. The argument for inner regularity is similar. □

2.1.17 Proposition. *Let θ be an $\mathfrak{L}(\mathcal{P}_E, \mathcal{Q}_F)$-valued measure such that for every $A \in \mathfrak{R}$ and $W \in \mathcal{W}$ there is a function space neighborhood $\mathfrak{v}$ such that $\|\theta\|_A^{(\mathfrak{v},W^\circ)} < +\infty$. If θ is regular, then*

$$\theta_A(C) = \lim_{O\supset A} \theta_O(C) = \lim_{K\subset A} \theta_K(C)$$

for all bounded sets $C \in \mathcal{P}_E$ and all $A \in \mathfrak{R}$ in the symmetric topology of $\mathcal{Q}_F$.

Proof. Suppose that θ is regular and satisfies our assumption for its variation. Let $A \in \mathfrak{R}$, $W \in \mathcal{W}$ and let C be a bounded element of $\mathcal{P}_E$. We choose an open set $O_0 \in \mathfrak{O}$ containing A and a function space neighborhood $\mathfrak{v}$ such that $\|\theta\|_{O_0}^{(\mathfrak{v},W^\circ)} < +\infty$. There is $V \in \mathcal{V}$ such that $\chi_{O_0} \cdot V \leqslant \mathfrak{v}$ and there is $\lambda > 0$ such that $C \subset \lambda V$. According to Proposition 2.1.16, given $\varepsilon > 0$ there is $O_1 \in \mathfrak{O}$ such that $O_0 \supset O_1 \supset A$ and $\|\theta\|_{O\setminus A}^{(\mathfrak{v},W^\circ)} \leqslant \varepsilon/\lambda$ for all $O \in \mathfrak{O}$ such that $O_1 \supset O \supset A$. Thus

$$\theta_{O\setminus A}(C)(\mu) = |\theta|_{O\setminus A}(C)(\mu) \leqslant \varepsilon$$

for all $\mu \in W^\circ$ by Lemma 2.1.5(a). Since $-\mu \in W^\circ$ whenever $\mu \in W^\circ$ and since $\theta_{O\setminus A}(C)(\mu) \geqslant -\theta_{O\setminus A}(C)(-\mu) \geqslant -\varepsilon$, we infer that both

$$\theta_O(C)(\mu) = \theta_A(C)(\mu) + \theta_{O\setminus A}(C)(\mu) \leqslant \theta_A(C)(\mu) + \varepsilon$$

and

$$\theta_A(C)(\mu) = \theta_O(C)(\mu) - \theta_{O\setminus A}(C)(\mu) \leqslant \theta_O(C)(\mu) + \varepsilon$$

holds for all $\mu \in W^\circ$ and all $O \in \mathfrak{O}$ such that $O_1 \supset O \supset A$. But this verifies that $\theta_O(C)$ is an element of the symmetric neighborhood $W^s(\theta_A(C))$ of $\theta_O(C)$ (see Section 1.4). The argument for the second equality regarding inner regularity is similar. □

2.1.18 Proposition. *The restriction $\theta|_B$ of an $\mathfrak{L}(\mathcal{P}_E, \mathcal{Q}_F)$-valued measure θ to a set $B \in \mathfrak{A}$ is outer (or inner) regular for a set $A \in \mathfrak{R}$ whenever θ is outer (or inner) regular for A.*

Proof. If θ is outer regular for a set $A \in \mathfrak{R}$, then for every $\mu \in F^*$ there is $V \in \mathcal{V}$ such that $|\theta|_A(V)(\mu) < +\infty$ and that $\lim_{O\supset A} |\theta|_{O\setminus A}(V)(\mu) = 0$. Since $|(\theta|_B)|_{O\setminus A}(V) \leqslant |\theta|_{O\setminus A}(V)$, this yields $\lim_{O\supset A} |(\theta|_B)|_{O\setminus A}(V)(\mu) = 0$ as well. The argument for inner regularity is similar. □

2.1.19 Proposition. *Let θ be a regular $\mathfrak{L}(\mathcal{P}_E, \mathcal{Q}_F)$-valued measure, which coincides with its modulus on $\mathcal{P}_E$, and suppose that all sets in $\mathcal{P}_E$ are bounded. If $\theta_K = 0$ for every compact subset K of a set $B \in \mathfrak{A}$, then B is of measure zero. The union of a family of open sets of measure zero is of measure zero.*

Proof. Suppose that θ satisfies our assumptions, and let $B \in \mathfrak{A}$ and let $A \in \mathfrak{R}$ be a subset of B. Together with our assumptions, Lemma 2.1.12 yields that $\theta_A = 0$, provided that $\theta_K = 0$ for every compact subset K of A. Our first claim follows. Now let $\mathcal{O}$ be a family of open sets of measure zero and let K be a compact subset of their union. There are finitely many sets $O_i \in \mathcal{O}$ such that $K \subset \bigcup_{i=1}^n O_i$, and their union is of measure zero by Proposition 2.1.10(c). Hence K is of measure zero by 2.1.10(a), and our second statement follows from the first one. □

If θ satisfies the assumptions of Proposition 2.1.19, then there exists a largest open subset of measure zero in X. Its complement is the smallest closed subset of X supporting θ, called the *support of* θ and denoted as $\operatorname{supp}(\theta)$.

Composition measures

If $(G,\mathcal{U})$ and $(H,\mathcal{R})$ are additional locally convex topological vector spaces such that H is Hausdorff and if θ is an $\mathfrak{L}(E,\mathcal{Q}_F)$-valued measure, then for any choice of operators $S \in \mathfrak{L}(G,E)$ and $T \in \mathfrak{L}(F,H)$ the formula

$$(T \circ \theta \circ S_A(g))(v) = (\theta_A(S(g)))(T^*(v))$$

defines an $\mathfrak{L}(G,\mathcal{Q}_H)$-valued measure $T \circ \theta \circ S$. In this expression, we have $A \in \mathfrak{R}$, $g \in G$, $v \in H^*$, and $T^* : H^* \to F^*$ is the adjoint of the operator $T \in \mathfrak{L}(F,H)$. We go forward to check he requirements for a measure for $T \circ \theta \circ S$. First, for $A \in \mathfrak{R}$ and $g \in G$ the function $v \mapsto ((T\circ\theta\circ S)_A(g))(v)$ is linear on H^* and bounded on the polars of the neighborhoods in $\mathcal{R}$. Indeed for $R \in \mathcal{R}$, there is $W \in \mathcal{W}$ such that $T(W) \subset R$, hence $T^*(R^\circ) \subset W^\circ$. There is $V \in \mathcal{V}$ such that $\theta_A(a) \leqslant \sigma_W$ for all $a \in V$ and there is $U \in \mathcal{U}$ such that $S(U) \subset V$. Then for all $u \in U$ we have $\theta_A(S(u)) \leqslant \sigma_W$. There is $\lambda \geqslant 0$ such that $g \in \lambda U$. Hence $(\theta_A(S(g)))(T^*(v)) \leqslant \lambda$ holds for all $v \in R^\circ$. We infer that $(T \circ \theta \circ S)_A(g) \in \mathcal{Q}_H$. Next, we realize that for every $A \in \mathfrak{R}$ the operator $g \mapsto (T\circ\theta\circ S)_A(g)$ is linear from G to $\mathcal{Q}_H$ and continuous, since for $R \in \mathcal{R}$ we choose $U \in \mathcal{U}$ as above and have $\theta_A(S(g)) \leqslant \sigma_W$, hence $(\theta_A(S(g)))(T^*(v)) \leqslant 1$ for all $g \in U$ and all $v \in R^\circ$. Summarizing, we have established that $(T\circ\theta\circ S)_A \in \mathfrak{L}(G,\mathcal{Q}_H)$. Because the countable additivity of $T \circ \theta \circ S$ follows immediately from the countable additivity of θ, all that is left to verify is boundedness. For this, we shall work out the modulus of this measure. For $A \in \mathfrak{R}$, a set $D \in \operatorname{Conv}(G)$ and $v \in H^*$ we have $S(D) \in \operatorname{Conv}(E)$ and calculate

$$\begin{aligned}
|T \circ \theta \circ S|_A(D)(v) &= \sup\left\{\sum_{i=1}^{n}(T \circ \theta \circ S)_{A_i}(d_i)(v) \;\middle|\; \begin{array}{l} d_i \in D,\ A_i \in \mathfrak{R} \text{ disjoint,} \\ \bigcup_{i=1}^{n} A_i = A \end{array}\right\} \\
&= \sup\left\{\sum_{i=1}^{n}\theta_{A_i}(c_i)(T^*(v)) \;\middle|\; \begin{array}{l} c_i \in S(D),\ A_i \in \mathfrak{R} \text{ disjoint,} \\ \bigcup_{i=1}^{n} A_i = A \end{array}\right\} \\
&= |\theta|_A(S(D))(T^*(v)).
\end{aligned}$$

Now for boundedness of the measure $T \circ \theta \circ S$, given $A \in \mathfrak{R}$ and $R \in \mathcal{R}$ there is $W \in \mathcal{W}$ such that $T^*(R^\circ) \subset W^\circ$ and $V \in \mathcal{V}$ such that $|\theta|_A(V) \leqslant \sigma_W$. There is $U \in \mathcal{U}$ such that $S(U) \subset V$, and using the above we realize that

$$|T \circ \theta \circ S|_A(U)(v) \leqslant |\theta|_A(V)(T^*(v)) \leqslant 1$$

for all $v \in R^{\circ}$. That is, $|T \circ \theta \circ S|_A(U) \leqslant \sigma_R$, as required. Regularity like countable additivity is determined pointwise on the elements of F^* and H^*, respectively. Thus regularity for θ implies regularity for $T \circ \theta \circ S$. Moreover, if the measure θ is indeed $\mathfrak{L}(E, F^{*\blacksquare})$-valued, that is,

$$\theta_A(\alpha a)(\mu) = \theta_A(a)(\alpha\mu)$$

holds for all $A \in \mathfrak{R}$, all $a \in E$, $\mu \in F^*$ and α in the common scalar field of the involved vector spaces, then this property transfers directly to the composition measure. That is, the measure $T \circ \theta \circ S$ is $\mathfrak{L}(E, H^{*\blacksquare})$-valued in this case.

2.2 Measurable functions

Measurability for vector-valued functions was introduced in various places (see, e. g., Dunford and Schwartz [26] III.2.10 or Diestel and Uhl [21] II.1). Different from most of these concepts, our forthcoming take on measurability will be quite restrictive and not refer to a particular measure. Rather than being a prerequisite for integrability, we shall establish a class of functions that are integrable with respect to any measure on a given σ-ring $\mathfrak{R}$. Integrability with respect to a specific measure will be developed in the following section and extend beyond this class.

We shall use the same broad assumptions and notation as in the preceding sections. The set X is generally not required to carry a topological structure, and $\mathfrak{R}$ is a σ-ring of subsets of X. Any measure, if referred to, is supposed to be defined on $\mathfrak{R}$, $\mathfrak{L}(\mathcal{P}_E, \mathcal{Q}_F)$-valued and bounded. The topological scenario, where X is locally compact and $\mathfrak{R}$ consists of all relatively compact Borel subsets of X, is included as a special case. We shall utilize step functions as introduced in Section 1.2, but it will turn out to be advantageous to have these available as not just E-valued, but indeed as $\mathrm{Conv}(E)$-valued functions.

Set-valued step functions

We consider $\mathrm{Conv}(E)$-valued step functions on X, that is, functions

$$h = \sum_{i=1}^{n} \chi_{A_i} \cdot C_i$$

for sets $A_i \in \mathfrak{R}$ and elements C_i in $\mathrm{Conv}(E)$. The subcone of $\mathcal{F}(X, \mathrm{Conv}(E))$ constituted by these functions is denoted by $\mathcal{F}_S(X, \mathrm{Conv}(E))$, the subspace of all E-valued step functions is denoted by $\mathcal{F}_S(X, E)$. (Recall that via the embedding $a \mapsto \{a\}$ we consider E as a subspace of $\mathrm{Conv}(E)$.) The above representation of a step function in $\mathcal{F}_S(X, \mathrm{Conv}(E))$ is of course not unique, but we may obtain a representation where the sets $A_i \in \mathfrak{R}$ are disjoint in the following way.

2.2.1 Lemma. *For every step function,* $h = \sum_{i=1}^{n} \chi_{A_i} \cdot C_i$ *there exists a representation* $\sum_{k=1}^{m} \chi_{B_k} \cdot D_k$ *such that the sets* $B_k \in \mathfrak{R}$ *are pairwise disjoint and such that* $\sum_{i=1}^{n} |\theta|_{A_i}(C_i) = \sum_{k=1}^{m} |\theta|_{B_k}(D_k)$ *holds for any* $\mathfrak{L}(\mathcal{P}_E, \mathcal{Q}_F)$*-valued measure* θ.

Proof. Let $h = \sum_{i=1}^{n} \chi_{A_i} \cdot C_i \in \mathcal{F}_S(X, \mathrm{Conv}(E))$ For a nonempty subset I of $\{1, \dots, n\}$, set

$$B_I = \bigcap_{i \in I} A_i \quad \bigcap_{i \in \{1,\dots,n\} \setminus I} X \setminus A_i.$$

The sets B_I are contained in $\mathfrak{R}$, pairwise disjoint for distinct index sets I, and their union is the union of the sets A_i. Let $\mathcal{I}$ be the family of all I such that $B_I \neq \emptyset$, and for every $I \in \mathcal{I}$ set $D_I = \sum_{i \in I} C_i$. One easily verifies that

$$h = \sum_{I \in \mathcal{I}} \chi_{B_I} \cdot D_I.$$

Moreover, for every i we have $A_i = \bigcup_{i \in I} B_I$. Hence $|\theta|_{A_i}(C_i) = \sum_{i \in I} |\theta|_{B_I}(C_i)$ holds for any $\mathfrak{L}(\mathcal{P}_E, \mathcal{Q}_F)$-valued measure θ and, therefore,

$$\sum_{i=1}^{n} |\theta|_{A_i}(C_i) = \sum_{i=1}^{n} \sum_{i \in I} |\theta|_{B_I}(C_i) = \sum_{I \in \mathcal{I}} |\theta|_{B_I}\Big(\sum_{i \in I} C_i\Big) = \sum_{I \in \mathcal{I}} |\theta|_{B_I}(D_I),$$

our claim. □

The representation of a step function $h = \sum_{i=1}^{n} \chi_{A_i} \cdot C_i$ with pairwise disjoint sets in $A_i \in \mathfrak{R}$ is unique, provided that the elements $C_i \in \mathrm{Conv}(E)$ are distinct and not zero. In this case, it is called the *standard representation* of h.

We recall from Section 1.4 the definition of boundedness for elements and subsets of $\mathcal{F}(X, \mathrm{Conv}(E))$. A function $f \in \mathcal{F}(X, \mathrm{Conv}(E))$ is bounded above (or below) if for every function space neighborhood $\mathfrak{v}$ there is $\lambda \geqslant 0$ such that $f \leqslant \lambda \mathfrak{v}$ (or $0 \leqslant f + \lambda \mathfrak{v}$). A subset of $\mathcal{F}(X, \mathrm{Conv}(E))$ is bounded above (or below) if these conditions hold with the same $\lambda \geqslant 0$ for all functions in this subset.

2.2.2 Lemma. *If* $f \in \mathcal{F}(X, \mathrm{Conv}(E))$ *is bounded above (or bounded below), then there is* $A \in \mathfrak{R}$ *such that* $f(x) = 0$ *(or* $f(x) \geqslant 0$*) for all* $x \in X \setminus A$, *and for every* $V \in \mathcal{V}$ *there is* $\lambda \geqslant 0$ *such that* $f \leqslant \lambda \chi_A \cdot V$ *(or* $0 \leqslant f + \lambda \chi_A \cdot V$*). The range of* f *is bounded above (or bounded below) in* $\mathrm{Conv}(E)$.

Proof. For our first statement, let us consider the case that $f \in \mathcal{F}(X, \mathrm{Conv}(E))$ is bounded below. For every $A \in \mathfrak{R}$, choose a neighborhood $V_A \in \mathcal{V}$. The convex hull $\mathcal{N}$ of the neighborhood functions $\chi_A \cdot V_A$ defines a function space neighborhood $\mathfrak{v}$ (see Example 1.2.7(f)). There is $n \in \mathcal{N}$ and $\lambda \geqslant 0$ such that $0 \leqslant f + \lambda n$. Because all neighborhood functions in $\mathcal{N}$ are supported by a set in $\mathfrak{R}$, $0 \leqslant f(x) + \lambda n(x)$ for all $x \in X$ implies that there is $A \in \mathfrak{R}$ such that $f(x) \geqslant 0$ for all $x \in X \setminus A$. The case of upper boundedness is similar, taking into account that $f(x) \leqslant 0$ means that $f(x) = 0$. Now for

$V \in \mathcal{V}$ consider the function space neighborhood $\mathfrak{v}$ defined by the constant neighborhood function $x \mapsto V$. If f is bounded above (or below), then there is $\lambda \geqslant 0$ such that

$$f \leqslant \lambda\chi_X \cdot V \quad (\text{or } 0 \leqslant f + \lambda\chi_X \cdot V).$$

That is to say,

$$f \leqslant \lambda\chi_A \cdot V \quad (\text{or } 0 \leqslant f + \lambda\chi_A \cdot V).$$

The latter also implies that $f(X) \leqslant \lambda V$ (or that $0 \leqslant f(X) + \lambda V$). Our claim follows. □

2.2.3 Lemma.

(a) *Every step function is bounded below.*

(b) *Let $f \in \mathcal{F}(X, \mathrm{Conv}(E))$. If for every function space neighborhood $\mathfrak{v}$ there is a bounded below function $g \in \mathcal{F}(X, \mathrm{Conv}(E))$ such that $g \leqslant f + \mathfrak{v}$, then f is bounded below.*

Proof. (a) Let $h = \sum_{i=1}^{n} \chi_{A_i} \cdot C_i$ be a $\mathrm{Conv}(E)$-valued step function and let $\mathfrak{v}$ be a function space neighborhood. We may assume that the sets A_i are disjoint and set $A = \bigcup_{i=1}^{n} A_i \in \mathfrak{R}$. There is $V \in \mathcal{V}$ such that $\chi_A \cdot V \leqslant \mathfrak{v}$, and in turn there is $\lambda \geqslant 0$ such that $0 \leqslant C_i + \lambda V$ for all $i = 1, \ldots, n$. This shows $0 \leqslant h + \lambda\chi_A \cdot V \leqslant h + \lambda\mathfrak{v}$, our claim in (a). Now suppose that the assumption of (b) holds for $f \in \mathcal{F}(X, \mathrm{Conv}(E))$. Given a function space neighborhood $\mathfrak{v}$, there is a bounded below function $g \in \mathcal{F}(X, \mathrm{Conv}(E))$ such that $g \leqslant f + \mathfrak{v}$ and $\lambda \geqslant 0$ such that $0 \leqslant g + \lambda\mathfrak{v}$. Hence $0 \leqslant f + (\lambda + 1)\mathfrak{v}$, our claim in Part (b). □

Measurable vector-valued functions

We shall first define measurability for E-valued functions. A function $f \in \mathcal{F}(X, E)$ is said to be *measurable* if for every function space neighborhood $\mathfrak{v}$ there is a step function $h \in \mathcal{F}_S(X, E)$ such that $h \leqslant f + \mathfrak{v}$, that is, $h - f \leqslant \mathfrak{v}$.

2.2.4 Proposition. *Every measurable function in $\mathcal{F}(X, E)$ is bounded and is supported by a set in $\mathfrak{R}$. Its range is a bounded subset of E. The measurable functions form a subspace of $\mathcal{F}(X, E)$.*

Proof. The first statements follow from Lemmas 2.2.3 and 2.2.2 and since for E-valued functions upper and lower boundedness coincide. The last statement is obvious, since $h - f \leqslant \mathfrak{v}$ and $l - g \leqslant \mathfrak{v}$ for step functions $h, l \in \mathcal{F}_S(X, E)$ and $f, g \in \mathcal{F}(X, E)$ implies that $(h + l) - (f + g) \in 2\mathfrak{v}$ and $ah - af \in |a|\mathfrak{v}$ for all a in $\mathbb{R}$ or $\mathbb{C}$. □

2.2.5 Proposition. *A function $f \in \mathcal{F}(X, E)$ is measurable if and only if it is supported by a set $A \in \mathfrak{R}$ and for every $V \in \mathcal{V}$ there is $h \in \mathcal{F}_S(X, E)$ such that $h - f \leqslant \chi_A \cdot V$. The step function h can be chosen such that all its values are in the range of f.*

Proof. Suppose that $f \in \mathcal{F}(X,E)$ is measurable and let $V \in \mathcal{V}$. Using the function space neighborhood $\mathfrak{v}$ defined by the constant neighborhood function $x \mapsto (1/2)V$ in our definition of measurability we find a step function $h \in \mathcal{F}_\mathcal{S}(X,E)$ such that $h - f \leqslant (1/2)\chi_X \cdot V$. According to Proposition 2.2.4, the function f is supported by a set $A \in \mathfrak{R}$ and we may assume the same for the step function h. That is, $h - f \leqslant (1/2)\chi_A \cdot V$. We have $h = \sum_{i=1}^n \chi_{A_i} \cdot a_i$ for disjoint sets $A_i \in \mathfrak{R}$ whose union is A and $a_i \in E$. We choose $x_i \in A_i$, set $b_i = f(x_i)$ and $g = \sum_{i=1}^n \chi_{A_i} \cdot b_i \in \mathcal{F}_\mathcal{S}(X,E)$. Then all values of g are in the range of f, and for all $x \in A_i$ we notice that

$$f(x) - g(x) = f(x) - f(x_i) = \big(f(x) - h(x)\big) + \big(h(x_i) - f(x_i)\big) \in V$$

since $h(x) = h(x_i)$. Hence $g - f \leqslant \chi_A \cdot V$, our claim. Conversely, if our condition holds for $f \in \mathcal{F}(X,E)$, there is $A \in \mathfrak{R}$ such that $f = \chi_A \cdot f$ and for every function space neighborhood $\mathfrak{v}$ there is $V \in \mathcal{V}$ such that $\chi_A \cdot V \leqslant \mathfrak{v}$. We find $h \in \mathcal{F}_\mathcal{S}(X,E)$ such that $h - f \leqslant \chi_A \cdot V$ and, therefore, $h - f \leqslant \mathfrak{v}$, as required. □

We shall consider two special cases of sufficient interest.

2.2.6 Proposition. *If E is a normed space, then a function in $\mathcal{F}(X,E)$ is measurable if and only if it is supported by a set in $\mathfrak{R}$ and the uniform limit of a sequence of step functions in $\mathcal{F}_\mathcal{S}(X,E)$.*

Proof. Let E be a normed space with unit ball $\mathbb{B}$. According to Proposition 2.2.5, a function $f \in \mathcal{F}(X,E)$ is measurable if and only if for every $n \in \mathbb{N}$ there is a step function $h_n \in \mathcal{F}_\mathcal{S}(X,E)$ such that $h_n - f \leqslant \chi_A \cdot (1/n)\mathbb{B}$, that is, $\|h_n - f\| \leqslant 1/n$. □

A real-valued function is measurable if it satisfies our condition of measurability with $E = \mathbb{R}$. We shall however also use the term $\mathfrak{A}$*-measurable* for an $\overline{\mathbb{R}}$-valued function φ if it is measurable in the conventional sense with respect to the σ-field $\mathfrak{A}$, that is, if for every $\alpha \in \mathbb{R}$ the set $\{x \in X \mid \varphi(x) > \alpha\}$ is contained in $\mathfrak{A}$ (see 11.2 in [53]).

2.2.7 Proposition. *A real-valued function in $\mathcal{F}(X)$ is measurable if and only if it is bounded, supported by a set in $\mathfrak{R}$ and $\mathfrak{A}$-measurable.*

Proof. If $\varphi \in \mathcal{F}(X)$ is measurable in our sense with $E = \mathbb{R}$, then by 2.2.4 it is bounded and supported by a set $A \in \mathfrak{R}$. Following Proposition 2.2.6, φ is the uniform limit of a sequence of real-valued step functions and, therefore, $\mathfrak{A}$-measurable (Theorem 6 in 11.2 of [53]). Conversely, suppose that φ is bounded, supported by a set in $A \in \mathfrak{R}$ and $\mathfrak{A}$-measurable. Let $\rho \geqslant 0$ be a bound for $|\varphi|$. Given a function space neighborhood $\mathfrak{v}$ for $\mathcal{F}(X)$ there is $\varepsilon > 0$ such that $\chi_A \cdot (\varepsilon\mathbb{B}) \leqslant \mathfrak{v}$, where $\mathbb{B} = [-1, +1]$. Choose $n > (1/\varepsilon)\rho$. Since φ is $\mathfrak{A}$-measurable, the sets

$$A_i = \big\{x \in A \mid (i-1)\varepsilon \leqslant \varphi(x) < i\varepsilon\big\}$$

for $i = -n, \ldots, n$ are contained in $\mathfrak{R}$ and their union is all of A. We choose $x_i \in A_i$, provided that $A_i \neq \emptyset$, and consider the real-valued step function. $h = \sum_{i=-n}^{n} \chi_{A_i} \cdot \varphi(x_i)$. Then for every i, we have

$$|\varphi(x) - h(x)| = |\varphi(x) - \varphi(x_i)| \leqslant \varepsilon$$

for all $x \in A_i$. This demonstrates that $h - f \leqslant \chi_A \cdot (\varepsilon \mathbb{B}) \leqslant \mathfrak{v}$, as required. □

2.2.8 Proposition. *If $\varphi \in \mathcal{F}(X)$ is measurable, then the function $\varphi \cdot a \in \mathcal{F}(X, E)$ is measurable for every $a \in E$.*

Proof. Suppose that $\varphi \in \mathcal{F}(X)$ is measurable and let $a \in E$. There is $A \in \mathfrak{R}$ such that φ is supported by A, and given a function space neighborhood $\mathfrak{v}$ there is $V \in \mathcal{V}$ such that $\chi_A \cdot V \leqslant \mathfrak{v}$, and there is $\lambda > 0$ such that $a \in \lambda V$. There is a real-valued step function ψ such that $|\psi(x) - \varphi(x)| \leqslant (1/\lambda)$ for all $x \in X$. We may assume that ψ is also supported by A and set $h = \psi \cdot a \in \mathcal{F}_{\mathcal{S}}(X, E)$. Then

$$h(x) - f(x) = (\psi(x) - \varphi(x))a \in V$$

for all $x \in A$ and $h(x) = f(x) = 0$ for all $x \in X \setminus A$. Therefore, $h - f \leqslant \chi_A \cdot V \leqslant \mathfrak{v}$, as required. □

The topological case offers the following observation, which is an immediate consequence of Proposition 1.2.5(ii) applied with the finest function space topology $\mathfrak{V}_*$ (see Example 1.2.7(f)).

2.2.9 Proposition. *Suppose that X is a locally compact Hausdorff space and that $\mathfrak{R}$ consists of all relatively compact Borel subsets of X. Then for every $f \in C(X, E)$ and every $A \in \mathfrak{R}$ the function $\chi_A \cdot f \in \mathcal{F}(X, E)$ is measurable. In particular, all functions in $C_{\mathcal{K}}(X, E)$ are measurable.*

Proof. For every function $f \in C(X, E)$ and $A \in \mathfrak{R}$, there is a function $g \in C_{\mathcal{K}}(X, E)$, which coincides with f on A. Because $C_{\mathcal{K}}(X, E) \subset C_{\mathfrak{V}_*}(X, E)$, according to Proposition 1.2.5(ii) for every function space neighborhood $\mathfrak{v}$ there is a step function $h \in \mathcal{F}_{\mathcal{S}}(X, E)$ such that $h - g \leqslant \mathfrak{v}$ and, therefore,

$$\chi_A \cdot h - \chi_A \cdot g = \chi_A \cdot h - \chi_A \cdot f \leqslant \mathfrak{v}.$$

Since $\chi_A \cdot h \in \mathcal{F}_{\mathcal{S}}(X, E)$, the definition of measurability is satisfied for the function $\chi_A \cdot f$. □

Our next proposition prepares the extension of the concept of measurability to set-valued functions.

2.2.10 Proposition. *A function $f \in \mathcal{F}(X, E)$ is measurable if and only if for every function space neighborhood $\mathfrak{v}$ there is a* Conv(E)*-valued step function h such that $f \leqslant h \leqslant f + \mathfrak{v}$.*

Proof. Suppose that $f \in \mathcal{F}(X,E)$ is measurable and let $\mathfrak{v}$ be a function space neighborhood. According to Proposition 2.2.5, f is supported by a set $A \in \mathfrak{R}$. There is $V \in \mathcal{V}$ such that $\chi_A \cdot V \leqslant (1/2)\mathfrak{v}$, and again using Proposition 2.2.5 we find an E-valued step function $g \in \mathcal{F}_{\mathcal{S}}(X,E)$ such that both $f \leqslant g + \chi_A \cdot V$ and $g \leqslant f + \chi_A \cdot V$. We choose the $\mathrm{Conv}(E)$-valued step function

$$h = g + \chi_A \cdot V \in \mathcal{F}_{\mathcal{S}}(X, \mathrm{Conv}(E))$$

and realize that $f \leqslant h$. Moreover, since both $g \leqslant f + \chi_A \cdot V \leqslant f + (1/2)\mathfrak{v}$ and $\chi_A \cdot V \leqslant (1/2)\mathfrak{v}$, we have $h \leqslant f + \mathfrak{v}$. Thus, the condition in our proposition holds for f with the step function $h \in \mathcal{F}_{\mathcal{S}}(X, \mathrm{Conv}(E))$. Conversely, let $f \in \mathcal{F}(X,E)$ and suppose that for every function space neighborhood $\mathfrak{v}$ there is a $\mathrm{Conv}(E)$-valued step function h such that $f \leqslant h \leqslant f + \mathfrak{v}$. For any choice of $\mathfrak{v}$, since step functions are supported by sets in $\mathfrak{R}$, the relation $f \leqslant h$ implies that f is supported by a set $A \in \mathfrak{R}$. Given a function space neighborhood $\mathfrak{v}$, there is $V \in \mathcal{V}$ such that $\chi_A \cdot V \leqslant \mathfrak{v}$. Now we use our assumption with the neighborhood function $\mathfrak{v}_V$ generated by the single neighborhood function of uniform convergence $x \mapsto V$ (Example 1.2.7(a)). There is a $\mathrm{Conv}(E)$-valued step function

$$h = \sum_{i=1}^{n} \chi_{A_i} \cdot C_i \in \mathcal{F}_{\mathcal{S}}(X, \mathrm{Conv}(E))$$

with $A_i \in \mathfrak{R}$ and $C_i \in \mathrm{Conv}(E)$ such that $f \leqslant h \leqslant f + \mathfrak{v}_V$. We may assume that h is also supported by A. We choose $a_i \in C_i$ and set $g = \sum_{i=1}^{n} \chi_{A_i} \cdot a_i \in \mathcal{F}_{\mathcal{S}}(X,E)$. Then $g \leqslant h \leqslant f + \chi_A \cdot V \leqslant f + \mathfrak{v}$, that is, $g - f \leqslant \mathfrak{v}$. Hence f is measurable. □

Measurable set-valued functions

We shall utilize the criterion from Proposition 2.2.10 in order to extend the notion of measurability to $\mathrm{Conv}(E)$-valued functions. This is however less than straightforward. Just applying the definition for E-valued functions to $\mathrm{Conv}(E)$-valued ones with $\mathrm{Conv}(E)$-valued step functions would prove unduly restrictive. For an unbounded set $C \in \mathrm{Conv}(E)$ and a measurable function $\varphi \in \mathcal{F}(X,E)$, for example, not even the mapping

$$x \mapsto \varphi(x)C : \mathbb{R} \to \mathrm{Conv}(E)$$

is guaranteed to be measurable if we would use this criterion (see Example 2.2.22(c) below). As a remedy, we shall employ sequences of step functions rather than single ones and implicitly use the *relative topologies* of the locally convex cone $\mathrm{Conv}(E)$. These were introduced and discussed in detail in Section I.4 of [48] to which we refer for details, motivation and clarification. We shall properly define and employ these topologies in a later section.

We shall say that a function $f \in \mathcal{F}(X, \mathrm{Conv}(E))$ is *measurable* if for every $A \in \mathfrak{R}$, every function space neighborhood $\mathfrak{v}$ and $\varepsilon > 0$ there is $1 \leqslant \gamma \leqslant 1+\varepsilon$ and a countable set $\mathcal{H}$ of step functions in $\mathcal{F}_{\mathcal{S}}(X, \mathrm{Conv}(E))$ such that $h \leqslant f + \mathfrak{v}$ for all $h \in \mathcal{H}$, and for every $x \in A$ there is $h \in \mathcal{H}$ such that $f(x) \leqslant \gamma h(x)$.

We need to verify that this definition of measurability for $\mathrm{Conv}(E)$-valued functions is consistent with the preceding one for E-valued functions. In fact, if $f \in \mathcal{F}(X,E)$ is measurable with respect to the earlier definition, then it satisfies the condition from above for every $A \in \mathfrak{R}$, and $\varepsilon > 0$ with the step function h from Proposition 2.2.10 and the singleton set $\mathcal{H} = \{h\}$. Conversely, if the above condition holds for a function $f \in \mathcal{F}(X,E)$, then for $\varnothing \in \mathfrak{R}$, for every function space neighborhood $\mathfrak{v}$ there is a step function

$$h = \sum_{i=1}^{n} \chi_{A_i} \cdot C_i \in \mathcal{F}_{\mathcal{S}}(X, \mathrm{Conv}(E))$$

such that $h \leqslant f + \mathfrak{v}$. We choose elements $c_i \in C_i$ and set $g = \sum_{i=1}^{n} \chi_{A_i} \cdot c_i$. Then $g \in \mathcal{F}_{\mathcal{S}}(X,E)$ and $g \leqslant h \leqslant f + \mathfrak{v}$. Hence the E-valued function f is measurable in the manner that was specified earlier in this section.

The presence of the only imprecisely specified factor γ in this definition of measurability is the cause for many of the technical difficulties in some of the upcoming arguments. Its involvement in dealing with set-valued functions appears however to be indispensable as simple examples (see 2.2.22(c)) can demonstrate.

2.2.11 Proposition. *Every measurable function $f \in \mathcal{F}(X, \mathrm{Conv}(E))$ is bounded below, and there is $A \in \mathfrak{R}$ such that $f(x) \geqslant 0$ for all $x \in X \setminus A$. The measurable functions form a subcone of $\mathcal{F}(X, \mathrm{Conv}(E))$.*

Proof. Let $f \in \mathcal{F}(X, \mathrm{Conv}(E))$ be measurable. Using the above definition for $\varnothing \in \mathfrak{R}$ and any $\varepsilon > 0$ concedes that there is a step function $h \in \mathcal{F}_{\mathcal{S}}(X, \mathrm{Conv}(E))$ such that $h \leqslant f + \mathfrak{v}$. Lemmas 2.2.3 and 2.2.2 then yield the first and second statements. For the final statement, it is obvious that λf is measurable for all $\lambda \geqslant 0$ whenever f is measurable. Suppose that both f and g are measurable, let $A \in \mathfrak{R}$, let $\mathfrak{v}$ be a function space neighborhood and let $\varepsilon > 0$. There is $\lambda \geqslant 0$ such that both $0 \leqslant f + \lambda\mathfrak{v}$ and $0 \leqslant g + \lambda\mathfrak{v}$. We set $\delta = \min\{\varepsilon, 1/(2 + 2\lambda)\}$. There are countable sets of step functions $\mathcal{H}$ and $\mathcal{L}$ of step functions satisfying the condition for measurability with A, the function space neighborhood $\delta\mathfrak{v}$ and $\delta > 0$ for f and g, respectively. In particular, there are $1 \leqslant \gamma, \rho \leqslant 1+\delta$ such that for every $x \in A$ there are $h \in \mathcal{H}$ and $l \in \mathcal{L}$ such that $f(x) \leqslant \gamma h(x)$ and $g(x) \leqslant \rho\, l(x)$. We may assume that $\gamma \geqslant \rho$ and observe that

$$0 \leqslant (1 - \rho/\gamma)g + (1 - \rho/\gamma)\lambda\mathfrak{v} \leqslant (1 - \rho/\gamma)g + \delta\lambda\mathfrak{v}.$$

The last inequality uses that $1 - \rho/\gamma = (\gamma - \rho)/\gamma \leqslant \delta$. We continue to calculate that

$$\begin{aligned}(\rho/\gamma)l &\leqslant (\rho/\gamma)g + (\rho/\gamma)\delta\mathfrak{v}\\ &\leqslant (\rho/\gamma)g + \delta\mathfrak{v} + ((1 - \rho/\gamma)g + \delta\lambda\mathfrak{v})\\ &= g + \delta(1 + \lambda)\mathfrak{v}\end{aligned}$$

$$\leqslant g + (1/2)\mathfrak{v}$$

for all $l \in \mathcal{L}$. This argument uses that $\rho/\gamma \leqslant 1$ and $\delta \leqslant 1/(2 + 2\lambda)$. Because $\delta \leqslant (1/2)$ we also have $h \leqslant f + (1/2)\mathfrak{v}$ for all $h \in \mathcal{H}$. Now we set

$$\mathcal{S} = \mathcal{H} + (\rho/\gamma)\mathcal{L}$$

and claim that this is a countable family of step functions hat fulfils the requirements of measurability for the function $f + g$. Indeed, we have

$$s = h + (\rho/\gamma)l \leqslant (f + (1/2)\mathfrak{v}) + (g + (1/2)\mathfrak{v}) = (f + g) + \mathfrak{v}$$

for all $s \in \mathcal{S}$ by the above, and because $\delta \leqslant \varepsilon$ we have $1 \leqslant \gamma \leqslant 1+\delta \leqslant 1+\varepsilon$. Moreover, for every $x \in A$ there is $s = h + (\rho/\gamma)l \in \mathcal{S}$ for $h \in \mathcal{H}$ and $l \in \mathcal{L}$ such that both $f(x) \leqslant \gamma h(x)$ and $g(x) \leqslant \rho\, l(x)$. That is to say,

$$(f + g)(x) \leqslant \gamma h(x) + \gamma(\rho/\gamma)l(x) = \gamma s(x).$$

This demonstrates that the function $f + g$ is also measurable. □

For the following, recall from Section 1.4 that for $f \in \mathcal{F}(X, \mathrm{Conv}(E))$ and an element α of the scalar field of E the function $\alpha f \in \mathcal{F}(X, \mathrm{Conv}(E))$ is defined as the mapping $x \mapsto \alpha f(x) \in \mathrm{Conv}(E)$.

2.2.12 Proposition. *Let $f \in \mathcal{F}(X, \mathrm{Conv}(E))$ and $C \in \mathrm{Conv}(E)$.*
(a) *If f is measurable, then αf is measurable for every α in $\mathbb{R}$ or $\mathbb{C}$.*
(b) *If f is measurable, then $\chi_B \cdot f$ is measurable for all $B \in \mathfrak{A}$.*
(c) *The function $\chi_A \cdot C$ is measurable for all $A \in \mathfrak{R}$.*
(d) *If $B \in \mathfrak{A} \setminus \mathfrak{R}$, then $\chi_B \cdot C$ is measurable if and only if $C \geqslant 0$.*

Proof. For (a), if the countable set $\mathcal{H}$ of step functions satisfies the conditions for the measurability of f and if α is in the scalar field of E, then the family $\mathcal{G} = \{\alpha h \mid h \in \mathcal{H}\}$ of step functions satisfies the conditions for the measurability of αf. For (b), if the countable set $\mathcal{H}$ of step functions satisfies the conditions for the measurability of f, then multiplying each of its elements by χ_B returns a countable set $\mathcal{G}$ of step functions that satisfies the conditions for the measurability of $\chi_B \cdot f$. For (c), we note that the conditions for the measurability of the function $\chi_A \cdot C$ are satisfied with the singleton set of step functions $\mathcal{H} = \{\chi_A \cdot C\}$. For (d), suppose that $B \in \mathfrak{A} \setminus \mathfrak{R}$. If $C \geqslant 0$, then for every $A \in \mathfrak{R}$ every function space neighborhood $\mathfrak{v}$ and $\varepsilon > 0$ we use the singleton set $\mathcal{H} = \{\chi_{B \cap A} \cdot C\}$ that satisfies the requirements for the measurability of $\chi_B \cdot C$. Conversely, if the function $\chi_B \cdot C$ is measurable, then according to Proposition 2.2.11 there is $A \in \mathfrak{R}$ such that $\chi_B \cdot C(x) \geqslant 0$ for all $x \in X \setminus A$. Since $B \notin \mathfrak{R}$, we infer that $B \not\subset A$, and there is a point $x \in B \cap (X \setminus A)$. Thus $(\chi_B \cdot C)(x) = C \geqslant 0$, our claim. □

The next proposition establishes that the condition for measurability can be formulated in a stronger way, which will prove to be useful in our upcoming explorations.

2.2.13 Proposition. *If $f \in \mathcal{F}(X, \mathrm{Conv}(E))$ is measurable, then for every $A \in \mathfrak{R}$, every function space neighborhood $\mathfrak{v}$ and $\varepsilon > 0$ there is $1 \leqslant \gamma \leqslant 1 + \varepsilon$ and an increasing sequence $(h_n)_{n \in \mathbb{N}}$ of step functions in $\mathcal{F}_S(X, \mathrm{Conv}(E))$ such that $h_n \leqslant f + \mathfrak{v}$ for all $n \in \mathbb{N}$ and for every $x \in A$ there is $n \in \mathbb{N}$ such that $f(x) \leqslant \gamma h_n(x.)$*

Proof. Suppose that $f \in \mathcal{F}(X, \mathrm{Conv}(E))$ is measurable. Let $A \in \mathfrak{R}$, let $\mathfrak{v}$ be a function space neighborhood and let $\varepsilon > 0$. According to Proposition 2.2.11, there is $B \in \mathfrak{R}$ such that $A \subset B$ and such that $f(x) \geqslant 0$ for all $x \in X \backslash B$. In turn, we find $V \in \mathcal{V}$ such that $\chi_B \cdot V \leqslant \mathfrak{v}$. We apply the above definition of measurability with the function space neighborhood $\mathfrak{w}$ generated by the single neighborhood function of uniform convergence $x \mapsto V$ in place of $\mathfrak{v}$. There is $1 \leqslant \gamma \leqslant 1 + \varepsilon$ and a countable bounded below set $\mathcal{H} = \{h_n\}_{n \in \mathbb{N}}$ of step functions such that $h_n \leqslant f + \chi_X \cdot V$, and for every $x \in A$ there is $h_n \in \mathcal{H}$ such that $f(x) \leqslant \gamma h_n(x)$. Next, for every $n \in \mathbb{N}$ we define a step function g_n such that $g_n(x)$ is the convex hull in E of the set $\bigcup_{i=1}^{n} h_i(x)$ for every $x \in X$. Clearly, $g_n \in \mathcal{F}_S(X, \mathrm{Conv}(E))$ and $h_i(x) \subset f(x) + V$ for all $i \in \mathbb{N}$ implies that $g_n(x) \subset f(x) + V$, thus $g_n \leqslant f + \chi_X \cdot V$. We set $l_n = \chi_B \cdot g_n$ and proceed to establish that the sequence $(l_n)_{n \in \mathbb{N}}$ in $\mathcal{F}_S(X, \mathrm{Conv}(E))$ answers our claim. Indeed, we have both $l_n = \chi_B \cdot g_n \leqslant \chi_B \cdot f + \chi_B \cdot V$ and $0 \leqslant \chi_{X \backslash B} \cdot f$ and, therefore,

$$l_n \leqslant (\chi_B \cdot f + \chi_B \cdot V) + \chi_{X \backslash B} \cdot f = f + \chi_B \cdot V \leqslant f + \mathfrak{v}$$

for all $n \in \mathbb{N}$. Moreover, our construction yields that $l_n \leqslant l_{n+1}$ and $\chi_B \cdot h_n \leqslant l_n$. All that is left to verify is that the sequence $(l_n)_{n \in \mathbb{N}}$ is bounded below. Let $\mathfrak{w}$ be a function space neighborhood. There is $\rho \geqslant 0$ such that $0 \leqslant h_1 + \rho \mathfrak{w}$ for all $n \in \mathbb{N}$. This shows

$$0 \leqslant \chi_B \cdot h_1 + \rho \mathfrak{w} \leqslant \chi_B \cdot g_n + \rho \mathfrak{w} = l_n + \rho \mathfrak{w}$$

for all $n \in \mathbb{N}$. □

2.2.14 Proposition. *Let $f \in \mathcal{F}(X, \mathrm{Conv}(E))$. If for every $A \in \mathfrak{R}$, every function space neighborhood $\mathfrak{v}$ and $\varepsilon > 0$ there is $1 \leqslant \gamma \leqslant 1 + \varepsilon$ and a countable set $\mathcal{G}$ of measurable functions in $\mathcal{F}(X, \mathrm{Conv}(E))$ such that $g \leqslant f + \mathfrak{v}$ for all $g \in \mathcal{G}$ and for every $x \in A$ there is $g \in \mathcal{G}$ such that $f(x) \leqslant \gamma g(x)$, then f is measurable.*

Proof. Suppose that $f \in \mathcal{F}(X, \mathrm{Conv}(E))$ satisfies our assumptions. Let $A \in \mathfrak{R}$, let $\mathfrak{v}$ be a function space neighborhood and $\varepsilon > 0$. According to Lemma 2.2.3(b) f is bounded below, hence there is $\lambda > 0$ such that $0 \leqslant f + \lambda \mathfrak{v}$. We set $\delta = \min\{1/(2\lambda), \sqrt{1+\varepsilon} - 1\}$ and find a countable family $\mathcal{G} = \{g_n\}_{n \in \mathbb{N}}$ of measurable functions and $1 \leqslant \gamma \leqslant 1 + \delta$ such that $g_n \leqslant f + (1/4)\mathfrak{v}$ for all $n \in \mathbb{N}$ and for every $x \in A$ there is $g_n \in \mathcal{G}$ such that $f(x) \leqslant \gamma g_n(x)$. In turn, using the definition of measurability for each of the functions g_n with the function space neighborhood $(1/4)\mathfrak{v}$, for every $n \in \mathbb{N}$ there is

$1 \leqslant \gamma_n \leqslant 1+\delta$ and a countable set $\{h_{(n,m)}\}_{m\in\mathbb{N}}$ of step functions in $\mathcal{F}_{\mathcal{S}}(X, \mathrm{Conv}(E))$ such that $h_{(n,m)} \leqslant g_n + (1/4)\mathfrak{v}$, hence $h_{(n,m)} \leqslant f + (1/2)\mathfrak{v}$ for all $m \in \mathbb{N}$, and for every $x \in A$ there is $h_{(n,m)}$ such that $g_n(x) \leqslant \gamma_n h_{(n,m)}(x)$. We set $l_{(n,m)} = (\gamma_n/(1+\delta))h_{(n,m)}$ and collect the following observations. Since

$$1/(1+\delta) \leqslant \gamma_n/(1+\delta) \leqslant 1$$

and

$$0 \leqslant (1-\gamma_n/(1+\delta)) \leqslant \delta/(1+\delta) \leqslant \delta,$$

we infer that

$$0 \leqslant (1-\gamma_n/(1+\delta))(f+\lambda\mathfrak{v}) \leqslant (1-\gamma_n/(1+\delta))f + \delta\lambda\mathfrak{v}.$$

By our choice of δ this renders

$$\begin{aligned} l_{(n,m)} &\leqslant (\gamma_n/(1+\delta))(f+(1/2)\mathfrak{v}) + (1-\gamma_n/(1+\delta))f + \delta\lambda\mathfrak{v} \\ &\leqslant f + ((1/2)\gamma_n/(1+\delta) + \delta\lambda)\mathfrak{v} \\ &\leqslant f + ((1/2)+(1/2))\mathfrak{v} \\ &= f + \mathfrak{v} \end{aligned}$$

for all $n, m \in \mathbb{N}$. For every $x \in A$ on the other hand, by the above there is $l_{(n,m)}$ such that

$$f(x) \leqslant \gamma g_n(x) \leqslant \gamma\gamma_n h_{(n,m)}(x) = (1+\delta)\gamma\, l_{(n,m)}(x).$$

Since $1 \leqslant (1+\delta)\gamma \leqslant (1+\delta)^2 \leqslant 1+\varepsilon$ by our choice of δ, the countable set $\mathcal{L} = \{l_{(n,m)}\}_{n,m\in\mathbb{N}}$ of step functions meets the requirements for the measurability of the function f with $(1+\delta)\gamma$ in place of γ. □

2.2.15 Proposition. *If $f \in \mathcal{F}(X, \mathrm{Conv}(E))$ is measurable, then for every $A \in \mathfrak{R}$, $V \in \mathcal{V}$ and $\varepsilon > 0$ there are $1 \leqslant \gamma \leqslant 1+\varepsilon$ and disjoint sets $A_n \in \mathfrak{R}$ for $n \in \mathbb{N}$ such that $\bigcup_{n\in\mathbb{N}} A_n = A$ and such that $f(x) \leqslant \gamma f(y) + V$ whenever $x, y \in A_n$ for the same $n \in \mathbb{N}$.*

Proof. Suppose that the function $f \in \mathcal{F}(X, \mathrm{Conv}(E))$ is measurable. For $A \in \mathfrak{R}$, $V \in \mathcal{V}$ and $\varepsilon > 0$, we consider the function space neighborhood $\mathfrak{v}$ generated by the single neighborhood function of uniform convergence $x \mapsto (1/2)V$. We set $\delta = \min\{1, \sqrt{1+\varepsilon}-1\}$ and use the definition of measurability for f with the neighborhood $\mathfrak{v}$ and δ in place of ε. There is $1 \leqslant \gamma \leqslant 1+\delta$ and a countable set $\mathcal{H} = \{h_n\}_{n\in\mathbb{N}}$ of step functions in $\mathcal{F}_{\mathcal{S}}(X, \mathrm{Conv}(E))$ as in the definition of measurability. In particular, we have $h_n(x) \leqslant f(x) + (1/2)V$ for all $x \in X$. For every $n \in \mathbb{N}$, we express $h_n = \sum_{i=1}^{m_n} \chi_{A_n^i} \cdot C_n^i$

with disjoint sets $A_n^i \in \mathfrak{R}$ whose union contains A (for this we may of course always add terms $\chi_{A_n^i} \cdot \{0\}$) and $C_n^i \in \mathrm{Conv}(E)$. For every $n \in \mathbb{N}$ and $i = 1, \ldots, m_n$, the sets

$$B_n^i = \{x \in A_n^i \cap A \mid h_m(x) \leqslant \gamma h_n(x) + (1/2)V, \text{ for all } m \in \mathbb{N}\}$$

are in $\mathfrak{R}$, and we have $\bigcup_{n\in\mathbb{N}} \bigcup_{i=1}^{m_n} B_n^i = A$. Indeed, for every $x \in A$ there is $n \in \mathbb{N}$ such that $f(x) \leqslant \gamma h_n(x)$, and $i \in \{1, \ldots, m_i\}$ such that $x \in A_n^i$. Thus

$$h_m(x) \leqslant f(x) + (1/2)V \leqslant \gamma h_n(x) + (1/2)V$$

for all $m \in \mathbb{N}$. Hence $x \in B_n^i$. Moreover, because for any $x \in A$ we have $f(x) \leqslant \gamma h_m(x)$ for some $m \in \mathbb{N}$, and since the step function h_n is constant on A_n^i, for any choice of x, y in the same set B_n^i we calculate

$$\begin{aligned} f(x) &\leqslant \gamma h_m(x) \\ &\leqslant \gamma^2 h_n(x) + (\gamma/2)V \\ &= \gamma^2 h_n(y) + (\gamma/2)V \\ &\leqslant \gamma^2 f(y) + (1/2)(1+\gamma^2)V. \end{aligned}$$

We have $1 \leqslant \gamma^2 \leqslant (1+\delta)^2 \leqslant 1+\varepsilon$ and $1+\gamma^2 \leqslant 2$ by our choice of δ. The family of the sets B_n^i is countable. Thus after renumbering, we obtain sets $B_k \in \mathfrak{R}$ for $k \in \mathbb{N}$ such that $f(x) \leqslant \gamma^2 f(y) + V$ whenever $x, y \in B_k$ for some $k \in \mathbb{N}$. Finally, we set $A_1 = B_1$ and $A_n = B_n \setminus \bigcup_{i=1}^{n-1} A_i$ for $n \geqslant 2$. The sets A_n are disjoint, we have $A_n \subset B_n$ and $\bigcup_{n\in\mathbb{N}} A_n = \bigcup_{n\in\mathbb{N}} B_n = A$. The family of the sets A_n then fulfills our claim. □

Corresponding to a neighborhood $V \in \mathcal{V}$ for sets $C, D \in \mathrm{Conv}(E)$ we denote that $C \preccurlyeq_V D$ if $C \leq D + \varepsilon V$ for all $\varepsilon > 0$. That is, C is contained in the closure of D with respect to the topology on E generated by the neighborhood system $\{\lambda V \mid \lambda > 0\}$.

2.2.16 Corollary. *For measurable functions $f, g \in \mathcal{F}(X, \mathrm{Conv}(E))$ and a neighborhood $V \in \mathcal{V}$, the set $\{x \in X \mid f(x) \preccurlyeq_V g(x)\}$ is contained in $\mathfrak{A}$.*

Proof. Let $f, g \in \mathcal{F}(X, \mathrm{Conv}(E))$ be measurable and let $V \in \mathcal{V}$. We denote

$$B = \{x \in X \mid f(x) \preccurlyeq_V g(x)\}$$

and proceed to verify that $B \in \mathfrak{A}$, that is, $B \cap A \in \mathfrak{R}$ for every set $A \in \mathfrak{R}$. For this, let $A \in \mathfrak{R}$, and for $\varepsilon > 0$ set

$$B_\varepsilon^A = \{x \in A \mid f(x) \leqslant \gamma g(x) + \varepsilon V \text{ for some } 1 \leqslant \gamma \leqslant 1+\varepsilon\}.$$

According to Proposition 2.2.15, for any choice of $0 < \varepsilon \leqslant 1$ there are $1 \leqslant \gamma_1, \gamma_2 \leqslant 1+\varepsilon$ and disjoint sets $A_i \in \mathfrak{R}$ for $i \in \mathbb{N}$ such that $\bigcup_{n\in\mathbb{N}} A_n = A$ and $f(x) \leqslant \gamma_1 f(y) + \varepsilon V$ as well as $g(x) \leqslant \gamma_2 g(y) + \varepsilon V$ whenever $x, y \in A_i$ for the same $i \in \mathbb{N}$. Set

$$C_\varepsilon^A = \bigcup \{A_i \mid f(x_i) \leqslant \gamma_3 g(x_i) + \varepsilon V \text{ for some } x_i \in A_i \text{ and } 1 \leqslant \gamma_3 \leqslant 1+\varepsilon\}.$$

Clearly, $C_\varepsilon^A \in \mathfrak{R}$ and $B_\varepsilon^A \subset C_\varepsilon^A$. If $x \in C_\varepsilon^A$, on the other hand, then $x \in A_i$ for one of the sets A_i forming this union, hence

$$f(x) \leqslant \gamma_1 f(x_i) + \varepsilon V, \quad f(x_i) \leqslant \gamma_3 g(x_i) + \varepsilon V \quad \text{and} \quad g(x_i) \leqslant \gamma_2 g(x) + \varepsilon V.$$

Combining these inequalities, we obtain that

$$f(x) \leqslant (\gamma_1\gamma_2\gamma_3) g(x) + (1 + \gamma_1 + \gamma_1\gamma_2)\varepsilon V \leqslant (\gamma_1\gamma_2\gamma_3) g(x) + 7\varepsilon V.$$

The latter follows since $0 < \varepsilon \leqslant 1$, hence $1 \leqslant \gamma_1, \gamma_2 \leqslant 2$. Moreover, since

$$1 \leqslant (\gamma_1\gamma_2\gamma_3) \leqslant (1+\varepsilon)^3 = 1 + (3 + 3\varepsilon + \varepsilon^2)\varepsilon \leqslant 1 + 7\varepsilon,$$

we infer that

$$B_\varepsilon^A \subset C_\varepsilon^A \subset B_{7\varepsilon}^A.$$

This yields

$$\bigcap_{n\in\mathbb{N}} C_{(1/n)}^A = \bigcap_{n\in\mathbb{N}} B_{(1/n)}^A \in \mathfrak{R}.$$

All that is left to show is that $\bigcap_{n\in\mathbb{N}} B_{(1/n)}^A = B \cap A$. The inclusion $B \cap A \subset \bigcap_{n\in\mathbb{N}} B_{(1/n)}^A$ is obvious. For the reverse, let $x \in \bigcap_{n\in\mathbb{N}} B_{(1/n)}^A$ and let $\varepsilon > 0$. For $a \in f(x)$, there is $\lambda \geqslant 0$ such that $a \in \lambda V$. Then for any choice of $n \geqslant (1+\lambda)/\varepsilon$ we have $f(x) \leqslant \gamma g(x) + (1/n)V$ with some $1 \leqslant \gamma \leqslant 1 + (1/n)$. Hence $a \in \gamma g(x) + (1/n)V$ and, therefore, $(1/\gamma)a \in g(x) + (1/n)V$. From $1 - (1/\gamma) = (\gamma - 1)/\gamma \leqslant 1/n$, we deduce that

$$a = (1/\gamma)a + (1 - (1/\gamma))a \in \big(g(x) + (1/n)V\big) + (\lambda/n)V \subset g(x) + \varepsilon V.$$

This yields $f(x) \leqslant g(x) + \varepsilon V$ for all $\varepsilon > 0$ and we infer that $f(x) \preccurlyeq_V g(x)$. Therefore, $x \in B$, and our claim follows. □

The reverse conclusion of Proposition 2.2.15 holds true under an additional assumption.

2.2.17 Proposition. *Let $f \in \mathcal{F}(X, \mathrm{Conv}(E))$ such that for every function space neighborhood $\mathfrak{v}$ there is a step function $h \in \mathcal{F}_S(X, \mathrm{Conv}(E))$ such that $h \leqslant f + \mathfrak{v}$. If for every $A \in \mathfrak{R}$, $V \in \mathcal{V}$ and $\varepsilon > 0$ there are $1 \leqslant \gamma \leqslant 1+\varepsilon$ and sets $A_n \in \mathfrak{R}$ for $n \in \mathbb{N}$ such that $\bigcup_{n\in\mathbb{N}} A_n = A$ and $f(x) \leqslant \gamma f(y) + V$ whenever $x, y \in A_n$ for the same $n \in \mathbb{N}$, then f is measurable.*

Proof. Suppose that $f \in \mathcal{F}(X, \mathrm{Conv}(E))$ satisfies our assumptions. Given $A \in \mathfrak{R}$, a function space neighborhood $\mathfrak{v}$ and $\varepsilon > 0$, there is $h \in \mathcal{F}_{\mathcal{S}}(X, \mathrm{Conv}(E))$ such that $h \leqslant f + (1/3)\mathfrak{v}$ and there is $V \in \mathcal{V}$ such that $\chi_A \cdot V \leqslant (1/3)\mathfrak{v}$. We set $\delta = \sqrt{1+\varepsilon} - 1$. There are $1 \leqslant \gamma \leqslant 1 + \delta$ and sets $A_n \in \mathfrak{R}$ as in our assumption. We may assume that the sets A_n are disjoint and not empty and choose points $x_n \in A_n$. For every $n \in \mathbb{N}$, we set

$$h_n = (1/\gamma)\chi_{A_n} \cdot (f(x_n) + V) + \chi_{X \setminus A_n} \cdot h.$$

Then $h_n \in \mathcal{F}_{\mathcal{S}}(X, \mathrm{Conv}(E))$, and for every $x \in A_n$ we have

$$f(x) \leqslant \gamma f(x_n) + V \leqslant \gamma(f(x_n) + V) = \gamma^2 h_n(x)$$

and

$$h_n(x) = (1/\gamma)(f(x_n) + V) \leqslant (1/\gamma)(\gamma f(x) + 2V) \leqslant f(x) + 2V.$$

The former shows that for every $x \in A$ there is $n \in \mathbb{N}$ such that $f(x) \leqslant \gamma^2 h_n(x)$, and our choice of δ guarantees that $1 \leqslant \gamma^2 \leqslant 1 + \varepsilon$. The latter inequality yields that

$$\chi_{A_n} \cdot h_n \leqslant \chi\chi_{A_n} \cdot f + \chi_{A_n} \cdot 2V \leqslant \chi_{A_n} \cdot f + (2/3)\mathfrak{v}.$$

Since $\chi_{X \setminus A_n} \cdot h \leqslant \chi_{X \setminus A_n} \cdot f + (1/3)\mathfrak{v}$ by Lemma 1.2.1(a), we infer that

$$h_n = \chi_{A_n} \cdot h_n + \chi_{X \setminus A_n} \cdot h_n \leqslant f + \mathfrak{v}.$$

The family $\mathcal{H} = \{h_n\}_{n \in \mathbb{N}}$ of step functions then satisfies the criterion for the measurability of the function f. □

For a positive function in $\mathcal{F}(X, \mathrm{Conv}(E))$, the existence of a step function h as required in Proposition 2.2.17, that is, $h = 0$, is of course guaranteed. Moreover, we may choose $\gamma = 1 + \varepsilon$ in the respective criteria of Propositions 2.2.15 and 2.2.17. We obtain the following.

2.2.18 Corollary. *A positive function $f \in \mathcal{F}(X, \mathrm{Conv}(E))$ is measurable if and only if for every $A \in \mathfrak{R}$, $V \in \mathcal{V}$ and $\varepsilon > 0$ there are sets $A_n \in \mathfrak{R}$ for $n \in \mathbb{N}$ such that $\bigcup_{n \in \mathbb{N}} A_n = A$ and $f(x) \leqslant (1 + \varepsilon) f(y) + V$ whenever $x, y \in A_n$ for the same $n \in \mathbb{N}$.*

2.2.19 Proposition. *Let $f \in \mathcal{F}(X, \mathrm{Conv}(E))$ be measurable and let φ be a nonnegative real-valued $\mathfrak{A}$-measurable function on X. If either f is positive or if φ is bounded on every set in $\mathfrak{R}$, then $\varphi \cdot f$ is measurable.*

Proof. It is obvious from our definition of measurability for $\mathrm{Conv}(E)$-valued functions that their restrictions to sets in $\mathfrak{R}$, that is, products with the characteristic functions of these sets, are also measurable (Proposition 2.2.12(b)). Suppose that $f \in \mathcal{F}(X, \mathrm{Conv}(E))$

is measurable and that φ is a positive real-valued $\mathfrak{A}$-measurable function on X with the stated properties. Let $A \in \mathfrak{R}$, let $\mathfrak{v}$ be a function space neighborhood and let $\varepsilon > 0$. We may assume that $f(x) \geqslant 0$ for all $x \in X \setminus A$. There is $V \in \mathcal{V}$ such that $\chi_A \cdot V \leqslant \mathfrak{v}$, and since f is bounded below, there is $\lambda \geqslant 0$ such that $0 \leqslant f + \lambda\chi_X \cdot V$. We shall verify the criterion of Proposition 2.2.14 in order to prove that the function $\varphi \cdot f$ is measurable. For any choice of $\gamma > 1$, the sets

$$A_n = \{x \in A \mid \gamma^n \leqslant \varphi(x) < \gamma^{n+1}\}$$

for all $n \in \mathbb{Z}$ are contained in $\mathfrak{R}$ by the $\mathfrak{A}$-measurability of φ. In our first case, if f is positive, we choose $\gamma = 1+\varepsilon$ and consider the countable set $\mathcal{G}$ of measurable functions

$$g_n = \gamma^n \chi_{A_n} \cdot f,$$

for all $n \in \mathbb{Z}$. Following Proposition 2.2.12(b), these functions are measurable. For all $x \in A_n$, since $\gamma^n \leqslant \varphi(x)$, we have $g_n(x) = \gamma^n f(x) \leqslant \varphi(x)f(x)$ and $0 = g_n(x) \leqslant \varphi(x)f(x)$ for all $x \in X \setminus A_n$. Thus $g_n \leqslant \varphi \cdot f \leqslant \varphi \cdot f + \mathfrak{v}$ for all $n \in \mathbb{Z}$. Let $x \in A$. If $\varphi(x) = 0$, then $\varphi(x)f(x) \leqslant \gamma g_n(x)$ for all $n \in \mathbb{Z}$. If $x \in A_n$ for some $n \in \mathbb{Z}$, then $\varphi(x)f(x) \leqslant \gamma^{n+1}f(x) = \gamma g_n(x)$. Thus the criterion of Proposition 2.2.14 applies for the function $\varphi \cdot f$ with the countable set $\mathcal{G} = \{g_n\}_{n\in\mathbb{N}}$ of measurable functions.

In our second case, if φ is bounded on A but f is not necessarily positive, there is $\rho > 0$ such that $0 \leqslant \varphi(x) \leqslant \rho$ for all $x \in A$. We shall first argue that there is a step function $h \in \mathcal{F}_S(X, \mathrm{Conv}(E))$ such that $h \leqslant \varphi \cdot f + \chi_X \cdot V$. Surely, since f is measurable there is $l \in \mathcal{F}_S(X, \mathrm{Conv}(E))$ such that $l \leqslant f + (1/2\rho)\chi_X \cdot V$, and since φ is $\mathfrak{A}$-measurable and bounded, there is a real-valued nonnegative step function ψ on X such that

$$\psi(x) \leqslant \varphi(x) \leqslant \psi(x) + \frac{1}{2\lambda}$$

for all $x \in X$. Then $h = \psi \cdot l \in \mathcal{F}_S(X, \mathrm{Conv}(E))$, and indeed

$$\begin{aligned} h &\leqslant \psi \cdot f + \frac{1}{2\rho}\psi \cdot V \\ &\leqslant \psi \cdot f + \frac{1}{2}\chi_X \cdot V + (\varphi - \psi) \cdot (f + \lambda V) \\ &\leqslant \varphi \cdot f + \frac{1}{2}\chi_X \cdot V + \frac{1}{2}\chi_X \cdot V = \varphi \cdot f + \chi_X \cdot V, \end{aligned}$$

our claim. Now we select $\gamma = 1 + 1/(2\rho\lambda)$ and use the sets A_n as specified for our first case. Since φ is bounded by ρ on A, the sets A_n are empty whenever $\gamma^n > \rho$. Hence for a nonempty set A_n, we have $\gamma^{n+1} - \gamma^n = \gamma^n(\gamma - 1) \leqslant 1/(2\lambda)$. We consider the countable set $\mathcal{G}$ of measurable functions

$$g_n = \chi_{A_n} \cdot (\gamma^n f + (1/2)V) + \chi_{A \setminus A_n} \cdot h,$$

for all $n \in \mathbb{Z}$ such that $A_n \neq \emptyset$. First, we shall verify that $g_n \leqslant \varphi \cdot f + \mathfrak{v}$. Indeed, for $x \in A_n$ we have

$$\begin{aligned} g_n(x) &= \gamma^n f(x) + (1/2)V \\ &\leqslant \gamma^n f(x) + (1/2)V + (\varphi(x) - \gamma^n)(f(x) + \lambda V) \\ &\leqslant \varphi(x)f(x) + V. \end{aligned}$$

The latter uses that $A_n \neq \emptyset$ and $\varphi(x) - \gamma^n \leqslant 1/(2\lambda)$ by the above. For $x \in A \setminus A_n$, we have $g_n(x) = h(x) \leqslant \varphi(x)f(x) + V$ by our choice of the step function h, and $0 = g_n(x) \leqslant \varphi(x)f(x)$ for all $x \in X \setminus A$. Hence indeed, $g_n \leqslant f + \chi_A \cdot V \leqslant f + \mathfrak{v}$ for all $n \in \mathbb{Z}$ such that $A_n \neq \emptyset$. Finally, let $x \in A$. If $\varphi(x) = 0$, then $\varphi(x)f(x) \leqslant \gamma g_n(x)$ for all $n \in \mathbb{Z}$. If $x \in A_n$ for some $n \in \mathbb{Z}$, then

$$\begin{aligned} \varphi(x)f(x) &\leqslant \varphi(x)f(x) + (\gamma^{n+1} - \varphi(x))(f(x) + \lambda V) \\ &\leqslant \gamma^{n+1} f(x) + (1/2)V \leqslant \gamma g_n(x). \end{aligned}$$

Therefore, Proposition 2.2.14 also applies in this case for the function $\varphi \cdot f$ with the family $\mathcal{G} = \{g_n\}_{n \in \mathbb{N}}$. □

Our first corollary broadens the statement of Proposition 2.2.7.

2.2.20 Corollary. *If $\varphi \in \mathcal{F}(X)$ is nonnegative and measurable, then the function $\varphi \cdot C \in \mathcal{F}(X, E)$ is measurable for every $C \in \mathrm{Conv}(E)$.*

Proof. Let φ be as stated. According to Proposition 2.2.7, then φ is bounded, supported by a set $A \in \mathfrak{R}$ and $\mathfrak{A}$-measurable. Hence $\varphi \cdot C = \varphi \cdot (\chi_A \cdot C)$ for every $C \in \mathrm{Conv}(E)$, and our claim follows from Propositions 2.2.19 and 2.2.12(c). □

For our second corollary recall that in Section 1.1, the product of a positive set $C \in \mathrm{Conv}(E)$ with $+\infty \in \overline{\mathbb{R}}$ was defined to be the subcone of E spanned by C. If C is absorbing, then $(+\infty)C = E$.

2.2.21 Corollary. *If φ is an $\overline{\mathbb{R}}_+$-valued $\mathfrak{A}$-measurable function on X and if $C \in \mathrm{Conv}(E)$ is positive, then the function $\varphi \cdot C$ is measurable.*

Proof. Because φ is nonnegative and $\mathfrak{A}$-measurable, the set $B = \{x \in X \mid \varphi(x) < +\infty\}$ is contained in $\mathfrak{A}$ and the real-valued function $\chi_B \cdot \varphi$ is $\mathfrak{A}$-measurable. Let D be the conic hull of C in E, that is, $D = (+\infty)C$. According to Proposition 2.2.12(d), since $C, D \geqslant 0$, the mappings $\chi_X \cdot C$ and $\chi_{X \setminus B} \cdot D$ are both measurable. Proposition 2.2.19 applies and renders that $(\chi_B \cdot \varphi) \cdot (\chi_X \cdot C) = \chi_B \cdot (\varphi \cdot C)$ is also measurable. Summarizing, the function

$$\varphi \cdot C = \chi_B \cdot (\varphi \cdot C) + \chi_{X \setminus B} \cdot (\varphi \cdot C) = \chi_B \cdot (\varphi \cdot C) + \chi_{X \setminus B} \cdot D$$

is measurable, our claim. □

Corollary 2.2.21 applies in particular to neighborhood functions $\varphi \cdot V$ where φ is an $\overline{\mathbb{R}}_+$-valued $\mathfrak{A}$-measurable function on X and $V \in \mathcal{V}_c$, in which $\mathcal{V}_c$ denotes the extended neighborhood system of E, that is, the subcone of $\mathrm{Conv}(E)$ spanned by $\mathcal{V} \cup \{E\}$ (see Section 1.2). All function space neighborhoods involved in our Examples 1.2.7(a) to (d) in Section 1.2 are convex combinations of functions of this type and, therefore, are measurable.

2.2.22 Examples. (a) There are various well-established definitions of measurability in case that E is a Banach space. Most involve a specific measure and as a precondition for integrability are considerably more generous than ours. (See, e. g., Chapter II.1 in [21].) Given a measure θ, an E-valued function f is said to be *strongly θ-measurable* if there exists a sequence $(h_n)_{n\in\mathbb{N}}$ of E-valued step functions such that the sequence $(\|h_n(x) - f(x)\|)_{n\in\mathbb{B}}$ converges θ-almost everywhere. This is clearly implied by our requirement for measurability for E-valued functions from the beginning of this section, which calls for the existence of such a sequence of step functions, which converges uniformly toward f. A function $f \in \mathcal{F}(X,E)$ is said to be *weakly θ-measurable* if for every linear functional $\mu \in E^*$ the real- or complex-valued function $x \mapsto \mu(f(x))$ is strongly θ-measurable. This notion is of course weaker than strong measurability and also implied by our requirements. We note that strong θ-measurability of a function $f \in \mathcal{F}(X,E)$ implies that the real-valued step functions $\|h_n\|$ converge pointwise θ-almost everywhere to the real-valued function $\|f\|$. Since the pointwise limit of $\mathfrak{A}$-measurable functions is again $\mathfrak{A}$-measurable (see Theorem 6 in 11.2 of [53]), there is a set $Z \in \mathfrak{A}$ of measure zero such that the function $\chi_{X\setminus Z} \cdot \|f\|$ is $\mathfrak{A}$-measurable.

(b) Let us consider the case that $E = \mathbb{R}$. According to Proposition 2.2.7, a real-valued function in $\mathcal{F}(X)$ is measurable if and only if it is bounded, supported by a set in $\mathfrak{R}$ and $\mathfrak{A}$-measurable. For $\mathrm{Conv}(\mathbb{R})$-valued, that is, interval-valued functions we observe the following. Let φ and ψ be $\overline{\mathbb{R}}_+$-valued $\mathfrak{A}$-measurable functions on X. We claim that the function $f \in \mathcal{F}(X, \mathrm{Conv}(\mathbb{R}))$ such that

$$f(x) = [-\varphi(x), \psi(x)]$$

for all $x \in X$ is measurable. Indeed, let $A \in \mathfrak{R}$ and let $\mathfrak{v}$ be a function space neighborhood for $\mathcal{F}(X)$. According to the definition of a function space neighborhood, there is $\rho > 0$ such that $\chi_A \cdot (\rho\mathbb{B}) \leqslant \mathfrak{v}$, where $\mathbb{B} = [-1, +1]$ denotes the unit ball in $\mathbb{R}$. For all $n \in \mathbb{N}$, the sets

$$A_n = \{x \in A \mid (n-1)\rho \leqslant \varphi(x) < n\rho\}$$

and

$$B_n = \{x \in A \mid (n-1)\rho \leqslant \psi(x) < n\rho\}$$

are contained in $\mathfrak{R}$ by the $\mathfrak{A}$-measurability of the functions φ and ψ. So, are the sets $A_{+\infty} = \{x \in A \mid \varphi(x) = +\infty\}$ and $B_{+\infty} = \{x \in A \mid \psi(x) = +\infty\}$. We consider the countable set $\mathcal{G}$ of Conv($\mathbb{R}$)-valued step functions

$$h_n^m = \chi_{A_n \cap B_m} \cdot [-n\rho, m\rho]$$

for $n, m \in \mathbb{N} \cup \{+\infty\}$. The sets $A_n \cap B_m$ cover all of A and for every $x \in A_n \cap B_m$ we have

$$f(x) \leqslant h_n^m(x) \leqslant f(x) + \rho\mathbb{B}.$$

Since $h_n^m(x) = 0 \leqslant f(x)$ for all $x \in X \setminus (A_n \cap B_m)$, the right-hand side of this inequality shows that $h_n^m \leqslant f + \chi_A \cdot (\rho\mathbb{B}) \leqslant f + \mathfrak{v}$, while its left-hand side demonstrates that for every $x \in A$ there is $h_n^m \in \mathcal{H}$ such that $f(x) \leqslant h_n^m(x)$. Hence f is measurable according to our definition.

(c) Let us illustrate our earlier remark that the inclusion of the unspecified factor γ in the definition of measurability for a Conv(E)-valued function is essential and that its omission would exclude some basic and indispensable cases. We consider the following simple example. Let $X = [0,1]$ and let $E = \mathbb{R}^2$ with the neighborhood system $\mathcal{V} = \{\rho\mathbb{B} \mid \rho > 0\}$, where $\mathbb{B}$ denotes the Euclidean unit ball in $\mathbb{R}^2$. Let $\varphi \in \mathcal{F}([0,1])$ such that $\varphi(x) = x$ and let

$$C = \{(x,y) \in \mathbb{R}^2 \mid y \geqslant x^2\} \in \operatorname{Conv}(\mathbb{R}^2).$$

We shall see that the function $\varphi{\cdot}C \in \mathcal{F}(X, \operatorname{Conv}(\mathbb{R}^2))$, which according to Corollary 2.2.21 is measurable and does however not satisfy the criterion of measurability with the fixed factor $\gamma = 1$, that is, $\varepsilon = 0$. Indeed, let $A = [0,1] \in \mathfrak{R}$ and let $\mathfrak{v}_{\mathbb{B}}$ be the function space neighborhood of uniform convergence defined by the constant neighborhood function $x \mapsto \mathbb{B}$ and assume that there is a countable set $\mathcal{H}$ of Conv($\mathbb{R}^2$)-valued step functions such that $h \leqslant \varphi \cdot C + \mathfrak{v}_{\mathbb{B}}$ for all $h \in \mathcal{H}$ and for every $x \in [0,1]$ there is $h \in \mathcal{H}$ such that $(\varphi \cdot C)(x) \leqslant h(x)$. Each of the step function in $\mathcal{H}$ is defined using finitely many disjoint sets in $\mathfrak{R}$, so only countably many are involved in the build of all the functions in $\mathcal{H}$. We find an element $x \in [0,1]$, which is not the infimum of any of these sets. There is $h \in \mathcal{H}$ such that $xC \leqslant h(x)$. The point x is contained in one of the disjoint sets in $\mathfrak{R}$ used for the definition of h, and there is a point $y < x$ contained in the same set. Thus

$$xC \leqslant h(x) = h(y) \leqslant yC + \mathbb{B}.$$

But a brief inspection of the geometry of the set $C \subset \mathbb{R}^2$ reveals that this can not hold true.

(d) A modification of the settings in (c) leads to the construction of a nonmeasurable function. With the same choices for X, E and $\mathfrak{V}$, we probe the function $f \in \mathcal{F}(X, \mathrm{Conv}(\mathbb{R}^2))$ defined as

$$f(t) = \{(x,y) \in \mathbb{R}^2 \mid y \geqslant |x|^{t+1}\}$$

for $t \in [0,1]$. It is again self-evident that for $s,t \in [0,1]$ such that $s < t$ there is no choice of $\gamma \geqslant 1$ such that $f(s) \leqslant \gamma f(t) + \mathbb{B}$. The measurability criterion in Proposition 2.2.15 can therefore not be satisfied.

Composition of measurable functions

If $(G, \mathcal{U})$ is an additional locally convex topological vector space and $S \in \mathfrak{L}(G,E)$, then for a function $f \in \mathcal{F}(X, \mathrm{Conv}(G))$ we define the function $S \circ f \in \mathcal{F}(X, \mathrm{Conv}(E))$ by

$$(S \circ f)(x) = S(f(x))$$

for all $x \in X$. We shall demonstrate that $S \circ f$ is measurable whenever f is measurable. Indeed, given $A \in \mathfrak{R}$, a function space neighborhood $\mathfrak{v}$ and $\varepsilon > 0$ there is $B \in \mathfrak{R}$ such that $A \subset B$ and $f(x) \geqslant 0$ for all $x \in X \setminus B$. There is $V \in \mathcal{V}$ such that $\chi_B \cdot V \leqslant \mathfrak{v}$ and we find $U \in \mathcal{U}$ such that $S(U) \subset V$. Using the function space neighborhood $\mathfrak{u}$ for $\mathcal{F}(X, G)$ defined by the single neighborhood function $\chi_X \cdot U$ and the measurability of f there is $1 \leqslant \gamma \leqslant 1 + \varepsilon$ and a countable set $\mathcal{H}$ of step functions in $\mathcal{F}_S(X, \mathrm{Conv}(G))$ such that $h \leqslant f + \mathfrak{u}$ for all $h \in \mathcal{H}$, and for every $x \in A$ there is $h \in \mathcal{H}$ such that $f(x) \leqslant \gamma h(x)$. We may assume that all functions in $\mathcal{H}$ are supported by B. Set

$$S \circ \mathcal{H} = \{S \circ h \mid h \in \mathcal{H}\} \subset \mathcal{F}(X, \mathrm{Conv}(E)).$$

We understand that $h \leqslant f + \mathfrak{u}$ implies that $S \circ h \leqslant S \circ f + \mathfrak{v}$. Indeed, we have $(S \circ h)(x) = 0 \leqslant (S \circ f)(x)$ for all $x \in X \setminus B$ and $h(x) \leqslant f(x) + U$, hence

$$(S \circ h)(x) \leqslant S(f(x) + U) = S(f(x)) + S(U) \leqslant (S \circ f)(x) + V$$

for all $x \in B$. Thus

$$S \circ h \leqslant S \circ f + \chi_B \cdot V \leqslant S \circ f + \mathfrak{v}$$

as claimed. Moreover, for every $x \in A$ there is $h \in \mathcal{H}$ such that $f(x) \leqslant \gamma h(x)$. Then $(S \circ f)(x) \leqslant \gamma (S \circ h)(x)$. The countable set $S \circ \mathcal{H}$ of $\mathrm{Conv}(E)$-valued step functions therefore satisfies the criterion for measurability of the function $S \circ f \in \mathcal{F}(X, \mathrm{Conv}(E))$.

2.3 Integrals of set-valued functions

We continue to use the general assumptions and notation from the preceding sections. We suppose that $\mathfrak{R}$ is a σ-ring of subsets of a set X, that $\mathcal{P}_E = E$ and that θ is a bounded $\mathfrak{L}(E, F^{*\blacksquare})$-valued measure defined on the sets in $\mathfrak{R}$. Recall from Sections 1.4 and 2.1 that the latter requires that

$$\theta_A(\alpha a)(\mu) = \theta_A(a)(\alpha\mu)$$

holds for all $A \in \mathfrak{R}$, all $a \in E$, all α in the common scalar field of E and F and all $\mu \in F^*$. This condition needs to be verified only in the complex case and then only for $\alpha = i$. The topological scenario, where X is locally compact, $\mathfrak{R}$ consists of all relatively compact Borel subsets of X and θ is a regular measure, will be treated as a special case. In a first step, we define integrals for step functions, then for measurable functions and finally for a wider class of θ-integrable functions. Having these integrals available for not just E-valued, but also for $\mathrm{Conv}(E)$-valued functions is of interest not only for their own sake, but essential for the development of our approach to integration theory. Integrals for set-valued functions had been established by Aumann [3] and others using integrable selections. They arise more naturally in the framework of our methodology. The values of the integrals of E-valued functions can be interpreted to be elements of F, F^{**} or $F^{*\blacksquare}$, those of $\mathrm{Conv}(E)$-valued functions to be convex subsets of these spaces (Propositions 2.3.10 and 2.3.23 below).

We start out with functions in $\mathcal{F}_S(X, \mathrm{Conv}(E))$, that is, $\mathrm{Conv}(E)$-valued step functions, and shall utilize the modulus $|\theta|$ of θ as its extension to an $\mathfrak{L}(\mathrm{Conv}(E), \mathcal{Q}_F)$-valued measure. The values of these integrals will be elements of $\mathcal{Q}_F$. In preparation, we observe the following.

2.3.1 Lemma. *Let $A_i, B_k \in \mathfrak{R}$ and $C_i, D_k \in \mathrm{Conv}(E)$ for $i = 1, \ldots, n$ and $k = 1, \ldots, m$. If*

$$\sum_{i=1}^{n} \chi_{A_i} \cdot C_i \leqslant \sum_{k=1}^{m} \chi_{B_k} \cdot D_k, \quad \textit{then} \quad \sum_{i=1}^{n} |\theta|_{A_i}(C_i) \leqslant \sum_{k=1}^{m} |\theta|_{B_k}(D_k).$$

Proof. Following Lemma 2.2.1, we may assume that both families of sets A_i and B_k are pairwise disjoint and suppose that $\sum_{i=1}^{n} \chi_{A_i} \cdot C_i \leqslant \sum_{k=1}^{m} \chi_{B_k} \cdot D_k$. By adding suitable terms $\chi_{A'} \cdot 0$ and $\chi_{B'} \cdot 0$ on the left- and right-hand sides, we may assume in addition that $\bigcup_{i=1}^{n} A_i = \bigcup_{k=1}^{m} B_k$. Under these assumptions, the sets $A_i \cap B_k$ form a disjoint partition of this union and we have either $A_i \cap B_k = \varnothing$ or $C_i \leqslant D_k$. This yields

$$\begin{aligned} \sum_{i=1}^{n} |\theta|_{A_i}(C_i) &= \sum_{i=1}^{n} \sum_{k=1}^{m} |\theta|_{(A_i \cap B_k)}(C_i) \\ &\leqslant \sum_{k=1}^{m} \sum_{i=1}^{n} |\theta|_{(A_i \cap B_k)}(D_k) = \sum_{k=1}^{m} |\theta|_{B_k}(D_k), \end{aligned}$$

as claimed. □

Integrals of step functions

We are now in a position to define the integral for a $\mathrm{Conv}(E)$-valued step function

$$h = \sum_{i=1}^{n} \chi_{A_i} \cdot C_i \in \mathcal{F}_{\mathcal{S}}(X, \mathrm{Conv}(E))$$

over a set $B \in \mathfrak{A}$ with respect to θ as an element of $\mathcal{Q}_F$ by

$$\int_B h\, d\theta = \sum_{i=1}^{n} |\theta|_{(A_i \cap B)}(C_i).$$

Lemma 2.3.1 implies that the sum on the right-hand side is independent of the particular representation for the step function h. The integral constitutes a monotone linear operator from $\mathcal{F}_{\mathcal{S}}(X, \mathrm{Conv}(E))$ into $\mathcal{Q}_F$.

2.3.2 Proposition. *Let $B \in \mathfrak{A}$, let $g, h \in \mathcal{F}_{\mathcal{S}}(X, \mathrm{Conv}(E))$ and $\alpha \geqslant 0$. Then*
(a) $\int_B \alpha g\, d\theta = \alpha \int_B g\, d\theta$.
(b) $\int_B (g + h)\, d\theta = \int_B g\, d\theta + \int_B h\, d\theta$.
(c) $\int_B g\, d\theta = \int_X \chi_B \cdot g\, d\theta$.
(d) $\int_B g\, d\theta \leqslant \int_B h\, d\theta$ *whenever* $g \underset{a.e.\,B}{\leqslant} h$.

Proof. Properties (a), (b) and (c) are obvious from the above definition of the integral. For (d), suppose that $g \underset{a.e.\,B}{\leqslant} h$ for $g, h \in \mathcal{F}_{\mathcal{S}}(X, \mathrm{Conv}(E))$. That is, $\chi_{B\setminus Z} \cdot g \leqslant \chi_{B\setminus Z} \cdot h$ with some set $Z \in \mathfrak{A}$ of measure zero. It then follows immediately from the definition of the integral for step functions that

$$\int_{B\setminus Z} g\, d\theta \leqslant \int_{B\setminus Z} h\, d\theta.$$

As $|\theta|_{(A\cap Z)} = 0$ holds for all $A \in \mathfrak{R}$ by Proposition 2.1.10(a) and (b), we infer that $\int_{B\cap Z} l\, d\theta = 0$ for all step functions $l \in \mathcal{F}_{\mathcal{S}}(X, \mathrm{Conv}(E))$. Moreover, parts (b) and (c) yield that

$$\int_B l\, d\theta = \int_{B\cap Z} d\theta + \int_{B\setminus Z} d\theta = \int_{B\setminus Z} l\, d\theta.$$

Combining all of the above, we obtain

$$\int_B g\, d\theta = \int_{B\setminus Z} g\, d\theta \leqslant \int_{B\setminus Z} h\, d\theta = \int_B h\, d\theta. \qquad \square$$

The integral of an E-valued step function is directly determined by the values of θ rather than those of its modulus because both measures coincide on the elements of E. The following lemma demonstrates how this observation relates to the integrals of Conv(E)-valued step functions.

2.3.3 Lemma. *Let* $B \in \mathfrak{A}$ *and* $g \in \mathcal{F}_{\mathcal{S}}(X, \mathrm{Conv}(E))$. *Then*

$$\int_B g\, d\theta = \sup\left\{ \int_B h\, d\theta \mid h \in \mathcal{F}_{\mathcal{S}}(X,E),\ h \leqslant g \right\}.$$

Proof. Following 2.3.2(c), we may assume that $B = X$. Let $g = \sum_{i=1}^{n} \chi_{A_i} \cdot C_i \in \mathcal{F}_{\mathcal{S}}(X, \mathrm{Conv}(E))$ and let $h = \sum_{k=1}^{m} \chi_{B_k} \cdot a_k \in \mathcal{F}_{\mathcal{S}}(X,E)$ such that $h \leqslant g$. Then

$$\int_X h\, d\theta = \sum_{k=1}^{m} \theta_{B_k}(a_k) \leqslant \sum_{i=1}^{n} |\theta|_{A_i}(C_i) = \int_X g\, d\theta$$

by Lemma 2.3.1. Conversely, for each of the sets A_i suppose that the sets $A_i^k \in \mathfrak{R}$ for $k = 1, \ldots, n_i$ form a disjoint partition of A_i and let $c_i^k \in C_i$. Then

$$\sum_{i=1}^{n} \sum_{k=1}^{n_i} \theta_{A_i^k}(c_i^k) = \int_h h\, d\theta,$$

where $h = \sum_{i=1}^{n} \sum_{k=1}^{n_i} \chi_{A_i^k} \cdot (c_i^k)$. Then $h \in \mathcal{F}_{\mathcal{S}}(X,E)$ and $h \leqslant g$. Thus

$$\sum_{i=1}^{n} \sum_{k=1}^{n_i} \theta_{A_i^k}(c_i^k) \leqslant \sup\left\{ \int_B h\, d\theta \mid h \in \mathcal{F}_{\mathcal{S}}(X,E),\ h \leqslant g \right\}$$

Taking the supremum over all choices for partitions of the sets A_i on the left-hand side returns

$$\int_B g\, d\theta = \sum_{i=1}^{n} |\theta|_{A_i}(C_i) \leqslant \sup\left\{ \int_B h\, d\theta \mid h \in \mathcal{F}_{\mathcal{S}}(X,E),\ h \leqslant g \right\},$$

hence our claim. □

Integrals of measurable functions

Now in a second step, for a measurable function $f \in \mathcal{F}(X, \mathrm{Conv}(E))$, a set $B \in \mathfrak{A}$ and a function space neighborhood $\mathfrak{v}$ we define the expression

$$\begin{aligned}\int_B^{(\mathfrak{v})} f\, d\theta &= \sup\left\{ \int_B h\, d\theta \mid h \in \mathcal{F}_{\mathcal{S}}(X, \mathrm{Conv}(E)),\ h \leqslant f + \mathfrak{v} \right\} \\ &= \sup\left\{ \int_B h\, d\theta \mid h \in \mathcal{F}_{\mathcal{S}}(X,E),\ h \leqslant f + \mathfrak{v} \right\}\end{aligned}$$

as a precursor of the integral. The supremum is meant pointwise on F^* and according to Proposition 1.4.7(a) defines an element of $\mathcal{Q}_F$. The equality of both expressions on the right-hand side of this definition results from Lemma 2.3.3. For our further investigations, with every neighborhood $W \in \mathcal{W}$ we associate the function space neighborhood $\mathfrak{v}_W$, defined by the family $\mathcal{N}_{\mathfrak{v}_W}$ of all neighborhood-valued step functions $n = \sum_{i=1}^{n} \chi_{A_i} \cdot V_i$, where $A_i \in \mathfrak{R}$ and $V_i \in \mathcal{V}$ such that $\int_X n\, d\theta \leqslant \sigma_W$. This clearly is a function space neighborhood in the sense of Section 1.2, as $\mathcal{N}_{\mathfrak{v}_W}$ is convex and for every set $A \in \mathfrak{R}$ there is $V \in \mathcal{V}$ such that $|\theta|_A(V) \leqslant \sigma_W$ by our premise that θ is bounded. Hence $\chi_A \cdot V \in \mathcal{N}_W$, that is, $\chi_A \cdot V \leqslant \mathfrak{v}_W$, thus satisfying the requirements. We note that all neighborhood functions in $\mathcal{N}_{\mathfrak{v}_W}$ are supported by some set $A \in \mathfrak{R}$. We continue with a simple observation for step functions in $\mathcal{F}_S(X, \mathrm{Conv}(E))$.

2.3.4 Lemma. *Let $B \in \mathfrak{A}$, $h \in \mathcal{F}_S(X, \mathrm{Conv}(E))$ and $W \in \mathcal{W}$. Then*

$$\int_B h\, d\theta \leqslant \int_B^{(\mathfrak{v}_W)} h\, d\theta \leqslant \int_B h\, d\theta + \sigma_W.$$

Proof. The first part of this inequality is trivial. For the second part, let $g \leqslant h + \mathfrak{v}_W$ for $g \in \mathcal{F}_S(X, \mathrm{Conv}(E))$. That is to say, $g \leqslant h + n$ for some neighborhood-valued function $n \in \mathcal{N}_{\mathfrak{v}_W}$. Following Proposition 2.3.2(b) and (c), this implies

$$\int_B g\, d\theta \leqslant \int_B h\, d\theta + \int_B n\, d\theta \leqslant \int_B h\, d\theta + \sigma_W$$

for each such step function $g \in \mathcal{F}_S(X, \mathrm{Conv}(E))$, hence $\int_B^{(\mathfrak{v}_W)} h\, d\theta \leqslant \int_B h\, d\theta + \sigma_W$ as claimed. □

For the following, we recall the definition of boundedness for subsets of $\mathcal{F}(X, \mathrm{Conv}(E))$ from Section 2.2.

2.3.5 Proposition. *Let $A \in \mathfrak{R}$, let $f \in \mathcal{F}(X, \mathrm{Conv}(E))$ be measurable, and let $(h_n)_{n\in\mathbb{N}}$ be a bounded below sequence of step functions in $\mathcal{F}_S(X, \mathrm{Conv}(E))$ such that for every $x \in A$ there is $n_0 \in \mathbb{N}$ such that $f(x) \leqslant h_n(x)$ for all $n \geqslant n_0$. Then*

$$\int_A^{(\mathfrak{v}_W)} f\, d\theta \leqslant \liminf_{n\to\infty} \int_A h_n\, d\theta + \sigma_W$$

for every $W \in \mathcal{W}$. The limit is taken pointwise on F^*.

Proof. Let $A \in \mathfrak{R}$, let $f \in \mathcal{F}(X, \mathrm{Conv}(E))$ be measurable and let $(h_n)_{n\in\mathbb{N}}$ be a sequence of step functions in $\mathcal{F}_S(X, \mathrm{Conv}(E))$ satisfying our assumptions. For $W \in \mathcal{W}$, let $l \in \mathcal{F}_S(X, \mathrm{Conv}(E))$ such that $l \leqslant f + \mathfrak{v}_W$, that is, $l \leqslant f + n$ for some $n \in \mathcal{N}_{\mathfrak{v}_W}$. Now we set

$$A_n = \{x \in A \mid l(x) \leqslant h_m(x) + n(x) \text{ for all } m \geqslant n\}.$$

Since all the functions involved are step functions, all the sets A_n are in $\mathfrak{R}$. We have $A_n \subset A_{n+1}$ and $A = \bigcup_{n\in\mathbb{N}} A_n$ by our assumption. There is $V \in \mathcal{V}$ such that $|\theta|_A(V) \leqslant \sigma_W$, and as the sequence $(h_n)_{n\in\mathbb{N}}$ is bounded below, there is $\lambda \geqslant 0$ such that $0 \leqslant h_n + \lambda\chi_X \cdot V$ for all $n \in \mathbb{N}$. Thus

$$\begin{aligned}\chi_{A_n} \cdot l &\leqslant \chi_{A_n} \cdot (h_n + n) + \chi_{A\setminus A_n}(h_n + \lambda\chi_X \cdot V)\\ &\leqslant \chi_A \cdot h_n + \chi_{A_n} \cdot n + \lambda\chi_{A\setminus A_n} \cdot V.\end{aligned}$$

Hence by Proposition 2.3.2, and as

$$\int_X \chi_{A_n} n\, d\theta \leqslant \int_X n\, d\theta \leqslant \sigma_W, \quad \text{and} \quad \int_X \chi_{A\setminus A_n} \cdot V = |\theta|_{(A\setminus A_n)}(V),$$

when taking the integrals over X in the above inequality, we obtain

$$\int_{A_n} l\, d\theta \leqslant \int_A h_n\, d\theta + \lambda|\theta|_{A\setminus A_n}(V) + \sigma_W.$$

Because $A_n \subset A_{n+1}$ and $\bigcup_{n\in\mathbb{N}} A_n = A$, Theorem 2.1.3 together with Lemma 2.1.2(a) yields that $|\theta|_{(B\cap A)}(C) = \lim_{n\to\infty} |\theta|_{(B\cap A_n)}(C)$ for all $B \in \mathfrak{R}$ and $C \in \operatorname{Conv}(E)$. Considering the definition of the integral of a step function, this renders

$$\lim_{n\to\infty} \int_{A_n} l\, d\theta = \int_A l\, d\theta,$$

and Lemma 2.1.2(b) yields that $\lim_{n\to\infty} |\theta|_{(A\setminus A_n)}(V)(\mu) = 0$ for all $\mu \in W^\circ$. Thus we obtain

$$\int_A l\, d\theta \leqslant \liminf_{n\to\infty} \int_A h_n\, d\theta + \sigma_W.$$

Our claim follows, since the above inequality holds true for all step functions $l \in \mathcal{F}_{\mathcal{S}}(X, \operatorname{Conv}(E))$ such that $l \leqslant f + \mathfrak{v}_W$. □

2.3.6 Corollary. *Let $B \in \mathfrak{A}$, let $f \in \mathcal{F}(X, \operatorname{Conv}(E))$ be measurable, let $W \in \mathcal{W}$ and let $\mathfrak{v}$ be a function space neighborhood. Then*

$$\int_B^{(\mathfrak{v}_W)} f\, d\theta \leqslant \int_B^{(\mathfrak{v})} f\, d\theta + \sigma_W.$$

Proof. Let $B \in \mathfrak{A}$, let $f \in \mathcal{F}(X, \operatorname{Conv}(E))$ be measurable, let $W \in \mathcal{W}$ and let $\mathfrak{v}$ be a function space neighborhood. Let $l \in \mathcal{F}_{\mathcal{S}}(X, \operatorname{Conv}(E))$ such that $l \leqslant f + \mathfrak{v}_W$, and according to Proposition 2.2.11 we choose $A \in \mathfrak{R}$ such that l is supported by A and $\chi_{X\setminus A} \cdot f \geqslant 0$. For the set $B \cap A \in \mathfrak{R}$, let $(h_n)_{n\in\mathbb{N}}$ be a sequence of step functions in

$\mathcal{F}_S(X, \mathrm{Conv}(E))$ approaching f as in Proposition 2.2.13 with the function space neighborhood $\mathfrak{v}$ and $\varepsilon > 0$. We may assume that the functions h_n are supported by A, because we may otherwise replace them by their product with the characteristic function of this set. Proposition 2.3.5 renders that there is $1 \leqslant \gamma \leqslant 1 + \varepsilon$ such that

$$\int\limits_{B\cap A}^{(\mathfrak{v}_W)} f\,d\theta \leqslant \gamma \liminf_{n\to\infty} \int\limits_{B\cap A} h_n\,d\theta + \sigma_W.$$

On the other hand, we have

$$\int\limits_{B} l\,d\theta = \int\limits_{B\cap A} l\,d\theta \leqslant \int\limits_{B\cap A}^{(\mathfrak{v}_W)} f\,d\theta,$$

since the step function l is supported by A. Similarly, for the functions $h_n \leqslant f + \mathfrak{v}$ we observe that

$$\int\limits_{(B\cap A)} h_n\,d\theta = \int\limits_{B} h_n\,d\theta \leqslant \int\limits_{B}^{(\mathfrak{v})} f\,d\theta.$$

Now combining all of the above returns

$$\int\limits_{B} l\,d\theta \leqslant \int\limits_{B\cap A}^{(\mathfrak{v}_W)} f\,d\theta \leqslant \gamma \liminf_{n\to\infty} \int\limits_{B\cap A} h_n\,d\theta + \sigma_W \leqslant \gamma \int\limits_{B}^{(\mathfrak{v})} f\,d\theta + \sigma_W.$$

Our claim follows because this last inequality holds true, pointwise on F^*, for all $\varepsilon > 0$, and, consequently, we have

$$\int\limits_{B} l\,d\theta \leqslant \int\limits_{B}^{(\mathfrak{n})} f\,d\theta + \sigma_W$$

for all step functions $l \in \mathcal{F}_S(X, \mathrm{Conv}(E))$ such that $h \leqslant f + \mathfrak{v}_W$. □

We are now in a position to define the *integral over a set* $B \in \mathfrak{A}$ for a measurable function $f \in \mathcal{F}(X, \mathrm{Conv}(E))$. Let $\mathfrak{V}_*$ denote the collection of all function space neighborhoods, downward directed in the order introduced in Section 1.2. Corollary 2.3.6 demonstrates that for a measurable function $f \in \mathcal{F}(X, \mathrm{Conv}(E))$ and a set $B \in \mathfrak{A}$ the net $(\int_B^{(\mathfrak{v})} f\,d\theta)_{\mathfrak{v}\in\mathfrak{V}_*}$ forms a Cauchy net in the symmetric topology of $\mathcal{Q}_F$. Indeed, given $W \in \mathcal{W}$ we have

$$\int\limits_{B}^{(\mathfrak{u})} f\,d\theta \in W^s\Bigg(\int\limits_{B}^{(\mathfrak{v})} f\,d\theta\Bigg),$$

that is,

$$\int_B^{(\mathfrak{u})} f\,d\theta \leqslant \int_B^{(\mathfrak{v})} f\,d\theta + \sigma_W \quad \text{and} \quad \int_B^{(\mathfrak{v})} f\,d\theta \leqslant \int_B^{(\mathfrak{u})} f\,d\theta + \sigma_W,$$

for all $\mathfrak{u}, \mathfrak{v} \leqslant \mathfrak{v}_W$. Thus according to Proposition 1.4.5 this net is convergent in the symmetric topology of $\mathcal{Q}_F$, and we set

$$\begin{aligned}\int_B f\,d\theta &= \lim_{\mathfrak{v}\in\mathfrak{V}_*} \int_B^{(\mathfrak{v})} f\,d\theta = \inf_{\mathfrak{v}\in\mathfrak{V}_*} \int_B^{(\mathfrak{v})} f\,d\theta \\ &= \lim_{\mathfrak{v}\in\mathfrak{V}_*} \sup\Big\{ \int_B h\,d\theta \;\Big|\; h \in \mathcal{F}_{\mathcal{S}}(X, \mathrm{Conv}(E)),\ h \leqslant f + \mathfrak{v} \Big\} \\ &= \lim_{\mathfrak{v}\in\mathfrak{V}_*} \sup\Big\{ \int_B h\,d\theta \;\Big|\; h \in \mathcal{F}_{\mathcal{S}}(X, E),\ h \leqslant f + \mathfrak{v} \Big\}.\end{aligned}$$

The limits in these expressions refer to the symmetric topology of $\mathcal{Q}_F$ and the index set $\mathfrak{V}_*$, whereas the infima and suprema are taken pointwise on F^*. Our earlier observation in Lemma 2.3.4 justifies that the above definition of the integral is consistent with the preceding one for step functions. Corollary 2.3.6 yields that

$$\int_B f\,d\theta \leqslant \int_B^{(\mathfrak{v}_W)} f\,d\theta \leqslant \int_B f\,d\theta + \sigma_W$$

holds for all $W \in \mathcal{W}$ and that the integral of a measurable function f is already determined using only the function space neighborhoods $\mathfrak{v}_W$. That is,

$$\int_B f\,d\theta = \lim_{W\in\mathcal{W}} \int_B^{(\mathfrak{v}_W)} f\,d\theta.$$

Obviously, the integral is monotone, and we shall go forward to verify that it determines a linear operator from the subcone of measurable functions in $\mathcal{F}(X, \mathrm{Conv}(E))$ into $\mathcal{Q}_F$. We begin with an obvious, but useful consequence of Proposition 2.3.5 in the light of our definition of the integral of a measurable function.

2.3.7 Corollary. *Let $A \in \mathfrak{R}$, let $f \in \mathcal{F}(X, \mathrm{Conv}(E))$ be measurable and let $(h_n)_{n\in\mathbb{N}}$ be a bounded below sequence of step functions in $\mathcal{F}_{\mathcal{S}}(X, \mathrm{Conv}(E))$ such that for every $x \in A$ there is $n_0 \in \mathbb{N}$ such that $f(x) \leqslant h_n(x)$ for all $n \geqslant n_0$. Then*

$$\int_A f\,d\theta \leqslant \liminf_{n\to\infty} \int_A h_n\,d\theta.$$

In Part (e) of the following theorem, we consider $\mathfrak{R}$ as the index set of a net, directed upward by set inclusion.

2.3.8 Theorem. *Let $B \in \mathfrak{A}$, let $f, g \in \mathcal{F}(X, \mathrm{Conv}(E))$ be measurable functions and let $\alpha \geqslant 0$. Then*

(a) $\int_B \alpha f \, d\theta = \alpha \int_B f \, d\theta$.

(b) $\int_B (f + g) \, d\theta = \int_B f \, d\theta + \int_B g \, d\theta$.

(c) $\int_B f \, d\theta \leqslant \int_B g \, d\theta$ *whenever* $f \underset{a.e.B}{\leqslant} g$.

(d) $\int_B f \, d\theta = \int_X \chi_B \cdot f \, d\theta$.

(e) $\int_B f \, d\theta = \lim_{A \in \mathfrak{R}} \int_{A \cap B} f \, d\theta$.

Proof. Part (a) follows straightaway from our definition of the integral of measurable functions. We defer the proof of Part (b). For Part (c), suppose that $f \underset{a.e.B}{\leqslant} g$ for measurable functions $f, g \in \mathcal{F}_S(X, \mathrm{Conv}(E))$. That is to say, $\chi_{B \setminus Z} \cdot f \leqslant \chi_{B \setminus Z} \cdot g$ holds with some set $Z \in \mathfrak{A}$ of measure zero. Let $\mathfrak{v} \in \mathfrak{V}_*$. The inequality

$$\int_{B \setminus Z}^{(\mathfrak{v})} f \, d\theta \leqslant \int_{B \setminus Z}^{(\mathfrak{v})} g \, d\theta$$

follows immediately from our definition. Proposition 2.3.2(d) renders $\int_B h \, d\theta = \int_{(B \setminus Z)} h \, d\theta$ for all step functions $h \in \mathcal{F}_S(X, \mathrm{Conv}(E))$ and, therefore,

$$\int_{B}^{(\mathfrak{v})} l \, d\theta = \int_{B \setminus Z}^{(\mathfrak{v})} l \, d\theta$$

holds for all measurable functions $l \in \mathcal{F}(X, \mathrm{Conv}(E))$. Combining with the above, we obtain

$$\int_{B}^{(\mathfrak{v})} f \, d\theta = \int_{B \setminus Z}^{(\mathfrak{v})} f \, d\theta \leqslant \int_{B \setminus Z}^{(\mathfrak{v})} g \, d\theta = \int_{B}^{(\mathfrak{v})} g \, d\theta.$$

Our claim now follows since the last inequality holds true for every $\mathfrak{v} \in \mathfrak{V}_*$. For Part (d), we recall from Proposition 2.2.12(b) that $\chi_B \cdot f$ is measurable. Let $\mathfrak{v} \in \mathfrak{V}_*$. We have

$$\int_{B}^{(\mathfrak{v})} f \, d\theta = \sup\left\{ \int_B h \, d\theta \mid h \in \mathcal{F}_S(X, \mathrm{Conv}(E)),\ h \leqslant f + \mathfrak{v} \right\}$$

and

$$\int_X^{(\mathfrak{v})} \chi_B \cdot f\, d\theta = \sup\left\{ \int_X h'\, d\theta \mid h' \in \mathcal{F}_{\mathcal{S}}(X, \mathrm{Conv}(E)),\ h' \leqslant \chi_B \cdot f + \mathfrak{v} \right\}$$

First, let $h \in \mathcal{F}_{\mathcal{S}}(X, \mathrm{Conv}(E))$ such that $h \leqslant f + \mathfrak{v}$. Then $h' = \chi_B \cdot h \leqslant \chi_B \cdot f + \mathfrak{v}$, and $\int_X h'\, d\theta = \int_B h\, d\theta$ by Proposition 2.3.2(c). This shows

$$\int_B^{(\mathfrak{v})} f\, d\theta \leqslant \int_X^{(\mathfrak{v})} \chi_B \cdot f\, d\theta.$$

Taking the infima over all $\mathfrak{v} \in \mathfrak{V}_*$ on both sides of this inequality then yields

$$\int_B f\, d\theta \leqslant \int_X \chi_B \cdot f\, d\theta.$$

For the converse inequality, let $W \in \mathcal{W}$ and let $\mathfrak{v} \in \mathfrak{V}_*$ such that $\mathfrak{v} \leqslant \mathfrak{v}_W$. There is $h_0 \in \mathcal{F}_{\mathcal{S}}(X, \mathrm{Conv}(E))$ such that $h_0 \leqslant f + \mathfrak{v}$ and, therefore, $\chi_{(X\setminus B)} \cdot h_0 \leqslant \chi_{(X\setminus B)} \cdot f + \mathfrak{v}$. For any $h' \in \mathcal{F}_{\mathcal{S}}(X, \mathrm{Conv}(E))$ such that $h' \leqslant \chi_B \cdot f + \mathfrak{v}$, we have both $\chi_B \cdot h' \leqslant \chi_B \cdot f + \mathfrak{v}$ and $\chi_{(X\setminus B)} \cdot h' \leqslant \mathfrak{v}$. We set

$$h = \chi_B \cdot h' + \chi_{(X\setminus B)} \cdot h_0.$$

Then $h \in \mathcal{F}_{\mathcal{S}}(X, \mathrm{Conv}(E))$ and

$$h \leqslant (\chi_B \cdot f + \mathfrak{v}) + (\chi_{(X\setminus B)} \cdot f + \mathfrak{v}) \leqslant f + 2\mathfrak{v}.$$

We have $\int_B h\, d\theta = \int_B h'\, d\theta$ by Proposition 2.3.2(d) and $\int_{(X\setminus B)} h'\, d\theta \leqslant \sigma_W$, as $\chi_{(X\setminus B)} \cdot h' \leqslant \mathfrak{v} \leqslant \mathfrak{v}_W$. Hence using both 2.3.2(b) and 2.3.2(d), we obtain

$$\int_X h'\, d\theta = \int_B h'\, d\theta + \int_{X\setminus B} h'\, d\theta \leqslant \int_B h\, d\theta + \sigma_W \leqslant \int_B^{(2\mathfrak{v})} f\, d\theta + \sigma_W$$

for all step functions $h' \in \mathcal{F}_{\mathcal{S}}(X, \mathrm{Conv}(E))$ such that $h' \leqslant \chi_B \cdot f + \mathfrak{v}$ and, therefore,

$$\int_X \chi_B \cdot f\, d\theta \leqslant \int_X^{(\mathfrak{v})} \chi_B \cdot f\, d\theta \leqslant \int_B^{(2\mathfrak{v})} f\, d\theta + \sigma_W.$$

Taking the infima first over all $\mathfrak{v} \leqslant \mathfrak{w}_W$, then over all $W \in \mathcal{W}$ on the right-hand side of this inequality yields

$$\int_X \chi_B \cdot f\, d\theta \leqslant \int_B f\, d\theta.$$

Part (d) follows. For Part (e), it is therefore sufficient to consider the case that $B = X$, because the function f may be replaced by its product with the characteristic function χ_B. Let $A_0 \in \mathfrak{R}$ such that $f(x) \geqslant 0$ for all $x \in X \setminus A_0$. Then $\chi_{A_1} \cdot f \leqslant \chi_{A_2} \cdot f \leqslant f$ whenever $A_0 \subset A_1 \subset A_2$ for $A_1, A_2 \in \mathfrak{R}$, hence $\int_{A_1} f\,d\theta \leqslant \int_{A_2} f\,d\theta \leqslant \int_X f\,d\theta$ by parts (c) and (d). This shows

$$\lim_{A \in \mathfrak{R}} \int_A f\,d\theta = \sup_{A_0 \subset A \in \mathfrak{R}} \int_A f\,d\theta \leqslant \int_X f\,d\theta.$$

For the converse inequality, let $W \in \mathcal{W}$ and $h \leqslant f + \mathfrak{v}_W$ for $h \in \mathcal{F}_{\mathcal{S}}(X, \mathrm{Conv}(E))$. Because h is supported by a set in $\mathfrak{R}$, there is $A_0 \subset A_1 \in \mathfrak{R}$ such that $\int_X h\,d\theta = \int_A h\,d\theta \leqslant \int_A^{(\mathfrak{v}_W)} f\,d\theta$ for all $A_1 \subset A \in \mathfrak{R}$. Moreover, Corollary 2.3.6 shows that $\int_A^{(\mathfrak{v}_W)} f\,d\theta \leqslant \int_A f\,d\theta + \sigma_W$. Thus

$$\int_X h\,d\theta \leqslant \sup_{A_1 \subset A \in \mathfrak{R}} \int_A f\,d\theta + \sigma_W = \lim_{A \in \mathfrak{R}} \int_A f\,d\theta + \sigma_W.$$

This shows

$$\int_X f\,d\theta \leqslant \int_X^{(\mathfrak{v}_W)} f\,d\theta \leqslant \lim_{A \in \mathfrak{R}} \int_A f\,d\theta + \sigma_W.$$

Since $W \in \mathcal{W}$ was arbitrarily chosen, our claim follows. Finally, in the light of Part (e) it suffices to verify the statement in (b) for the sets in $\mathfrak{R}$. Let $A \in \mathfrak{R}$, let $\mathfrak{v} \in \mathfrak{V}_*$ and let $h \leqslant f + \mathfrak{v}$ and $l \leqslant g + \mathfrak{v}$ for $h, l \in \mathcal{F}_{\mathcal{S}}(X, E)$. Then

$$(h + l) \leqslant (f + g) + 2\mathfrak{v} \leqslant \int_A^{(2\mathfrak{v})} (f + g)\,d\theta.$$

From this, we infer that

$$\int_A^{(\mathfrak{v})} f\,d\theta + \int_A^{(\mathfrak{v})} g\,d\theta \leqslant \int_A^{(2\mathfrak{v})} (f + g)\,d\theta.$$

Taking the infima over all $\mathfrak{v} \in \mathfrak{V}_*$ renders

$$\int_A f\,d\theta + \int_A g\,d\theta \leqslant \int_A (f + g)\,d\theta.$$

For the reverse inequality, let $\varepsilon > 0$, let $\mathfrak{v} \in \mathcal{V}_*$ and let $(f_n)_{n \in \mathbb{N}}$ and $(g_n)_{n \in \mathbb{N}}$ be increasing sequences of step functions in $\mathcal{F}_{\mathcal{S}}(X, \mathrm{Conv}(E))$ as in Proposition 2.2.13 for the measurable functions f and g, respectively. Let $1 \leqslant \gamma, \sigma \leqslant 1 + \varepsilon$ be the respective constants involved in this statement and set $h_n = \gamma f_n + \sigma g_n \in \mathcal{F}_S(X, \mathrm{Conv}(E))$ Then for every $x \in A$ there is $n_0 \in \mathbb{N}$ such that $(f+g)(x) \leqslant h_n(x)$ holds for all $n \geqslant n_0$. Thus

$$\int_A (f+g)\, d\theta \leqslant \liminf_{n\to\infty} \int_A h_n\, d\theta$$

by Corollary 2.3.7. The sequences $(f)_{n\in\mathbb{N}}$, $(g)_{n\in\mathbb{N}}$ and $(h)_{n\in\mathbb{N}}$ are increasing, so are the respective sequences of their integrals, and we conclude that

$$\liminf_{n\to\infty} \int_A h_n\, d\theta = \lim_{n\to\infty} \int_A h_n\, d\theta = \gamma \lim_{n\to\infty} \int_A f_n\, d\theta + \sigma \lim_{n\to\infty} \int_A g_n\, d\theta.$$

Moreover, $f_n \leqslant f + \mathfrak{v}$ and $g_n \leqslant g + \mathfrak{v}$ implies that $\int_A f_n\, d\theta \leqslant \int_A^{(\mathfrak{v})} f\, d\theta$ and $\int_A g_n\, d\theta \leqslant \int_A^{(\mathfrak{v})} g\, d\theta$ for all $n \in \mathbb{N}$. Combining all of the above then yields

$$\int_A (f+g)\, d\theta \leqslant \gamma \int_A^{(\mathfrak{v})} f\, d\theta + \sigma \int_A^{(\mathfrak{v})} g\, d\theta.$$

The latter holds for all $\mathfrak{v} \in \mathfrak{V}_*$ and $\varepsilon > 0$ with γ and σ depending on both. First, with $\mathfrak{v} \in \mathfrak{V}_*$ fixed, passing to the limit over ε tending to 0, and then to the downward directed limits over $\mathfrak{v} \in \mathfrak{V}_*$, we obtain

$$\int_A (f+g)\, d\theta \leqslant \int_A f\, d\theta + \int_A g\, d\theta.$$

This completes our argument. □

In the expression below, for the calculation of the integral of a measurable function $f \in \mathcal{F}(X, \mathrm{Conv}(E))$ we consider only step functions with values in the range of f. This will turn out to be advantageous for our interpretation of the value of such an integral.

2.3.9 Proposition. *Let $B \in \mathfrak{A}$ and let $f \in \mathcal{F}(X, \mathrm{Conv}(E))$ be a measurable function. Then*

$$\int_B f\, d\theta = \lim_{W\in\mathcal{W}} \sup \left\{ \int_B h\, d\theta \;\middle|\; \begin{array}{l} h \in \mathcal{F}_S(X,E),\ h \leqslant f + \mathfrak{v}_W \\ h(B) \subset \bigcup_{x\in B} f(x) \end{array} \right\}.$$

Proof. Let $B \in \mathfrak{A}$ and let $f \in \mathcal{F}(X, \mathrm{Conv}(E))$ be a measurable function. Given $W \in \mathcal{W}$, for the sake of our argument we abbreviate

$$C_W = \left\{ \int_B h\, d\theta \;\middle|\; \begin{array}{l} h \in \mathcal{F}_S(X,E),\ h \leqslant f + \mathfrak{v}_W \\ h(B) \subset \bigcup_{x\in B} f(x) \end{array} \right\}.$$

For the corresponding element $\varphi_{C_W} = \sup C_W \in \mathcal{Q}_F$, we have

$$\varphi_{C_W} \leqslant \int\limits_B^{(\mathfrak{v}_w)} f\,d\theta \leqslant \int\limits_B f\,d\theta + \sigma_W.$$

Conversely, let $h = \sum_{i=1}^n \chi_{A_i} \cdot a_i \in \mathcal{F}_\mathcal{S}(X,E)$ such that $h \leqslant f + \mathfrak{v}_{(1/2)W}$. That is to say, $h \leqslant f + n$ for a neighborhood function $n = \sum_{k=1}^m \chi_{C_k} \cdot V_k$ with disjoint sets $C_k \in \mathfrak{R}$ and $V_k \in \mathcal{V}$ such that $\int_X n\,d\theta = \sum_{k=1}^m |\theta|_{C_k}(V_k) \leqslant \sigma_{(1/2)W}$. We may also assume that the sets $A_i \in \mathfrak{R}$ are disjoint, either contained in B or in $X \setminus B$, and that their union contains the union C of the sets C_k. The sets $C_i^k = A_i \cap C_k$ form a disjoint cover of C. If $C_i^k \neq \varnothing$, we choose $x_i^k \in C_i^k$ and have

$$a_i = h(x_i^k) \leqslant f(x_i^k) + V_k$$

hence there is $c_i^k \in f(x_i^k)$ such that $a_i - c_i^k \in V_k$. If $C_i^k = \varnothing$, we assign $c_i^k = 0$. Using these choices, we set

$$\tilde{h} = \sum_{i,k} \chi_{C_i^k} \cdot c_i^k + \chi_{X \setminus C} \cdot h \in \mathcal{F}_\mathcal{S}(X,E)$$

and observe the following. For every $x \in B \cap C$, we have $x \in C_i^k$ for exactly one choice of i and k and

$$\tilde{h}(x) = c_i^k \in f(x_i^k) \subset \bigcup_{x \in B} f(x).$$

For $x \in B \cap (X \setminus C)$, on the other hand, we have $n(x) = 0$ and

$$\tilde{h}(x) = h(x) \leqslant f(x) + n(x) = f(x).$$

Thus $\tilde{h}(x) \in \bigcup_{x \in B} f(x)$ also holds. We conclude that $\tilde{h}(B) \subset \bigcup_{x \in B} f(x)$. Moreover, for $x \in C$ we have $x \in C_i^k$ for some i and k. Thus

$$\tilde{h}(x) = c_i^k \leqslant a_i + V_k = h(x) + n(x) \leqslant f(x) + 2n(x)$$

and

$$h(x) = a_i \leqslant c_i^k + V_k = \tilde{h}(x) + n(x).$$

Since $\tilde{h}(x) = h(x)$ for all $x \in X \setminus C$, we infer that $\tilde{h} \leqslant f + 2n$, that is, $\tilde{h} \leqslant f + \mathfrak{v}_W$ and $h \leqslant \tilde{h} + n$. Summarizing, we have $\tilde{h} \in C_W$ and

$$\int\limits_B h\,d\theta \leqslant \int\limits_B \tilde{h}\,d\theta + \int\limits_B n\,d\theta \leqslant \int\limits_B \tilde{h}\,d\theta + \sigma_{(1/2)W} \leqslant \varphi_{C_W} + \sigma_W.$$

The latter holds true for all $h \in \mathcal{F}_\mathcal{S}(X,E)$ such that $h \leqslant f + \mathfrak{v}_{(1/2)W}$, and we conclude that

$$\int_B f\,d\theta \leqslant \int_B^{(\mathfrak{v}_{(1/2)W})} f\,d\theta \leqslant \varphi_{C_W} + \sigma_W.$$

Together with

$$\varphi_{C_W} \leqslant \int_B^{(\mathfrak{v}_W)} f\,d\theta \leqslant \int_B f\,d\theta + \sigma_W$$

this yields $\varphi_{C_W} \in W^s(\int_B f\,d\theta)$, and our claim follows. □

The following result allows us to interpret the value of the integral of a measurable $\mathrm{Conv}(E)$-valued function as an element of $\mathrm{Conv}(F^{**})$ or of $\mathrm{Conv}(F)$. Proposition 1.4.3(a) establishes a correspondence between elements of $\mathcal{Q}_F$ that are $\sigma(F^*, F^{**})$-lower semicontinuous and convex subsets of E^{**}, and Proposition 1.4.3(d) states that an element of $\mathcal{Q}_F$ represents a weakly compact convex subset of F if and only it is continuous in the Mackey topology $\tau(F^*, F)$.

2.3.10 Proposition.

(a) *Suppose that F is a normed space. If $f \in \mathcal{F}(X, \mathrm{Conv}(E))$ is measurable, then for every $B \in \mathfrak{A}$ the integral $\int_B f\,d\theta$ represents a convex subset of F^{**}.*

(b) *Suppose that the measure θ is $\mathfrak{L}(E, F)$-valued and weakly compact. Let $f \in \mathcal{F}(X, \mathrm{Conv}(E))$ be a measurable function and let $A \in \mathfrak{R}$ such that $\bigcup_{x\in A} f(x)$ is bounded in E. Then $\int_A f\,d\theta$ represents a weakly compact convex subset of F.*

Proof. (a) If F is a normed space, then $F^{*\blacksquare} = F^{**}$, and our measures are $\mathfrak{L}(E, F^{**})$-valued. For a set $B \in \mathfrak{A}$, the integral of an E-valued step function over B is an element of $F^{**} \subset \mathcal{Q}_F$, and the integral of a $\mathrm{Conv}(E)$-valued step function evaluates as a $\sigma(F^*, F^{**})$-lower semicontinuous element of $\mathcal{Q}_F$. The same observation holds true for the expression $\int_B^{(\mathfrak{v})} f\,d\theta$ for a measurable function $f \in \mathcal{F}(X, \mathrm{Conv}(E))$, a set $B \in \mathfrak{A}$ and a function space neighborhood $\mathfrak{v} \in \mathfrak{V}_*$. Moreover, the integral $\int_B f\,d\theta$ of a measurable function is the limit of these expressions in the symmetric topology of $\mathcal{Q}_F$, hence also $\sigma(F^*, F^{**})$-lower semicontinuous by Proposition 1.4.6(b). For the latter, we recall from 1.4.4 that a sublinear functional on F^* is $\sigma(F^*, F^{**})$-lower semicontinuous if and only if it is $\beta(F^*, F)$-lower semicontinuous, provided that F is a normed space. Thus in the light of Proposition 1.4.3(a), which states a correspondence between $\sigma(F^*, F^{**})$-lower semicontinuous sublinear functionals on F^* and convex subsets of F^{**}, the integral of $\int_B f\,d\theta$ may be considered to be an element of $\mathrm{Conv}(F^{**})$. For Part (b), let θ and $f \in \mathcal{F}(X, \mathrm{Conv}(E))$ be as stated. For $A \in \mathfrak{R}$, let

$$D = \left\{ \sum_{i=1}^{n} \theta_{A_i}(c_i) \;\middle|\; c_i \in \bigcup_{x\in A} f(x),\ A_i \in \mathfrak{R} \text{ disjoint},\ \bigcup_{i=1}^{n} A_i = A \right\}$$

As $\bigcup_{x\in A} f(x)$ is bounded and θ is weakly compact, this set is relatively weakly compact in F. For every $W \in \mathcal{W}$, we have

$$C_W = \left\{ \int_A h\,d\theta \;\middle|\; \begin{array}{l} h \in \mathcal{F}_S(X,E),\ h \leqslant f + \mathfrak{v}_W \\ h(B) \subset \bigcup_{x\in A} f(x) \end{array} \right\} \subset D.$$

Hence the sets C_W are also relatively weakly compact and the corresponding sublinear functionals $\varphi_{C_W} \in \mathcal{Q}_F$ are $\tau(F^*, F)$-continuous. According to Proposition 2.3.9, we have

$$\int_A f\,d\theta = \lim_{W\in\mathcal{W}} \varphi_{C_W} = \inf_{W\in\mathcal{W}} \varphi_{C_W}.$$

Thus following Proposition 1.4.3(c), the functional $\int_A f\,d\theta \in \mathcal{Q}_F$ is also $\tau(F^*, F)$-continuous and in the light of 1.4.3(d) may be interpreted to be a weakly compact convex subset of F. □

For the following, recall from Section 1.2 that a single neighborhood function n defines a function space neighborhood if and only if for every $A \in \mathfrak{R}$ there is $V \in \mathcal{V}$ such that $\chi_A \cdot V \leqslant n$. For an $\mathfrak{L}(E, F^{*\blacksquare})$-valued measure, a fixed neighborhood $V \in \mathcal{V}$ and $\mu \in F^*$ we introduce the positive $\overline{\mathbb{R}}$-valued measure $[|\theta|(V)(\mu)]$, that is, the countably additive set-function

$$A \mapsto |\theta|_A(V)(\mu) : \mathfrak{R} \to \overline{\mathbb{R}}_+.$$

This measure is easily extended to the σ-field $\mathfrak{A}$ if we set

$$\$[|\theta|(V)(\mu)]_B = \sup_{A\in\mathfrak{R}} [|\theta|(V)(\mu)]_A$$

for all $B \in \mathfrak{A}$. For an $\overline{\mathbb{R}}_+$-valued $\mathfrak{A}$-measurable function ρ and $B \in \mathfrak{A}$, we have

$$\int_B \rho\,d[|\theta|(V)(\mu)] = \sup\left\{ \sum_{i=1}^{n} \alpha_i [|\theta|(V)(\mu)]_{A_i} \;\middle|\; \begin{array}{l} \alpha_i \geqslant 0,\ A_i \in \mathfrak{R},\ A_i \subset B, \\ \sum_{i=1}^{n} \alpha_i \chi_{B_i} \leqslant \rho \end{array} \right\}.$$

Using this, we formulate the next result.

2.3.11 Proposition. *Suppose that the function space neighborhood* $\mathfrak{v}$ *is defined by a single measurable neighborhood function* n. *Then*

$$\left(\int_B n\,d\theta \right)(\mu) = \sup_{A\in\mathfrak{R}} \|\theta\|_{A\cap B}^{(\mathfrak{v},\{\mu\})}$$

for every $\mu \in F^*$ *and* $B \in \mathfrak{A}$. *If* $n = \rho \cdot V$ *for* $V \in \mathcal{V}$ *and an* $\overline{\mathbb{R}}_+$*-valued* $\mathfrak{A}$*-measurable function* ρ *such that* $\inf\{\rho(x) \mid x \in A\} > 0$ *for every* $A \in \mathfrak{R}$, *then the above expressions equal* $\int_B \rho\,d[|\theta|(V)(\mu)]$.

Proof. Let $\mathfrak{v}$ and n be as stated and let $B \in \mathfrak{A}$ and $\mu \in F^*$. According to our definition of the variation of a measure in Section 2.1, we have

$$\|\theta\|_A^{(\mathfrak{v},\{\mu\})} = \sup\left\{ \sum_{i=1}^{n} \theta_{A_i}(a_i)(\mu) \;\middle|\; \begin{array}{l} A_i \in \mathfrak{R} \text{ disjoint}, A_i \subset A, \\ a_i \in E, \chi_{A_i} \cdot a_i \leqslant n \end{array} \right\}$$

for every $A \in \mathfrak{R}$. Using Lemma 1.2.1(b), we realize that $\chi_{A_i} \cdot a_i \leqslant n$ for disjoint sets $A_i \in \mathfrak{R}$ implies that $\sum_{i=1}^n \chi_{A_i} \cdot a_i \leqslant n$ and, therefore, $\sum_{i=1}^n \theta_{A_i}(a_i) \leqslant \int_B n\, d\theta$, provided that all $A_i \subset B$. The inequality

$$\sup_{A \in \mathfrak{R}} \|\theta\|_{A \cap B}^{(\mathfrak{v},\{\mu\})} \leqslant \left(\int_B n\, d\theta \right)(\mu)$$

is therefore evident. For the reverse relation, let $\varepsilon > 0$. For any step function $h \in \mathcal{F}_S(X,E)$ such that $h \leqslant n + \varepsilon\mathfrak{v}$, we have $h \leqslant (1+\varepsilon)n$. This yields

$$\int_B^{(\varepsilon\mathfrak{v})} n\, d\theta \leqslant (1+\varepsilon) \sup\left\{ \int_B h\, d\theta \;\middle|\; h \in \mathcal{F}_S(X,E),\; h \leqslant n \right\}$$

for all $\varepsilon > 0$. Thus

$$\int_B n\, d\theta \leqslant \sup\left\{ \int_B h\, d\theta \;\middle|\; h \in \mathcal{F}_S(X,E),\; h \leqslant n \right\}.$$

Our claim follows, because for every step function $h \in \mathcal{F}_S(X,E)$ that is included in the supremum on the right-hand side of the last inequality, its product $\chi_B \cdot h$ with χ_B is involved in the definition of $\|\theta\|_{A \cap B}^{(\mathfrak{v},\{\mu\})}$ for some set $A \in \mathfrak{R}$. This completes the argument for the first part of our claim. Now suppose that $n = \rho \cdot V$ for an $\overline{\mathbb{R}}_+$-valued $\mathfrak{A}$-measurable function ρ such that $\inf\{\rho(x) \mid x \in A\} > 0$ for every $A \in \mathfrak{R}$, and $V \in \mathcal{V}$. We calculate that

$$\begin{aligned}
\int_B \rho\, d[|\theta|(V)(\mu)] &= \sup\left\{ \sum_{i=1}^{n} \alpha_i [|\theta|(V)(\mu)]_{A_i} \;\middle|\; \begin{array}{l} A_i \in \mathfrak{R}, A_i \subset B, \alpha_i \geqslant 0, \\ \sum_{i=1}^n \alpha_i \chi_{A_i} \leqslant \rho \end{array} \right\} \\
&= \sup\left\{ \sum_{i=1}^{n} |\theta|_{A_i}(\alpha_i V)(\mu) \;\middle|\; \begin{array}{l} A_i \in \mathfrak{R}, A_i \subset B, \alpha_i \geqslant 0, \\ \sum_{i=1}^n \chi_{A_i} \cdot (\alpha_i V) \leqslant n \end{array} \right\} \\
&\leqslant \left(\int_B n\, d\theta \right)(\mu).
\end{aligned}$$

For the reverse inequality, let $\gamma > 1$, let $A_i \in \mathfrak{R}$ be disjoint subsets of B and $a_i \in E$ such that $\chi_{A_i} \cdot a_i \leqslant n$ and set $\alpha_i = \inf\{\rho(x) \mid x \in A_i\} > 0$. Then $a_i \in \rho(x)V$ for all $x \in A_i$ implies that $a_i \in \gamma\alpha_i V$ and, therefore,

$$\sum_{i=1}^{n} \chi_{A_i} \cdot a_i \leqslant \sum_{i=1}^{n} \chi_{A_i} \cdot (\gamma a_i V) \leqslant \gamma n.$$

Hence

$$\sum_{i=1}^{n} \theta_{A_i}(a_i)(\mu) \leqslant \sum_{i=1}^{n} |\theta|_{A_i}(\gamma a_i V)(\mu) \leqslant \gamma \int_B \rho \, d[|\theta|(V)(\mu)].$$

Taking the supremum over all such choices for $A_i \in \mathfrak{R}$ and $a_i \in E$ yields

$$\Big(\int_B n \, d\theta\Big)(\mu) = \sup_{A \in \mathfrak{R}} \|\theta\|_{A \cap B}^{(\mathfrak{v}, \{\mu\})} \leqslant \gamma \int_B \rho \, d[|\theta|(V)(\mu)]$$

for all $\gamma > 1$. Hence our claim. □

If the measure θ is $\mathfrak{L}(E, G)$-valued for a subspace G of $F^{*\blacksquare}$, then the integral of a neighborhood function n satisfying the assumptions of Proposition 2.3.11 may be considered to be an element of $\mathrm{Conv}(G)$. Indeed, it follows immediately from the definition of the variation of a measure that the mapping

$$\mu \mapsto \sup_{A \in \mathfrak{R}} \|\theta\|_{A \cap B}^{(\mathfrak{v}, \{\mu\})} \ : \ F^* \to \overline{\mathbb{R}}$$

is sublinear and lower semicontinuous in the topology $\sigma(F^*, G)$. Hence it represents a closed convex subset of G. (See Sections 1.4 and 3.1 below.)

Integrals of set-valued functions

We are now in a position to extend our notion of integrability to $\mathrm{Conv}(E)$-valued functions that are not necessarily measurable in the sense of Section 2.2. First, for a set $A \in \mathfrak{R}$ we shall say that a function $f \in \mathcal{F}(X, \mathrm{Conv}(E))$ is *integrable over A with respect to an $\mathfrak{L}(E, F^{*\blacksquare})$-valued measure θ* if for every $W \in \mathcal{W}$ there are a measurable function f_W and a measurable neighborhood function n_W in $\mathcal{F}(X, \mathrm{Conv}(E))$ such that

$$f \underset{a.e.A}{\leqslant} f_W \underset{a.e.A}{\leqslant} f + n_W \quad \text{and} \quad \int_A n_W \, d\theta \leqslant \sigma_W.$$

Recall from Section 1.2 that neighborhood functions are $\mathcal{V}_c$-valued, where $\mathcal{V}_c$ denotes the extended function space neighborhood system of E, that is, the convex cone generated by $\mathcal{V} \cup \{E\}$ in $\mathrm{Conv}(E)$.

We consider nets $(f_W)_{W \in \mathcal{W}}$ of measurable functions as in this definition. The index set $\mathcal{W}$ is endowed with the reverse set inclusion as its order. For the following, we recall from Section 1.4 that convergence of a net in the symmetric topology of $\mathcal{Q}_F$ implies pointwise convergence on the elements of F^* with respect to the symmetric topology of $\overline{\mathbb{R}}$.

2.3.12 Lemma. *For a function $f \in \mathcal{F}(X, \mathrm{Conv}(E))$ that is integrable over $A \in \mathfrak{R}$ and measurable functions f_W as in the definition of integrability, the net $(\int_A f_W \, d\theta)_{W \in \mathcal{W}}$ converges in the symmetric topology of $\mathcal{Q}_F$ and its limit is independent of the particular choice for the functions f_W.*

Proof. Suppose that $f \in \mathcal{F}(X, \mathrm{Conv}(E))$ is integrable over $A \in \mathfrak{R}$. Let $W \in \mathcal{W}$. For any choice of $U, V \in \mathcal{W}$ such that $U, V \subset W$ and any choice of the measurable functions f_U and f_V from our definition of integrability, we have

$$f_U \leqslant f + n_U \leqslant f_V + n_U \leqslant f_V + n_W.$$

Hence using the statements of Theorem 2.3.8, we obtain

$$\int_A f_U \, d\theta \leqslant \int_A f_V \, d\theta + \sigma_W.$$

The net $(\int_A f_W \, d\theta)_{W \in \mathcal{W}}$ is therefore a Cauchy net in the symmetric topology of $\mathcal{Q}_F$, hence convergent according to Proposition 1.4.5. Our argument also implies that its limit is independent of the particular choice for the functions f_W. □

Integrals over sets in $\mathfrak{R}$

Consequently, for a function $f \in \mathcal{F}(X, \mathrm{Conv}(E))$ that is integrable over a set $A \in \mathfrak{R}$ we define its integral over A as

$$\int_A f \, d\theta = \lim_{W \in \mathcal{W}} \int_A f_W \, d\theta,$$

where $(f_W)_{W \in \mathcal{W}}$ is a net of measurable functions in $\mathcal{F}(X, \mathrm{Conv}(E))$ as in the above definition of integrability for the function f, and the limit refers to the symmetric topology of $\mathcal{Q}_F$. This notion is of course consistent with the earlier definition of the integral of a measurable function.

2.3.13 Proposition. *If $f \in \mathcal{F}(X, \mathrm{Conv}(E))$ is integrable over a set $A \in \mathfrak{R}$, then for every $W \in \mathcal{W}$ there is a measurable neighborhood function n and $\lambda \geqslant 0$ such that $0 \leqslant f + n$ and $\int_A n \, d\theta \leqslant \lambda \sigma_W$.*

Proof. Suppose that $f \in \mathcal{F}(X, \mathrm{Conv}(E))$ is integrable over $A \in \mathfrak{R}$ and let $W \in \mathcal{W}$. According to the definition of integrability, there are measurable functions $g \in \mathcal{F}(X, \mathrm{Conv}(E))$ and a measurable neighborhood function n such that $f \underset{a.e.A}{\leqslant} g \underset{a.e.A}{\leqslant} f + n$

and $\int_A n\,d\theta \leqslant \sigma_W$. We choose $V \in \mathcal{V}$ such that $|\theta|_A(V) \leqslant \sigma_W$. Following Proposition 2.2.11, there is $\lambda \geqslant 0$ such that $0 \leqslant g + \lambda\chi_X \cdot V$. There is a set $Z \in \mathfrak{A}$ of measure zero such that $g \leqslant f + n + \chi_{(X\setminus A)\cup Z} \cdot E$. The combination of the above renders

$$0 \leqslant g + \lambda\chi_X \cdot V \leqslant f + (n + \lambda\chi_X \cdot V + \chi_{(X\setminus A)\cup Z} \cdot E).$$

As $m = n + \lambda\chi_X \cdot V + \chi_{(X\setminus A)\cup Z} \cdot E$ is a measurable neighborhood function, and as

$$\int_A m\,d\theta = \int_A n\,d\theta + \lambda|\theta|_A(V) + \int_Z \chi \cdot E\,d\theta \leqslant (1+\lambda)\sigma_W$$

holds by 2.3.8, our claim follows. □

Integrals over sets in $\mathfrak{A}$

Obviously, integrability for a function $f \in \mathcal{F}(X, \mathrm{Conv}(E))$ over a set $A \in \mathfrak{R}$ implies integrability over all subsets $B \in \mathfrak{R}$ of A. This observation, together with Theorem 2.3.8(e) shows that we may consistently define integrability over sets in the σ-field $\mathfrak{A}$ in the following way: We shall say that a function $f \in \mathcal{F}(X, \mathrm{Conv}(E))$ is *integrable over* $B \in \mathfrak{A}$ *with respect to* θ if f is integrable over the sets $A \cap B$ for all $A \in \mathfrak{R}$ and if the limit

$$\int_B f\,d\theta = \lim_{A\in\mathfrak{R}} \int_{A\cap B} f\,d\theta$$

exists, pointwise on F^*, in $\mathcal{Q}_F$. The set of all functions in $\mathcal{F}(X, \mathrm{Conv}(E))$ or in $\mathcal{F}(X,E)$ that are integrable over B with respect to θ shall be denoted by $\mathcal{F}_{(\theta,B)}(X, \mathrm{Conv}(E))$ or $\mathcal{F}_{(\theta,B)}(X,E)$, respectively. Obviously, if $f \in \mathcal{F}_{(\theta,B)}(X, \mathrm{Conv}(E))$ and $f \underset{a.e.B}{=} g$ for $g \in \mathcal{F}(X, \mathrm{Conv}(E))$, then $g \in \mathcal{F}_{(\theta,B)}(X, \mathrm{Conv}(E))$. Theorem 2.3.8(e) implies that all measurable functions in $\mathcal{F}(X, \mathrm{Conv}(E))$ are contained in $\mathcal{F}_{(\theta,B)}(X, \mathrm{Conv}(E))$ for every set $B \in \mathfrak{A}$ and every $\mathfrak{L}(E, F^{*\blacksquare})$-valued measure θ on $\mathfrak{R}$.

We recall from Section 1.4 that for $f \in \mathcal{F}(X, \mathrm{Conv}(E))$ and an element α of the scalar field of E the function $\alpha f \in \mathcal{F}(X, \mathrm{Conv}(E))$ is defined as the mapping $x \mapsto \alpha f(x) \in \mathrm{Conv}(E)$. An element of $\mathcal{Q}_F$ represents a convex subset of F, F^{**} or $F^{*\blacksquare}$, if it is lower semicontinuous as a sublinear functional on F^* in a topology that is consistent with the duality (F^*, F), (F^*, F^{**}) or $(F^*, F^{*\blacksquare})$, respectively.

2.3.14 Proposition. *If $f \in \mathcal{F}_{(\theta,B)}(X, \mathrm{Conv}(E))$ for $B \in \mathfrak{A}$, then $\alpha f \in \mathcal{F}_{(\theta,B)}(X, \mathrm{Conv}(E))$ for all α in $\mathbb{R}$ or $\mathbb{C}$ and*

$$\Big(\int_B \alpha f\,d\theta\Big)(\mu) = \Big(\int_B f\,d\theta\Big)(\alpha\mu)$$

holds for all $\mu \in F^*$. *If* $\int_B f\,d\theta$ *represents a convex subset* C *of* F, F^{**} *or* $F^{*\blacksquare}$, *then* $\int_B af\,d\theta$ *represents* αC.

Proof. We recollect that $af \leqslant ag$ holds whenever $f \leqslant g$, and $af \leqslant ag + |\alpha|\mathfrak{v}$ holds whenever $f \leqslant g + \mathfrak{v}$ for $f, g \in \mathcal{F}(X, \mathrm{Conv}(E))$, an element α of the common scalar field of E and F and a function space neighborhood $\mathfrak{v}$. Moreover, the function af is measurable whenever f is measurable (Proposition 2.2.12(a)). Because θ is supposed to be $\mathfrak{L}(E, F^{*\blacksquare})$-valued, we have

$$\theta_A(\alpha a)(\mu) = \theta_A(a)(\alpha\mu)$$

for all $A \in \mathfrak{R}$, all $a \in E$, all α in $\mathbb{R}$ or $\mathbb{C}$ and all $\mu \in F^*$. The latter yields immediately that our claim holds for integrals of step functions in $\mathcal{F}_{\mathcal{S}}(X, \mathrm{Conv}(E))$, and since they are pointwise suprema of integrals of step functions on F^*, also for the precursory expressions $\int_A^{\mathfrak{v}} f\,d\theta$ for measurable functions. Because they are pointwise limits of these expressions, this property then further transfers to integrals of measurable functions. We continue to use the same argument for a function $f \in \mathcal{F}(X, \mathrm{Conv}(E))$ that is integrable over a set $A \in \mathfrak{R}$. Indeed, if $(f_W)_{W \in \mathcal{W}}$ is a net of measurable functions in $\mathcal{F}(X, \mathrm{Conv}(E))$ approaching f as in our definition of integrability over A, then since $\alpha n_W = |\alpha| n_W$, the net $(\alpha f_W)_{W \in \mathcal{W}}$ approaches af in the prescribed way with the corresponding neighborhood functions $|\alpha| n_W$. Pointwise convergence of the integrals on F^* then yields our claim. The extension to integrals over a set $B \in \mathfrak{A}$ is argued in the same way. Finally, if $\int_B f\,d\theta$ represents a convex subset C of F, F^{**} or $F^{*\blacksquare}$, then

$$\Big(\int_B f\,d\theta\Big)(\mu) = \sup\{\mathfrak{Re}\,\mu(c) \mid c \in C\}$$

for all $\mu \in F^*$, and

$$\Big(\int_B af\,d\theta\Big)(\mu) = \Big(\int_B af\,d\theta\Big)(\alpha\mu) = \sup\{\mathfrak{Re}\,\mu(c) \mid c \in \alpha C\}$$

Hence $\int_B af\,d\theta$ represents αC. □

2.3.15 Theorem. *Let* $B \in \mathfrak{A}$. *Then* $\mathcal{F}_{(\theta,B)}(X, \mathrm{Conv}(E))$ *is a subcone of the cone* $\mathcal{F}(X, \mathrm{Conv}(E))$ *that contains all measurable functions, and for* $f, g \in \mathcal{F}_{(\theta,B)}(X, \mathrm{Conv}(E))$ *and* $0 \leqslant \alpha \in \mathbb{R}$ *we have*

(a) $\int_B (\alpha f)\,d\theta = \alpha \int_B f\,d\theta$ *for all* $\alpha \geqslant 0$.

(b) $\int_B (f + g)\,d\theta = \int_B f\,d\theta + \int_B g\,d\theta$.

(c) $\int_B f\,d\theta \leqslant \int_B g\,d\theta$ *whenever* $f \underset{a.e.B}{\leqslant} g$.

Proof. Part (a) follows immediately from Proposition 2.3.14 and the fact that the integral is a sublinear functional on F^*. Thus for $f \in \mathcal{F}_{(\theta,B)}(X, \mathrm{Conv}(E))$ and $\alpha \geqslant 0$, we calculate that

$$\Big(\int_B \alpha f \, d\theta\Big)(\mu) = \Big(\int_B f \, d\theta\Big)(\alpha\mu) = \Big(\alpha \int_B f \, d\theta\Big)(\mu)$$

holds for all $\mu \in F^*$. Hence our claim in (a). We shall argue for the remaining parts in steps. First, we shall verify our statements for integrals over sets in $\mathfrak{R}$. For this, let $A \in \mathfrak{R}$ and let $f, g \in \mathcal{F}_{(\theta,A)}(X, \mathrm{Conv}(E))$. Let $(f_W)_{W\in\mathcal{W}}, (n_W)_{W\in\mathcal{W}}$ and $(g_W)_{W\in\mathcal{W}}, (m_W)_{W\in\mathcal{W}}$ be nets of measurable functions as in the definition of the integral for the functions f and g over A, respectively. For Part (b), we set

$$h_W = f_W + g_W \quad \text{and} \quad s_W = n_W + m_W$$

and observe that

$$(f+g) \underset{a.e.A}{\leqslant} h_W \underset{a.e.A}{\leqslant} (f+g) + s_W$$

and

$$\int_A s_W \, d\theta = \int_A n_W \, d\theta + \int_A m_W \, d\theta \leqslant 2\sigma_W$$

holds for all $W \in \mathcal{W}$ by Theorem 2.3.8(b). The measurable functions h_W and s_W therefore satisfy the definition of integrability over A with $2W$ in place of W for the function $f+g$. Now the corresponding statement for measurable functions in 2.3.8(b) and the usual limit rules yield

$$\begin{aligned}
\int_A (f+g) \, d\theta &= \lim_{W\in\mathcal{W}} \int_A h_W \, d\theta \\
&= \lim_{W\in\mathcal{W}} \int_A f_W \, d\theta + \lim_{W\in\mathcal{W}} \int_A g_W \, d\theta \\
&= \int_A f \, d\theta + \int_A g \, d\theta.
\end{aligned}$$

For Part (c), suppose that $f \underset{a.e.A}{\leqslant} g$. Then

$$f_W \underset{a.e.A}{\leqslant} f + n_W \underset{a.e.A}{\leqslant} g + n_W \underset{a.e.A}{\leqslant} g_W + m_W + n_W$$

holds for all $W \in \mathcal{W}$. Hence using Theorem 2.3.8(a), (b) and (c) for measurable functions, we infer that

$$\int_A f_W\, d\theta \leqslant \int_A g_W\, d\theta + 2\sigma_W$$

for all $W \in \mathcal{W}$. According to the limit rules, this yields

$$\int_A f\, d\theta = \lim_{W \in \mathcal{W}} \int_A f_W\, d\theta \leqslant \lim_{W \in \mathcal{W}} \int_A g_W\, d\theta = \int_A g\, d\theta.$$

For the general case, let $B \in \mathfrak{A}$ and $f, g \in \mathcal{F}_{(\theta,B)}(X, \mathrm{Conv}(E))$. Then the claims in parts (b) and (c) hold for integrals over all sets $A \cap B$ for $A \in \mathfrak{R}$. The definition of the respective integrals over B together with the validity of our claims for integrals over sets in $\mathfrak{R}$ and the usual limit rules yield their validity for the integrals over $B \in \mathfrak{A}$ as well. Parts (a) and (b) imply that $\mathcal{F}_{(\theta,B)}(X, \mathrm{Conv}(E))$ forms a subcone of $\mathcal{F}(X, \mathrm{Conv}(E))$. □

For the following corollaries, recall the definition of the order $\preccurlyeq_V$ in $\mathrm{Conv}(E)$ from Section 2.2. That is to say, for a neighborhood $V \in \mathcal{V}$ and $C, D \in \mathrm{Conv}(E)$ we denote $C \preccurlyeq_V D$ if $C \leqslant D + \varepsilon V$ for all $\varepsilon > 0$.

2.3.16 Corollary. *Let $f, g \in \mathcal{F}_{(\theta,B)}(X, \mathrm{Conv}(E))$ for $B \in \mathfrak{A}$ and let $V \in \mathcal{V}$. If $f(x) \preccurlyeq_V g(x)$ θ-almost everywhere on B, then $(\int_B f\, d\theta)(\mu) \leqslant (\int_B g\, d\theta)(\mu)$ for all $\mu \in F^*$ such that $(\int_B \chi_X \cdot V\, d\theta)(\mu) < +\infty$.*

Proof. Let $f, g \in \mathcal{F}_{(\theta,B)}(X, \mathrm{Conv}(E))$ for $B \in \mathfrak{A}$ and let $V \in \mathcal{V}$. If $f(x) \preccurlyeq_V g(x)$ holds θ-almost everywhere on B, then $f \underset{a.e.\,B}{\leqslant} g + \varepsilon \chi_X \cdot V$ for all $\varepsilon > 0$. Thus

$$\int_B f\, d\theta \leqslant \int_B g\, d\theta + \varepsilon \int_B \chi_X \cdot V\, d\theta$$

by Proposition 2.3.15(c). Hence our claim. □

We note that $\int_B \chi_X \cdot V\, d\theta = \sup_{A \in \mathfrak{R}} |\theta|_{A \cap B}(V)$ for $V \in \mathcal{V}$ and $B \in \mathfrak{A}$. Other than for measurable functions $f, g \in \mathcal{F}(X, \mathrm{Conv}(E))$ and a neighborhood $V \in \mathcal{V}$ (see Corollary 2.2.16), for integrable functions it can no longer be guaranteed that the set $\{x \in X \mid f(x) \preccurlyeq_V g(x)\}$ is contained in $\mathfrak{A}$. However, we have the following.

2.3.17 Corollary. *Let $f, g \in \mathcal{F}_{(\theta,A)}(X, \mathrm{Conv}(E))$ for $A \in \mathfrak{R}$ and let $W \in \mathcal{W}$ and $V \in \mathcal{V}$ such that $|\theta|_A(V) \leqslant \sigma_W$. There is a subset $B \in \mathfrak{R}$ of A that contains all points $x \in A$ such that $f(x) \preccurlyeq_V g(x)$, and $(\int_C f\, d\theta)(\mu) \leqslant (\int_C g\, d\theta)(\mu)$ holds for all $\mu \in F^*$ such that $\sigma_W(\mu) < +\infty$ and every subset $C \in \mathfrak{R}$ of A.*

Proof. Let $f, g \in \mathcal{F}_{(\theta,A)}(X, \mathrm{Conv}(E))$ for $A \in \mathfrak{R}$ and let $W \in \mathcal{W}$ and $V \in \mathcal{V}$ such that $|\theta|_A(V) \leqslant \sigma_W$. For every $n \in \mathbb{N}$, there are measurable functions $f_n, g_n \in \mathcal{F}(X, \mathrm{Conv}(E))$, measurable $\mathcal{V}_c$-valued functions n_n and m_n such that

$$f \underset{a.e.\,A}{\leqslant} f_n \underset{a.e.\,A}{\leqslant} f + n_n \quad \text{and} \quad g \underset{a.e.\,A}{\leqslant} g_n \underset{a.e.\,A}{\leqslant} g + m_n,$$

and both $\int_A n_n \, d\theta \leqslant (1/n)\sigma_W$ and $\int_A m_n \, d\theta \leqslant (1/n)\sigma_W$. We may assume that the above order relations hold everywhere on A instead of θ-almost everywhere. Indeed, the relations for each of the functions f_n, g_n, n_n and m_n hold for all $x \in A \setminus Z_n$ with some set $Z_n \in \mathfrak{R}$ of measure zero. According to Proposition 2.1.10(c), the set $Z = \bigcup_{n\in\mathbb{N}} Z_n \in \mathfrak{R}$ is again of measure zero, and the products of all of the functions involved with $\chi_{X\setminus Z}$ are again integrable and satisfy the above order relations everywhere on A. Their respective integrals coincide with those of the given functions. Let $x \in A$. If $f(x) \preccurlyeq_V g(x)$, that is, $f(x) \leqslant g(x) + \varepsilon V$ holds for all $\varepsilon > 0$, then for every choice of $n, m \in \mathbb{N}$, we compute

$$\begin{aligned} f_n(x) &\leqslant f(x) + n_n(x) \\ &\leqslant g(x) + (1/m)V + n_n(x) \\ &\leqslant g_n(x) + n_n(x) + (2/m)V. \end{aligned}$$

According to Corollary 2.2.16, all the sets

$$B_n^m = \{x \in A \mid f_n(x) \preccurlyeq_V g_n(x) + n_n(x) + (2/m)V\}$$

are contained in $\mathfrak{R}$ and so is their intersection $B = \bigcap_{n,m\in\mathbb{N}} B_n^m$. Moreover, B contains all points $x \in B$ such that $f(x) \preccurlyeq_V g(x)$, and following Corollary 2.3.16 for every subset $C \in \mathfrak{R}$ of B, we have

$$\begin{aligned} \int_C f_n \, d\theta &\leqslant \int_C g_n \, d\theta + \int_C n_n \, d\theta + (2/m)\sigma_W \\ &\leqslant \int_C g_n \, d\theta + (1/n + 2/m)\sigma_W. \end{aligned}$$

Therefore,

$$\begin{aligned} \int_C f \, d\theta &\leqslant \int_C g_n \, d\theta + (1/n + 2/m)\sigma_W \\ &\leqslant \int_C (g + m_n) \, d\theta + (1/n + 2/m)\sigma_W \\ &\leqslant \int_C g \, d\theta + (2/n + 2/m)\sigma_W \end{aligned}$$

holds for all $n, m \in \mathbb{N}$. Let $\mu \in F^*$ such that $\sigma_W(\mu) < +\infty$. Evaluating the above inequality at μ and taking the limit over n and m now returns our claim. □

According to our definition, integrability for a function in $\mathcal{F}(X, \mathrm{Conv}(E))$ over a set in $B \in \mathfrak{A}$ with respect to a measure θ requires integrability over all subsets $A \in \mathfrak{R}$ of B. However, the following simple example shows that this claim does not generally extend to all subsets $C \in \mathfrak{A}$ of B. For this, let $X = [0, +\infty)$, let $\mathfrak{R}$ be the collection of all relatively compact Borel subsets of X and let $E = F = \mathbb{R}$. Let θ be the Lebesgue measure on X. Consider the function f on X such that $f(x) = (-1)^{n+1}/n$ for $x \in [n-1, n)$ and $n \in \mathbb{N}$. Then f is integrable over X in the sense of our definition, and $\int_X f\,d\theta = \sum_{n=1}^{\infty}(-1)^{n+1}/n = \ln 2$. It is however not integrable over the set $C = \bigcup_{n\in\mathbb{N}}[2n-1, 2n) \in \mathfrak{A}$, since $\lim_{A\in\mathfrak{R}} \int_{A\cap C} f\,d\theta = -\infty$.

2.3.18 Theorem. *Let $f, g \in \mathcal{F}(X, \mathrm{Conv}(E))$ and $B, C \in \mathfrak{A}$.*

(a) *The function f is integrable over $B \cap C$ if and only if $\chi_C \cdot f$ is integrable over B, if and only if $\chi_B \cdot f$ is integrable over C. In this case, we have* $\int_{B\cap C} f\,d\theta = \int_B \chi_C \cdot f\,d\theta = \int_C \chi_B \cdot f\,d\theta.$

(b) *If B and C are disjoint and f is integrable over B and C, then f is integrable over $B \cup C$ and $\int_{B\cup C} f\,d\theta = \int_B f\,d\theta + \int_C f\,d\theta.$*

(c) *If $B \subset C$, if f and g are integrable over B and C and if $f \underset{a.e.\,C}{\leqslant} g$, then* $\int_C f\,d\theta + \int_B g\,d\theta \leqslant \int_B f\,d\theta + \int_C g\,d\theta.$

Proof. Let $f \in \mathcal{F}(X, \mathrm{Conv}(E))$. For Part (a), in a first step we shall verify that f is integrable over a set $A \in \mathfrak{R}$ if and only $\chi_A \cdot f$ is integrable over X and that $\int_A f\,d\theta = \int_X \chi_A \cdot f\,d\theta$ holds in this case. Indeed, if $\chi_A \cdot f$ is integrable over X, our definition of integrability over in $X \in \mathfrak{A}$ yields that $\chi_A \cdot f$ is integrable over all sets in $\mathfrak{R}$. But it is immediate from our definition of integrability over a set in $\mathfrak{R}$ that $\chi_A \cdot f$ is integrable over A if and only if f is integrable over A and that $\int_A f\,d\theta = \int_A \chi_A \cdot f\,d\theta$. Conversely, assume that f is integrable over A. We shall verify that $\chi_A \cdot f$ is integrable over X. For this, let $D \in \mathfrak{R}$. Given $W \in \mathcal{W}$, there are measurable functions f_W and n_W as in our definition of integrability over A, that is, that $f \underset{a.e.\,A}{\leqslant} f_W \underset{a.e.\,A}{\leqslant} f + n_W$ and $\int_A n_W\,d\theta \leqslant \sigma_W$. Then we have

$$\chi_A \cdot f \underset{a.e.\,D}{\leqslant} \chi_A \cdot f_W \underset{a.e.\,D}{\leqslant} \chi_A \cdot f + \chi_A \cdot n_W$$

and $\int_D \chi_A \cdot n_W\,d\theta \leqslant \sigma_W$ as well. Because the functions $\chi_A \cdot f_W$ and $\chi_A \cdot n_W$ are also measurable (Proposition 2.2.12(b)), we conclude that the function $\chi_A \cdot f$ is integrable over D and using Theorem 2.3.8(d) for measurable functions, that

$$\int_D \chi_A \cdot f\,d\theta = \lim_{W\in\mathcal{W}} \int_D \chi_A \cdot f_W\,d\theta = \lim_{W\in\mathcal{W}} \int_X \chi_{D\cap A} \cdot f_W\,d\theta.$$

The above holds for all sets $D \in \mathfrak{R}$, hence using our definition for the integrability over a set in $\mathfrak{A}$, we realize that the function $\chi_A \cdot f$ is indeed integrable over X and that

$$\begin{aligned}\int_X \chi_A \cdot f\,d\theta &= \lim_{D\in\mathfrak{R}} \int_D \chi_A \cdot f\,d\theta \\ &= \lim_{D\in\mathfrak{R}} \lim_{W\in\mathcal{W}} \int_X \chi_{D\cap A} \cdot f_W\,d\theta\end{aligned}$$

$$= \lim_{W\in\mathcal{W}} \int_X \chi_A \cdot f_W \, d\theta$$
$$= \lim_{W\in\mathcal{W}} \int_A f_W \, d\theta$$
$$= \int_A f \, d\theta.$$

This demonstrates our claim in the first step. Now in a second step we shall verify the statement of our first step for the sets in $\mathfrak{A}$. For this, let $B \in \mathfrak{A}$. By our first step, the function f is integrable over all sets $A \cap B$ for $A \in \mathfrak{R}$, if and only if all the functions $\chi_{(A\cap B)} \cdot f = \chi_A \cdot (\chi_B \cdot f)$ are integrable over X. In this case,

$$\int_{A\cap B} f \, d\theta = \int_X \chi_{A\cap B} \cdot f \, d\theta = \int_X \chi_A \cdot (\chi_B \cdot f) \, d\theta = \int_A \chi_B \cdot f \, d\theta$$

holds by our first step. According to our definition of integrability over sets in $\mathfrak{A}$, therefore f is integrable over B if and only $\chi_B \cdot f$ is integrable over X, and in this case we have

$$\int_B f \, d\theta = \lim_{A\in\mathfrak{R}} \int_{A\cap B} f \, d\theta = \lim_{A\in\mathfrak{R}} \int_A \chi_B \cdot f \, d\theta = \int_X \chi_B \cdot f \, d\theta.$$

In a third and final step for Part (a), let $B, C \in \mathfrak{A}$. From the preceding, we conclude that $\chi_C \cdot f$ is integrable over B if and only if $\chi_B \cdot (\chi_C \cdot f) = \chi_{(B\cap C)} \cdot f$ is integrable over X, that is, f is integrable over $B \cap C$, and all the integrals coincide.

For Part (b), suppose that $B, C \in \mathfrak{A}$, that $B \cap C = \varnothing$ and that f is integrable over both B and C. Then both functions $\chi_B \cdot f$ and $\chi_C \cdot f$ are integrable over X by Part (a), hence $\chi_{(B\cup C)} \cdot f = \chi_B \cdot f + \chi_C \cdot f$ is also integrable over X by Theorem 2.3.15(b). Thus f is indeed integrable over $B \cup C$ and

$$\int_{B\cup C} f \, d\theta = \int_X \chi_{B\cup C} \cdot f = \int_X \chi_B \cdot f + \int_X \chi_C \cdot f = \int_B f \, d\theta + \int_C f \, d\theta$$

by 2.3.15(b). Part (c) is obvious if both $B, C \in \mathfrak{R}$ such that $B \subset C$, since both f and g are integrable over $C \backslash B \in \mathfrak{R}$ in this case and Proposition 2.3.15(b) yields that $\int_{C\backslash B} f \, d\theta \leqslant \int_{C\backslash B} g \, d\theta$. Adding $\int_B f \, d\theta + \int_B g \, d\theta$ to both sides of this inequality renders our claim with Part (b). For the general case, that is, $B, C \in \mathfrak{A}$, let $A \in \mathfrak{R}$. Then both $A \cap B, A \cap C \in \mathfrak{R}$ and $A \cap B \subset A \cap C$. Both functions f and g are integrable over these sets, and we have

$$\int_{A\cap C} f \, d\theta + \int_{A\cap B} g \, d\theta \leqslant \int_{A\cap B} f \, d\theta + \int_{A\cap C} g \, d\theta$$

by the prior argument for sets in $\mathfrak{R}$. Taking the limit over all sets $A \in \mathfrak{R}$ renders our claim by the definition of the integral over B and C, respectively. □

We take note that Part (b) of Theorem 2.3.18 implies that a function f in $\mathcal{F}(X, \mathrm{Conv}(E))$ that is integrable over each of two not necessarily disjoint sets $B, C \in \mathfrak{R}$ is also integrable over their union. Indeed, since the sets $B \setminus C$, $C \setminus B$ and $B \cap C$ are in $\mathfrak{R}$ and subsets of B and C, respectively, f is integrable over each of these sets and, therefore, over their disjoint union $B \cup C$.

For the following observation, recall the definition of the restriction $\theta|_W$ of a measure θ to a set $B \in \mathfrak{A}$ from Section 2.1 together with the results from Proposition 2.1.9.

2.3.19 Proposition. *A function $f \in \mathcal{F}(X, \mathrm{Conv}(E))$ is integrable over $B \in \mathfrak{A}$ with respect to θ if and only if it is integrable over X with respect to $\theta|_B$. In this case, we have $\int_B f\, d\theta = \int_X f\, d\theta|_B$.*

Proof. Our claim is immediate for step functions and, therefore, also for measurable functions. Since the relations $\underset{a.e.B}{\leqslant}$ for θ and $\underset{a.e.X}{\leqslant}$ for $\theta|_B$ coincide, the respective definitions of integrability agree and the integrals are equal. □

2.3.20 Proposition. *Let $f \in \mathcal{F}_{(\theta,A)}(X, \mathrm{Conv}(E))$ for $A \in \mathfrak{R}$. For every $W \in \mathcal{W}$ and every $\varepsilon > 0$, there is an increasing sequence $(h_n)_{n\in\mathbb{N}}$ of step functions in $\mathcal{F}_{\mathcal{S}}(X, \mathrm{Conv}(E))$, $1 \leqslant \gamma \leqslant 1 + \varepsilon$ and a measurable neighborhood function $\mathfrak{n}$ such that:*
(i) *$h_n \underset{a.e.A}{\leqslant} f + \mathfrak{n}$ for all $n \in \mathbb{N}$ and $\int_A \mathfrak{n}\, d\theta \leqslant \sigma_W$.*
(ii) *θ-almost everywhere on A, for $x \in A$ there is $n \in \mathbb{N}$ such that $f(x) \leqslant \gamma h_n(x)$.*
(iii) *$\int_B f\, d\theta \leqslant \gamma \liminf_{n\to\infty} \int_B h_n\, d\theta$, pointwise on F^*, for all $B \in \mathfrak{R}$ such that $B \subset A$.*

Proof. Suppose that $f \in \mathcal{F}_{(\theta,A)}(X, \mathrm{Conv}(E))$ for $A \in \mathfrak{R}$. Let $W \in \mathcal{W}$ and $\varepsilon > 0$. Following our definition of integrability, there are a measurable function g and a measurable neighborhood function $\mathfrak{m}$ such that $f \underset{a.e.A}{\leqslant} g \underset{a.e.A}{\leqslant} f + \mathfrak{m}$ and $\int_A \mathfrak{m}\, d\theta \leqslant (1/2)\sigma_W$. There is $V \in \mathcal{V}$ such that $|\theta|_A(V) \leqslant (1/2)\sigma_W$. We use Proposition 2.2.13 for the measurable function g with the function space neighborhood $\mathfrak{v}$ generated by the single neighborhood function $\chi_X \cdot V$. There is $1 \leqslant \gamma \leqslant 1+\varepsilon$ and an increasing sequence $(h_n)_{n\in\mathbb{N}}$ of step functions in $\mathcal{F}_{\mathcal{S}}(X, \mathrm{Conv}(E))$ such that $h_n \leqslant g + \chi_X \cdot V$ for all $n \in \mathbb{N}$, and for every $x \in A$ there is $n \in \mathbb{N}$ such that $g(x) \leqslant \gamma h_n(x)$. We proceed to establish that the sequence $(h_n)_{n\in\mathbb{N}}$ of step functions and the measurable neighborhood function $\mathfrak{n} = \mathfrak{m} + \chi_X \cdot V$ fulfill our claim. Indeed, we have $\int_A \mathfrak{n}\, d\theta = \int_A \mathfrak{m}\, d\theta + |\theta|_A(V) \leqslant \sigma_W$ by Theorem 2.3.8(b), and for all $n \in \mathbb{N}$ we calculate that

$$h_n \leqslant g + \chi_X \cdot V \underset{a.e.A}{\leqslant} f + (\mathfrak{m} + \chi_X \cdot V) = f + \mathfrak{n}$$

holds, as required in (i). Moreover, θ-almost everywhere on A, for $x \in A$ there is $n \in \mathbb{N}$ such that $f(x) \leqslant g(x) \leqslant \gamma h_n(x)$ as stated in (ii). Finally, since g is measurable, Corollary 2.3.7 yields that

$$\int_B f\, d\theta \leqslant \int_B g\, d\theta \leqslant \gamma \liminf_{n\to\infty} \int_B h_n\, d\theta$$

holds for all $B \in \mathfrak{R}$ such that $B \subset A$. □

The following observation extends the result from Proposition 2.3.10(a) from measurable to integrable functions, but restricts the domain to sets in $\mathfrak{R}$.

2.3.21 Proposition. *Suppose that F is a normed space and let $A \in \mathfrak{R}$. For every $f \in \mathcal{F}_{(\theta,A)}(X, \mathrm{Conv}(E))$, the integral $\int_A f\, d\theta$ represents a convex subset of F^{**}.*

Proof. Let $f \in \mathcal{F}_{(\theta,A)}(X, \mathrm{Conv}(E))$ for $A \in \mathfrak{R}$. According to Lemma 2.3.12, the integral $\int_A f\, d\theta$ is the symmetric limit in $\mathcal{Q}_F$ of a net of integrals of measurable functions, which according to Proposition 2.3.10(a) are $\sigma(F^*, F^{**})$-lower semicontinuous elements of $\mathcal{Q}_F$. Hence $\int_A f\, d\theta$ itself is also $\sigma(F^*, F^{**})$-lower semicontinuous by Proposition 1.4.6(b). Thus in the light of Proposition 1.4.3(a), which states a correspondence between $\sigma(F^*, F^{**})$-lower semicontinuous sublinear functionals on F^* and convex subsets of E^{**}, the integral $\int_A f\, d\theta$ represents an element of $\mathrm{Conv}(F^{**})$. □

Integrals of vector-valued functions

For E-valued functions, the last proposition leads to a straightforward characterization of integrability.

2.3.22 Corollary. *A function $f \in \mathcal{F}(X, E)$ is integrable over a set $A \in \mathfrak{R}$ if and only if for every $W \in \mathcal{W}$ there are a step function $h \in \mathcal{F}_{\mathcal{S}}(X, E)$ and a measurable neighborhood function n such that $h - f \underset{a.e.A}{\leqslant} n$ and $\int_A n\, d\theta \leqslant \sigma_W$.*

Proof. Suppose that $f \in \mathcal{F}(X, E)$ is integrable over $A \in \mathfrak{R}$. Given $W \in \mathcal{W}$ and any $\varepsilon > 0$ according to Proposition 2.3.20(i), there is a step function $h \in \mathcal{F}_{\mathcal{S}}(X, \mathrm{Conv}(E))$ and the neighborhood function n such that $h \underset{a.e.A}{\leqslant} f + n$ and $\int_A n\, d\theta \leqslant \sigma_W$. That is, $h = \sum_{i=1}^{n} \chi_{A_i} \cdot C_i$ for disjoint sets $A_i \in \mathfrak{R}$ and $C_i \in \mathrm{Conv}(E)$. We choose elements $c_i \in C_i$ and set $g = \sum_{i=1}^{n} \chi_{A_i} \cdot c_i \in \mathcal{F}_{\mathcal{S}}(X, E)$. Then $g \leqslant h \underset{a.e.A}{\leqslant} f + n$, that is, $g - f \underset{a.e.A}{\leqslant} n$, our claim. Conversely, if for every $W \in \mathcal{W}$ there is $h \in \mathcal{F}_{\mathcal{S}}(X, E)$ and a measurable neighborhood function n such that $h - f \underset{a.e.A}{\leqslant} n$ and $\int_A n\, d\theta \leqslant \sigma_W$, the function $g = h + n \in \mathcal{F}(X, \mathrm{Conv}(E))$ is measurable and we have both $f \underset{a.e.A}{\leqslant} g$ and $g \underset{a.e.A}{\leqslant} f + 2n$. Hence the definition of integrability over A is satisfied for the function f. □

Corollary 2.3.22 implies that for a function $f \in \mathcal{F}_{(\theta,A)}(X, E)$ there exists a net $(h_W)_{W \in \mathcal{W}}$ in $\mathcal{F}_{\mathcal{S}}(X, E)$ such that $\int_A f\, d\theta = \lim_{W \in \mathcal{W}} \int_A h_W\, d\theta$ in the symmetric topology of $\mathcal{Q}_F$. We continue with some observations about integrals of E-valued functions. For the following, recall the definition of a weakly compact measure from Section 2.1.

2.3.23 Proposition. *Let $B \in \mathfrak{A}$.*

(a) *$\mathcal{F}_{(\theta,B)}(X, E)$ is a real- or complex-linear subspace of $\mathcal{F}(X, E)$. For all $f \in \mathcal{F}_{(\theta,B)}(X, E)$, the integral $\int_B f\, d\theta \in \mathcal{Q}_F$ is a linear functional on F^*, that is, an element of $F^{*\blacksquare}$. If F is a normed space, then $\int_B f\, d\theta \in F^{**}$.*

(b) *The mapping $f \mapsto \int_B f\, d\theta : \mathcal{F}_{(\theta,B)}(X, E) \to F^{*\blacksquare}$ is real or complex linear.*

(c) *If* θ *is* $\mathfrak{L}(E,F)$*-valued, and if for a function* $f \in \mathcal{F}_{(\theta,B)}(X,E)$*, the net* $(\int_{A\cap B} f\,d\theta)_{A\in\mathfrak{R}}$ *converges in the symmetric topology of* $\mathcal{Q}_F$ *to* $\int_B f\,d\theta$*, then* $\int_B f\,d\theta \in \widehat{F}$*, the completion of* F*.*

(d) *If* θ *is* $\mathfrak{L}(E,F)$*-valued and weakly compact and if* $f \in \mathcal{F}(X,E)$ *is measurable, then* $\int_B f\,d\theta \in F$*.*

Proof. Let $B \in \mathfrak{R}$ and let $f \in \mathcal{F}_{(\theta,B)}(X,E)$. Then $af \in \mathcal{F}_{(\theta,B)}(X,E)$ for all a in the scalar field of E by Proposition 2.3.14. Hence $\mathcal{F}_{(\theta,B)}(X,E)$, which by Proposition 2.3.15 is known to be a subcone, is indeed a subspace of $\mathcal{F}(X,E)$. We have $\int_B(-f)\,d\theta = -\int_B f\,d\theta \in \mathcal{Q}_F$ by 2.3.15(b). The latter implies that $\int_B f\,d\theta$ is a real-linear functional on F^*, hence an element of the embedding of $F^{*\blacksquare}$ into $\mathcal{Q}_F$. If F is a normed space, then $F^{*\blacksquare} = F^{**}$ (see Section 1.4). Hence our claim in (a). Our claim in Part (b) is obvious in the real case, since we already established that $\int_B(-f)\,d\theta = -\int_B f\,d\theta$. In the complex case, we recover the integral as an element $\omega_{I(f)}$ of $F^{*\blacksquare}$ from its embedding in $\mathcal{Q}_F$ as

$$\omega_{I(f)}(\mu) = \Big(\int_B f\,d\theta\Big)(\mu) - i\Big(\int_B f\,d\theta\Big)(i\mu)$$

for all $\mu \in F^*$ (see Section 1.4). Thus $\omega_{I(f)}$ is a complex-valued complex-linear functional on F^* and for $\alpha \in \mathbb{C}$, using Proposition 2.3.14, we calculate that

$$\omega_{I(af)}(\mu) = \Big(\int_B f\,d\theta\Big)(\alpha\mu) - i\Big(\int_B f\,d\theta\Big)(i\alpha\mu) = \omega_{I(f)}(\alpha\mu) = \alpha\omega_{I(f)}(\mu)$$

holds for all $\mu \in F^*$. Hence $\omega_{I(af)} = a\omega_{I(f)}$, our claim in (b). For Part (c), suppose that θ is $\mathfrak{L}(E,F)$-valued. In a first step, we consider a set $A \in \mathfrak{R}$ such that $f \in \mathcal{F}_{(\theta,A)}(X,E)$. Following Corollary 2.3.22, there exists a net $(h_W)_{W\in\mathcal{W}}$ of step functions in $\mathcal{F}_S(X,E)$ such that $\int_A f\,d\theta = \lim_{W\in\mathcal{W}} \int_A h_W\,d\theta$ in the symmetric topology of $\mathcal{Q}_F$. The integrals of the functions h_W are of course F-valued. The net $(\int_A h_W\,d\theta)_{W\in\mathcal{W}}$ is therefore a Cauchy net in F and its limit $\int_A f\,d\theta$ is contained inits completion $\widehat{F} \subset F^{*\blacksquare}$. Now in a second step, let $B \in \mathfrak{A}$ and suppose that the net $(\int_{A\cap B} f\,d\theta)_{A\in\mathfrak{R}}$ converges in the symmetric topology of $\mathcal{Q}_F$ to $\int_B f\,d\theta$. All integrals $\int_{A\cap B} f\,d\theta$ are elements of $\widehat{F}$ by our first step, and since $(\int_{A\cap B} f\,d\theta)_{A\in\mathfrak{R}}$ is a Cauchy net, its limit is also contained in $\widehat{F}$. For Part (d), suppose that θ is $\mathfrak{L}(E,F)$-valued and weakly compact and assume that the function $f \in \mathcal{F}(X,E)$ is measurable. Then f is supported by a set $D \in \mathfrak{R}$, and $f(X)$ is a bounded subset of E (Proposition 2.2.4). Thus

$$\int_B f\,d\theta = \int_{B\cap D} f\,d\theta$$

and Proposition 2.3.10(b) applies. The integral corresponds to a weakly compact subset of F, that is, a singleton subset by Part (a), that is, an element of F. □

If θ is $\mathfrak{L}(E,F)$-valued, Part (c) of Proposition 2.3.23 applies of course to all functions in $\mathcal{F}_{(\theta,A)}(X,E)$ when A is a set in $\mathfrak{R}$. It also applies to all measurable functions in $\mathcal{F}(X,E)$ and all sets $B \in \mathfrak{A}$ because they are integrable over all sets in $\mathfrak{A}$ and are supported by a set in $\mathfrak{R}$. The topological scenario, where X is locally compact and $\mathfrak{R}$ consists of all relatively compact Borel subsets of X allows for some additional observations.

2.3.24 Proposition. *Suppose that X is a locally compact Hausdorff space and that $\mathfrak{R}$ consists of all relatively compact Borel subsets of X. Then every function $f \in C(X,E)$ is integrable over every set $A \in \mathfrak{R}$ with respect to θ. If θ is regular, then*

$$\int_A f\,d\theta = \lim_{O \supset A} \int_O f\,d\theta = \lim_{K \subset A} \int_K f\,d\theta.$$

The limits are meant pointwise on F^* and are taken over all open sets $O \in \mathfrak{O}$ containing A and all compact subsets K of A, respectively. If θ is $\mathfrak{L}(E,F)$-valued and weakly compact, then the integral is an element of F.

Proof. Let $f \in C(X,E)$ and $A \in \mathfrak{R}$. Following Proposition 2.2.9, the function $\chi_A \cdot f$ is measurable, hence f is integrable over A. Now suppose that θ is regular. For every $\mu \in F^*$, there is $V \in \mathcal{V}$ such that $\lim_{K \subset A} |\theta|_{A\setminus K}(V)(\mu) = 0$. Since A is relatively compact in X and f is continuous, the set $f(A)$ is bounded in E. Hence there is $\lambda > 0$ such that $f(A) \subset \lambda V$. Given $\varepsilon > 0$, we find a compact subset K_0 of A such that $|\theta|_{A\setminus K_0}(V)(\mu) \leqslant \varepsilon/\lambda$. Thus for every compact set $K_0 \subset K \subset A$, this renders

$$\begin{aligned}\left|\left(\int_A f\,d\theta\right)(\mu) - \left(\int_K f\,d\theta\right)(\mu)\right| &= \left|\left(\int_{A\setminus K} f\,d\theta\right)(\mu)\right| \\ &\leqslant |\theta|_{A\setminus K}(\lambda V)(\mu) \\ &\leqslant \lambda|\theta|_{A\setminus K_0}(V)(\mu) \leqslant \varepsilon.\end{aligned}$$

The second part of the equality in our claim, which involves outer regularity and relatively compact open sets containing A is verified in a similar fashion. Since $\chi_A \cdot f$ is measurable, if θ is $\mathfrak{L}(E,F)$-valued and weakly compact, the last statement follows from Proposition 2.3.23(d). □

For the following, recall from Section 1.2 that a single neighborhood function n defines a function space neighborhood if and only if for every $A \in \mathfrak{R}$ there is $V \in \mathcal{V}$ such that $\chi_A \cdot V \leqslant n$.

2.3.25 Proposition. *Suppose that X is a locally compact Hausdorff space and that $\mathfrak{R}$ consists of all relatively compact Borel subsets of X. Let n be a measurable weakly lower continuous neighborhood function that defines a function space neighborhood. If θ is regular, then*

$$\int_B n\,d\theta = \sup\left\{\int_B f\,d\theta \,\middle|\, f \in \mathcal{C}_{\mathcal{K}}(X,E),\ \operatorname{supp}(f) \subset O,\ f \leqslant n\right\}$$

for every $B \in \mathfrak{A}$, whereby O is any open subset of X containing B.

Proof. Let $\mathfrak{v}$ be the function space neighborhood defined by $\mathcal{N}_{\mathfrak{v}} = \{n\}$. By Proposition 2.3.11, we have

$$\Big(\int_B n\,d\theta\Big)(\mu) = \sup_{A \in \mathfrak{R}} \|\theta\|_{A\cap B}^{(\mathfrak{v},\{\mu\})}$$

for every $\mu \in F^*$ and all $B \in \mathfrak{A}$. In detail, and using Proposition 2.1.14 this reads as

$$\Big(\int_B n\,d\theta\Big)(\mu) = \sup\left\{\sum_{i=1}^{n} \theta_{K_i}(a_i)(\mu) \,\middle|\, \begin{array}{l} K_i \text{ disjoint compact, } K_i \subset B, \\ a_i \in E,\ \chi_{K_i}\cdot a_i \leqslant n \end{array}\right\}.$$

Let O be an open subset of X such that $B \subset O$. The inequality

$$\sup\left\{\int_B f\,d\theta \,\middle|\, f \in \mathcal{C}_{\mathcal{K}}(X,E),\ \operatorname{supp}(f) \subset O,\ f \leqslant n\right\} \leqslant \int_B n\,d\theta$$

is obvious. For the converse, let $\mu \in F^*$ and $\varepsilon > 0$. For $i = 1,\dots,n$ let K_i be disjoint compact subsets of B and let $a_i \in E$ such that $\chi_{K_i}\cdot a_i \leqslant n$. There is $V \in \mathcal{V}$ such that $\lim_{O\supset K_i} |\theta|_{O\setminus K_i}(V)(\mu) = 0$, and there is $\lambda \geqslant 0$ such that $a_i \in \lambda V$ for all $i = 1,\dots,n$. Next, we find disjoint open sets $O_i \in \mathfrak{O}$ such that $K_i \subset O_i \subset O$, $|\theta|_{O_i\setminus K_i}(V)(\mu) \leqslant \varepsilon/(n\lambda)$ and $\chi_{O_i}\cdot a_i \leqslant (1+\varepsilon)n$. The last mentioned property uses Lemma 1.2.4(a). There are functions $\varphi_i \in \mathcal{C}_{\mathcal{K}}(X)$ such that $0 \leqslant \varphi_i \leqslant 1$, $\operatorname{supp}(\varphi_i) \subset O_i$ and $\varphi_i(x) = 1$ for all $x \in K_i$ (see Proposition 9.16 in [53]). We set

$$f = \sum_{i=1}^{n} \varphi_i \cdot a_i$$

and observe that $f \in \mathcal{C}_{\mathcal{K}}(X,E)$, $\operatorname{supp}(f) \subset O$ and $f \leqslant (1+\varepsilon)n$. The latter is easily established pointwise on X. We calculate that

$$\int_B f\,d\theta = \sum_{i=1}^{n} \int_{O_i} \varphi_i \cdot a_i\,d\theta = \sum_{i=1}^{n} \int_{K_i} \varphi_i \cdot a_i\,d\theta + \sum_{i=1}^{n} \int_{O_i\setminus K_i} \varphi_i \cdot a_i\,d\theta.$$

We have $\sum_{i=1}^{n} \int_{K_i} \varphi_i \cdot a_i\,d\theta = \sum_{i=1}^{n} \theta_{K_i}(a_i)$. Moreover, for each $i = 1,\dots,n$, we observe that $\gamma\varphi_i \cdot a_i \leqslant \lambda\chi_X \cdot V$ for all $|\gamma| \leqslant 1$ and, therefore,

$$\left|\Big(\int_{O_i\setminus K_i} \varphi_i \cdot a_i\,d\theta\Big)(\mu)\right| \leqslant \lambda\Big(\int_{O_i\setminus K_i} \chi_X \cdot V\,d\theta\Big)(\mu) = \lambda|\theta|_{O_i\setminus K_i}(V)(\mu) \leqslant \varepsilon/n.$$

Thus $|(\sum_{i=1}^{n} \int_{O_i\setminus K_i} \varphi_i \cdot a_i\,d\theta)(\mu)| \leqslant \varepsilon$, and consequently,

$$\left(\sum_{i=1}^{n} \theta_{K_i}(a_i)\right)(\mu) \leqslant \left(\int_B f\,d\theta\right)(\mu) + \varepsilon.$$

Furthermore, because

$$\left(\int_B f\,d\theta\right)(\mu) \leqslant \sup\left\{\left(\int_B f\,d\theta\right)(\mu) \;\middle|\; \begin{array}{l} f \in \mathcal{C}_{\mathcal{K}}(X,E), \\ \operatorname{supp}(f) \subset O,\ f \leqslant (1+\varepsilon)\mathsf{n} \end{array}\right\}$$

$$= (1+\varepsilon)\sup\left\{\left(\int_B f\,d\theta\right)(\mu) \;\middle|\; \begin{array}{l} f \in \mathcal{C}_{\mathcal{K}}(X,E), \\ \operatorname{supp}(f) \subset O,\ f \leqslant \mathsf{n} \end{array}\right\},$$

we have

$$\left(\sum_{i=1}^{n} \theta_{K_i}(a_i)\right)(\mu) \leqslant (1+\varepsilon)\sup\left\{\left(\int_B f\,d\theta\right)(\mu) \;\middle|\; \begin{array}{l} f \in \mathcal{C}_{\mathcal{K}}(X,E), \\ \operatorname{supp}(f) \subset O,\ f \leqslant \mathsf{n} \end{array}\right\} + \varepsilon.$$

This argument holds true for all such choices of disjoint compact sets $K_i \subset B$, all such $a_i \in E$ and also for all $\varepsilon > 0$ and, therefore, yields that

$$\left(\int_B \mathsf{n}\,d\theta\right)(\mu) = \sup_{A \in \mathfrak{R}} \|\theta\|_{A \cap B}^{(\mathfrak{v},\{\mu\})} \leqslant \sup\left\{\left(\int_B f\,d\theta\right)(\mu) \;\middle|\; \begin{array}{l} f \in \mathcal{C}_{\mathcal{K}}(X,E), \\ \operatorname{supp}(f) \subset O,\ f \leqslant \mathsf{n} \end{array}\right\}.$$

Our claim follows. □

2.3.26 Examples. (a) In case that $E = F$ is a Banach space over $\mathbb{R}$ or $\mathbb{C}$, real- or complex-valued measures are indeed $\mathfrak{L}(E,E)$-valued, and there is a variety of established concepts for the integration of vector-valued functions with respect to scalar-valued measures. The most prominent is the following (see Chapter II.2 in [21]). A function $f \in \mathcal{F}(X,E)$ is called *Bochner integrable* over a set $A \in \mathfrak{R}$ if it is strongly θ-measurable (see Example 2.2.22(a)) and if there exists a sequence $(h_n)_{n\in\mathbb{N}}$ of E-valued step functions such that $\lim_{n\to\infty}\int_A \|h_n - f\|\,d|\theta| = 0$. In this context, $|\theta|$ is the usual modulus of a real- or complex-valued measure. (See Example 2.1.8(f).) We shall verify that a Bochner-integrable function $f \in \mathcal{F}(X,E)$ is integrable in our sense. Indeed, for every $\varepsilon > 0$, there is a step function $h \in \mathcal{F}_{\mathcal{S}}(X,E)$ such that $\int_A \|h - f\|\,d|\theta| \leqslant \varepsilon$. The strong θ-measurability of the function $h - f \in \mathcal{F}(X,E)$ implies that there is a set $Z \in \mathfrak{A}$ of measure zero such that the real-valued function $\chi_{A\setminus Z}\cdot\|h - f\|$ is $\mathfrak{A}$-measurable (see Example 2.2.22), and according to Corollary 2.2.21 the neighborhood function $\mathsf{n} = (\chi_{A\setminus Z}\cdot\|h - f\|)\cdot\mathbb{B}$, where $\mathbb{B}$ denotes the unit ball of E, is measurable in the vein of our definition. We have $h - f \underset{a.e.A}{\leqslant} h + \mathsf{n}$, and since we shall argue below that $\int_A (2\mathsf{n})\,d\theta \leqslant 2\varepsilon\sigma_{\mathbb{B}}$, following Corollary 2.3.22 the function f is integrable over A. Conversely, if a function $f \in \mathcal{F}(X,E)$ is strongly θ-measurable and integrable over a set $A \in \mathfrak{R}$ in our sense, then for every $\varepsilon > 0$ according to Corollary 2.3.22 there is a step function $h \in \mathcal{F}_S(X, E)$ and a measurable neighborhood function n such that $h - f \underset{a.e.A}{\leqslant} n$ and $\int_A n\, d\theta \leqslant \varepsilon\sigma_{\mathbb{B}}$. Hence $\|h - f\| \cdot \mathbb{B} \leqslant n$ and $\int_A \|h - f\|\, d|\theta| \leqslant \varepsilon$. Hence f is Bochner integrable over A.

In order to complete this argument, let us consider a function $\varphi \cdot C \in \mathcal{F}(X, \mathrm{Conv}(E))$, where φ is an $\overline{\mathbb{R}}_+$-valued $\mathfrak{A}$-measurable function on X and C is a balanced set in $\mathrm{Conv}(E)$. According to 2.2.21 this function is measurable, hence integrable over every set in $\mathfrak{A}$, and using an approximation of φ by $\overline{\mathbb{R}}_+$-valued step functions, its integral over a set $B \in \mathfrak{A}$ as an element of $\mathcal{Q}_F$ can be evaluated as

$$\Big(\int_B \varphi \cdot C\, d\theta\Big)(\mu) = \lim_{A \in \mathfrak{R}} \Big(\int_{A \cap B} \varphi\, d|\theta|\Big) p_C(\mu),$$

for $\mu \in F^*$, where $|\theta|$ denotes the usual modulus of a real- or complex-valued measure and $p_C(\mu) = \sup\{|\mu(c)| \mid c \in C\} \in \overline{\mathbb{R}}$ (see Example 2.1.8(f)). If $C = \mathbb{B}$, that is, the unit ball of E, then $\varphi \cdot \mathbb{B}$ is a neighborhood function and $p_{\mathbb{B}}(\mu) = \|\mu\|$, the dual norm of the functional $\mu \in F^*$. This observation validates the preceding argument. Moreover, Proposition 2.3.23(c) guarantees that the values of the integrals of E-valued functions are indeed elements of E.

(b) In case that $E = F = \mathbb{R}$, integrals of real-valued functions are linear functionals on $\mathbb{R}^* = \mathbb{R}$, that is, real numbers via their evaluations at $+1$, whereas the modulus of a real-valued measure and integrals of $\mathrm{Conv}(\mathbb{R})$-valued functions are $\overline{\mathbb{R}}$-valued sublinear functionals on $\mathbb{R}$ and determined by their evaluations at the elements $+1$ and -1 of $\mathbb{R}^*$. A real-valued measure θ defined on the σ-field $\mathfrak{A}$ associated with $\mathfrak{R}$ is the difference of two positive measures θ_+ and θ_-, which are supported by disjoint sets (see Proposition 21 in Section 11.5 of [53]). For a set $C = [a, b] \in \mathrm{Conv}(\mathbb{R})$, where $a \leqslant b$ we have

$$|\theta|_A(C)(+1) = b\,(\theta_+)_A - a\,(\theta_-)_A$$

and

$$|\theta|_A(C)(-1) = b\,(\theta_-)_A - a\,(\theta_+)_A.$$

For nonnegative $\overline{\mathbb{R}}$-valued $\mathfrak{A}$-measurable functions φ and ψ on X, the set-valued function $f \in \mathcal{F}(X, \mathrm{Conv}(\mathbb{R}))$ such that $x \mapsto [-\varphi(x), \psi(x)]$ is measurable (Example 2.2.22(b)), hence integrable over all sets $B \in \mathfrak{R}$. We calculate that

$$\Big(\int_B f\, d\theta\Big)(+1) = \lim_{A \in \mathfrak{R}} \int_{A \cap B} \psi\, d\theta_+ + \lim_{A \in \mathfrak{R}} \int_{A \cap B} \varphi\, d\theta_-$$

and

$$\Big(\int_B f\,d\theta\Big)(-1) = \lim_{A\in\mathfrak{R}} \int_{A\cap B} \varphi\,d\theta_+ + \lim_{A\in\mathfrak{R}} \int_{A\cap B} \psi\,d\theta_-.$$

In other words, the integral $\int_B f\,d\theta$ represents the interval

$$\Big[\Big(\int_B f\,d\theta\Big)(+1),\ -\Big(\int_B f\,d\theta\Big)(-1)\Big]$$

in $\mathcal{F}(X, \mathrm{Conv}(\mathbb{R}))$. Because the function $-f$, that is, the mapping $x \mapsto [-\psi(x), \varphi(x)]$, is defined in the same way by the functions φ and ψ in reverse order, we verify that $(\int_B(-f)\,d\theta)(+1) = (\int_B f\,d\theta)(-1)$ and $(\int_B(-f)\,d\theta)(-1) = (\int_B f\,d\theta)(+1)$ holds, as stated in Proposition 2.3.14. We note that the integral of a set-valued function may take the value $+\infty$ on $\mathbb{R}^* = \mathbb{R}$.

Further into this example, let φ be an $\overline{\mathbb{R}}_+$-valued $\mathfrak{A}$-measurable function on X such that both

$$\int_A \varphi\,d\theta_+,\ \int_A \varphi\,d\theta_- < +\infty$$

for $A \in \mathfrak{R}$. A simple argument returns that for every $\varepsilon > 0$ there is $\rho \geqslant 0$ such that both

$$\int_A (\varphi\vee\rho-\varphi)\,d\theta_+ \leqslant \varepsilon \quad\text{and}\quad \int_A (\varphi\vee\rho-\varphi)\,d\theta_- \leqslant \varepsilon.$$

The real-valued functions $\chi_A\cdot(\varphi\vee\rho)$ and $\chi_A\cdot(\varphi\wedge\rho)$ are measurable in our sense by Proposition 2.2.7 and so by the above is the $\mathrm{Conv}(\mathbb{R})$-valued neighborhood function $n = (\varphi\vee\rho-\varphi)\cdot\mathbb{B}$, where $\mathbb{B} = [-1,+1]$ is the unit ball of $\mathbb{R}$. We set $g = \chi_A\cdot(\varphi\wedge\rho)+n$. By Proposition 2.2.11, the function $g \in \mathcal{F}(X, \mathrm{Conv}(\mathbb{R}))$ is also measurable, and because $\varphi = (\varphi\wedge\rho)+(\varphi\vee\rho-\varphi)$, we have

$$\varphi \underset{a.e.A}{\leqslant} g \underset{a.e.A}{\leqslant} \varphi + 2n.$$

Moreover, using the above we compute that both $(\int_A n\,d\theta)(+1) \leqslant 2\varepsilon$ and $(\int_A n\,d\theta)(-1) \leqslant 2\varepsilon$. This observation can be reformulated as $\int_A n\,d\theta \leqslant 2\varepsilon\sigma_{\mathbb{B}}$, where $\sigma_{\mathbb{B}}$ is the sublinear functional $\alpha \mapsto |\alpha|$ on $\mathbb{R}^* = \mathbb{R}$. The function φ is therefore integrable over every $A \in \mathfrak{R}$ in the understanding of our definition.

Integrals with respect to composition measures

In the last paragraph of Section 2.1, we considered the following situation. If $(G, \mathcal{U})$ and $(H, \mathcal{R})$ are additional locally convex topological vector spaces such that H is Hausdorff and if $S \in \mathfrak{L}(G, E)$ and $T \in \mathfrak{L}(F, H)$, then the formula

$$((T \circ \theta \circ S)_A(g))(v) = (\theta_A(S(g)))(T^*(v)),$$

where $A \in \mathfrak{R}$, $g \in G$ and $v \in H^*$, defines an $\mathfrak{L}(G, H^{*\blacksquare})$-valued measure. In Section 2.2, we established that the function $S \circ f \in \mathcal{F}(X, \mathrm{Conv}(E))$ is measurable provided that $f \in \mathcal{F}(X, \mathrm{Conv}(G))$ is measurable. We shall now investigate integrability. We note that the measures $\theta \circ S$ and $T \circ \theta$ are $\mathfrak{L}(G, F^{*\blacksquare})$-valued and $\mathfrak{L}(E, H^{*\blacksquare})$-valued, respectively. In preparation, we take note of the following.

2.3.27 Lemma. *Let G be a subspace of E and let $B \in \mathfrak{R}$. If $\theta_A(a) = 0$ for all subsets $A \in \mathfrak{R}$ of B and all $a \in G$, then $\int_B f\, d\theta = 0$ for all $f \in \mathcal{F}_{(\theta,B)}(X, \mathrm{Conv}(E))$ such that $f(x) \subset G$ for all $x \in A$.*

Proof. Let G, θ and $B \in \mathfrak{A}$ be as stated. Because $|\theta|_A(C) = 0$ whenever $A \subset B$ for $A \in \mathfrak{R}$ and $C \subset G$, our claim is obvious for step functions. Suppose that $f \in \mathcal{F}(X, \mathrm{Conv}(E))$ is measurable and $\mathrm{Conv}(G)$-valued on B. Given $W \in \mathcal{W}$, let $h = \sum_{i=1}^n \chi_{A_i} \cdot C_i$ be a step function in $\mathcal{F}_S(X, \mathrm{Conv}(E))$ such that $h \leqslant f + \mathfrak{v}_W$. Then $h \leqslant f + n$ with some neighborhood-valued step function $n \in \mathcal{N}_{\mathfrak{v}_W}$. Thus $h \leqslant \sum_{i=1}^n \chi_{A_i} \cdot G + n$ and, therefore, $\int_B h\, d\theta \leqslant \int_B n\, d\theta \leqslant \sigma_W$. Hence $\int_B f\, d\theta \leqslant \int_B^{(\mathfrak{v}_W)} f\, d\theta \leqslant \sigma_W$. This holds for all $W \in \mathcal{W}$. We infer that $\int_B f\, d\theta \leqslant 0$ and, therefore, $\int_B f\, d\theta = 0$ since $\int_B f\, d\theta$ is a sublinear functional on F^*. Now suppose that $f \in \mathcal{F}_{(\theta,A)}(X, \mathrm{Conv}(E))$ for $A \in \mathfrak{R}$ such that $A \subset B$ and that f is $\mathrm{Conv}(G)$-valued on A. For $W \in \mathcal{W}$, let f_W and n_W be measurable functions in $\mathcal{F}(X, \mathrm{Conv}(E))$ as in the definition of integrability for f. Then $f_W \underset{a.e.A}{\leqslant} f + n_W$ implies that $f_W \underset{a.e.A}{\leqslant} \chi_A \cdot G + n_W$ and $\int_A f_W\, d\theta \leqslant \int_A n_W\, d\theta \leqslant \sigma_W$. Hence

$$\int_A f\, d\theta = \lim_{W \in \mathcal{W}} \int_A f_W\, d\theta = 0.$$

Finally, if $f \in \mathcal{F}_{(\theta,B)}(X, \mathrm{Conv}(E))$ for $B \in \mathfrak{A}$ such that f is $\mathrm{Conv}(G)$-valued on B, we recognize that

$$\int_B f\, d\theta = \lim_{A \in \mathfrak{R}} \int_{A \cap B} f\, d\theta = 0,$$

our claim. □

2.3.28 Theorem. *Let $S \in \mathfrak{L}(G, E)$ and $T \in \mathfrak{L}(F, H)$. Let $f \in \mathcal{F}(X, \mathrm{Conv}(G))$ and $B \in \mathfrak{A}$. If both f and $S \circ f$ are integrable over B with respect to $\theta \circ S$ and θ, respectively, then f is integrable over B with respect to $T \circ \theta \circ S$ and*

$$\left(\int_B f\, d(T \circ \theta \circ S) \right)(v) = \left(\int_B (S \circ f)\, d\theta \right)(T^*(v))$$

holds for all $v \in H^*$.

Proof. Let $h = \sum_{i=1}^{n} \chi_{A_i} \cdot G_i$ be a $\mathrm{Conv}(G)$-valued step function. Using the evaluation of the modulus of the involved measures from Section 2.1, for every $B \in \mathfrak{A}$ and $v \in H^*$ we calculate

$$\Big(\int_B h\, d(T \circ \theta \circ S)\Big)(v) = \sum_{i=1}^{n} |T \circ \theta \circ S|_{A_i \cap B}(G_i)(v)$$
$$= \sum_{i=1}^{n} |\theta|_{A_i \cap B}(S(G_i))(T^*(v))$$
$$= \Big(\int_B (S \circ h)\, d\theta\Big)(T^*(v)).$$

Thus our statement holds for G-valued step functions.

Now let $f \in \mathcal{F}(X, \mathrm{Conv}(G))$ be measurable and let $A \in \mathfrak{R}$. Then the function $S \circ f \in \mathcal{F}(X, \mathrm{Conv}(E))$ was also understood to be measurable in the preceding section. Given $v \in H^*$ and $\varepsilon > 0$, there is $R \in \mathcal{R}$ such that $v \in \varepsilon R^*$. In turn, there is $W \in \mathcal{W}$ such that $T^*(R^\circ) \subset W^\circ$ and $V \in \mathcal{V}$ such that $|\theta|_A(V) \leqslant \sigma_W$. Then there is $U \in \mathcal{U}$ such that $S(U) \subset V$. Hence both

$$\int_A \chi_X \cdot U\, d(T \circ \theta \circ S) \leqslant \sigma_R \quad \text{and} \quad \int_A \chi_X \cdot V\, d\theta \leqslant \sigma_W$$

by the above. We choose the function space neighborhoods $\mathfrak{v}$ for $\mathcal{F}(X, E)$ and $\mathfrak{u}$ for $\mathcal{F}(X, G)$ defined by the single neighborhood functions $\chi_X \cdot V$ and $\chi_X \cdot U$, respectively. There is an increasing sequence $(h_n)_{n \in \mathbb{N}}$ of step functions in $\mathcal{F}_S(X, \mathrm{Conv}(G))$ for the function f as in Proposition 2.2.13. That is, $h_n \leqslant f + \mathfrak{u}$ for all $n \in \mathbb{N}$, and there is $1 \leqslant \gamma \leqslant 1 + \varepsilon$ such that for every $x \in A$ there is $n \in \mathbb{N}$ such that $f(x) \leqslant \gamma h_n(x.)$ The sequence $(S \circ h_n)_{n \in \mathbb{N}}$ in $\mathcal{F}_S(X, \mathrm{Conv}(E))$ has the same properties for the function $S \circ f \in \mathrm{Conv}(E)$ with $\mathfrak{v}$ in place of $\mathfrak{u}$. According to Corollary 2.3.7, this implies both

$$\int_A f\, d(T \circ \theta \circ S) \leqslant \gamma \liminf_{n \to \infty} \int_A h_n\, d(T \circ \theta \circ S)$$

and

$$\int_A (S \circ f)\, d\theta \leqslant \gamma \liminf_{n \to \infty} \int_A (S \circ h_n)\, d\theta,$$

pointwise on F^* and G^*, respectively. Recall that the measures $T \circ \theta \circ S$ and θ are $\mathfrak{L}(G, H^{*\blacksquare})$-valued and $\mathfrak{L}(E, F^{*\blacksquare})$-valued, respectively. On the other hand, we have both

$$\int_A h_n\, d(T \circ \theta \circ S) \leqslant \int_A f\, d(T \circ \theta \circ S) + \sigma_R$$

and

$$\int_A (S \circ h_n)\, d\theta \leqslant \int_A (S \circ f)\, d\theta + \sigma_W$$

for all $n \in \mathbb{N}$. Because

$$\Big(\int_A h_n\, d(T \circ \theta \circ S)\Big)(v) = \Big(\int_A (S \circ h_n)\, d\theta\Big)(T^*(v))$$

by the above, and since $v \in \varepsilon R^\circ$ and $T^*(v) \in \varepsilon W^\circ$, we conclude that

$$\begin{aligned}\Big(\int_A f\, d(T \circ \theta \circ S)\Big)(v) &\leqslant \gamma \liminf_{n\to\infty}\Big(\int_A (S \circ h_n)\, d\theta\Big)(T^*(v)) \\ &\leqslant \gamma\Big(\int_A (S \circ f)\, d\theta\Big)(T^*(v)) + \varepsilon\end{aligned}$$

and, likewise,

$$\begin{aligned}\Big(\int_A (S \circ f)\, d\theta\Big)(T^*(v)) &\leqslant \gamma \liminf_{n\to\infty}\Big(\int_A h_n\, d(S \circ \theta \circ S)\Big)(v) \\ &\leqslant \gamma\Big(\int_A f\, d(T \circ \theta \circ S)\Big)(v) + \varepsilon.\end{aligned}$$

These inequalities hold true for all $\varepsilon > 0$ and, therefore, validate that

$$\Big(\int_A f\, d(T \circ \theta \circ S)\Big)(v) = \Big(\int_B (S \circ f)\, d\theta\Big)(T^*(v))$$

holds for every measurable function $f \in \mathcal{F}(X, \mathrm{Conv}(G))$ and all $A \in \mathfrak{R}$. In case that $H = F$ and T is the identity operator, this reads $\int_A f\, d(\theta \circ S) = \int_B (S \circ f)\, d\theta$.

Next, we shall argue that every function $f \in \mathcal{F}(X, \mathrm{Conv}(G))$ that is integrable with respect to $\theta \circ S$ over a set $A \in \mathfrak{R}$ is also integrable over A with respect to $T \circ \theta \circ S$. Recall that the measures $\theta \circ S$ and $T \circ \theta \circ S$ are $\mathfrak{L}(G, F^{*\blacksquare})$-valued and $\mathfrak{L}(G, H^{*\blacksquare})$-valued, respectively. Given $R \in \mathcal{R}$, there is $W \in \mathcal{W}$ such that $T(W) \subset R$, and by the integrability of f with respect to $\theta \circ S$ there are measurable functions f_W and n_W in $\mathcal{F}(X, \mathrm{Conv}(G))$ such that n_W is $\mathcal{U}_c$-valued and

$$f \underset{a.e.A}{\leqslant} f_W \underset{a.e.A}{\leqslant} f + n_W \quad \text{and} \quad \int_A n_W\, d(\theta \circ S) = \int_A (S \circ n_W)\, d\theta \leqslant \sigma_W.$$

Any set $Z \in \mathfrak{A}$ of measure zero with respect to $\theta \circ S$ is also of measure zero with respect to $T \circ \theta \circ S$, and we have

$$\Big(\int_A n_W\, d(T \circ \theta \circ S)\Big)(v) = \Big(\int_B (S \circ n_W)\, d\theta\Big)(T^*(v)) \leqslant 1$$

for all $v \in R^{\circ}$ by our prior argument for measurable functions. Hence

$$\int_A n_W \, d(T \circ \theta \circ S) \leqslant \sigma_R.$$

As a consequence, the same measurable functions $f_R = f_W$ and $n_R = n_W$ from the definition of integrability of f with respect to $\theta \circ S$ can also be used to demonstrate its integrability with respect to $T \circ \theta \circ S$.

Next, we assume that $f \in \mathcal{F}(X, \mathrm{Conv}(G))$ is integrable over $A \in \mathfrak{R}$ with respect to $\theta \circ S$, hence with respect to $T \circ \theta \circ S$ by the above, and that $S \circ f$ is integrable over A with respect to θ. Given $W \in \mathcal{W}$, we choose the functions f_W and n_W as in the definition of integrability with respect to $\theta \circ S$. That is,

$$f \underset{a.e.A}{\leqslant} f_W \underset{a.e.A}{\leqslant} f + n_W \quad \text{and} \quad S \circ f \underset{a.e.A}{\leqslant} S \circ f_W \underset{a.e.A}{\leqslant} S \circ f + S \circ n_W.$$

We find a subset $Z \in \mathfrak{R}$ of A of measure zero with respect to $\theta \circ S$ such that the stated inequalities hold everywhere on $A \setminus Z$. By our assumption, the function $S \circ f$ is integrable over A with respect to θ and so are the measurable functions $S \circ f_W$ and $S \circ n_W$. Then the preceding shows

$$\int_{A \setminus Z} (S \circ f) \, d\theta \leqslant \int_{A \setminus Z} (S \circ f_W) \, d\theta \leqslant \int_{A \setminus Z} (S \circ f) \, d\theta + \int_{A \setminus Z} (S \circ n_W) \, d\theta.$$

The values of all functions involved in these integrals are contained in the subspace $S(G)$ of E, and Z being of measure zero with respect to $\theta \circ S$ implies and $\theta_C(a) = 0$ for all $C \in \mathfrak{R}$ such that $C \subset Z$ and all $a \in G$. Thus Lemma 2.3.27 applies and we infer that the integrals over Z with respect to θ of all the functions involved equals zero. We therefore conclude that

$$\int_A (S \circ f) \, d\theta \leqslant \int_A (S \circ f_W) \, d\theta \leqslant \int_A (S \circ f) \, d\theta + \int_A (S \circ n_W) \, d\theta.$$

Because $\int_A (S \circ n_W) \, d\theta = \int_A n_W \, d(S \circ \theta) \leqslant \sigma_W$ and because for every $R \in \mathcal{R}$ there is $W \in \mathcal{W}$ such that $T^*(R) \subset W$, for every $v \in H^*$ passing to the limit over $W \in \mathcal{W}$ in this inequality leads to

$$\begin{aligned}
\left(\int_A (S \circ f) \, d\theta \right)(T^*(v)) &= \lim_{W \in \mathcal{W}} \left(\int_A (S \circ f_W) \, d\theta \right)(T^*(v)) \\
&= \lim_{W \in \mathcal{W}} \left(\int_A f_W \, d(T \circ \theta \circ S) \right)(v) \\
&= \left(\int_A f \, d(T \circ \theta \circ S) \right)(v).
\end{aligned}$$

Finally, if $f \in \mathcal{F}(X, \mathrm{Conv}(G))$ is integrable over a set $B \in \mathfrak{A}$ with respect to $\theta \circ S$ and $S \circ f$ is integrable over B with respect to θ. Then the limit

$$\int_B (S \circ f)\, d\theta = \lim_{A \in \mathfrak{R}} \int_{A \cap B} (S \circ f)\, d\theta$$

exists pointwise on F^*. Hence for every $\nu \in H^*$ the limit

$$\lim_{A \in \mathfrak{R}} \Big(\int_{A \cap B} (S \circ f)\, d\theta \Big)(T^*(\nu)) = \lim_{A \in \mathfrak{R}} \Big(\int_{A \cap B} f\, d(T \circ \theta \circ S) \Big)(\nu)$$

also exists and equals

$$\Big(\int_B f\, d(T \circ \theta \circ S) \Big)(\nu) = \Big(\int_B (S \circ f)\, d\theta \Big)(T^*(\nu)).$$

This observation completes our argument. □

We shall consider a few special cases of interest. First, suppose that $G = E$ with the identity operator for S. Theorem 2.3.28 states that if a function $f \in \mathcal{F}(X, \mathrm{Conv}(E))$ is integrable over $B \in \mathfrak{A}$ with respect to θ, then it is integrable over B with respect to $T \circ S$ and we have

$$\Big(\int_B f\, d(T \circ \theta) \Big)(\nu) = \Big(\int_B f\, d\theta \Big)(T^*(\nu))$$

for all $\nu \in H^*$. Now suppose in addition that G equals either $\mathbb{R}$ or $\mathbb{C}$ and that T is a linear functional $\mu \in F^*$. The adjoint operator T^* then maps the scalars $\alpha \in G^*$ into $\alpha\mu \in F^*$, and Theorem 2.3.28 returns that

$$\Big(\int_B f\, d(\mu \circ \theta) \Big)(\alpha) = \Big(\int_B f\, d\theta \Big)(\alpha\mu)$$

holds for all α in $\mathbb{R}$ or $\mathbb{C}$ in this case. We shall formulate this observation for $\alpha = 1$ as a corollary.

2.3.29 Corollary. *If $f \in \mathcal{F}_{(\theta,B)}(X, \mathrm{Conv}(E))$ for $B \in \mathfrak{A}$, then $f \in \mathcal{F}_{(\mu\circ\theta,B)}(X, \mathrm{Conv}(E))$ for every $\mu \in F^*$ and*

$$\Big(\int_B f\, d\theta \Big)(\mu) = \Big(\int_B f\, d(\mu \circ \theta) \Big)(+1).$$

Now suppose that $H = F$ with the identity operator for T. Theorem 2.3.28 states that if both functions $f \in \mathcal{F}(X, \mathrm{Conv}(G))$ and $S \circ f \in \mathcal{F}(X, \mathrm{Conv}(E))$ are integrable over B with respect to $\theta \circ S$ and θ, respectively, then

$$\int_B f\,d(\theta \circ S) = \int_B (S \circ f)\,d\theta$$

In particular, if G is a subspace of E and S is the imbedding operator, then $S \circ f = f$ for a $\mathrm{Conv}(G)$-valued function and $\theta \circ S$ arises from the restriction of the values of θ to the subspace G of E. Hence if f is integrable both with respect to θ and to $\theta \circ S$, then the values of these integrals coincide.

Let us consider the case that $E = F$ and that θ is real- or complex-valued. (We shall formulate the complex-valued case.) Given a linear functional $\mu \in E^*$, we choose E in place of G and $\mathbb{C}$ for E, F and H in Theorem 2.3.28. The functional μ stands for the operator S and T is the identity on $\mathbb{C}$. The measure θ is $\mathfrak{L}(\mathbb{C}, \mathbb{C})$-valued. For $f \in \mathcal{F}(X, \mathrm{Conv}(E))$, then $\mu \circ f$ is the $\mathrm{Conv}(\mathbb{C})$-valued function $x \mapsto \mu(f(x))$. The measure $\theta \circ \mu$ is $\mathfrak{L}(E, \mathbb{C})$-valued, that is, E^*-valued. Theorem 2.3.28 states that

$$\int_B f\,d(\theta \circ \mu) = \int_B (\mu \circ f)\,d\theta$$

holds for all functions $f \in \mathcal{F}(X, \mathrm{Conv}(E))$ such that f is integrable with respect to $\theta \circ \mu$ and $\mu \circ f$ is integrable with respect to θ. The integrals in the preceding equation are $\mathcal{Q}_{\mathbb{C}}$-valued.

Alternatively, in the same setting, where $E = F$, we may consider the complex-valued measure θ as being $\mathfrak{L}(E, E)$-valued (see Example 2.3.26(a)). Corollary 2.3.29 then states that $(\int_B f\,d\theta)(\mu) = (\int_B f\,d(\mu \circ \theta))(+1)$ holds for every $f \in \mathcal{F}_{(\theta,B)}(X, \mathrm{Conv}(E))$. We realize that the measures $\theta \circ \mu$ and $\mu \circ \theta$ are both $\mathfrak{L}(E, \mathbb{C})$-valued and coincide. Indeed, we have

$$(\theta \circ \mu)_A(a) = (\mu \circ \theta)_A(a) = \mu(a)\,\theta_A$$

for all $A \in \mathfrak{R}$ and $a \in E$. Moreover, integrability over B for an $\mathrm{Conv}(E)$-valued function with respect to θ implies integrability with respect to $\theta \circ \mu = \mu \circ \theta$ by the above. Hence if $f \in \mathcal{F}(X, \mathrm{Conv}(E))$ and if both f and $\mu \circ f$ are integrable over B with respect to θ (see 2.3.26(a)), then

$$\Big(\int_B f\,d\theta\Big)(\mu) = \Big(\int_B f\,d(\mu \circ \theta)\Big)(+1) = \Big(\int_B (\mu \circ f)\,d\theta\Big)(+1).$$

2.4 The convergence theorems

We continue to use the previous assumptions and go forward to establish a range of general convergence results for sequences of functions and their respective integrals. These results are modeled after the dominated convergence theorem from classical measure theory. However, the presence of unbounded elements in the range of $\mathrm{Conv}(E)$-valued functions and the general absence of negatives will considerably complicate some technical aspects of the approach, which is modeled after the corresponding method for locally convex cones in Chapter II.5 of [50]. As in the preceding section, we shall assume that θ is a bounded $\mathfrak{L}(E, F^{*\blacksquare})$-valued measure. Subsequently, we shall set up suitable notions for convergence of sequences and series of Conv(E)-valued functions. Convergence for sequences of integrals will generally refer to the topology of pointwise convergence in $\mathcal{Q}_F$.

Theorem 2.3.18(b) states that a function in $\mathcal{F}(X, \mathrm{Conv}(E))$ that is integrable over each of finitely many disjoint sets in $\mathfrak{A}$ is also integrable over their union. If the sets are indeed in $\mathfrak{R}$, then this holds true as well if they are not necessarily disjoint (see the remark following 2.3.18). Simple examples show that these statements do however generally not extent to countable unions. For the following, we recall from Section 1.4 that a function $f \in \mathcal{F}(X, \mathrm{Conv}(E))$ is bounded above (or below) if for every function space neighborhood $\mathfrak{v}$ there is $\lambda \geqslant 0$ such that $f \leqslant \lambda\mathfrak{v}$ (or $0 \leqslant f + \lambda\mathfrak{v}$). It is bounded, if it is both bounded above and below. A subset of $\mathcal{F}(X, \mathrm{Conv}(E))$ is bounded above (or below) if these conditions hold with the same $\lambda \geqslant 0$ for all functions in this subset.

2.4.1 Lemma. *Suppose that $|\theta|$ is strongly additive at all neighborhoods $V \in \mathcal{V}$. Let $A_n \in \mathfrak{R}$ such that $A = \bigcup_{n\in\mathbb{N}} A_n \in \mathfrak{R}$. If $f \in \mathcal{F}(X, \mathrm{Conv}(E))$ is bounded and integrable over all sets A_n, then it is integrable over A.*

Proof. Set $B_n = \bigcup_{i=1}^{n} A_i \in \mathfrak{R}$. Then $B_n \subset B_{n+1}$ and $\bigcup_{n\in\mathbb{N}} B_n = \bigcup_{n\in\mathbb{N}} A_n = A$. If the function $f \in \mathcal{F}(X, \mathrm{Conv}(E))$ is integrable over all sets A_n, then it is integrable over all sets B_n. If in addition $|\theta|$ is strongly additive and f is bounded, we shall verify its integrability over A. For this, let $W \in \mathcal{W}$. There is $V \in \mathcal{V}$ such that $|\theta|_A(V) \leqslant (1/2)\sigma_W$, and there is $\lambda \geqslant 0$ such that both $f \leqslant \lambda\chi_X \cdot V$ and $0 \leqslant f + \lambda\chi_X \cdot V$. As $|\theta|$ is supposed to be strongly additive at V, there is $n \in \mathbb{N}$ such that $|\theta|_{A\setminus B_n} \leqslant (1/(6\lambda))\sigma_W$. By the definition of integrability of f over B_n, we find measurable functions g and $\mathfrak{n}$ satisfying

$$f \underset{a.e. B_n}{\leqslant} g \underset{a.e. B_n}{\leqslant} f + \mathfrak{n} \quad \text{and} \quad \int_{B_n} \mathfrak{n}\, d\theta \leqslant (1/2)\sigma_W.$$

There is a subset $B \in \mathfrak{R}$ of B_n such that $B_n \setminus B$ is of measure zero, and the above inequalities hold everywhere on B. We have $|\theta|_{A\setminus B} = |\theta|_{A\setminus B_n} + |\theta|_{B_n\setminus B} = |\theta|_{A\setminus B_n} \leqslant (1/(6\lambda))\sigma_W$. Since

$$\chi_A \cdot f = \chi_B \cdot f + \chi_{A\setminus B} \cdot f \leqslant \chi_B \cdot f + \lambda\chi_{A\setminus B} \cdot V$$

and $\chi_B \cdot f \leqslant \chi_B \cdot g \leqslant \gamma\chi_B \cdot f + \mathfrak{n}$, we conclude that

$$\begin{aligned}
\chi_A \cdot f &\leqslant (\chi_B \cdot g + \lambda\chi_{A\setminus B} \cdot V) \\
&\leqslant \gamma\chi_B \cdot f + \lambda\chi_{A\setminus B} \cdot V + \mathfrak{n} \\
&\leqslant \gamma\chi_B \cdot f + \lambda\chi_{A\setminus B} \cdot V + \mathfrak{n} + \gamma\chi_{A\setminus B} \cdot (f + \lambda\chi_X \cdot V) \\
&\leqslant \gamma\chi_A \cdot f + (\mathfrak{n} + \lambda(1+\gamma)\chi_{A\setminus B} \cdot V).
\end{aligned}$$

The function $\tilde{g} = \chi_B \cdot g + \lambda\chi_{A\setminus B} \cdot V$ is measurable, likewise the neighborhood function

$$\tilde{n} = n + \lambda(1+\gamma)\chi_{A\setminus B} \cdot V,$$

and we calculate that

$$\int_A \tilde{n}\, d\theta \leqslant \int_A n\, d\theta \leqslant \sigma_W + (1+\gamma)\lambda|\theta|_{A\setminus B}(V).$$

Here, we use that both $\int_A n\, d\theta \leqslant (1/2)\sigma_W$ and

$$(1+\gamma)\lambda|\theta|_{A\setminus B}(V) \leqslant 3\lambda|\theta|_{A\setminus B}(V) \leqslant (1/2)\sigma_W.$$

Our definition of integrability for f over A is therefore satisfied with the measurable function $\tilde{g} \in \mathcal{F}(X, \mathrm{Conv}(E))$ and the measurable neighborhood function $\tilde{n}$. □

2.4.2 Proposition. *Let $f \in \mathcal{F}_{(\theta,A)}(X, \mathrm{Conv}(E))$ for $A \in \mathfrak{R}$.*
(a) *If $A_n \in \mathfrak{R}$ are disjoint sets and $A = \bigcup_{n\in\mathbb{N}} A_n$, then*

$$\int_A f\, d\theta = \sum_{n=1}^{\infty} \int_{A_n} f\, d\theta.$$

(b) *If $A_n \in \mathfrak{R}$ such that $A_n \subset A_{n+1}$ for all $n \in \mathbb{N}$, and $A = \bigcup_{n\in\mathbb{N}} A_n$, then*

$$\int_A f\, d\theta = \lim_{n\to\infty} \int_{A_n} f\, d\theta.$$

(c) *If $A_n \in \mathfrak{R}$ such that $A \supset A_n \supset A_{n+1}$ for all $n \in \mathbb{N}$, and $\bigcap_{n\in\mathbb{N}} A_n = \emptyset$, then*

$$\lim_{n\to\infty} \left(\int_{A_n} f\, d\theta \right)(\mu) = 0$$

for all $\mu \in F^*$ such that $(\int_A f\, d\theta)(\mu) < +\infty$.

Proof. We recall that a function that is integrable over a set in $\mathfrak{A}$ is integrable over all its intersections with sets in $\mathfrak{R}$. Therefore, integrability over $A = \bigcup_{n\in\mathbb{N}} A_n \in \mathfrak{R}$ implies integrability over each of the sets A_n. Parts (a) and (b) of the proposition are clearly equivalent, and we shall verify the second. Let $A_n \in \mathfrak{R}$ such that $A_n \subset A_{n+1}$ for all $n \in \mathbb{N}$, and $A = \bigcup_{n\in\mathbb{N}} A_n \in \mathfrak{R}$. We shall first assume that the function $f \in \mathcal{F}_{(\theta,A)}(X, \mathrm{Conv}(E))$ is measurable. For $W \in \mathcal{W}$, there is a neighborhood $V \in \mathcal{V}$ such that $\theta_A(V) \leqslant \sigma_W$ and by Lemma 2.2.2 there is $\lambda \geqslant 0$ such that $0 \leqslant f + \lambda\chi_X \cdot V$. This implies

$$\chi_{A_n} \cdot f \leqslant \chi_{A_n} \cdot f + \chi_{A\setminus A_n} \cdot (f + \lambda\chi_X \cdot V) = \chi_A \cdot f + \lambda\chi_{A\setminus A_n} \cdot V.$$

Thus

$$\int_{A_n} f\,d\theta \leqslant \int_A (f + \lambda\chi_{A\setminus A_n} \cdot V)\,d\theta = \int_A f\,d\theta + \lambda|\theta|_{A\setminus A_n}(V).$$

Following Lemma 2.1.2(b), this yields

$$\limsup_{n\to\infty}\Big(\int_{A_n} f\,d\theta\Big)(\mu) \leqslant \Big(\int_A f\,d\theta\Big)(\mu)$$

for all $\mu \in W^\circ$. For any $U \in \mathcal{W}$ let $h \in \mathcal{F}_S(X, \mathrm{Conv}(E))$ be a step function such that $h \leqslant f + \mathfrak{v}_U$, that is, $h \leqslant f + n$ for some neighborhood-valued step function n such that $\int_X n\,d\theta \leqslant \sigma_U$. Then $\int_{A_n} h\,d\theta \leqslant \int_{A_n} f\,d\theta + \sigma_U$ by Theorem 2.3.8(b) and (c). Hence using the countable additivity of $|\theta|$ for the step function h, we conclude that

$$\int_A h\,d\theta = \lim_{n\to\infty}\int_{A_n} h\,d\theta \leqslant \liminf_{n\to\infty}\int_{A_n} f\,d\theta + \sigma_U,$$

holds pointwise on F^*. Taking the supremum over all such step functions $h \leqslant f + \mathfrak{v}_U$ renders

$$\int_A^{(U)} f\,d\theta \leqslant \liminf_{n\to\infty}\int_{A_n} f\,d\theta + \sigma_U.$$

Taking the infimum over all $U \in \mathcal{W}$ in the preceding inequality yields

$$\int_A f\,d\theta \leqslant \liminf_{n\to\infty}\int_{A_n} f\,d\theta,$$

and combining all of the above, we infer that

$$\begin{aligned}\Big(\int_A f\,d\theta\Big)(\mu) &\leqslant \liminf_{n\to\infty}\Big(\int_{A_n} f\,d\theta\Big)(\mu)\\ &\leqslant \limsup_{n\to\infty}\Big(\int_{A_n} f\,d\theta\Big)(\mu)\\ &\leqslant \Big(\int_A f\,d\theta\Big)(\mu)\end{aligned}$$

holds for all $\mu \in W^\circ$. Since every $\mu \in F^*$ is contained in the polar of some neighborhood $W \in \mathcal{W}$, this delivers our claim for a measurable function $f \in \mathcal{F}(X, \mathrm{Conv}(E))$. Now for the general case, let $f \in \mathcal{F}_{(\theta,A)}(X, \mathrm{Conv}(E))$. Given $W \in \mathcal{W}$, we choose the measurable functions f_W and n_W as in our definition of integrability over a set in $A \in \mathfrak{R}$. That is,

$$f \underset{a.e.A}{\leqslant} f_W \underset{a.e.A}{\leqslant} f + n_W \quad \text{and} \quad \int_A n_W \, d\theta \leqslant \sigma_W.$$

Using the previously established claim for measurable functions for each of the functions f_W, we obtain

$$\int_A f \, d\theta \leqslant \int_A f_W \, d\theta = \lim_{n\to\infty} \int_{A_n} f_W \, d\theta \leqslant \liminf_{n\to\infty} \int_{A_n} f \, d\theta + \sigma_W,$$

and

$$\limsup_{n\to\infty} \int_{A_n} f \, d\theta \leqslant \limsup_{n\to\infty} \int_{A_n} f_W \, d\theta = \int_A f_W \, d\theta \leqslant \int_A f \, d\theta + \sigma_W,$$

pointwise on F^*. Given $\mu \in F^*$, there is $W \in \mathcal{W}$ such that $\mu \in \varepsilon W^\circ$. Then a combination of the above inequalities returns

$$\limsup_{n\to\infty} \Big(\int_{A_n} f \, d\theta \Big)(\mu) \leqslant \Big(\int_A f \, d\theta \Big)(\mu) + \varepsilon \leqslant \liminf_{n\to\infty} \Big(\int_{A_n} f \, d\theta \Big)(\mu) + 2\varepsilon.$$

Letting $\varepsilon > 0$ tend to 0 renders our claim in Part (b). For Part (c), let $A_n \in \mathfrak{R}$ such that $A \supset A_n \supset A_{n+1}$ for all $n \in \mathbb{N}$ and $\bigcap_{n\in\mathbb{N}} A_n = \emptyset$. Using Part (a) for the sets $A\backslash A_n$ yields

$$\int_A f \, d\theta = \lim_{n\to\infty} \int_{A\backslash A_n} f \, d\theta.$$

We have $\int_A f \, d\theta = \int_{A\backslash A_n} f \, d\theta + \int_{A_n} f \, d\theta$ by Theorem 2.3.15(b), and consequently,

$$\lim_{n\to\infty} \Big(\int_{A_n} f \, d\theta \Big)(\mu) = \Big(\int_A f \, d\theta \Big)(\mu) - \lim_{n\to\infty} \Big(\int_{A\backslash A_n} f \, d\theta \Big)(\mu) = 0,$$

provided that $(\int_A f \, d\theta)(\mu) < +\infty$. □

The convergence statements in Proposition 2.4.2 do generally not apply to the stronger symmetric topology of $\mathcal{Q}_F$. This consolidation can however be established under some additional assumptions. In this vein, for a function $f \in \mathcal{F}(X, \mathrm{Conv}(E))$ we shall say that its integral over a set $A \in \mathfrak{R}$ is *strongly additive* if it is integrable over A, if $\int_A f \, d\theta$ is a bounded element of the locally convex cone $\mathcal{Q}_F$ and if the convergence claims in Proposition 2.4.2 hold for the symmetric topology of $\mathcal{Q}_F$. We note that boundedness of the integral implies that $(\int_A f \, d\theta)(\mu) < +\infty$ for all $\mu \in F^*$, and the statement in 2.4.2(c) implies those in 2.4.2(a) and (b). The integral of a function f over a set $B \in \mathfrak{R}$ is *strongly additive* if for all $A \in \mathfrak{R}$ its integrals over the sets $A \cap B$ are strongly additive and if the limit $\int_B f\,d\theta = \lim_{A \in \mathfrak{R}} \int_{A \cap B} f\,d\theta$ refers to the symmetric topology of $\mathcal{Q}_F$.

If θ is $\mathfrak{L}(E,F)$-valued, then according to Proposition 2.1.1 it is strongly additive at all elements of E, hence the integral of any E-valued step function over a set in $\mathfrak{R}$ is strongly additive. The same statement holds for integrable E-valued functions in general.

2.4.3 Corollary. *Let $f \in \mathcal{F}_{(\theta,A)}(X,E)$ for $A \in \mathfrak{R}$. If θ is $\mathfrak{L}(E,F)$-valued, then the integral of f over A is strongly additive.*

Proof. Suppose that θ is $\mathfrak{L}(E,F)$-valued and let $f \in \mathcal{F}_{(\theta,A)}(X,E)$ for $A \in \mathfrak{R}$. Proposition 2.3.23(a) states that $\int_A f\,d\theta \in F^{*\blacksquare}$, hence is a bounded element of $\mathcal{Q}_F$. We shall verify the statement in 2.4.2(c) for the symmetric topology of $\mathcal{Q}_F$, which in this case implies the remaining parts. Let $A_n \in \mathfrak{R}$ such that $A \supset A_n \supset A_{n+1}$ for all $n \in \mathbb{N}$ and $\bigcap_{n\in\mathbb{N}} A_n = \emptyset$. Given $W \in \mathcal{W}$, according to Corollary 2.3.22, there is a step function $h \in \mathcal{F}_S(X,E)$ and a measurable neighborhood function n such that $h - f \underset{a.e.A}{\leqslant} n$ and $\int_A n\,d\theta \leqslant \sigma_W$. Because $\int_A h\,d\theta$ is strongly additive, there is n_0 such that both

$$\int_{A_n} h\,d\theta \leqslant \sigma_W \quad \text{and} \quad 0 \leqslant \int_{A_n} h\,d\theta + \sigma_W$$

for all $n \geqslant n_0$. We infer that

$$\int_{A_n} f\,d\theta \leqslant \int_{A_n} h\,d\theta + \sigma_W \leqslant 2\sigma_W$$

and

$$0 \leqslant \int_{A_n} h\,d\theta + \sigma_W \leqslant \int_{A_n} f\,d\theta + 2\sigma_W$$

holds for all $n \geqslant n_0$. Our claim follows. □

2.4.4 Corollary. *Let $f \in \mathcal{F}_{(\theta,A)}(X,\mathrm{Conv}(E))$ for $A \in \mathfrak{R}$. If f is bounded and if $|\theta|$ is strongly additive at all neighborhoods $V \in \mathcal{V}$, then the integral of f over A is strongly additive.*

Proof. Let $A \in \mathfrak{R}$ and $f \in \mathcal{F}_{(\theta,A)}(X,\mathrm{Conv}(E))$. Given $W \in \mathcal{W}$, there is $V \in \mathcal{V}$ such that $|\theta|_A(V) \leqslant \sigma_W$, and there is $\lambda > 0$ such that both $0 \leqslant f + \lambda\chi_X \cdot V$ and $f \leqslant \lambda\chi_X \cdot V$. Hence $0 \leqslant \int_A f\,d\theta + \lambda\sigma_W$ and $\int_A f\,d\theta \leqslant \lambda\sigma_W$. That is, $\int_A f\,d\theta \in \lambda W^s(0)$, and $\int_A f\,d\theta$ is seen to be a bounded element of the locally convex cone $\mathcal{Q}_F$ (see Section 1.4). Next, we shall verify the statement in 2.4.2(c) for the symmetric topology of $\mathcal{Q}_F$, which implies the remaining parts. Let $A_n \in \mathfrak{R}$ such that $A \supset A_n \supset A_{n+1}$ for all $n \in \mathbb{N}$ and $\bigcap_{n\in\mathbb{N}} A_n = \emptyset$.

Given $W \in \mathcal{W}$, we choose $V \in \mathcal{V}$ and $\lambda \geqslant 0$ from above. Because $|\theta|$ is supposed to be strongly additive at V, we find n_0 such that $|\theta|_{A_n} \leqslant (1/\lambda)\sigma_W$ for all $n \geqslant n_0$. Hence $0 \leqslant \int_{A_n} f\,d\theta + \sigma_W$ and $\int_{A_n} f\,d\theta \leqslant \sigma_W$. Our claim follows. □

Recall from Example 2.1.8(e) that every $F^{*\blacksquare}$-valued measure θ may be considered to be an operator-valued measure if we choose $E = \mathbb{R}$ or $E = \mathbb{C}$.

2.4.5 Corollary. *If $f \in \mathcal{F}(X, \mathrm{Conv}(E))$ is integrable over all sets in $\mathfrak{R}$, then the mapping*

$$A \mapsto \int_A f\,d\theta : \mathfrak{R} \to F^{*\blacksquare}$$

defines an $F^{*\blacksquare}$-valued measure. If all integrals are F-valued, then this measure is F-valued and strongly additive. If f is bounded, then this measure is bounded.

Proof. Let $f \in \mathcal{F}(X, \mathrm{Conv}(E))$ be as stated and set $\Theta_A = \int_A f\,d\theta$ for all $A \in \mathfrak{R}$. Then Θ is $F^{*\blacksquare}$-valued, that is, $\mathfrak{L}(\mathbb{R}, F^{*\blacksquare})$- or $\mathfrak{L}(\mathbb{C}, F^{*\blacksquare})$-valued, and countably additive as stated in Proposition 2.4.2(a). If Θ is in fact F-valued, then according to Proposition 2.1.1 it is strongly additive. If f is bounded, then given $A \in \mathfrak{R}$ and $W \in \mathcal{W}$, there is $V \in \mathcal{V}$ such that $|\theta|_A(V) \leqslant \sigma_W$, and there is $\lambda > 0$ such that both $0 \leqslant f + \lambda\chi_X \cdot V$ and $f \leqslant \lambda\chi_X \cdot V$, that is, $0 \leqslant \int_B f\,d\theta + \lambda|\theta|_B(V)$ and $\int_B f\,d\theta \leqslant \lambda|\theta|_B(V)$ for every $B \in \mathfrak{R}$. For disjoint subsets $A_i \in \mathfrak{R}$ of A and $\mu \in F^*$ this yields

$$\sum_{i=1}^{n} \left| \int_{A_i} f\,d\theta\,(\mu) \right| \leqslant \lambda \sum_{i=1}^{n} |\theta|_{A_i}(V)(\mu) \leqslant \lambda|\theta|_A(V)(\mu).$$

Hence for the unit ball $\mathbb{B}$ of $\mathbb{C}$ or $\mathbb{R}$ we compute using 2.1.8(e)

$$|\Theta|_A(\mathbb{B})(\mu) = \sup\left\{ \sum_{i=1}^{n} |\Theta_{A_i}(\mu)| \;\middle|\; A_i \in \mathfrak{R} \text{ disjoint } A_i \subset A \right\} \leqslant \lambda\sigma_W(\mu).$$

Thus $|\Theta|_A((1/\lambda)\mathbb{B}) \leqslant \sigma_W$. The $F^{*\blacksquare}$-valued measure Θ is therefore bounded as claimed. □

The relative topologies

Chapter II.5 in [50] presents a variety of convergence theorems simultaneously involving sequences of measures and integrable functions. They involve the (generally coarser) relative rather that the given topologies of a locally convex cone which we will now establish for $\mathrm{Conv}(E)$. For a neighborhood $V \in \mathcal{V}$ and $\varepsilon > 0$, we define the *upper, lower* and *symmetric relative neighborhoods* for an element $C \in \mathrm{Conv}(E)$ by

$$V_\varepsilon(C) = \{D \in \mathrm{Conv}(E) \mid B \leqslant \gamma C + \varepsilon V \text{ for some } 1 \leqslant \gamma \leqslant 1 + \varepsilon\}$$

$$(C)V_\varepsilon = \{D \in \mathrm{Conv}(E) \mid C \leqslant \gamma D + \varepsilon V \text{ for some } 1 \leqslant \gamma \leqslant 1+\varepsilon\}$$

and

$$V_\varepsilon^s(C) = V_\varepsilon(C) \cap (C)V_\varepsilon,$$

respectively. These neighborhoods are convex subsets of $\mathrm{Conv}(E)$ and one easily checks the conditions for neighborhood bases defining a topology (see Theorem 4.5 in [63] and Lemma I.4.1 in [50]). They create the *upper, lower* and *symmetric relative topologies* on $\mathrm{Conv}(E)$. These are generally coarser than the given upper, lower and symmetric topologies, which are generated by the neighborhoods $V(C) = \{D \in \mathrm{Conv}(E) \mid D \subset C + V\}$, $(C)V = \{D \in \mathrm{Conv}(E) \mid C \subset D + V\}$ and $V^s(C) = V(C) \cap (C)V$, respectively (see Section 1.4). The relative topologies are however locally equivalent to the given topologies on bounded sets $C \in \mathrm{Conv}(E)$. Indeed, for $V \in \mathcal{V}$ and $\varepsilon > 0$ we have $(\varepsilon V)(C) \subset V_\varepsilon(C)$ for all $C \in \mathrm{Conv}(E)$. If C is bounded, then $C \subset \lambda V$ for some $\lambda \geqslant 0$, and we have $V_{1/(1+\lambda)}(C) \subset V(C)$. On the subspace E of $\mathrm{Conv}(E)$, all of the above topologies coincide with the given one. We had implicitly employed the relative topologies in the earlier sections about measurability and integrability.

The main justification for the use of the relative rather than the given topologies on a locally convex cone is that they render continuity not only for the addition but also the multiplication by positive scalars (see Proposition I.4.2 in [48]). This property is indispensable for the investigation of continuous cone-valued functions, as we shall illustrate in an example below.

Sequences of set-valued functions

We shall use different patterns of pointwise convergence for sequences of functions in $\mathcal{F}(X, \mathrm{Conv}(E))$. For sequences $(C_n)_{n\in\mathbb{N}}$ in $\mathrm{Conv}(E)$ and $C \in \mathrm{Conv}(E)$, we shall write

$$C_n \searrow C, \quad C_n \nearrow C, \quad \text{or} \quad C_n \to C$$

if $(C_n)_{n\in\mathbb{N}}$ converges to C with respect to the upper, lower or symmetric relative topology of $\mathrm{Conv}(E)$, that is, if for every $V \in \mathcal{V}$ and $\varepsilon > 0$ there is $n_0 \in \mathbb{N}$ such that $C_n \in V_\varepsilon(C)$, $C_n \in (C)V_\varepsilon$ or $C_n \in V_\varepsilon^s(C)$ for all $n \geqslant n_0$, respectively. Similarly, for functions $(f_n)_{n\in\mathbb{N}}$ and f in $\mathcal{F}(X, \mathrm{Conv}(E))$, and a set $B \in \mathfrak{A}$ we shall write

$$f_n \,{}_B\!\!\searrow f, \quad f_n \underset{B}{\nearrow} f, \quad \text{or} \quad f_n \underset{B}{\to} f,$$

if $(f_n)_{n\in\mathbb{N}}$ converges to f pointwise on B with respect to the upper, lower or symmetric relative topology of $\mathrm{Conv}(E)$, respectively. We shall also use these notions for double sequences $(f_n)_{n\in\mathbb{N}}$ and $(g_n)_{n\in\mathbb{N}}$ in $\mathcal{F}(X, \mathrm{Conv}(E))$. That is to say, we shall write $f_n \,{}_B\!\!\searrow g_n$

if for every $x \in B$, $V \in \mathcal{V}$ and $\varepsilon > 0$ there is $n_0 \in \mathbb{N}$ such that $f_m(x) \in V_\varepsilon(g_n(x))$ for all $m, n \geqslant n_0$. Correspondingly, we shall denote

$$f_n \ {}_{a.e.B}\!\searrow f, \quad f_n \nearrow_{a.e.B} f, \quad f_n \xrightarrow[a.e.B]{} f \quad \text{or} \quad f_n \ {}_{a.e.B}\!\searrow g_n,$$

if this convergence holds θ-almost everywhere on B, that is, pointwise everywhere on a subset $B \setminus Z$ with some set $Z \in \mathfrak{A}$ of measure zero. All the above notions of convergence are compatible with the algebraic operations in $\mathrm{Conv}(E)$, that is, the limit rules apply with sums and multiples by positive scalars.

2.4.6 Lemma. *Let $a, a_n \geqslant 0$ and let $C \in \mathrm{Conv}(E)$.*
(a) *If $\limsup_{n\to\infty} a_n \leqslant a$, if $C \geqslant 0$, and if either $a > 0$ or C is bounded, then $a_nC \searrow aC$.*
(b) *If $\liminf_{n\to\infty} a_n \geqslant a$ and if $C \geqslant 0$, then $a_nC \nearrow aC$.*
(c) *If $\lim_{n\to\infty} a_n = a$ and if either $a > 0$ or C is bounded, then $a_nC \to aC$.*

Proof. Let $a, a_n \geqslant 0$ and $C \in \mathrm{Conv}(E)$ Let $V \in \mathcal{V}$ and $\varepsilon > 0$. There is $\lambda > 0$ such that $0 \leqslant C + \lambda V$. For those statements in (a) and (c) which require C to be bounded, we shall also assume that $C \leqslant \lambda V$. For (a), suppose that $\limsup_{n\to\infty} a_n \leqslant a$ and that $C \geqslant 0$. In case that $a > 0$, we set $\delta = a\varepsilon$. There is $n_0 \in \mathbb{N}$ such that $a_n \leqslant a + \delta$ for all $n \geqslant n_0$. Since $C \geqslant 0$, this implies

$$a_nC \leqslant (a + \delta)C \leqslant (1 + \varepsilon)aC$$

for all $n \geqslant n_0$. In case that $a = 0$ and C is bounded, we set $\delta = \varepsilon/\lambda$. There is $n_0 \in \mathbb{N}$ such that $0 \leqslant a_n \leqslant \delta$ for all $n \geqslant n_0$. Then

$$a_nC \leqslant \delta C \leqslant \delta\lambda V \leqslant 0\,C + \varepsilon V.$$

Thus $a_nC \in V_\varepsilon(aC)$ holds for all $n \geqslant n_0$ in both cases. For (b), suppose that $\limsup_{n\to\infty} a_n \geqslant a$ and that $C \geqslant 0$. In case that $a > 0$, we set $\delta = a\varepsilon/(1 + \varepsilon)$. There is $n_0 \in \mathbb{N}$ such that $a \leqslant a_n + \delta$ for all $n \geqslant n_0$. Then $\delta(1 + \varepsilon) \leqslant \varepsilon a \leqslant \varepsilon(a_n + \delta)$, hence $\delta \leqslant \varepsilon a_n$ and

$$aC \leqslant (a_n + \delta)C \leqslant (1 + \varepsilon)a_nC$$

for all $n \geqslant n_0$. In case that $a = 0$, we have $0\,C \leqslant a_nC$ for all $n \in \mathbb{N}$. Thus $aC \in V_\varepsilon(a_nC)$, that is, $a_nC \in (aC)V_\varepsilon$, holds for all $n \geqslant n_0$ in both cases. For (c), suppose that $\lim_{n\to\infty} a_n = a$. In case that $a > 0$, we set $\delta = \min\{\varepsilon/\lambda, a\varepsilon/(1 + \varepsilon)\}$. There is $n_0 \in \mathbb{N}$ such that $|a_n - a| \leqslant \delta$ for all $n \geqslant n_0$. We claim that this implies that $a_nC \in V_\varepsilon^s(aC)$. Indeed, if $a_n \leqslant a$, then

$$a_nC \leqslant a_nC + (a - a_n)(C + \lambda V) \leqslant aC + \delta\lambda V \leqslant aC + \varepsilon V$$

and $\alpha C = \gamma \alpha_n C$ with some $1 \leqslant \gamma \leqslant 1 + \varepsilon$. The former follows since $\alpha - \delta \leqslant \alpha_n \leqslant \alpha$, hence $0 \leqslant \alpha - \alpha_n \leqslant \delta \leqslant \varepsilon/\lambda$ by our choice for δ, and the latter since

$$\alpha - \delta \geqslant \alpha(1 - \varepsilon/(1 + \varepsilon)) = \alpha/(1 + \varepsilon),$$

hence

$$1 \leqslant \gamma = \alpha/\alpha_n \leqslant \alpha/(\alpha - \delta) \leqslant 1 + \varepsilon.$$

If $\alpha < \alpha_n$, then $\alpha_n C = \gamma \alpha C$ and

$$\alpha C \leqslant \alpha C + (\alpha_n - \alpha)(C + \lambda V) \leqslant \alpha_n C + \varepsilon V,$$

with some $1 \leqslant \gamma \leqslant 1 + \varepsilon$. The former follows since

$$\alpha < \alpha_n \leqslant \alpha + \delta \leqslant \alpha(1 + \varepsilon/(1 + \varepsilon)) \leqslant \alpha(1 + \varepsilon),$$

hence $1 \leqslant \gamma = \alpha_n/\alpha \leqslant 1 + \varepsilon$, and the latter since $0 < \alpha_n - \alpha \leqslant \delta \leqslant \varepsilon/\lambda$. We therefore have $\alpha_n C \in V_\varepsilon^s(\alpha C)$ for all $n \geqslant n_0$ in both instances. Finally, in case that $\alpha = 0$ and C is bounded, we set $\delta = \varepsilon/\lambda$. There is $n_0 \in \mathbb{N}$ such that $0 \leqslant \alpha_n \leqslant \delta$ for all $n \geqslant n_0$. Then

$$\alpha_n C \leqslant \alpha_n \lambda V \leqslant \delta \lambda V \leqslant 0\,C + \varepsilon V$$

and

$$0\,C \leqslant \alpha_n(C + \lambda V) \leqslant \alpha_n C + \delta \lambda V \leqslant \alpha_n C + \varepsilon V.$$

Thus $\alpha_n C \in V_\varepsilon^s(0\,C)$ for all $n \geqslant n_0$. □

For bounded sets $C \in \mathrm{Conv}(E)$, the statements of Lemma 2.4.6 hold indeed with the given rather than the relative topologies of $\mathrm{Conv}(E)$. But they fail with the given topologies for unbounded sets as simple examples can demonstrate. Indeed, let $E = \mathbb{R}^2$ with the Euclidean unit ball $\mathbb{B}$. Let

$$C = \{(x, y) \mid y \geqslant x^2 + 1\} \in \mathrm{Conv}(\mathbb{R}^2).$$

Then $\alpha C = \{(x, y) \mid y \geqslant (1/\alpha)x^2 + \alpha\}$ for $\alpha > 0$, and we realize that $\alpha C \not\subset C + \mathbb{B}$, that is, $\alpha C \notin \mathbb{B}(C)$ for any choice of $\alpha > 1$. However, given $0 < \varepsilon \leqslant 1$, a brief calculation shows that $\alpha C \in \mathbb{B}_\varepsilon^s(C)$ holds whenever $|\alpha - 1| < \varepsilon$.

The following statement is an immediate consequence of the preceding lemma. Recall from Section 1.4 that a subscript to the symbol of a relation between functions signifies that this relation holds pointwise on the specified subset.

2.4.7 Proposition. *Let $B \in \mathfrak{A}$. Let $\varphi, \varphi_n \in \mathcal{F}(X)$ be nonnegative functions and let $C \in$* Conv(E).

(a) *If* $\limsup_{n\to\infty} \varphi_n \leqslant_B \varphi$, *if $C \geqslant 0$ and if either $\varphi \geqslant_B 0$ or C is bounded, then* $\varphi_n \cdot C \;{}_B\!\searrow\; \varphi \cdot C$.

(b) *If* $\liminf_{n\to\infty} \varphi_n \geqslant_B \varphi$ *and if $C \geqslant 0$, then* $\varphi_n \cdot C \nearrow_B \varphi \cdot C$.

(c) *If* $\lim_{n\to\infty} \varphi_n =_B \varphi$ *and if either $\varphi >_B 0$ or C is bounded, then* $\varphi_n \cdot C \underset{B}{\to} \varphi \cdot C$.

The next result is our main convergence theorem for sequences of integrable functions in $\mathcal{F}(X, \mathrm{Conv}(E))$. Recall from Section 1.4 that we consider the upper, lower and symmetric locally convex cone topologies on $\mathcal{Q}_F$. That is, $\varphi \in W(\psi)$ for $\varphi, \psi \in \mathcal{Q}_F$ and $W \in \mathcal{W}$ means that $\varphi \leqslant \psi + \sigma_W$.

2.4.8 Theorem. *Let $f_n, g_n, f^*, g_*, g_{**} \in \mathcal{F}_{(\theta,B)}(X, \mathrm{Conv}(E))$ for $B \in \mathfrak{A}$ and suppose that $f_n \underset{a.e.B}{\leqslant} f^*$ and $g_{**} \underset{a.e.B}{\leqslant} g_n + g_*$ for all $n \in \mathbb{N}$. If $f_n \;{}_{a.e.B}\!\searrow\; g_n$, then for every $\mu \in F^*$ such that both $(\int_B f^*\, d\theta)(\mu) < +\infty$ and $(\int_B g_*\, d\theta)(\mu) < +\infty$ and every $\varepsilon > 0$ there is n_0 such that*

$$\left(\int_B f_m\, d\theta\right)(\mu) \leqslant \left(\int_B g_n\, d\theta\right)(\mu) + \varepsilon$$

holds for all $m, n \geqslant n_0$. If the integrals of f^*, g_* and g_{**} over B are strongly additive, then the above holds uniformly on the polars of all neighborhoods in $\mathcal{W}$. That is, for every $W \in \mathcal{W}$ there is n_0 such that $\int_B f_m\, d\theta \leqslant \int_B g_n\, d\theta + \sigma_W$ for all $m, n \geqslant n_0$.

Proof. Without loss of generality, we may assume that $B = X$. Indeed, the respective integrals over $B \in \mathfrak{A}$ equal the integrals over X for the products of the concerned functions with χ_B, and these products satisfy the conditions of the theorem with X in place of B. Also, we may assume that the required convergence and boundedness properties hold everywhere on X instead of θ-almost everywhere. Indeed, the convergence condition for f_m and g_n holds for all $x \in X \setminus Z_1$ with some set $Z_1 \in \mathfrak{A}$ of measure zero. Using the fact that countable unions of sets of measure zero are again of measure zero (see Proposition 2.1.10(c)), we can find a set $Z_2 \in \mathfrak{A}$ of measure zero such that both $\chi_{X\setminus Z_2} \cdot f_n \leqslant \chi_{X\setminus Z_2} \cdot f^*$ and $0 \leqslant \chi_{X\setminus Z_2} \cdot (g_n + g_*)$ holds for all $n \in \mathbb{N}$ everywhere on X. The set $Z = Z_1 \cup Z_2 \in \mathfrak{A}$ is also of measure zero, and the functions

$$f_n' = \chi_{X\setminus Z} \cdot f_n, \quad g_n' = \chi_{X\setminus Z} \cdot g_n,$$
$$f^{*\prime} = \chi_{X\setminus Z} \cdot f^*, \quad g_*{}' = \chi_{X\setminus Z} \cdot g_* \quad \text{and} \quad g_{**}{}' = \chi_{X\setminus Z} \cdot g_{**}$$

fulfill all the assumptions of the theorem everywhere on X, and their respective integrals coincide with those of the given functions.

Let $W \in \mathcal{W}$. In order to argue the general and the strongly additive instance of our statement simultaneously, we choose $\Omega = \{\mu\}$, where μ is an element of W° such that both $(\int_B f^* \, d\theta)(\mu) < +\infty$ and $(\int_B g_* \, d\theta)(\mu) < +\infty$, for the general case and $\Omega = W^\circ$ for the strongly additive case. For any $\mu \in \Omega$, there is nothing to prove in case that $(\int_X g_n \, d\theta)(\mu) = +\infty$ for all $n \in \mathbb{N}$. Otherwise the relation $g_{**} \underset{a.e.B}{\leqslant} g_n + g_*$ guarantees that $(\int_B g_{**} \, d\theta)(\mu) < +\infty$ and that the sequence $((\int_X g_n \, d\theta)(\mu))_{n\in\mathbb{N}}$ is bounded below in $\mathbb{R}$. The same observation applies to the sequences $((\int_A g_n \, d\theta)(\mu))_{n\in\mathbb{N}}$ for all $A \in \mathfrak{R}$ since Theorem 2.3.18(b) guarantees that $(\int_A g_* \, d\theta)(\mu) < +\infty$ also holds. In a first step of our argument, we consider a set $A \in \mathfrak{R}$. Let $\varepsilon > 0$. There is $V \in \mathcal{V}$ such that $|\theta|_A(V) \leqslant \sigma_W$. According to Proposition 2.3.13, there is a measurable neighborhood function n and $\lambda \geqslant 0$ such that $0 \leqslant f^* + n$ and $\int_A n \, d\theta \leqslant \lambda\sigma_W$. We choose $\delta > 0$ such that both $\delta(1+\lambda) \leqslant \varepsilon$ and $\delta(\int_A f^* \, d\theta)(\mu) \leqslant \varepsilon$ for all $\mu \in \Omega$. For the latter, we recall that in the strongly additive case $\int_A f^* \, d\theta$ is bounded in $\mathcal{Q}_F$. If $f_m(x) \in V_\delta(g_n(x))$ for $x \in A$, then

$$f_m(x) \leqslant \gamma g_n(x) + \delta V, \quad \text{hence} \quad (1/\gamma)f_m(x) \leqslant g_n(x) + \delta V,$$

with some $1 \leqslant \gamma \leqslant 1+\delta$. Using that $f_m \leqslant f^*$ and that both $n(x) \geqslant 0$ and $n(x)+f^*(x) \geqslant 0$, we continue to argue that

$$\begin{aligned} f_m(x)(x) &= (1/\gamma)f_m(x) + (1-1/\gamma)f_m(x) \\ &\leqslant (g_n(x) + \delta V) + (1-1/\gamma)(f^*(x) + n(x)) \\ &\leqslant g_n(x) + \delta(f^*(x) + n(x) + V). \end{aligned}$$

The latter follows since $(\gamma - 1)/\gamma \leqslant \delta$. According to Corollary 2.3.17, there are subsets $B_m^n \in \mathfrak{R}$ of A that contain all points $x \in A$ such that

$$f_m(x) \leqslant g_n(x) + \delta(f^*(x) + n(x) + V),$$

therefore all points $x \in A$ such that $f_m(x) \in V_\delta(g_n(x))$, and we have

$$\begin{aligned} \int_C f_m \, d\theta &\leqslant \int_C (g_n + \delta\,(f^* + n + \chi_X \cdot V)) \, d\theta \\ &\leqslant \int_C g_n \, d\theta + \delta \int_C f^* \, d\theta + \delta(\lambda+1)\sigma_W \, d\theta \end{aligned}$$

for every subset $C \in \mathfrak{R}$ of B_m^n. In fact, Corollary 2.3.17 states that the above inequality holds at all $\mu \in F^*$ such that $\sigma_W(\mu) < +\infty$, but it obviously also holds if $\sigma_W(\mu) = +\infty$. We set $A_k = \bigcap\{B_m^n \mid m, n \geqslant k\} \in \mathfrak{R}$ and observe that $A_k \subset A_{k+1}$, and $\bigcup_{k\in\mathbb{N}} A_k = A$ by our assumption. By our choice of δ, we have

$$\int_{A_k} f_m \, d\theta \leqslant \int_{A_k} g_n \, d\theta + \delta \int_{A_k} f^* \, d\theta + \varepsilon\sigma_W \, d\theta$$

for all $m, n \geqslant k$. From $f_m \leqslant f^*$ and $g_{**} \leqslant g_n + g_*$ for all $m, n \in \mathbb{N}$, we infer that

$$\int_{A \setminus A_k} f_m \, d\theta \leqslant \int_{A \setminus A_k} f^* d\theta \quad \text{and} \quad \int_{A \setminus A_k} g_{**} \, d\theta \leqslant \int_{A \setminus A_k} g_m \, d\theta + \int_{A \setminus A_k} g_* \, d\theta.$$

Adding these two inequalities to the prior one, we obtain

$$\int_A f_m \, d\theta + \int_{A \setminus A_k} g_{**} \, d\theta \leqslant \int_A g_n \, d\theta + \delta \int_A f^* \, d\theta + \int_{A \setminus A_k} (f^* + g_*) \, d\theta + \varepsilon \sigma_W.$$

This inequality remains true for all $k \in \mathbb{N}$ and $m, n \geqslant k$. Following Proposition 2.4.2(c), there is k_0 such that both

$$\Big(\int_{A \setminus A_k} g_{**} \, d\theta \Big)(\mu) \geqslant -\varepsilon \quad \text{and} \quad \Big(\int_{A \setminus A_k} (f^* + g_*) \, d\theta \Big)(\mu) \leqslant \varepsilon$$

holds for all $\mu \in \Omega$ and all $k \geqslant k_0$. Hence, using our choice for δ, we conclude that

$$\Big(\int_A f_m \, d\theta \Big)(\mu) \leqslant \Big(\int_A g_n \, d\theta \Big)(\mu) + 4\varepsilon$$

holds for all $m, n \geqslant k_0$ and all $\mu \in \Omega$.

Now in the second step of our argument, we shall extend the preceding inequality from integrals over sets $A \in \mathfrak{R}$ to the corresponding integrals over X. Given $\varepsilon > 0$, there is $A \in \mathfrak{R}$ such that

$$\Big(\int_X f^* \, d\theta \Big)(\mu) \leqslant \Big(\int_A f^* \, d\theta \Big)(\mu) + \varepsilon, \quad \Big(\int_X g_* \, d\theta \Big)(\mu) \leqslant \Big(\int_A g_* \, d\theta \Big)(\mu) + \varepsilon$$

and

$$\Big(\int_A g_{**} \, d\theta \Big)(\mu) \leqslant \Big(\int_X g_{**} \, d\theta \Big)(\mu) + \varepsilon$$

for all $\mu \in \Omega$. Recall that all these evaluations are finite. By our first step, there is n_0 such that $(\int_A f_n \, d\theta)(\mu) \leqslant (\int_A g_m \, d\theta)(\mu) + \varepsilon$ holds for all $m, n \geqslant n_0$ and all $\mu \in \Omega$. We have $f_m \leqslant f^*$ for all $m \in \mathbb{N}$ and, therefore,

$$\int_X f_m \, d\theta + \int_A f^* \, d\theta \leqslant \int_A f_m \, d\theta + \int_X f^* \, d\theta,$$

by Theorem 2.3.18(c). By our choice for $A \in \mathfrak{R}$, this yields

$$\Big(\int_X f_m \, d\theta\Big)(\mu) \leqslant \Big(\int_A f_m \, d\theta\Big)(\mu) + \varepsilon,$$

for all $m \in \mathbb{N}$ and $\mu \in \Omega$. Likewise, $g_{**} \leqslant g_n + g_*$ for all $n \in \mathbb{N}$, implies that

$$\int_X g_{**} \, d\theta + \int_A (g_n + g_*) \, d\theta \leqslant \int_A g_{**} \, d\theta + \int_X (g_n + g_*) \, d\theta$$

by 2.3.18(c) and

$$\Big(\int_A g_n \, d\theta\Big)(\mu) \leqslant \Big(\int_X g_n \, d\theta\Big)(\mu) + 2\varepsilon$$

for all $n \in \mathbb{N}$ and $\mu \in \Omega$. Now combining all of the above, we calculate that

$$\begin{aligned}\Big(\int_X f_m \, d\theta\Big)(\mu) &\leqslant \Big(\int_A f_m \, d\theta\Big)(\mu) + \varepsilon \\ &\leqslant \Big(\int_A g_n \, d\theta\Big)(\mu) + 2\varepsilon \\ &\leqslant \Big(\int_X g_n \, d\theta\Big)(\mu) + 4\varepsilon\end{aligned}$$

holds for all $m, n \geqslant n_0$ and all $\mu \in \Omega$. Hence our claim. □

We take notice that the claim in the general case of Theorem 2.4.8 can be reformulated, stating that

$$\limsup_{m\to\infty}\Big(\int_B f_m \, d\theta\Big)(\mu) \leqslant \liminf_{n\to\infty}\Big(\int_B g_n \, d\theta\Big)(\mu)$$

holds for all $\mu \in F^*$ such that both $(\int_B f^* \, d\theta)(\mu) < +\infty$ and $(\int_B g_* \, d\theta)(\mu) < +\infty$. The result of this theorem becomes more recognizable if one of the sequences involved is constant. We formulate these special cases as our second convergence theorem.

2.4.9 Theorem. *Let* $f_n, f, f_*, f^* \in \mathcal{F}_{(\theta,B)}(X, \mathrm{Conv}(E))$ *for* $B \in \mathfrak{A}$.

(a) *If* $f_n \underset{a.e.B}{\leqslant} f^*$ *for all* $n \in \mathbb{N}$ *and* $f_n \, {}_{a.e.B}\!\searrow f$, *then for every* $\mu \in F^*$ *such that* $(\int_B f^* \, d\theta)(\mu) < +\infty$ *and every* $\varepsilon > 0$ *there is* n_0 *such that*

$$\Big(\int_B f_n \, d\theta\Big)(\mu) \leqslant \Big(\int_B f \, d\theta\Big)(\mu) + \varepsilon$$

for all $n \geqslant n_0$.

(b) *If* $0 \underset{a.e.B}{\leqslant} f_n + f_*$ *for all* $n \in \mathbb{N}$, *and* $f_n \nearrow_{a.e.B} f$, *then for every* $\mu \in F^*$ *such that both* $(\int_B f\, d\theta)(\mu) < +\infty$ *and* $(\int_B f_*\, d\theta)(\mu) < +\infty$ *and every* $\varepsilon > 0$ *there is* n_0 *such that*

$$\left(\int_B f\, d\theta\right)(\mu) \leqslant \left(\int_B f_n\, d\theta\right)(\mu) + \varepsilon$$

for all $n \geqslant n_0$.

(c) *If* $0 \underset{a.e.B}{\leqslant} f_n + f_*$ *and* $f_n \underset{a.e.B}{\leqslant} f^*$ *for all* $n \in \mathbb{N}$ *and* $f_n \underset{a.e.B}{\longrightarrow} f$, *then*

$$\left(\int_B f\, d\theta\right)(\mu) = \lim_{n\to\infty} \left(\int_B f_n\, d\theta\right)(\mu)$$

for all $\mu \in F^*$ *such that both* $(\int_B f_*\, d\theta)(\mu) < +\infty$ *and* $(\int_B f^*\, d\theta)(\mu) < +\infty$.

If the integrals of f, f_* *and* f^* *over* B *are strongly additive, then the inequalities in (a) and (b) hold uniformly on the polars of all neighborhoods in* $\mathcal{W}$, *and the limit in (c) holds with respect to the symmetric topology of* $\mathcal{Q}_F$.

Proof. We use Theorem 2.4.8 with the following insertions. For Part (a), we set $g_n = g_{**} = f$, $g_* = 0$, and our claim follows from 2.4.8. For Part (b), we insert f for f_n, rename f_n for g_n, f_* for g_* and choose $g_{**} = 0$. Part (c) combines both Parts (a) and (b). □

Again, in the general case the claims of Theorem 2.4.9(a) and (b) can be reformulated, stating that

$$\limsup_{m\to\infty} \left(\int_B f_m\, d\theta\right)(\mu) \leqslant \left(\int_B f\, d\theta\right)(\mu)$$

holds for all $\mu \in F^*$ such that $(\int_B f^*\, d\theta)(\mu) < +\infty$ in (a), and

$$\left(\int_B f\, d\theta\right)(\mu) \leqslant \liminf_{n\to\infty} \left(\int_B g_n\, d\theta\right)(\mu)$$

holds for all $\mu \in \Omega$ such that both $(\int_B f\, d\theta)(\mu) < +\infty$ and $(\int_B f_*\, d\theta)(\mu) < +\infty$ in Part (b).

2.4.10 Examples. (a) In case that $E = F$ is a Banach space over $\mathbb{R}$ or $\mathbb{C}$ and θ is a real- or complex-valued measure, then according to Corollary 2.4.3 the integrals for integrable E-valued functions over sets in $\mathfrak{R}$ are strongly additive. For every absorbing set $C \in \mathrm{Conv}(E)$ and an $\overline{\mathbb{R}}_+$-valued $\mathfrak{A}$-measurable function φ on X, the function $\varphi \cdot C \in \mathcal{F}(X, \mathrm{Conv}(E))$ is integrable over every set $B \in \mathfrak{A}$ and its integral is evaluated as

$$\left(\int_B \varphi \cdot C\, d\theta\right)(\mu) = \lim_{A\in\mathfrak{R}}\left(\int_{A\cap B} \varphi\, d|\theta|\right) p_C(\mu),$$

for $\mu \in F^*$, where $|\theta|$ denotes the usual modulus of a real- or complex-valued measure and $p_C(\mu) = \sup\{|\mu(c)| \mid c \in C\} \in \overline{\mathbb{R}}$ (see Example 2.3.26(a)). If C is bounded, then p_C is bounded in $\mathcal{Q}_F$ and if $\lim_{A\in\mathfrak{R}}(\int_{A\cap B} \varphi\, d|\theta|) < +\infty$ for $B \in \mathfrak{A}$, then the integral of $\varphi \cdot C$ over B is strongly additive. Theorem 2.4.9(c) therefore implies the dominated convergence theorem (Theorem 3 in II.2 of [21]) for Bochner integrable functions.

(b) For a concrete example, let $X = \mathbb{R}$, let $\mathfrak{R}$ be the family of all relatively compact Borel subsets of X and let θ be the Lebesgue measure on X. For $E = F$, we choose the space of all continuous real-valued functions on $[0, +\infty)$ that vanish at infinity, endowed with the norm of uniform convergence. Its dual F^* then consists of all bounded regular Borel measures on $[0, +\infty)$. We consider the balanced set $C \in \mathrm{Conv}(E)$ consisting of all functions $\rho \in E$ such that $\rho(x) \leqslant x^2$ for all $x \in [0, +\infty)$. According to Example 2.1.8(f), we have

$$|\theta|_A(C)(\mu) = \sup\left\{\sum_{i=1}^{n} |\theta_{A_i}| \,\middle|\, A_i \in \mathfrak{R} \text{ disjoint}, A_i \subset A\right\} p_C(\mu) = \theta_A\, p_C(\mu)$$

for $A \in \mathfrak{R}$ and $\mu \in F^*$, where

$$p_C(\mu) = \sup\{|\mu(\rho)| \mid \rho \in C\} = \int_0^\infty x^2\, d|\mu| \in \overline{\mathbb{R}}_+,$$

according to Example 2.1.8(f). In this context, $|\mu|$ denotes the usual modulus of the bounded real-valued measure $\mu \in F^*$. Following Example 2.3.26(a) for an $\overline{\mathbb{R}}_+$-valued measurable function φ on X and a Borel subset B of X, we compute

$$\begin{aligned}\left(\int_B \varphi \cdot C\, d\theta\right)(\mu) &= \lim_{A\in\mathfrak{R}}\left(\int_{A\cap B} \varphi\, d|\theta|\right) p_C(\mu)\\ &= \int_B \varphi\, dx \int_0^\infty x^2\, d|\mu| \in \overline{\mathbb{R}}_+\end{aligned}$$

for all $\mu \in F^*$. This integral is generally neither a bounded element of $\mathcal{Q}_F$ nor is it strongly additive. Now consider a sequence of θ-almost everywhere positive measurable functions $\varphi_n \in \mathcal{F}(X)$ converging pointwise to $\varphi \in \mathcal{F}(X)$. For every Borel subset B of X, we have $\varphi_n \cdot C \underset{a.e.B}{\longrightarrow} \varphi \cdot C$ by Proposition 2.4.7(c). If there is a measurable function $\varphi^* \in F$ such that $0 \underset{a.e.B}{<} \varphi_n \underset{a.e.B}{\leqslant} \varphi^*$ for all $n \in \mathbb{N}$, then Theorem 2.4.9(c) applies with $f_* = 0$ and $f^* = \varphi^* \cdot C$. If for example we choose $\varphi_n(x) = |x|^n$, then $\varphi(x) = 0$ for $|x| < 1$, $\varphi(\pm 1) = 1$ and $\varphi(x) = +\infty$ for $|x| > 1$, and we choose $\varphi^*(x) = 1$ for $|x| < 1$, $\varphi^*(\pm 1) = 1$ and $\varphi^*(x) = +\infty$ for $|x| > 1$. Thus Theorem 2.4.9(c) applies to all Borel subsets B of $[-1, +1]$ and $\mu \in F^*$ that as measures on $[0, +\infty)$ have a compact support.

2.5 Measures as linear operators

Using the general assumptions from the previous sections, we proceed to investigate the integral as a linear operator from the integration cones $\mathcal{F}_{(\theta,B)}(X, \mathrm{Conv}(E))$ into $\mathcal{Q}_F$. For an $\mathfrak{L}(E, F^{*\blacksquare})$-valued measure θ and a set $B \in \mathfrak{A}$, we shall denote the operator

$$f \mapsto \int_B f\, d\theta : \mathcal{F}_{(\theta,B)}(X, \mathrm{Conv}(E)) \to \mathcal{Q}_F$$

by $\mathcal{T}_{(\theta,B)}$. According to Theorem 2.3.15, this operator is linear. For every $W \in \mathcal{W}$, we denote by $\mathfrak{v}_{(\theta,W)}$ the function space neighborhood, which is generated by the family $\mathcal{N}_{\mathfrak{v}_{(\theta,W)}}$ of all measurable neighborhood functions n such that $\int_X n\, d\theta \leqslant \sigma_W$. The requirements for a function space neighborhood are readily verified. By $\mathfrak{V}_\theta$, we denote the function space neighborhood system in the sense of Section 1.2 comprised of the neighborhoods $\mathfrak{v}_{(\theta,W)}$ for all $W \in \mathcal{W}$. We elaborated earlier (see Section 2.2) that the concept of function space neighborhoods canonically extends to the set-valued functions in $\mathcal{F}(X, \mathrm{Conv}(E))$. Locally convex cones were introduced in Section 1.4.

2.5.1 Proposition. *For every $A \in \mathfrak{R}$, the cone $\mathcal{F}_{(\theta,A)}(X, \mathrm{Conv}(E))$ endowed with $\underset{a.e.A}{\leqslant}$ as its order and the neighborhood system $\mathfrak{V}_\theta$ forms a locally convex cone. The subcone of measurable functions is dense in its symmetric topology. $\mathcal{T}_{(\theta,A)}$ is a continuous linear operator from $\mathcal{F}_{(\theta,A)}(X, \mathrm{Conv}(E))$ into $\mathcal{Q}_F$.*

Proof. Let $A \in \mathfrak{R}$. We shall establish that $\mathcal{F}_{(\theta,A)}(X, \mathrm{Conv}(E))$ is a subcone of a full locally convex cone $(\mathcal{P}, \mathfrak{V}_\theta)$, which contains the neighborhoods in $\mathfrak{V}_\theta$ as its elements. Sums, positive multiples and the order for function space neighborhoods were defined in Section 1.2. We denote by $\widetilde{\mathfrak{V}}_\theta$ the cone generated by the sums of elements of $\mathfrak{V}_\theta$ adjoined by a formal zero element o and set

$$\mathcal{P} = \{(f, \mathfrak{v}) \mid f \in \mathcal{F}_{(\theta,A)}(X, \mathrm{Conv}(E)),\ \mathfrak{v} \in \widetilde{\mathfrak{V}}_\theta\}.$$

Endowed with the componentwise algebraic operations $\mathcal{P}$ forms a cone. Its canonical order is given by $(f, \mathfrak{v}) \leqslant (g, \mathfrak{u})$ if for every $n \in \mathcal{N}_\mathfrak{v}$ there is $m \in \mathcal{N}_\mathfrak{u}$ such that $f + n \underset{a.e.A}{\leqslant} g + m$. If either $\mathfrak{v} = \mathrm{o}$ or $\mathfrak{u} = \mathrm{o}$, the latter means $f \underset{a.e.A}{\leqslant} g + m$ or $f + n \underset{a.e.A}{\leqslant} g$, respectively. The embeddings $f \mapsto (f, \mathrm{o})$ of $\mathcal{F}_{(\theta,A)}(X, \mathrm{Conv}(E))$ and $\mathfrak{v}_{(\theta,W)} \mapsto (0, \mathfrak{v}_{(\theta,W)})$ of $\mathfrak{V}_\theta$ into $\mathcal{P}$ preserve the respective algebraic and order structures. Furthermore, every element $(f, \mathfrak{u}) \in \mathcal{P}$ is bounded below with respect to the neighborhood system $\mathfrak{V}_\theta$. Indeed, let $\mathfrak{v}_{(\theta,W)} \in \mathfrak{V}_\theta$ with $W \in \mathcal{W}$. According to Proposition 2.3.13, there is a measurable neighborhood function n and $\lambda \geqslant 0$ such that $0 \leqslant f + n$ and $\int_A n\, d\theta \leqslant \lambda \sigma_W$. Hence $m = \chi_A \cdot n \in \mathcal{N}_{\lambda \mathfrak{v}_{(\theta,W)}}$. We choose any neighborhood function $u \in \mathcal{N}_\mathfrak{u}$ and set $s = u + m \in \mathcal{N}_{\mathfrak{u} + \lambda \mathfrak{v}_{(\theta,W)}}$. Then $0 \underset{a.e.A}{\leqslant} f + s$ and, therefore,

$$(0, \mathrm{o}) \leqslant (f, \mathfrak{u}) + \lambda(0, \mathfrak{v}_{(\theta,W)}) = (f, \mathfrak{u} + \lambda \mathfrak{v}_{(\theta,W)}).$$

In this way, $(\mathcal{P},\mathcal{V}_\theta)$ becomes a full locally convex cone, which contains the cone $\mathcal{F}_{(\theta,A)}(X,\mathrm{Conv}(E))$ of integrable functions as a subcone. A brief inspection of our definition of integrability shows that every symmetric neighborhood of a function in $\mathcal{F}_{(\theta,A)}(X,\mathrm{Conv}(E))$ contains measurable functions, which therefore form a dense subcone. Finally, the operator

$$\mathcal{T}_{(\theta,A)}:\mathcal{F}_{(\theta,A)}(X,\mathrm{Conv}(E))\to\mathcal{Q}_F$$

is real-linear and continuous. For the latter, let $W\in\mathcal{W}$. If $f\underset{\text{a.e.}A}{\leqslant} g+\mathfrak{v}_{(\theta,W)}$ for $f,g\in\mathcal{F}_{(\theta,A)}(X,\mathrm{Conv}(E))$, then there exists a neighborhood function $n\in\mathcal{N}_{\mathfrak{v}_{(\theta,W)}}$ such that $f\underset{\text{a.e.}A}{\leqslant} g+n$. Following Theorem 2.3.15(b) and (c), this renders

$$\mathcal{T}_{(\theta,A)}(f)=\int_A f\,d\theta\leqslant\int_A g\,d\theta+\int_A n\,d\theta\leqslant\mathcal{T}_{(\theta,A)}(g)+\sigma_W.$$ □

We note that the order $\underset{\text{a.e.}A}{\leqslant}$ induces the usual equivalence relation for integration spaces. The locally convex cone $\mathcal{F}_{(\theta,B)}(X,\mathrm{Conv}(E))$ is closed in $\mathcal{F}(X,\mathrm{Conv}(E))$ in the following sense.

2.5.2 Proposition. *Let $f\in\mathcal{F}(X,\mathrm{Conv}(E))$ and let $B\in\mathfrak{A}$. If for every $W\in\mathcal{W}$, there is $g\in\mathcal{F}_{(\theta,B)}(X,\mathrm{Conv}(E))$ such that $f\underset{\text{a.e.}B}{\leqslant} g+\mathfrak{v}_{(\theta,W)}$ and $g\underset{\text{a.e.}B}{\leqslant} f+\mathfrak{v}_{(\theta,W)}$, then $f\in\mathcal{F}_{(\theta,B)}(X,\mathrm{Conv}(E))$.*

Proof. Suppose that $f\in\mathcal{F}(X,\mathrm{Conv}(E))$ satisfies our assumptions. Using Theorem 2.3.18(a), we may assume that $B=X$. Given $W\in\mathcal{W}$, we set $U=(1/3)W\in\mathcal{W}$. There are $g\in\mathcal{F}_{(\theta,X)}(X,\mathrm{Conv}(E))$ and neighborhood functions $n_1,n_2\in\mathcal{N}_{\mathfrak{v}_{(\theta,U)}}$ such that

$$f\underset{\text{a.e.}X}{\leqslant} g+n_1,\quad g\underset{\text{a.e.}B}{\leqslant} f+n_2\quad\text{and}\quad\int_X n_1\,d\theta,\int_X n_2\,d\theta\leqslant\sigma_U.$$

Given $A\in\mathfrak{R}$, there are a measurable functions g_U and n_U in $\mathcal{F}(X,\mathrm{Conv}(E))$ as in the definition of the integrability of g over A. That is, $g\underset{\text{a.e.}A}{\leqslant} g_U\underset{\text{a.e.}A}{\leqslant} g+n_U$ and $\int_A n_U\,d\theta\leqslant\sigma_U$. Hence

$$f\underset{\text{a.e.}A}{\leqslant} g_U+n_1\underset{\text{a.e.}A}{\leqslant} g+n_U+n_1\underset{\text{a.e.}A}{\leqslant} f+n_U+n_1+n_2.$$

Now setting $f_W=g_U+n_1$ and $n_W=n_U+n_1+n_2$, we have

$$f\underset{\text{a.e.}A}{\leqslant} f_W\underset{\text{a.e.}A}{\leqslant} f+n_W,$$

where f_W is measurable and n_W is a measurable neighborhood function such that

$$\int_A n_W\,d\theta\leqslant 3\sigma_U=\sigma_W.$$

In this way, f satisfies the definition of integrability over every set $A \in \mathfrak{R}$. Furthermore, the above yields that

$$\int_A f\,d\theta \leqslant \int_A g\,d\theta + \sigma_U \quad \text{and} \quad \int_A g\,d\theta \leqslant \int_A f\,d\theta + \sigma_U$$

holds for all $A \in \mathfrak{R}$. Since g is integrable over X, the limit $\int_X g\,d\theta = \lim_{A\in\mathfrak{R}} \int_A g\,d\theta$ exist pointwise on F^* and is an element of $\mathcal{Q}_F$. This yields

$$\begin{aligned}\limsup_{A\in\mathfrak{R}}\Big(\int_A f\,d\theta\Big)(\mu) &\leqslant \Big(\int_X g\,d\theta\Big)(\mu) + \sigma_U(\mu)\\ &\leqslant \liminf_{A\in\mathfrak{R}}\Big(\int_A f\,d\theta\Big)(\mu) + 2\sigma_U(\mu)\end{aligned}$$

for all $\mu \in F^*$. Since this last inequality holds for any choice of W, we can infer that the limit $\lim_{A\in\mathfrak{R}}(\int_A f\,d\theta)(\mu) \in \overline{\mathbb{R}}$ exists for all $\mu \in F^*$. Moreover, for every $W \in \mathcal{W}$ the integral $\int_X g\,d\theta$ as an element of $\mathcal{Q}_F$ is bounded below on W°. Hence the above inequality also implies that the pointwise limit $\lim_{A\in\mathfrak{R}} \int_A f\,d\theta$ is similarly bounded below on W° and, therefore, also an element of $\mathcal{Q}_F$. We infer that $f \in \mathcal{F}_{(\theta,X)}(X, \mathrm{Conv}(E))$, our claim. □

The approximation procedure described in Proposition 2.5.2 uses the symmetric neighborhoods created by $\mathfrak{V}_\theta$. Together with the order $\underset{a.e.A}{\leqslant}$ they define a topology on all of $\mathcal{F}(X, \mathrm{Conv}(E))$, which does however not render a locally convex cone since the elements need not be bounded below. Propositions 2.5.1 and 2.5.2 yield that $\mathcal{F}_{(\theta,A)}(X, \mathrm{Conv}(E))$ is the closure in this topology of the cone of measurable functions in $\mathcal{F}(X, \mathrm{Conv}(E))$.

The comparison of function space neighborhood systems was defined in Section 1.2. In accordance with this definition, $\mathfrak{V}$ is *finer* than $\mathfrak{U}$ if for every $\mathfrak{u} \in \mathfrak{U}$ there is $\mathfrak{v} \in \mathfrak{V}$ such that $\mathfrak{v} \leqslant \mathfrak{u}$. In this sense, the earlier (see Section 2.3) specified function space neighborhood system $\{\mathfrak{v}_W \mid W \in \mathcal{W}\}$ is finer than $\mathfrak{V}_\theta$.

2.5.3 Proposition. *Suppose that $\mathfrak{V}$ is finer than $\mathfrak{V}_\theta$ and let $A \in \mathfrak{R}$. If for every $\mathfrak{v} \in \mathfrak{V}$, there is $n \in \mathcal{N}_\mathfrak{v}$ such that $n(x) = E$ for all $x \in A$, then A is of measure zero.*

Proof. Suppose that $\mathfrak{V}$ is finer that $\mathfrak{V}_\theta$. Let $A \in \mathfrak{R}$ and let $B \in \mathfrak{R}$ be any subset of A. For $W \in \mathcal{W}$, choose a function space neighborhood $\mathfrak{v} \in \mathfrak{V}$ such that $\mathfrak{v} \leqslant \mathfrak{v}_{(\theta,W)}$. Following our assumption, there is $n \in \mathcal{N}_\mathfrak{v}$, such that $n(x) = E$ for all $x \in A$. Then for every $a \in E$ the function $\chi_B \cdot a$ is measurable (Proposition 2.2.9), hence integrable, and we have $\chi_B \cdot a \leqslant n$, that is, $\chi_B \cdot a \leqslant \mathfrak{v}$. Thus $\theta_B(a) = \int_X \chi_B \cdot a\,d\theta \leqslant \sigma_W$. As $W \in \mathcal{W}$ was arbitrarily chosen, this shows $\theta_B(a) = 0$. Hence $\theta_B = 0$. □

Properties of operators defined by an integral

We shall now return to the investigation of function spaces, that is, spaces of E-valued functions as introduced in Section 1.2 and to the topological case. Throughout the remainder of this section and indeed of this text, we shall assume that X is a locally compact Hausdorff space and that $\mathfrak{R}$ consists of all relatively compact Borel subsets of X. Any measure θ on $\mathfrak{R}$ is presumed to be $\mathfrak{L}(E, F^{*\blacksquare})$-valued and bounded. Recall that integrals of E-valued functions are elements of $F^{*\blacksquare}$ (Proposition 2.3.23(a)) and, therefore, finitely-valued on F^*. The function space neighborhood system $\mathfrak{V}$ is assumed to be measurable. As a reference, we shall use the function space neighborhood systems $\mathfrak{V}_\theta$ and the system $\mathfrak{V}_c$ of compact convergence from Example 1.2.7(b). Using the conventions from Section 1.3, the space of all continuous linear operators from $\mathcal{C}_{\mathfrak{V}}(X, E)$ to $F, \widehat{F}, F^{**}$ or $F^{*\blacksquare}$ is denoted by

$$\mathfrak{L}(\mathcal{C}_{\mathfrak{V}}(X,E),F), \quad \mathfrak{L}(\mathcal{C}_{\mathfrak{V}}(X,E),\widehat{F}), \quad \mathfrak{L}(\mathcal{C}_{\mathfrak{V}}(X,E),F^{**}) \quad \text{or} \quad \mathfrak{L}(\mathcal{C}_{\mathfrak{V}}(X,E),F^{*\blacksquare}),$$

respectively. Proposition 2.5.1 implies that the operator $\mathcal{T}_{(\theta,B)}$ is continuous on every function space contained in $\mathcal{F}_{(\theta,B)}(X, \mathrm{Conv}(E))$ provided that its topology is finer than $\mathfrak{V}_\theta$. In our first set of results, we shall derive properties of the operator $\mathcal{T}_{(\theta,B)}$ from those of the measure θ.

2.5.4 Proposition. *Suppose that $\mathfrak{V}$ is finer than $\mathfrak{V}_\theta$, let $f \in \mathcal{C}_{\mathfrak{V}}(X,E)$ and $B \in \mathfrak{A}$.*

(a) *The function f is integrable over B with respect to θ and the net $(\int_{A\cap B} f\, d\theta)_{A\in\mathfrak{R}}$ converges in the symmetric topology of $\mathcal{Q}_F$ to $\int_B f\, d\theta$.*
Hence $\mathcal{T}_{(\theta,B)} \in \mathfrak{L}(\mathcal{C}_{\mathfrak{V}}(X,E),F^{*\blacksquare})$.

(b) *If θ is $\mathfrak{L}(E,F)$-valued, then $\int_B f\, d\theta$ is strongly additive and an element of $\widehat{F}$, the completion of F. Hence $\mathcal{T}_{(\theta,B)} \in \mathfrak{L}(\mathcal{C}_{\mathfrak{V}}(X,E),\widehat{F})$.*

(c) *If θ is $\mathfrak{L}(E,F)$-valued and weakly compact, then $\int_A f\, d\theta \in F$ for every $A \in \mathfrak{R}$. Hence $\mathcal{T}_{(\theta,A)} \in \mathfrak{L}(\mathcal{C}_{\mathfrak{V}}(X,E),F)$.*

Proof. Suppose that $\mathfrak{V}$ is finer than $\mathfrak{V}_\theta$ and let $f \in \mathcal{C}_{\mathfrak{V}}(X,E)$. According to Proposition 2.3.24, every continuous E-valued function is integrable over all sets $A \in \mathfrak{R}$ and the integral is an element of $F^{*\blacksquare}$ by 2.3.23(a). Let $B \in \mathfrak{A}$. Given $W \in \mathcal{W}$, since $\mathfrak{V}$ is finer than $\mathfrak{V}_\theta$, there is $A \in \mathfrak{R}$ such that $\chi_{X\setminus D} \cdot f \leqslant \mathfrak{v}_{(\theta,W)}$ Then for any choice of $C, D \in \mathfrak{R}$, both containing A we have

$$C \cap B = (C \cap D \cap B) \cup (C \setminus (C \cap D \cap B))$$

and

$$D \cap B = (C \cap D \cap B) \cup (D \setminus (C \cap D \cap B)).$$

Since both sets $C\setminus(C\cap D\cap B)$ and $D\setminus(C\cap D\cap B)$ are contained in $X\setminus A$, the integral of f over each of these sets is smaller than σ_W. Consequently, with 2.3.18(b), we realize that both

$$\int_{C\cap B} f\,d\theta \leqslant \int_{D\cap B} f\,d\theta + \sigma_W \quad \text{and} \quad \int_{D\cap B} f\,d\theta \leqslant \int_{C\cap B} f\,d\theta + \sigma_W.$$

Hence $(\int_{A\cap B} f\,d\theta)_{A\in\mathfrak{R}}$ is a Cauchy net in the symmetric topology of $\mathcal{Q}_F$ and, therefore, convergent in this topology to $\int_B f\,d\theta$. For the latter, we recall from Section 2.2 that $\mathcal{Q}_F$ is complete in this topology. The integral defines a real- or complex-linear operator from $\mathcal{F}_{(\theta,A)}(X,E)$ into $F^{*\blacksquare}$ by 2.3.23(b). Part (a) follows. Part (b) is a consequence of Corollary 2.4.3 and Proposition 2.3.23(c), and Part (c) follows from Proposition 2.3.23(d). □

Part (c) of Proposition 2.5.4 can be further strengthened in case that the neighborhood system $\mathfrak{V}$ is also finer than $\mathfrak{V}_c$.

2.5.5 Proposition. *If $\mathfrak{V}$ is finer than both $\mathfrak{V}_\theta$ and $\mathfrak{V}_c$ and if θ is $\mathfrak{L}(E,F)$-valued and compact (or weakly compact), then for every $A\in\mathfrak{R}$ the operator $\mathcal{T}_{(\theta,A)}\in\mathfrak{L}(C_{\mathfrak{V}}(X,E),F)$ is compact (or weakly compact).*

Proof. Suppose that $\mathfrak{V}$ is finer than both $\mathfrak{V}_\theta$ and $\mathfrak{V}_c$ and let $A\in\mathfrak{R}$. According to Propositions 2.5.1 and 2.5.4, $\mathcal{T}_{(\theta,A)}$ is a continuous linear operator from $C_{\mathfrak{V}}(X,E)$ to F. Let K be a compact subset of X that contains the closure of A in its interior. There is $\varphi\in C_K(X)$ such that $\chi_A\leqslant\varphi\leqslant\chi_K$. Let $\mathcal{B}$ be a bounded subset of $C_{\mathfrak{V}}(X,E)$. Following Lemma 1.2.1(a), the set $\widetilde{\mathcal{B}}=\{\varphi\cdot f \mid f\in\mathcal{B}\}$ is also bounded and since

$$\mathcal{T}_{(\theta,A)}(f)=\int_A f\,d\theta=\int_A \varphi\cdot f\,d\theta=\mathcal{T}_{(\theta,A)}(\varphi\cdot f),$$

we have $\mathcal{T}_{(\theta,A)}(\mathcal{B})=\mathcal{T}_{(\theta,A)}(\widetilde{\mathcal{B}})$. By our assumption, for every $V\in\mathcal{V}$ there is $\mathfrak{v}\in\mathcal{V}$ such that $\mathfrak{v}\leqslant\mathfrak{v}_{(K,V)}$ (see Example 1.2.7(b)). There is $\lambda\geqslant 0$ such that $\widetilde{\mathcal{B}}\leqslant\lambda\mathfrak{v}$, that is, $f\leqslant\lambda\mathfrak{v}$, hence $f(x)\in\lambda V$ for all $x\in K$ and all $f\in\widetilde{\mathcal{B}}$. The set $C=\bigcup\{f(K)\mid f\in\widetilde{\mathcal{B}}\}$ is therefore seen to be bounded in E, and so is its convex hull $\widetilde{C}$. By the definition of (weak) compactness of a measure, the convex set

$$D=\left\{\sum_{i=1}^{n}\theta_{A_i}(c_i)\;\middle|\; c_i\in\widetilde{C},\ A_i\in\mathfrak{R}\text{ disjoint, } \bigcup_{i=1}^{n}A_i=K\right\}$$

is relatively compact (or relatively weakly compact) in F. Every function $f\in\widetilde{\mathcal{B}}$ is $\widetilde{C}$-valued and can be approximated in the topology of $C_{\mathfrak{V}}(X,E)$ by $\widetilde{C}$-valued step functions that are supported by K. (Proposition 1.2.5(ii)). The integrals of these step functions are therefore elements of D, and as the pointwise limit of these integrals, the value $\mathcal{T}_{(\theta,A)}(f)=\int_A f\,d\theta$ is contained in the weak closure of D. As D is convex, its closure in

F coincides with its weak closure. We infer that $\mathcal{T}_{(\theta,A)}(\mathcal{B}) = \mathcal{T}_{(\theta,A)}(\widetilde{\mathcal{B}})$ is indeed relatively compact (or relatively weakly compact) in F. □

We shall supply a reverse of the assertion of Proposition 2.5.5 below, but continue with some technical, but nevertheless highly practicable considerations. For the following statements, recall the definition of boundedness for subsets of $\mathcal{F}(X, \mathrm{Conv}(E))$ and $\mathcal{F}(X,E)$ from Section 1.4.

2.5.6 Lemma. *For every set $A \in \mathfrak{R}$, there is a net $(\varphi_i)_{i\in\mathcal{I}}$ in $\mathcal{C}_\mathcal{K}(X)$ such that for every compact subset K of A and every open set $O \in \mathfrak{O}$ containing A there is $i_0 \in \mathcal{I}$ such that $\chi_K \leqslant \varphi_i \leqslant \chi_O$ for all $i \geqslant i_0$. Every net $(\varphi_i)_{i\in\mathcal{I}}$ in $\mathcal{C}_\mathcal{K}(X)$ with this property converges pointwise on X to χ_A, and for every choice of E there is $i_0 \in \mathcal{I}$ such that for every $f \in \mathcal{C}(X,E)$ the set $\{\varphi_i \cdot f \mid i \in \mathcal{I},\ i \geqslant i_0\}$ is bounded in $\mathcal{F}(X,E)$. For every choice of F and every regular $\mathfrak{L}(E, F^{*\blacksquare})$-valued measure θ, we have*

$$\lim_{i\in\mathcal{I}} \mathcal{T}_{(\theta,X)}(\varphi_i \cdot f) = \int_A f \, d\theta.$$

Convergence is meant pointwise on F^*.

Proof. Given a set $A \in \mathfrak{R}$, we use the index set $\mathcal{I}$ consisting of all pairs (O,K), where $O \in \mathfrak{O}$ is an open set containing A and K is a compact subset of A, ordered by $(O,K) \leqslant (O',K')$ if $O \supset O'$ and $K \subset K'$. We choose functions $\varphi_{(O,K)} \in \mathcal{C}_\mathcal{K}(X)$ such that $\chi_K \leqslant \varphi_{(O,K)} \leqslant \chi_O$. The net $(\varphi_i)_{i\in\mathcal{I}}$ then satisfies our first statement. If $(\varphi_i)_{i\in\mathcal{I}}$ is any such net in $\mathcal{C}_\mathcal{K}(X)$, it converges pointwise to χ_A since for every $x \in A$, $K = \{x\}$ is compact subset of A, and for $x \notin A$ there is $O \in \mathfrak{O}$ containing A but not x. For the latter, consider any set $U \in \mathfrak{O}$ containing A and choose $O = U \setminus \{x\}$. Next, for any choice of O and K choose $i_0 \in \mathcal{I}$ such that $\chi_K \leqslant \varphi_i \leqslant \chi_O$ for all $i \geqslant i_0$. Let E be a locally convex topological vector space and let $f \in \mathcal{C}(X,E)$. We claim that the set

$$\mathcal{F} = \{\varphi_i \cdot f \mid i \in \mathcal{I},\ i \geqslant i_0\}$$

is bounded in $\mathcal{F}(X,E)$. Indeed, given a function space neighborhood $\mathfrak{v}$ by Lemma 1.2.2 there is $V \in \mathcal{V}$ and $\lambda \geqslant 0$ such that $\chi_O \cdot f \leqslant \lambda\mathfrak{v}$. Thus $\mathcal{F} \subset \lambda\mathfrak{v}$ by 1.2.1(a) and our claim follows. For our final statement, let F be a Hausdorff locally convex topological vector space and let θ be a regular $\mathfrak{L}(E, F^{*\blacksquare})$-valued measure. According to Proposition 2.3.24, every function $f \in \mathcal{C}(X,E)$ is integrable over A with respect to θ. Let $\mu \in F^*$. Since θ is regular, there is $V \in \mathcal{V}$ such that

$$\lim_{O\supset A} |\theta|_{O\setminus A}(V)(\mu) = \lim_{K\subset A} |\theta|_{A\setminus K}(V)(\mu) = 0.$$

We choose $U \in \mathfrak{O}$ such that $A \subset U$. Since U is relatively compact in X and f is continuous, the set $f(U)$ is bounded in E. Hence there is $\lambda > 0$ such that $f(U) \subset \lambda V$. Given $\varepsilon > 0$, we find a subset $O \in \mathfrak{O}$ of U containing A and a compact subset K

of A such that $|\theta|_{O\setminus K}(V)(\mu) \leqslant \varepsilon/(2\lambda)$. There is $i_0 \in \mathcal{I}$ such that $\chi_K \leqslant \varphi_i \leqslant \chi_O$ for all $i \geqslant i_0$. Since $(\varphi_i \cdot f)(U) \subset \lambda V$ holds as well, we infer that

$$\left|\left(\int_{A\setminus K} f\,d\theta\right)(\mu)\right| \leqslant \frac{\varepsilon}{2} \quad\text{and}\quad \left|\left(\int_{O\setminus K} \varphi_i \cdot f\,d\theta\right)(\mu)\right| \leqslant \frac{\varepsilon}{2}$$

holds for all $i \geqslant i_0$. Now combining the latter and using that $\mathcal{T}_{(\theta,X)}(\varphi_i \cdot f) = \int_X \varphi_i \cdot f\,d\theta = \int_O \varphi_i \cdot f\,d\theta$ and $\int_K f\,d\theta = \int_K \varphi_i \cdot f\,d\theta$, we conclude that

$$\left|\left(\int_A f\,d\theta\right)(\mu) - \mathcal{T}_{(\theta,X)}(\varphi_i \cdot f)(\mu)\right| = \left|\left(\int_{A\setminus K} f\,d\theta\right)(\mu) - \left(\int_{O\setminus K} \varphi_i \cdot f\,d\theta\right)(\mu)\right| \leqslant \varepsilon.$$

Our claim follows. □

A similar but slightly weaker statement is available for all sets in the σ-field $\mathfrak{A}$ associated with $\mathfrak{R}$ and functions in $\mathcal{C}_{\mathfrak{V}}(X,E)$. Following Proposition 2.5.4(a), every function $f \in \mathcal{C}_{\mathfrak{V}}(X,E)$ is integrable over every set $B \in \mathfrak{A}$ with respect to a measure θ, provided that $\mathfrak{V}$ is finer than $\mathfrak{V}_\theta$, and its integral is an element of $F^{*\blacksquare}$. We shall construct a diagonal net for the proof of the following lemma.

2.5.7 Lemma. *For every set $B \in \mathfrak{A}$, there is a net $(\varphi_i)_{i\in\mathcal{I}}$ in $\mathcal{C}_{\mathcal{K}}(X)$ such that $(\varphi_i)_{i\in\mathcal{I}}$ converges pointwise to χ_B and for every compact subset K of B and every open set U containing B there is $i_0 \in \mathcal{I}$ such that $\chi_K \leqslant \varphi_i \leqslant \chi_U$ for all $i \geqslant i_0$. For every choice of E, of $\mathcal{C}_{\mathfrak{V}}(X,E)$ and F, for every regular $\mathfrak{L}(E,F^{*\blacksquare})$-valued measure θ such that $\mathfrak{V}$ is finer than $\mathfrak{V}_\theta$ and every $f \in \mathcal{C}_{\mathfrak{V}}(X,E)$ we have*

$$\lim_{i\in\mathcal{I}} \mathcal{T}_{(\theta,X)}(\varphi_i \cdot f) = \int_B f\,d\theta$$

Convergence is meant pointwise on F^*.

Proof. Let $B \in \mathfrak{A}$. For every $A \in \mathfrak{R}$, let $(\varphi_i^A)_{i\in\mathcal{I}_A}$ be a net in $\mathcal{C}_{\mathcal{K}}(X)$ converging to $B \cap A \in \mathfrak{R}$ as in Lemma 2.5.6. We define the index set for a diagonal net as follows. Let

$$\mathcal{I} = \{(A,\xi) \mid A \in \mathfrak{R} \text{ and } \xi \in \Pi_{A\in\mathfrak{R}}\,\mathcal{I}_A\}.$$

We notice that the elements of $\Pi_{A\in\mathfrak{R}}\,\mathcal{I}_A$ are functions ξ on $\mathfrak{R}$ such that $\xi(A) \in \mathcal{I}_A$ for all $A \in \mathfrak{R}$. The order on $\mathcal{I}$ is defined by $(A_1,\xi_1) \leqslant (A_2,\xi_2)$ if $A_1 \subset A_2$ and $\xi_1(A) \leqslant \xi_2(A)$ for all $A \in \mathfrak{R}$. We shall verify that the net $(\varphi^A_{\xi(A)})_{(A,\xi)\in\mathcal{I}}$ satisfies our requirements. Indeed, let $K \subset B \subset U$ for a compact set K and an open set U. We choose $\xi_0 \in \Pi_{A\in\mathfrak{R}}\mathcal{I}_A$ such that for all sets $A \in \mathfrak{R}$ that contain K we have $\chi_K \leqslant \varphi_i^A \leqslant \chi_U$ for all $i \in \mathcal{I}_A$ such that $i \geqslant \xi_0(A)$. Then $\chi_K \leqslant \varphi^A_{\xi(A)} \leqslant \chi_U$ for all $(A,\xi) \geqslant (K,\xi_0)$. This implies pointwise convergence to χ_B since for every $x \in X$ we have either $K = \{x\} \subset B$ or $x \notin B$ and $U = X\setminus\{x\}$ is an open set containing B, but not x. Hence our first claim. For our second claim, let E, $C_{\mathfrak{V}}(X,E)$, F and θ be as stated and let $f \in C_{\mathfrak{V}}(X,E)$. Then $\varphi_i \cdot f \in C_{\mathfrak{V}}(X,E)$ for all $i \in \mathcal{I}$, hence $\varphi_i \cdot f$ is integrable over X with respect to θ by 2.5.4. Let $\mu \in F^*$ and $\varepsilon > 0$. Choose $A_0 \in \mathfrak{R}$ such that $|(\int_B f\,d\theta)(\mu) - (\int_{B\cap A} f\,d\theta)(\mu)| \leqslant \varepsilon/2$ for all $A \in \mathfrak{R}$ such that $A_0 \subset A$. Then according to Lemma 2.5.6, we choose $\xi_0 \in \Pi_{A\in\mathfrak{R}}\mathcal{I}_A$ such that for every $A \in \mathfrak{R}$ we have $|(\int_{B\cap A} f\,d\theta)(\mu) - (\mathcal{T}_{(\theta,X)}(\varphi_i^A \cdot f))(\mu)| \leqslant \varepsilon/2$ for all $i \in \mathcal{I}_A$ such that $i \geqslant \xi_0(A)$. Thus

$$\left|\left(\int_B f\,d\theta\right)(\mu) - (\mathcal{T}_{(\theta,X)}(\varphi_{\xi(A)}^A \cdot f))(\mu)\right| \leqslant \varepsilon$$

holds for all $(A,\xi) \geqslant (A_0,\xi_0)$, and our claim follows. □

Properties of measures representing an operator

Lemmas 2.5.6 and 2.5.7 permit to conveniently retrieve properties of a regular measure from those of the operator $\mathcal{T}_{(\theta,X)}$ that it induces on function spaces. Using Lemma 2.5.6, we can directly compute the values of the measure θ in terms $\mathcal{T}_{(\theta,X)}$. Indeed, let $A \in \mathfrak{R}$ and $a \in E$. We apply 2.5.6 with a net $(\varphi_i)_{i\in\mathcal{I}}$ in $C_{\mathcal{K}}(X)$ as stated and the constant function $x \mapsto a$ for f. This renders

$$\theta_A(a) = \lim_{i\in\mathcal{I}} \mathcal{T}_{\theta,X}(\varphi_i \cdot a) \in F^{*\blacksquare}.$$

As an immediate consequence, we obtain the following statements.

2.5.8 Proposition. *Let θ and ϑ be regular measures. If $\mathcal{T}_{(\theta,X)}(\varphi \cdot a) = \mathcal{T}_{(\vartheta,X)}(\varphi \cdot a)$ for all $\varphi \in C_{\mathcal{K}}(X)$ and $a \in E$, then $\theta = \vartheta$.*

2.5.9 Proposition. *Let θ be a regular measure and let $B \in \mathfrak{A}$ be a closed set. If $\mathcal{T}_{(\theta,B)}(\varphi \cdot a) = 0$ for all $\varphi \in C_{\mathcal{K}}(X)$ and $a \in E$ such that φ is supported by $X \setminus B$, then θ is supported by B.*

Proof. We shall verify that $X\setminus B$ is of measure zero. Let $A \in \mathfrak{R}$ be a subset of $X\setminus B$ and let $a \in E$. We choose the net $(\varphi_i)_{i\in\mathcal{I}}$ in $C_{\mathcal{K}}(X)$ from Lemma 2.5.6 for the set A with $K = \varnothing$ and $U = X \setminus B$. For $f \in C(X,E)$, we choose the constant function $x \mapsto a$. Then all the functions $\varphi_i \cdot a$ vanish on B, hence $\mathcal{T}(\varphi_i \cdot a) = 0$. Thus $\theta_A(a) = \int_A f\,d\theta = 0$ by Lemma 2.5.6. Our claim follows. □

2.5.10 Proposition. *Suppose that $\mathfrak{V}$ is finer than $\mathfrak{V}_\theta$, that θ is a regular measure and that $\mathcal{T}_{(\theta,X)} \in \mathfrak{L}(C_{\mathfrak{V}}(X,E),F)$.*

(a) *Then θ is $\mathfrak{L}(E,F^{**})$-valued and $\mathcal{T}_{(\theta,A)} \in \mathfrak{L}(C_{\mathfrak{V}}(X,E),F^{**})$ for all $A \in \mathfrak{R}$.*
(b) *If for every $A \in \mathfrak{R}$ and $W \in \mathcal{W}$, there is a function space neighborhood $\mathfrak{v}$ such that $\|\theta\|_A^{(\mathfrak{v},W^\circ)} < +\infty$, then θ is $\mathfrak{L}(E,\widehat{F})$-valued and $\mathcal{T}_{(\theta,A)} \in \mathfrak{L}(C_{\mathfrak{V}}(X,E),\widehat{F})$ for all $A \in \mathfrak{R}$.*

(c) *If $\mathcal{T}_{(\theta,X)}$ is compact (or weakly compact), then θ is $\mathfrak{L}(E,F)$-valued and compact (or weakly compact) and $\mathcal{T}_{(\theta,A)} \in \mathfrak{L}(\mathcal{C}_{\mathfrak{V}}(X,E),F)$ for all $A \in \mathfrak{R}$.*

Proof. Suppose that $\mathfrak{V}$ is finer than $\mathfrak{V}_\theta$, that θ is regular and that $\mathcal{T}_{(\theta,X)}(f) \in F$ for all $f \in \mathcal{C}_{\mathfrak{V}}(X,E)$. Let $A \in \mathfrak{R}$ and $f \in \mathcal{C}(X,E)$. Choose a net $(\varphi_i)_{i\in\mathcal{I}}$ in $\mathcal{C}_{\mathcal{K}}(X)$ as in Lemma 2.5.6. Then $\lim_{i\in\mathcal{I}} \mathcal{T}_{(\theta,X)}(\varphi_i \cdot f) = \int_A f\,d\theta$ pointwise on F^*, and there is $i_0 \in \mathcal{I}$ such that the set

$$\mathcal{F} = \{\varphi_i \cdot f \mid i \geqslant i_0\}$$

is bounded in $\mathcal{F}(X,E)$. For Part (a) we continue as follows. Since every continuous operator is bounded, the set $\mathcal{T}_{(\theta,X)}(\mathcal{F})$ is bounded in F and its polar is a neighborhood in the strong topology of F^*. For every μ in this neighborhood, we have $\mathfrak{Re}\,\mu(c) \leqslant 1$ for all $c \in \mathcal{T}_{(\theta,X)}(\mathcal{F})$. Hence $\mathfrak{Re}(\int_A f\,d\theta)(\mu) \leqslant 1$ holds as well. But this shows that the linear functional $\int_A f\,d\theta$ is bounded on a neighborhood of F^*, hence is an element of its dual F^{**}. If for $a \in E$, we use the constant function $x \mapsto a$ for f, this yields $\theta_A(a) \in F^{**}$. Moreover, for every α in the common scalar field of E and F, we have

$$\begin{aligned}\theta_A(\alpha a)(\mu) &= \lim_{i\in\mathcal{I}} \mathcal{T}_{(\theta,X)}(\varphi_i \cdot (\alpha a))(\mu)\\ &= \lim_{i\in\mathcal{I}} \mathcal{T}_{(\theta,X)}(\varphi_i \cdot a)(\alpha\mu)\\ &= \theta_A(a)(\alpha\mu)\ .\end{aligned}$$

Thus θ is indeed $\mathfrak{L}(E,F^{**})$-valued, as claimed. Furthermore, if $\mathfrak{V}$ is finer than $\mathfrak{V}_\theta$, then the operator $\mathcal{T}_{(\theta,A)}$ is continuous on $\mathcal{C}_{\mathfrak{V}}(X,E)$ and F^{**}-valued by the above. This yields Part (a). Under the assumptions of (b), we shall verify that $(\mathcal{T}_{(\theta,X)}(\varphi_i \cdot f))_{i\in\mathcal{I}}$ is a Cauchy net in F. For this, let $W \in \mathcal{W}$ and choose an open set $O_0 \in \mathfrak{O}$ containing A and a function space neighborhood $\mathfrak{v}$ such that $\|\theta\|_{O_0}^{(\mathfrak{v},W^\circ)} < +\infty$. There is $V \in \mathcal{V}$ such that $\chi_A \cdot V \leqslant \mathfrak{v}$ and there is $\rho > 0$ such that $\varphi_i \cdot f \leqslant \rho\chi_X \cdot V$ for all $i \geqslant i_0$. Now according to Proposition 2.1.16, we find a compact set K and an open set $O \in \mathfrak{O}$ such that $K \subset A \subset O$ and $\|\theta\|_{O\setminus K}^{(\mathfrak{v},W^\circ)} \leqslant 1/(2\rho)$. According to 2.5.6, there is $i_1 \geqslant i_0$ such that $\chi_K \leqslant \varphi_i \leqslant \chi_O$ for all $i \geqslant i_1$. Thus $|\varphi_i - \varphi_j| \leqslant \chi_{O\setminus K}$ for all $i,j \geqslant i_1$. We have $((\varphi_i - \varphi_j)\cdot f)(x) \in 2\rho V$ for all $x \in O \setminus K$, therefore

$$\int_{O\setminus K} (\varphi_i - \varphi_j)\cdot f\,d\theta \leqslant 2\rho|\theta|_{O\setminus K}(V)$$

pointwise on F^*. Using Lemma 2.1.5(a), for all $\mu \in W^\circ$ we calculate that

$$|\theta|_{O\setminus K}(V)(\mu) \leqslant \|\theta\|_{O\setminus K}^{(\mathfrak{v},W^\circ)} \leqslant 1/(2\rho).$$

Hence

$$(\mathcal{T}_{(\theta,X)}(\varphi_i \cdot f) - \mathcal{T}_{(\theta,X)}(\varphi_j \cdot f))(\mu) = \Big(\int_{O\setminus K} (\varphi_i - \varphi_j) \cdot f \, d\theta \Big)(\mu) \leqslant 1.$$

This implies that $\mathcal{T}_{(\theta,X)}(\varphi_i \cdot a) - \mathcal{T}_{(\theta,X)}(\varphi_j \cdot a) \in W$ for all $i,j \geqslant i_1$, our claim. The net $(\mathcal{T}_{(\theta,X)}(\varphi_i \cdot a))_{i\in\mathcal{I}}$ therefore converges to an element of $\widehat{F} \subset F^{*\blacksquare}$. As in the argument for (a), we infer that $\mathcal{T}_{(\theta,A)}(f) \in \widehat{F}$, and consequently, $\theta_A(a) \in \widehat{F}$ for all $a \in E$. For (c), if the operator $\mathcal{T}_{(\theta,X)}$ is compact (or weakly compact), then the set $\mathcal{T}_{(\theta,X)}(\mathcal{F})$ is relatively weakly compact in F and its polar is a neighborhood consistent with the duality (F^*,F). The argument from (a) then shows that $\int_A f \, d\theta \in F$, hence $\theta_A(a) \in F$ for all $a \in E$ and $\mathcal{T}_{(\theta,A)} \in \mathfrak{L}(\mathcal{C}_{\mathfrak{V}}(X,E),F)$. All left to show for Part (c) is that the measure θ is compact (or weakly compact). For this, let $A \in \mathfrak{R}$, let $O \in \mathfrak{O}$ be an open set containing A and let $C \in \mathrm{Conv}(E)$ be bounded. We may also assume that C is balanced. We shall establish that the set

$$D = \Big\{ \sum_{i=1}^{n} \theta_{A_i}(c_i) \;\Big|\; c_i \in C,\ A_i \in \mathfrak{R} \text{ disjoint}, \bigcup_{i=1}^{n} A_i = A \Big\}$$

is relatively compact (or relatively weakly compact) in F. Let $\mathcal{H}$ be the subset of all functions $f \in \mathcal{C}_{\mathfrak{V}}(X,E)$ that are supported by O and C-valued. For every $\mathfrak{v} \in \mathfrak{V}$, there is $V \in \mathcal{V}$ such that $\chi_A \cdot V \leqslant \mathfrak{v}$ and there is $\lambda \geqslant 0$ such that $C \subset \lambda V$. Thus $f \leqslant \lambda\mathfrak{v}$ for all $f \in \mathcal{H}$, and the set $\mathcal{H}$ is understood to be bounded in $\mathcal{C}_{\mathfrak{V}}(X,E)$. Its image $\mathcal{T}_{(\theta,X)}(\mathcal{H})$ is therefore relatively compact (or relatively weakly compact) in F. Since $\mathcal{T}_{(\theta,X)}(\mathcal{H})$ is convex, its closure $\overline{\mathcal{T}_{(\theta,X)}(\mathcal{H})}$ in the given and the weak topologies of F coincide and is compact (or weakly compact). We shall verify that $D \subset \overline{\mathcal{T}_{(\theta,X)}(\mathcal{H})}$. For this, let $\sum_{k=1}^{n} \theta_{A_k}(c_k) \in D$ and choose compact subsets K_k of A_k. There are disjoint open sets $O_k \subset O$ such that $K_k \subset O_k$. Now for each of the sets K_k, we choose a net $(\varphi^k_{i_k})_{i_k\in\mathcal{I}_k}$ in $\mathcal{C}_{\mathcal{K}}(X)$ with K_k for K and O_k for O in Lemma 2.5.6. We consider the index set $\mathcal{I} = \Pi_{k=1}^{n}\mathcal{I}_k$ with the componentwise order and for $i = (i_1,\dots,i_n) \in \mathcal{I}$ set $f_i = \sum_{k=1}^{n} \varphi^k_{i_k} \in \mathcal{H}$. Then according to Lemma 2.5.6 the net $(T_{(\theta,X)}(f_i))_{i\in\mathcal{I}}$ in $\mathcal{T}_{(\theta,X)}(\mathcal{H})$ converges pointwise on F^*, that is, in the weak topology of F to $\sum_{k=1}^{n} \theta_{K_k}(c_k)$. Hence

$$\sum_{k=1}^{n} \theta_{K_k}(c_k) \in \overline{\mathcal{T}_{(\theta,X)}(\mathcal{H})}.$$

Next, we construct an index set $\mathcal{K} = \{(K_1,\dots,K_n)\}$, where K_k are compact subsets of A_k, ordered componentwise by set inclusion. Then finally, using the regularity of θ and Lemma 2.1.12, we conclude that

$$\sum_{k=1}^{n} \theta_{A_k}(c_k) = \sum_{k=1}^{n} \lim_{K_k\subset A_k} \theta_{K_k}(c_k) = \lim_{(K_1,\dots,K_n)\in\mathcal{K}} \sum_{k=1}^{n} \theta_{K_k}(c_k) \in \overline{\mathcal{T}_{(\theta,X)}(\mathcal{H})}.$$

Our claim in (c) follows. □

2.5.11 Example. The following simple example will demonstrate that for a regular $\mathfrak{L}(E,F)$-valued measure θ with bounded variation the operator $\mathcal{T}_{(\theta,X)}$ is not necessarily weakly compact (compare VI.3 in [21]). Let $X = [0,1]$ and let $E = F$ be any normed space with unit ball $\mathbb{B}$. Suppose that $\mathcal{C}_{\mathfrak{V}}(X,E)$ is endowed with the topology of uniform convergence (Example 1.2.7(a)). We consider the point evaluation measure δ_0^I, where I is the identity operator on E. That is, $(\delta_0^I)_A = I$ if $0 \in A$ and $(\delta_0^I)_A = 0$ if $0 \notin A$. Then $\mathcal{T}_{(\delta_0^I,X)}(f) = f(0)$ for $f \in \mathcal{C}(X,E)$. We choose the unit ball $\mathbb{B}_{\mathcal{C}}$ of $\mathcal{C}(X,E)$ for $\mathfrak{v}$ and the dual unit ball $\mathbb{B}^\circ$ of $E = F$ for Π and calculate that $\|\theta\|_A^{(\mathbb{B}_{\mathcal{C}},\mathbb{B}^\circ)} = 1$. But $\mathcal{T}(\mathbb{B}_{\mathcal{C}}) = \mathbb{B}$, which in general is not relatively weakly compact.

2.6 Integral representation

The Riesz representation theorem for linear functionals on a space of real-valued continuous functions on a compact space is the starting point for a variety of subsequent results. In the preceding section, we investigated linear operators defined by measures on function spaces and the interplay of their properties. In this section, we shall formulate and prove an integral representation theorem, which establishes that every continuous linear operator on a function space is of this type. It is an adaptation (and in some aspects, strengthening) of a more general result for locally convex cones, that is, Theorem III.5.1 together with Corollary III.6.10 in [48]. There are numerous earlier versions of this variety. For the special case of a compact space X, for $E = \mathbb{R}$, a Banach space F and weakly compact operators, for example, a similar result is due to Bartle, Dunford and Schwartz [7].

Our approach to integral representations will use the assumptions and notation from the previous sections while considering only the topological case. Accordingly, throughout the remainder of this text, X represents a locally compact Hausdorff space and the σ-ring $\mathfrak{R}$ consists of all relatively compact Borel subsets of X. All measures are supposed to be bounded. A function space neighborhood system is said to be *measurable* if all neighborhood functions involved in the formation of its neighborhoods are measurable. It is weakly lower continuous (see Section 1.2) if these neighborhood functions are weakly lower continuous. All function space neighborhood systems in our Examples 1.2.7(a) to (d) in Section 1.2 are both measurable (see Corollary 2.2.21) and weakly lower continuous. The space $\mathfrak{L}(\mathcal{C}_{\mathfrak{V}}(X,E),F)$ consists of all continuous linear operators from the function space $\mathcal{C}_{\mathfrak{V}}(X,E)$ into the locally convex space F. We recall from Section 1.3 that in case that F is semireflexive, every operator $\mathcal{T} \in \mathfrak{L}(\mathcal{C}_{\mathfrak{V}}(X,E),F)$ is weakly compact.

2.6.1 Theorem. *If the function space neighborhood system $\mathfrak{V}$ is measurable and weakly lower continuous, then every operator $\mathcal{T} \in \mathfrak{L}(\mathcal{C}_{\mathfrak{V}}(X,E),F)$ can be represented as an integral on X. More precisely, there exists a unique regular $\mathfrak{L}(E,F^{**})$-valued measure θ*

on $\mathfrak{R}$ such that $\mathfrak{V}$ is finer than $\mathfrak{V}_\theta$, all functions in $\mathcal{C}_{\mathfrak{V}}(X,E)$ are integrable over every set in $\mathfrak{A}$ with respect to θ, their integrals are F^{**}-valued and

$$\int_X f\,d\theta = \mathcal{T}(f) \quad \textit{for all} \quad f \in \mathcal{C}_{\mathfrak{V}}(X,E).$$

If $\mathcal{T}$ is compact (or weakly compact), then the measure θ is $\mathfrak{L}(E,F)$-valued and compact (or weakly compact), countably additive with respect to the strong operator topology of $\mathfrak{L}(E,F)$, and the integrals of functions in $\mathcal{C}_{\mathfrak{V}}(X,E)$ over sets in $\mathfrak{R}$ are F-valued.

Proof. Let $\mathcal{T} \in \mathfrak{L}(\mathcal{C}_{\mathfrak{V}}(X,E),F)$. In our construction of a representation measure θ for the operator $\mathcal{T}$ on the σ-ring $\mathfrak{R}$, we shall use a version of the classical Riesz representation theorem, which states that every real-valued continuous linear functional Φ on the space $\mathcal{C}_{\mathcal{K}}(X)$ of continuous real-valued functions with compact support on a locally compact Hausdorff space X, endowed with the topology of uniform convergence, can be uniquely represented by a regular real-valued Borel measure ϑ. That is, $\Phi(\varphi) = \int_X \varphi\,d\vartheta$ holds for all $\varphi \in \mathcal{C}_{\mathcal{K}}(X)$ and $\|\Phi\| = |\vartheta|_X$. This version can for instance be found in Rudin's book [55] (Theorem 6.19). We shall go forward with a step-by-step construction of our representation measure for the operator $\mathcal{T}$.

Every open subset O of X is itself a locally compact Hausdorff space, and we denote by $\mathfrak{B}_O$ the σ-field of all Borel subsets of O. We claim that

$$\mathfrak{B}_O = \{B \in \mathfrak{B}_X \mid B \subset O\}.$$

Indeed, the family $\{B \in \mathfrak{B}_X \mid B \subset O\}$ forms a σ-field in O that contains all open subsets of O. Hence this family contains $\mathfrak{B}_O$. For the reverse inclusion, consider the family $\mathfrak{C} = \{B \subset X \mid B \cap O \in \mathfrak{B}_O\}$. This is a σ-field in X that contains all open subsets of X. Hence $\mathfrak{B}_X \subset \mathfrak{C}$, and for every $B \in \mathfrak{B}_X$ such that $B \subset O$ we have $B \in \mathfrak{C}$ and, therefore, $B = B \cap O \in \mathfrak{B}_O$. Moreover, if a Borel subset of O is relatively compact in O, then its closure in O, which is contained in its closure in X, is compact, hence also closed in X. Thus this set is also relatively compact in X. In summary, the family $\mathfrak{R}_O$ of all relatively compact Borel subsets of O is a subfamily of $\mathfrak{R}$. We shall use this observation in our construction.

In a first step, we fix elements $a \in E$ and $\mu \in F^*$ and choose a neighborhood $W \in \mathcal{W}$ such that $\mu \in W^\circ$. There is $\mathfrak{v} \in \mathfrak{V}$ such that $\mathcal{T}(\mathfrak{v}) \subset W$. For every $O \in \mathfrak{O}$, there is $V \in \mathcal{V}$ such that $\chi_O \cdot V \leqslant \mathfrak{v}$, and there is $\lambda_W \geqslant 0$ such that $a \in \lambda_W V$. Since $O \subset X$ is itself locally compact, we can apply the above version of the Riesz representation theorem to every continuous linear functional on $\mathcal{C}_{\mathcal{K}}(O)$, and because O is open, every function $\varphi \in \mathcal{C}_{\mathcal{K}}(O)$ if extended by zero on $X \setminus O$ is continuous on X. In this way, $\mathcal{C}_{\mathcal{K}}(O)$ is a subspace of $\mathcal{C}_{\mathcal{K}}(X)$. We define a linear functional $\Phi_O^{(a,\mu)}$ on $\mathcal{C}_{\mathcal{K}}(O)$ setting

$$\Phi_O^{(a,\mu)}(\varphi) = (\mathcal{T}(\varphi \cdot a))(\mu)$$

for $\varphi \in \mathcal{C}_{\mathcal{K}}(O)$. As $|\varphi| \leqslant 1$ implies that $\varphi \cdot a \leqslant \lambda_W \chi_O \cdot V \leqslant \lambda_W \mathfrak{v}$, we have $\mathcal{T}(\varphi \cdot a) \in \lambda_W W$ and, therefore, $\Phi_O^{(a,\mu)}(\varphi) \leqslant \lambda_W$. The functional $\Phi_O^{(a,\mu)}$ is therefore continuous on $\mathcal{C}_{\mathcal{K}}(O)$ and can be represented by a unique regular Borel measure $\vartheta_O^{(a,\mu)}$ on O such that $|\vartheta_O^{(a,\mu)}|_O = \|\Phi_O^{(a,\mu)}\| \leq \lambda_W$. If $O, U \in \mathfrak{O}$, then $\mathfrak{B}_{O\cap U} \subset \mathfrak{B}_O \cap \mathfrak{B}_U$ by the above, and

$$\mathcal{C}_{\mathcal{K}}(O \cap U) \subset \mathcal{C}_{\mathcal{K}}(O) \cap \mathcal{C}_{\mathcal{K}}(U).$$

For every function $\varphi \in \mathcal{C}_{\mathcal{K}}(O \cap U)$, we have

$$\Phi_{O\cap U}^{(a,\mu)}(\varphi) = \Phi_O^{(a,\mu)}(\varphi) = \Phi_U^{(a,\mu)}(\varphi),$$

and consequently,

$$\begin{aligned}\int_{O\cap U} \varphi \, d\vartheta_{O\cap U}^{(a,\mu)} &= \int_O \varphi \, d\vartheta_O^{(a,\mu)} = \int_{O\cap U} \varphi \, d\vartheta_O^{(a,\mu)}|_{O\cap U} \\ &= \int_U \varphi \, d\vartheta_U^{(a,\mu)} = \int_{O\cap U} \varphi \, d\vartheta_U^{(a,\mu)}|_{O\cap U}.\end{aligned}$$

The measures $\vartheta_{O\cap U}^{(a,\mu)}$, $\vartheta_O^{(a,\mu)}|_{O\cap U}$ and $\vartheta_U^{(a,\mu)}|_{O\cap U}$ all are regular on $O \cap U$ (Proposition 2.1.18) and their integrals coincide on $\mathcal{C}_{\mathcal{K}}(O \cap U)$. By the uniqueness statement of the Riesz representation theorem (see also Proposition 2.5.8) they are therefore equal. Thus for any set $A \in \mathfrak{B}_{O\cap U}$ we have

$$(\vartheta_O^{(a,\mu)})_A = (\vartheta_O^{(a,\mu)}|_{O\cap U})_A = (\vartheta_U^{(a,\mu)}|_{O\cap U})_A = (\vartheta_U^{(a,\mu)})_A.$$

For every $A \in \mathfrak{R}$ and $a \in E$, we can therefore consistently define a real-valued function $\theta_A(a)$ on F^* by

$$\theta_A(a)(\mu) = (\vartheta_O^{(a,\mu)})_A,$$

where $\mu \in F^*$ and O is any set in $\mathfrak{O}$ containing A.

We shall proceed to demonstrate that for every fixed element $a \in E$ the mapping $\mu \mapsto (\theta_A(a))(\mu)$ on F^* is an element of $\mathcal{Q}_F$. For $A \in \mathfrak{R}$ and $a \in E$, choose $O \in \mathfrak{O}$ such that $A \subset O$. Then

$$\Phi_O^{(a,\lambda\mu)} = \lambda \Phi_O^{(a,\mu)} \quad \text{and} \quad \Phi_O^{(a,\mu+\nu)} = \Phi_O^{(a,\mu)} + \Phi_O^{(a,\nu)}$$

for $\mu, \nu \in F^*$ and $\lambda \in \mathbb{R}$. We agree that sums and scalar multiples of real-valued measures are defined setwise and using the uniqueness of the corresponding real-valued representation measures conclude that

$$\vartheta_O^{(a,\lambda\mu)} = \lambda \vartheta_O^{(a,\mu)} \quad \text{and} \quad \vartheta_O^{(a,\mu+\nu)} = \vartheta_O^{(a,\mu)} + \vartheta_O^{(a,\nu)}.$$

Hence $\theta_A(a)$ is seen to be a linear functional on F^*. We still need to verify that it is bounded on the polars of the neighborhoods in $\mathcal{W}$. For this, let $W \in \mathcal{W}$. For every $\mu \in W^\circ$, we have

$$(\theta_A(a))(\mu) = (\vartheta_O^{(a,\mu)})_A \leqslant |\vartheta_O^{(a,\mu)}|_A \leqslant |\vartheta_O^{(a,\mu)}|_O \leqslant \lambda_W.$$

Thus indeed $\theta_A(a) \in \mathcal{Q}_F$.

Next, we shall establish that $\theta_A \in \mathfrak{L}(E, \mathcal{Q}_F)$. Again, we choose a set $O \in \mathfrak{O}$ containing A. For every $\mu \in F^*$, we have

$$\Phi_O^{(\lambda a,\mu)} = \lambda \Phi_O^{(a,\mu)} \quad \text{and} \quad \Phi_O^{(a+b,\mu)} = \Phi_O^{(a,\mu)} + \Phi_O^{(b,\mu)}$$

for $a, b \in E$ and $\lambda \in \mathbb{R}$. Hence

$$\vartheta_O^{(\lambda a,\mu)} = \lambda \vartheta_O^{(a,\mu)} \quad \text{and} \quad \vartheta_O^{(a+b,\mu)} = \vartheta_O^{(a,\mu)} + \vartheta_O^{(b,\mu)}.$$

This yields

$$\theta_A(\lambda a) = \lambda \theta_A \quad \text{and} \quad \theta_A(a+b) = \theta_A(a) + \theta_A(b).$$

Furthermore, given $W \in \mathcal{W}$, there is $\mathfrak{v} \in \mathfrak{V}$ such that $\mathcal{T}(\mathfrak{v}) \subset W$ and $V \in \mathcal{V}$ such that $\chi_O \cdot V \leqslant \mathfrak{v}$. Then for all $a \in V$ we have $\lambda_W = 1$ and, therefore, $(\theta_A(a))(\mu) \leqslant 1$ for all $\mu \in W^\circ$ by the above. Hence $\theta_A(a) \leqslant \sigma_W$ (see Section 1.4). We infer that θ_A is indeed a continuous linear operator from E into $\mathcal{Q}_F$.

Next, we shall reason that θ is countably additive, hence an $\mathfrak{L}(E, \mathcal{Q}_F)$-valued measure on $\mathfrak{R}$. For this, suppose that $A_i \in \mathfrak{R}$ are disjoint sets and that $A = \bigcup_{i \in \mathbb{N}} A_i \in \mathfrak{R}$. Again, we choose $O \in \mathfrak{O}$ containing A, an element $a \in E$ and $\mu \in F^*$. Using that the real-valued measure $\vartheta_O^{(a,\mu)}$ is countably additive, we compute that

$$(\theta_A(a))(\mu) = (\vartheta_O^{(a,\mu)})_A = \sum_{i=1}^{\infty} (\vartheta_O^{(a,\mu)})_{A_i} = \sum_{i=1}^{\infty} (\theta_{A_i}(a))(\mu).$$

The arguments for boundedness and for regularity of the $\mathfrak{L}(E, \mathcal{Q}_F)$-valued measure θ are somewhat more elaborate. Some complications arise from the feature that the expression defining its modulus involves different real-valued measures $\vartheta_O^{(a,\mu)}$. In preparation, for a set $A \in \mathfrak{R}$ and a neighborhood $V \in \mathcal{V}$, we choose an open set $O \in \mathfrak{O}$ containing A and calculate that

$$\begin{aligned} |\theta|_A(V)(\mu) &= \sup\left\{ \sum_{i=1}^{n} \theta_{A_i}(a_i)(\mu) \;\middle|\; a_i \in V,\ A_i \in \mathfrak{R} \text{ disjoint},\ A_i \subset A \right\} \\ &= \sup\left\{ \sum_{i=1}^{n} (\vartheta_O^{(a_i,\mu)})_{A_i} \;\middle|\; a_i \in V,\ A_i \in \mathfrak{R} \text{ disjoint},\ A_i \subset A \right\} \end{aligned}$$

$$= \sup\left\{\sum_{i=1}^{n}(\vartheta_O^{(a_i,\mu)})_{K_i} \;\middle|\; a_i \in V,\ K_i \text{ disjoint compact, } K_i \subset A\right\}$$

The last line in this equality, that is, the fact that we only need to consider compact subsets K_i of A, results from the regularity of the measures $\vartheta_O^{(a_i,\mu)}$. Indeed, let $A_i \in \mathfrak{R}$ be disjoint subsets of A, and let $a_i \in V$ for $i = 1,\dots,n$ and let $\mu \in F^*$. Because the real-valued measures $\vartheta_O^{(a_i,\mu)}$ are regular, and we have

$$(\vartheta_O^{(a_i,\mu)})_{A_i} = \lim_{K_i \subset A_i}(\vartheta_O^{(a_i,\mu)})_{K_i},$$

where K_i are compact subsets of A_i, for $i = 1,\dots,n$, we infer that

$$\sum_{i=1}^{n}(\vartheta_O^{(a_i,\mu)})_{A_i} = \lim_{K_i \subset A_i}\sum_{i=1}^{n}(\vartheta_O^{(a_i,\mu)})_{K_i}.$$

This demonstrates that the supremum in the last line of the precedent equation using only compact subsets of A equals the supremum using all sets in $\mathfrak{R}$ in the second line.

Now for the boundedness of θ let $A \in \mathfrak{R}$ and choose an open set $O \in \mathfrak{O}$ containing A. Let $W \in \mathcal{W}$. There is $\mathfrak{v} \in \mathfrak{V}$ such that $\mathcal{T}(\mathfrak{v}) \subset W$ and there is $V \in \mathcal{V}$ such that $\chi_O \cdot V \leqslant \mathfrak{v}$. We shall verify that $|\theta|_A(V) \leqslant \sigma_W$. For this, let $\mu \in W^\circ$. Let K_k be disjoint compact subsets of A, let $a_k \in V$ for $k = 1,\dots,n$ and let $\mu \in F^*$. There are disjoint open sets $O_k \in \mathfrak{R}$ such that $K_k \subset O_k \subset O$. For each of the sets K_k, there is a net $(\varphi_{i_k}^k)_{i \in \mathcal{I}_k}$ in $\mathcal{C}_\mathcal{K}(X)$ as in Lemma 2.5.6 with O_k in place of O. Hence

$$\lim_{i_k \in \mathcal{I}_k}\int_X(\varphi_{i_k}^k \cdot a_k)\, d\vartheta_O^{(a_k,\mu)} = (\vartheta_O^{(a_k,\mu)})_{K_k}.$$

For different choices of k, the functions $\varphi_{i_k}^k$ are supported by disjoint sets and, therefore,

$$\sum_{k=1}^{n}\varphi_{i_k}^k \cdot a_k \leqslant \chi_O \cdot V \leqslant \mathfrak{v}$$

for any choice of the indexes $i_k \in \mathcal{I}_k$. Since $\int_X(\varphi_{i_k}^k \cdot a_k)\, d\vartheta_O^{(a_k,\mu)} = \mathcal{T}(\varphi_{i_k}^k \cdot a_k)(\mu)$, this yields

$$\sum_{k=1}^{n}\int_X(\varphi_{i_k}^k \cdot a_k)\, d\vartheta_O^{(a_k,\mu)} = \mathcal{T}\left(\sum_{k=1}^{n}\varphi_{i_k}^k \cdot a_k\right)(\mu) \leqslant 1.$$

Thus taking the limits over all index sets $\mathcal{I}_k$ yields $\sum_{k=1}^{n}(\vartheta_O^{(a_k,\mu)})_{K_k} \leqslant 1$. Our claim now follows from our preparatory remarks.

The argument for the regularity of θ is more straightforward. Following Proposition 2.1.13, it suffices to verify inner regularity for all sets in $\mathfrak{R}$. Let $A \in \mathfrak{R}$ and choose

$O \in \mathfrak{O}$ containing A. The already proven boundedness of θ implies that for every $\mu \in F^*$ there is $V \in \mathcal{V}$ such that $|\theta|_O(V)(\mu) < +\infty$. The inequality

$$|\theta|_A(V)(\mu) \geqslant \lim_{K \subset A} |\theta|_K(V)(\mu) = \sup_{K \subset A} |\theta|_K(V)(\mu)$$

is obvious since $|\theta|_B(V)(\mu) \leqslant |\theta|_C(V)(\mu)$ whenever $B \subset C$ for $B, C \in \mathfrak{R}$. For the reverse inequality, let K_i be disjoint compact subsets of A, let $a_i \in V$ for $i = 1, \ldots, n$ and let $\mu \in F^*$. Set $K = \bigcup_{i=1}^n K_i \subset A$. Then

$$\sum_{i=1}^{n} (\vartheta_O^{(a_i,\mu)})_{K_i} \leqslant |\theta|_K(V) \leqslant \lim_{K \subset A} |\theta|_K(V)(\mu).$$

This renders $|\theta|_A(V)(\mu) \leqslant \lim_{K \subset A} |\theta|_K(V)(\mu)$, our claim.

Next, we shall argue that $\mathfrak{V}$ is finer than $\mathfrak{V}_\theta$ and proceed in several steps. Let $W \in \mathcal{W}$. There is $\mathfrak{v} \in \mathfrak{V}$ such that $\mathcal{T}(\mathfrak{v}) \subset W$. Let $h = \sum_{k=1} \chi_{A_k} \cdot a_k$ be an E-valued step function such that $h \leqslant \mathfrak{v}$. We may assume that the sets A_k are disjoint and choose compact subsets K_k of A_k. Given $\varepsilon > 0$, according to Lemma 1.2.4(a) there are disjoint open sets $O_k \in \mathfrak{O}$ such that $K_k \subset O_k$ and $\sum_{k=1} \chi_{O_k} \cdot a_k \leqslant (1 + \varepsilon)\mathfrak{v}$. We choose nets $(\varphi_{i_k}^k)_{i \in \mathcal{I}_k}$ in $\mathcal{C}_\mathcal{K}(X)$ as in the prior argument for boundedness. Hence

$$\sum_{k=1}^{n} \varphi_{i_k}^k \cdot a_k \leqslant (1 + \varepsilon)\mathfrak{v}, \quad \text{that is} \quad \mathcal{T}\left(\sum_{k=1}^{n} \varphi_{i_k}^k \cdot a_k\right) \leqslant (1 + \varepsilon)\sigma_W$$

for any choice of the indexes $i_k \in \mathcal{I}_k$. Taking the limits first over the indexes $i_k \in \mathcal{I}_k$ as in Lemma 2.5.6 and over all $\varepsilon > 0$ then returns

$$\int_X \left(\sum_{k=1}^{n} \chi_{A_k} \cdot a_k\right) d\theta = \sum_{k=1}^{n} \theta_{K_k}(a_k) \leqslant \sigma_W.$$

Finally, now using the previously established regularity of θ and passing to the limits over the compact subsets K_k of A_k in the last inequality renders

$$\int_X h \, d\theta = \sum_{k=1}^{n} \theta_{A_k}(a_k) \leqslant \sigma_W.$$

In the light of Lemma 2.3.3, this observation extends to $\mathrm{Conv}(E)$-valued step functions $h \leqslant \mathfrak{v}$ since the integral of a set-valued step function h was seen to be the supremum of the integrals of all vector-valued step functions smaller than h.

Now let $n \in \mathcal{N}_\mathfrak{v}$ be any of the measurable weakly lower continuous neighborhood functions that define $\mathfrak{v}$. We shall proceed to demonstrate that $\int_X n \, d\theta \leqslant \sigma_W$. For this, we recall the definition of the integral of a measurable function from Section 2.3. Let

$\varepsilon > 0$ and let $h \in \mathcal{F}_S(X, \operatorname{Conv}(E))$ be a step function such that $h \leqslant n + \varepsilon\mathfrak{v} \leqslant (1+\varepsilon)\mathfrak{v}$. Then $\int_X h\,d\theta \leqslant (1+\varepsilon)\sigma_W$ by the above. Hence

$$\int_X n\,d\theta \leqslant \int_X^{(\varepsilon\mathfrak{v})} n\,d\theta \leqslant (1+\varepsilon)\sigma_W$$

The latter holds for all $\varepsilon > 0$ and proves that $\int_X n\,d\theta \leqslant \sigma_W$, our claim. From this, we infer that $n \in \mathcal{N}_{\mathfrak{v}_{(\theta,W)}}$. In summary, we have established that, given $W \in \mathcal{W}$, there is $\mathfrak{v} \in \mathfrak{V}$ such that $\mathcal{N}_{\mathfrak{v}} \subset \mathcal{N}_{\mathfrak{v}_{(\theta,W)}}$. Accordingly, the neighborhood system $\mathfrak{V}$ is finer that $\mathfrak{V}_\theta$.

We are now in a position to reason that θ represents the operator $\mathcal{T}$. First, we shall prove that $\mathcal{T}(\varphi \cdot a) = \int_X \varphi \cdot a\,d\theta$ holds for all $\varphi \in \mathcal{C}_{\mathcal{K}}(X)$ and $a \in E$. Let $\mu \in F^*$. We shall use Theorem 2.3.28 with $G = H = \mathbb{R}$ and the operators $S \in \mathfrak{L}(\mathbb{R}, E)$ and $T \in \mathfrak{L}(F, \mathbb{R})$ such that $S(\alpha) = \alpha a$ and $T(c) = \mathfrak{Re}\,\mu(c)$ for $\alpha \in \mathbb{R}$ and $c \in F$. The real-valued composition measure $T \circ \theta \circ S$ then is given by

$$(T \circ \theta \circ S)_A = ((T \circ \theta \circ S)_A(1))(1) = \theta_A(a)(\mu) = (\vartheta_O^{(a,\mu)})_A,$$

where $O \in \mathfrak{O}$ is any open set containing A. The $\mathcal{Q}_F$-valued measure $\theta \circ S$, on the other hand is defined by

$$(\theta \circ S)_A(\mu) = (\theta_A(a))(\mu) = (\vartheta_O^{(a,\mu)})_A$$

for all $\mu \in F^*$. We choose $f = \varphi$ in 2.3.28 and observe that φ is integrable over X with respect to $\theta \circ S$ (Proposition 2.3.24). The function $S \circ \varphi$, that is, $\varphi \cdot a$, is integrable over X with respect to θ (also see 2.3.24). Theorem 2.3.28 then yields that

$$\mathcal{T}(\varphi \cdot a)(\mu) = \int_X \varphi\,d\vartheta_O^{(a,\mu)} = \Big(\int_X \varphi\,d(T \circ \theta \circ S)\Big)(1) = \Big(\int_X (\varphi \cdot a)\,d\theta\Big)(\mu)$$

holds for all $\mu \in F^*$. Hence our claim.

The preceding implies that $\mathcal{T}_{(X,\theta)}$, that is, the linear operator defined by the integral over X with respect to θ with values in $\mathcal{Q}_F$, is continuous on the subspace $\mathcal{C}_{\mathfrak{V}}(X, E)$ of $\mathcal{F}_{(\theta,X)}(X, \operatorname{Conv}(E))$. By the prior argument, it coincides with the given operator $\mathcal{T} \in \mathfrak{L}(\mathcal{C}_{\mathfrak{V}}(X,E), F)$ on the dense subspace of $\mathcal{C}_{\mathfrak{V}}(X, E)$ spanned by the functions $\varphi \cdot a$ for all $\varphi \in \mathcal{C}_{\mathcal{K}}(X)$ and $a \in E$ (Proposition 1.2.5(i)). Hence

$$\int_X f\,d\theta = \mathcal{T}(f)$$

holds for all functions $f \in \mathcal{C}_{\mathfrak{V}}(X, E)$ since both operators $\mathcal{T}$ and $\mathcal{T}_{(X,\theta)}$ are continuous.

There are a few more loose ends and claims that we need to justify. Proposition 2.5.10(a) yields that the measure θ is indeed $\mathfrak{L}(E, F^{**})$-valued and the integrals of functions in $\mathcal{C}_{\mathfrak{V}}(X,E)$ over sets in $\mathfrak{A}$ exist and are F^{**}-valued. The uniqueness of θ as a representation measure for the operator $\mathcal{T} \in \mathfrak{L}(\mathcal{C}_{\mathfrak{V}}(X,E),F)$ follows from Proposition 2.5.8. If $\mathcal{T}$ is compact (or weakly compact), the measure θ is $\mathfrak{L}(E,F)$-valued and compact (or weakly compact), and the integrals of functions in $\mathcal{C}_{\mathfrak{V}}(X,E)$ over sets in $\mathfrak{R}$ are F-valued by Proposition 2.5.10(c). Moreover, θ is countably additive with respect to the strong operator topology of $\mathfrak{L}(E,F)$ in this case by Proposition 2.1.1. □

Theorem 2.6.1 permits us to consider the elements of $\mathfrak{L}(\mathcal{C}_{\mathfrak{V}}(X,E),F)$ both as linear operators on a function space and as operator-valued measures, thus permitting to employ techniques from either perspective. This interchangeability will be used to a great extend in the following chapter. Several special cases are worth mentioning.

Positive operators

If both E and F are ordered topological vector spaces, then an operator $T \in \mathfrak{L}(E,F)$ is called *monotone* or *positive* if it maps positive elements of E into positive elements of F. $\mathfrak{L}(E,F)_+$ denotes the subcone of all positive operators in $\mathfrak{L}(E,F)$. Recall that the order relations transfer naturally to the function space $\mathcal{C}_{\mathfrak{V}}(X,E)$ (see Proposition 1.2.8(a)) and to the respective dual spaces of E and F. In this context, F^*_+ is the positive cone of F^*, which consists of all $\mu \in F^*$ such that $\mathfrak{Re}\,\mu(c) \geqslant 0$ for all $c \in F_+$, that is, F^*_+ is the negative of the polar of F_+. Similarly, F^{**}_+ is the negative of the polar of F^*_+. That is, an element of F^{**} is positive if its embedding into $\mathcal{Q}_F$ takes nonnegative values on the elements of $F^*_+ \subset F^*$.

2.6.2 Corollary. *If in addition to the assumptions of Theorem* 2.6.1*, both* $(E,\mathcal{V})$ *and* $(F,\mathcal{W})$ *are ordered topological vector spaces, then the operator* $\mathcal{T}$ *is positive if and only if all values of its representation measure* θ *are positive operators in* $\mathfrak{L}(E,F^{**})$.

Proof. First, suppose that the operator $\mathcal{T} \in \mathfrak{L}(\mathcal{C}_{\mathfrak{V}}(X,E),F)$ is positive, that is, $\mathcal{T}$ maps E_+-valued functions into F_+. Let θ be the representation measure for $\mathcal{T}$. Let $a \in E_+$ and $A \in \mathfrak{R}$. Choose a net $(\varphi_i)_{i\in\mathcal{I}}$ in $\mathcal{C}_{\mathcal{K}}(X)$ as in Lemma 2.5.6. Then for every $\mu \in F^*_+$ we have

$$\theta_A(a)(\mu) = \lim_{i\in\mathcal{I}} \mathcal{T}(\varphi_i \cdot a)(\mu) \geqslant 0.$$

Hence $\theta_A(a) \in F^{**}_+$. Conversely, if all values of θ are positive operators in $\mathfrak{L}(E,F^{**})$ and if $f \in \mathcal{C}_{\mathfrak{V}}(X,E)$ is E_+-valued, then f can be approximated with respect to the neighborhood system $\mathcal{V}$ by E_+-valued step functions (Proposition 1.2.5(ii)). The integrals of these step functions with respect to θ are elements of F^{**}_+, and as F^{**}_+ is closed in $\mathcal{Q}_F$ with respect to its topology of pointwise convergence, the integral of f is also contained in F^{**}_+. Finally, $\mathcal{T}(\mathcal{C}_{\mathfrak{V}}(X,E)_+) \subset F^{**}_+$ follows since $\mathcal{C}_{\mathfrak{V}}(X,E)_+$ is the closure of the subcone of all E_+-valued functions in $\mathcal{C}_{\mathfrak{V}}(X,E)$ (Proposition 1.2.8(a)). □

Lattice homomorphisms

If both E and F are indeed topological vector lattices over $\mathbb{R}$, and if $C_{\mathfrak{V}}(X,E)$ satisfies the conditions of Proposition 1.2.8(b), then we may inquire about additional properties of the representation measure for a lattice homomorphism in $\mathfrak{L}(C_{\mathfrak{V}}(X,E),F)$. As usual, we denote the respective lattice operations by $\vee$ and $\wedge$. A *lattice homomorphism* is an operator in $\mathfrak{L}(C_{\mathfrak{V}}(X,E),F)$ that is compatible with these operations. We recall that the dual and second dual of a topological vector lattice are again a topological vector lattices if endowed with their respective strong topologies (Theorem V.7.4 in [56]).

2.6.3 Corollary. *Suppose that in addition to the assumptions of Theorem* 2.6.1 *both* $(E,\mathcal{V})$ *and* $(F,\mathcal{W})$ *are topological vector lattices over* $\mathbb{R}$ *and that* $C_{\mathfrak{V}}(X,E)$ *fulfills the conditions of Proposition 1.2.8(b). If* $T \in \mathfrak{L}(C_{\mathfrak{V}}(X,E),F)$ *is a lattice homomorphism, then its representation measure* θ *satisfies*

$$\theta_A(a \vee b)(\delta) = \theta_A(a)(\delta) \vee \theta_A(b)(\delta)$$

and

$$\theta_A(a)(\delta) \vee \theta_B(b)(\delta) = \theta_A(a)(\delta) + \theta_B(b)(\delta)$$

for all real-valued lattice homomorphisms $\delta \in F^*$, *all* $a,b \in E_+$ *and disjoint sets* $A,B \in \mathfrak{R}$.

Proof. Let $A \in \mathfrak{R}$ and $a,b \in E$. We choose the net $(\varphi_i)_{i\in\mathcal{I}}$ in $C_{\mathcal{K}}(X)$ as in Proposition 2.5.6 and understand that

$$\theta_A(a) = \lim_{i\in\mathcal{I}} T(\varphi_i \cdot a), \quad \theta_A(b) = \lim_{i\in\mathcal{I}} T(\varphi_i \cdot b)$$

and

$$\theta_A(a \vee b) = \lim_{i\in\mathcal{I}} T(\varphi_i \cdot (a \vee b)).$$

We have $\varphi_i \cdot (a \vee b) = (\varphi_i \cdot a) \vee (\varphi_i \cdot b)$, hence

$$T(\varphi_i \cdot (a \vee b)) = T(\varphi_i \cdot a) \vee T(\varphi_i \cdot b) \in F,$$

since T is a lattice homomorphism in $\mathfrak{L}(C_{\mathfrak{V}}(X,E),F)$. Therefore,

$$T(\varphi_i \cdot (a \vee b))(\delta) = T(\varphi_i \cdot a)(\delta) \vee T(\varphi_i \cdot b)(\delta)$$

holds for every real-valued lattice homomorphism $\delta \in F^*$. Now passing to the limits over $i \in \mathcal{I}$ returns

$$\theta_A(a \vee b)(\delta) = \theta_A(a)(\delta) \vee \theta_A(b)(\delta),$$

our first claim. For the second claim, let $A, B \in \mathfrak{R}$ be disjoint sets and let $a, b \in E_+$. For any choice of compact subsets K and L of A and B, respectively, there are disjoint open sets $O, U \in \mathfrak{O}$ such that $K \subset O$ and $L \subset U$. We choose nets $(\varphi_i)_{i\in\mathcal{I}}$ and $(\psi_j)_{j\in\mathcal{J}}$ in $\mathcal{C}_\mathcal{K}(X)$ for K and L, respectively. That is, $\varphi_i \leqslant \chi_O$ and $\psi_j \leqslant \chi_U$ and, therefore,

$$(\varphi_i \cdot a) \vee (\psi_j \cdot b) = \varphi_i \cdot a + \psi_j \cdot b$$

since $a, b \geqslant 0$. Hence, again passing to the limits over $i \in \mathcal{I}$ and $j \in \mathcal{J}$ yields

$$\theta_K(a)(\delta) \vee \theta_L(b)(\delta) = \theta_K(a)(\delta) + \theta_L(b)(\delta)$$

for every lattice homomorphism $\delta \in F^*$. Now using the inner regularity of θ and passing to the limits over the compact sets $K \subset A$ and $L \subset B$ yields our claim. □

An element of F^* is known to be a real-valued lattice homomorphism if and only if it generates an extreme ray in F^*_+ (see V.1.7 in [56]). If the closure of the span of all lattice homomorphisms with respect to the strong topology of F^* is all of F^*, then any two elements of F^{**} that coincide on all lattice homomorphisms are equal. That is, the statements of Corollary 2.6.3 read as

$$\theta_A(a \vee b) = \theta_A(a) \vee \theta_A(b) \quad \text{and} \quad \theta_A(a) \vee \theta_B(b) = \theta_A(a) + \theta_B(b).$$

In case that the operator $\mathcal{T}$ is weakly compact, its representation measure θ is $\mathfrak{L}(E, F)$-valued, and it suffices to consider the closure of the span of all lattice homomorphisms in the weak* topology of F^* for the preceding observation.

Algebra homomorphisms

If both E and F are topological algebras and if $\mathcal{C}_\mathfrak{V}(X, E)$ satisfies the condition of Proposition 1.2.8(c), then we may ask how multiplicativity of an operator $\mathcal{T} \in \mathfrak{L}(\mathcal{C}_\mathfrak{V}(X, E), F)$ transfers to its representation measure θ. We also consider the case that both E and F contain an involutions $a \mapsto a^*$ and $c \mapsto c^*$, respectively. Other than in the case of topological vector lattices the continuity of the operations with respect to the given topology of a Hausdorff topological algebra implies weak continuity (see Section 1.3) and allows a more efficient use of Lemma 2.5.6. We shall formulate our result for the case that the representation measure of an algebra homomorphism is $\mathfrak{L}(E, F)$-valued. A more general and more technical result can be found in Corollary III.6.13 from [48].

2.6.4 Corollary. *Suppose that in addition to the assumptions of Theorem 2.6.1 both $(E, \mathcal{V})$ and $(F, \mathcal{W})$ are topological algebras and that $\mathcal{C}_\mathfrak{V}(X, E)$ fulfills the condition of Proposition 1.2.8(c). If $\mathcal{T} \in \mathfrak{L}(\mathcal{C}_\mathfrak{V}(X, E), F)$ is an algebra homomorphisms and its representation measure θ is $\mathfrak{L}(E, F)$-valued, then it satisfies*

$$\theta_A(ab) = \theta_A(a)\,\theta_A(b) \quad \textit{and} \quad \theta_A(a)\,\theta_B(b) = 0$$

for all $a, b \in E$ *and disjoint sets* $A, B \in \mathfrak{R}$. *If both* E *and* F *contain involutions satisfying the condition in 1.2.8(c) and if* $\mathcal{T}(f^*) = (\mathcal{T}(f))^*$ *for all* $f \in C_{\mathfrak{V}}(X, E)$, *then* $\theta_A(a^*) = \theta_A(a)^*$ *holds in addition.*

Proof. Let $A \in \mathfrak{R}$ and $a, b \in E$. Let K be compact subset of A and choose the net $(\varphi_i)_{i\in\mathcal{I}}$ in $C_K(X)$ for the set K as in Lemma 2.5.6. We may assume that $\chi_K \leqslant \varphi_i$ for all $i \in \mathcal{I}$. For every fixed $k \in \mathcal{I}$, the net $(\varphi_i\varphi_k)_{i\in\mathcal{I}}$ also satisfies the assumption of Lemma 2.5.6 and with $f = \varphi_k \cdot ab$ we infer that

$$\lim_{i\in\mathcal{I}} \mathcal{T}(\varphi_i \cdot (\varphi_k \cdot ab)) = \int_K \varphi_k \cdot ab\, d\theta = \theta_K(ab).$$

On the other hand, we have

$$\mathcal{T}(\varphi_i \cdot (\varphi_k \cdot ab)) = \mathcal{T}((\varphi_i \cdot a)(\varphi_k \cdot b)) = \mathcal{T}(\varphi_i \cdot a)\mathcal{T}(\varphi_k \cdot b)$$

since $\mathcal{T}$ is an algebra homomorphism. For every $d \in F$, the linear operators $c \mapsto cd$ and $c \mapsto dc$ are supposed to be continuous with the given and, therefore, also with respect to the weak topology of F (see IV.7.4 in [56]). For this, we recall that the space F is generally supposed to be Hausdorff. We therefore conclude that

$$\lim_{i\in\mathcal{I}}(\mathcal{T}(\varphi_i \cdot a)\mathcal{T}(\varphi_k \cdot b)) = \Big(\lim_{i\in\mathcal{I}} \mathcal{T}(\varphi_i \cdot a)\Big)\mathcal{T}(\varphi_k \cdot b) = \theta_K(a)\mathcal{T}(\varphi_k \cdot b)$$

pointwise on F^* and, therefore, $\theta_K(ab) = \theta_K(a)\mathcal{T}(\varphi_k \cdot b)$. Now passing to the limit over $k \in \mathcal{I}$ on the right-hand side yields $\theta_K(ab) = \theta_K(a)\,\theta_K(b)$. Our first claim follows with the inner regularity of θ. For the second claim, let $A, B \in \mathfrak{R}$ be disjoint sets and let $a, b \in E$. For any choice of compact subsets K and L of A and B, respectively, there are disjoint open sets $O, U \in \mathfrak{O}$ such that $K \subset O$ and $L \subset U$. We choose nets $(\varphi_i)_{i\in\mathcal{I}}$ and $(\psi_j)_{j\in\mathcal{J}}$ in $C_K(X)$ for K and L, respectively. That is, $\varphi_i \leqslant \chi_O$ and $\psi_j \leqslant \chi_U$ and, therefore, $(\varphi_i \cdot a)(\psi_j \cdot b) = 0$. Hence passing to the limits pointwise on F^* over $i \in \mathcal{I}$ and $j \in \mathcal{J}$ renders $\theta_K(a)\,\theta_L(b) = 0$. Now again using the inner regularity of θ, passing to the limits over $K \subset A$ and $L \subset B$ yields our second claim. Finally, suppose that E contains an involution satisfying the condition in Proposition 1.2.8(c) and that $\mathcal{T}(f^*) = (\mathcal{T}(f))^*$ holds for all $f \in C_{\mathfrak{V}}(X, E)$. Let $A \in \mathfrak{R}$ and $a \in E$ and choose the net $(\varphi_i)_{i\in\mathcal{I}}$ in $C_K(X)$ for A as in Lemma 2.5.6. Then

$$\theta_A(a^*) = \lim_{i\in\mathcal{I}} \mathcal{T}(\varphi_i \cdot a^*) = \lim_{i\in\mathcal{I}}(\mathcal{T}(\varphi_i \cdot a)^*) = \Big(\lim_{i\in\mathcal{I}} \mathcal{T}(\varphi_i \cdot a)\Big)^* = \theta_A(a)^*.$$

This last equality uses once more the fact that the involution in F is continuous, hence also weakly continuous. □

If F itself is an operator space $\mathfrak{L}_b(H, K)$ endowed with the strong operator topology, if K is reflexive and if an operator $\mathcal{T} \in \mathfrak{L}(C_{\mathfrak{V}}(X, E), \mathfrak{L}_b(H, K))$ is indeed bounded with respect to the uniform operator topology of $\mathfrak{L}_b(H, K)$, then we may employ Proposition 1.3.6 and Proposition 1.3.3 with the uniform operator topology for τ. We infer that $\mathcal{T}$ is weakly compact and summarize the following.

2.6.5 Corollary. *Suppose that $\mathfrak{L}_b(H,K)$ is endowed with the strong operator topology and that K is reflexive. If an operator in $\mathfrak{L}(C_{\mathfrak{V}}(X,E),\mathfrak{L}_b(H,K))$ is bounded with respect to the uniform operator topology of $\mathfrak{L}_b(H,K)$, then it is weakly compact. Hence its representation measure is $\mathfrak{L}(E,\mathfrak{L}_b(H,K))$-valued and weakly compact, countably additive with respect to the strong operator topology of $\mathfrak{L}(E,\mathfrak{L}_b(H,K))$ and the integrals of functions in $C_{\mathfrak{V}}(X,E)$ over sets in $\mathfrak{R}$ are $\mathfrak{L}_b(H,K)$-valued.*

The case that $E=\mathbb{R}$ or $E=\mathbb{C}$ is of particular interest. In the case of Corollary 2.6.5, then $\mathcal{T}$ is an operator from $\mathcal{C}(X)$ or $\mathcal{C}(X,\mathbb{C})$ into $\mathfrak{L}_b(H,K)$, its representation measure is $\mathfrak{L}_b(H,K)$-valued, countably additive and regular with respect to the strong operator topology of $\mathfrak{L}_b(H,K)$.

The spectral theorem

Corollaries 2.6.4 and 2.6.5 yield the classical Spectral theorem (see II.44 in [30] or X.2 in [26]) as an immediate application. Indeed, let B be a normal operator on a Hilbert space H, let $X\subset\mathbb{C}$ be its compact spectrum, and consider $\mathcal{C}(X,\mathbb{C})$ with the topology of uniform convergence. Let $F=\mathfrak{L}(H,H)$ be endowed with the strong operator topology. According to the Gelfand–Naimark theorem (Theorem IX.3.7 in [26]), there exists a multiplicative linear operator

$$\mathcal{T}:\mathcal{C}(X,\mathbb{C})\to\mathfrak{L}(H,H),$$

compatible with the involution and continuous with respect to the uniform operator topology of $\mathfrak{L}(H,H)$. It maps the identity function $x\mapsto x$ in $\mathcal{C}(X,\mathbb{C})$ to the operator B. According to Corollary 2.6.5, the operator

$$\mathcal{T}\in\mathfrak{L}(C_{\mathfrak{V}}(X,E),\mathfrak{L}_b(H,H))$$

is weakly compact, its representation measure θ is $\mathfrak{L}(H,H)$-valued, weakly compact and countably additive with respect to the strong operator topology of $\mathfrak{L}(H,H)$. Furthermore, according to Corollary 2.6.4 with $E=\mathbb{C}$ and $F=\mathfrak{L}(H,H)$, we have

$$\theta_A=\theta_A(1)=\theta_A(1\cdot 1)=\theta_A\circ\theta_A\quad\text{and}\quad\theta_A=\theta_A^*$$

for all $A\in\mathfrak{R}$. Hence the values of θ are indeed projections on H, and θ is understood to be the spectral measure of the operator B.

From now on, we shall always assume that $C_{\mathfrak{V}}(X,E)$ satisfies the assumptions of Theorem 2.6.1 and, therefore, every operator in $\mathfrak{L}(C_{\mathfrak{V}}(X,E),F)$ admits a unique integral representation with the stated properties.

Compositions of operators with real- or complex-valued functions

We continue with some observations about compositions of operators with scalar-valued functions. Let $\mathcal{T} \in \mathfrak{L}(\mathcal{C}_{\mathfrak{V}}(X,E),F)$ and let φ be a bounded real- or complex-valued (depending on the scalar field of E) continuous function on X. For every $\mathfrak{v} \in \mathfrak{V}$ and $f \leqslant \mathfrak{v}$, we have $\varphi \cdot f \in \rho\mathfrak{v}$, where $\rho \geqslant 0$ is a bound for $|\varphi|$ (see Lemma 1.2.1(a)). In particular, we have $\varphi \cdot f \in \mathcal{C}_{\mathfrak{V}}(X,E)$ for all $f \in \mathcal{C}_{\mathfrak{V}}(X,E)$. Hence the formula $\mathcal{T}_\varphi(f) = \mathcal{T}(\varphi \cdot f)$ defines an operator $\mathcal{T}_\varphi \in \mathfrak{L}(\mathcal{C}_{\mathfrak{V}}(X,E),F)$.

2.6.6 Proposition. *If θ is the representation measure for $\mathcal{T} \in \mathfrak{L}(\mathcal{C}_{\mathfrak{V}}(X,E),F)$ and φ and ψ are bounded real- or complex-valued continuous functions on X, then $T_\varphi = T_\psi$ if and only $\varphi(x) = \psi(x)$ for all x in the support of θ.*

Proof. Our claim is equivalent to the statement that $T_\varphi = 0$ if and only $\varphi(x) = 0$ for all x in the support of θ. We shall prove the latter. Let A be the support of θ. Then $\mathcal{T}(f) = \int_A f\,d\theta$ for all $f \in \mathcal{C}_{\mathfrak{V}}(X,E)$, hence $\mathcal{T}_\varphi(f) = 0$, provided that $\varphi(x) = 0$ for all $x \in A$. Conversely, if $\varphi(x) \neq 0$ for some $x \in A$, then there are $\varepsilon > 0$ and an open neighborhood U of x such that $|\varphi(y)| \geqslant \varepsilon$ for all $y \in U$. Since $A \setminus U$ is no longer the support of θ, using inner regularity we find a compact subset K of U and an element $a \in E$ such that $\theta_K(a) \neq 0$. We choose $\mu \in F^*$ such that $\mu(\theta_K(a)) = 1$. By outer regularity, there is $V \in \mathcal{V}$ such that $|\theta|_A(V)(\mu) < +\infty$ and an open set $K \subset O \subset U$ such that $|\theta|_{O\setminus K}(V)(\mu) \leqslant 1/(2\lambda)$, where $\lambda > 0$ is such that $a \in \lambda V$. Moreover, according to Urysohn's lemma there is $\psi \in \mathcal{C}(X)$ such that $0 \leqslant \psi \leqslant 1$, $\psi(x) = 1$ for all $x \in K$ and $\operatorname{supp}(\psi) \subset O$. Set $f(x) = (\psi(x)/\varphi(x))a$ for $x \in O$ and $f(x) = 0$, else. Then $f \in \mathcal{C}_{\mathfrak{V}}(X,E)$ and

$$\mathcal{T}_\varphi(f) = \mathcal{T}(\varphi f) = \int_O \psi \cdot a\,d\theta = \int_{O\setminus K} \psi \cdot a\,d\theta + \theta_K(a).$$

Finally, since $\psi \cdot a \leq \lambda\chi_X \cdot V$, hence

$$\Big|\Big(\int_{O\setminus K} \psi \cdot a\Big)(\mu)\Big| \leqslant \lambda|\theta|_{O\setminus K}(V)(\mu) \leqslant 1/2,$$

we infer that $\mathcal{T}_\varphi(f)(\mu) \neq 0$ and, therefore, that $T_\varphi \neq 0$. □

Lemma 2.5.6 facilitates the identification of the representation measure θ_φ of the operator $\mathcal{T}_\varphi$ associated with $\mathcal{T} \in \mathfrak{L}(\mathcal{C}_{\mathfrak{V}}(X,E),F)$, its representation measure θ and a bounded continuous real- or complex-valued function φ. For a set $A \in \mathfrak{R}$, we use a net $(\varphi_i)_{i\in\mathcal{I}}$ in $\mathcal{C}_{\mathcal{K}}(X)$ as in 2.5.6. Given $a \in E$, we choose the constant function $x \mapsto a$ for f and calculate

$$(\theta_\varphi)_A(a) = \int_A f\,d\theta_\varphi = \lim_{i\in\mathcal{I}}(\mathcal{T}_\varphi)_{\varphi_i}(f) = \lim_{i\in\mathcal{I}} \mathcal{T}_{\varphi_i}(\varphi \cdot f) = \int_A \varphi \cdot a\,d\theta.$$

Thus, according to Theorem 2.6.1 $(\theta_\varphi)_A(a)$ is indeed an element of F^{**}, as required. If the operator $\mathcal{T} \in \mathfrak{L}(\mathcal{C}_{\mathfrak{V}}(X,E),F)$ is compact (or weakly compact), we may use the characteristic function χ_A of a set $A \in \mathfrak{R}$ in place of φ and obtain the *restriction* $\mathcal{T}_{\chi_A}$ of $\mathcal{T}$ to A. For $f \in \mathcal{C}_{\mathfrak{V}}(X,E)$, we set

$$\mathcal{T}_{\chi_A}(f) = \int_A f\,d\theta = \int_X f\,d\theta|_A.$$

According to Theorem 2.6.1, $\mathcal{T}_{\chi_A}(f)$ is an element of F. Moreover, given $W \in \mathcal{W}$ there is $\mathfrak{v} \in \mathfrak{V}$ such that $|\int_X f\,d\theta(\mu)| \leqslant 1$ for all integrable functions $f \in \mathcal{F}(X,E)$ and all $\mu \in W^\circ$. Since $\chi_A \cdot f \leqslant \mathfrak{v}$, whenever $f \leqslant \mathfrak{v}$ (Lemma 1.2.1(a)) we have $\mathcal{T}_{\chi_A}(f) \in W$ for all $f \leqslant \mathfrak{v}$. Hence $\mathcal{T}_{\chi_A}$ is continuous and an element of $\mathfrak{L}(\mathcal{C}_{\mathfrak{V}}(X,E),F)$. As observed in Proposition 2.1.9, the restriction of a regular compact (or weakly compact) measure is again regular and compact (or weakly compact). Thus according to Proposition 2.5.5, the operator $\mathcal{T}_{\chi_A}$ is also compact (or weakly compact). Finally, because $\mathcal{T}_{\chi_{(X\setminus A)}}(f) = \mathcal{T}(f) - \mathcal{T}_{\chi_A}(f) \in F$, the same conclusions hold for the operator $\mathcal{T}_{\chi_{X\setminus A}}$. We summarize the following.

2.6.7 Proposition. *If θ is the representation measure for $\mathcal{T} \in \mathfrak{L}(\mathcal{C}_{\mathfrak{V}}(X,E),F)$ and φ is a bounded real-or complex-valued continuous function on X, then the representation measure θ_φ for $\mathcal{T}_\varphi \in \mathfrak{L}(\mathcal{C}_{\mathfrak{V}}(X,E),F)$ is defined by*

$$(\theta_\varphi)_A(a) = \int_A \varphi \cdot a\,d\theta$$

for all $A \in \mathfrak{R}$ and $a \in E$. If $\mathcal{T} \in \mathfrak{L}(\mathcal{C}_{\mathfrak{V}}(X,E),F)$ is compact (or weakly compact), if either $A \in \mathfrak{R}$ or $X\setminus A \in \mathfrak{R}$, then the operator $\mathcal{T}_{\chi_A}$ is also in $\mathfrak{L}(\mathcal{C}_{\mathfrak{V}}(X,E),F)$, compact (or weakly compact) and its representation measure θ_{χ_A} is the restriction $\theta|_A$ of θ to A.

2.6.8 Proposition. *If θ is the representation measure for $\mathcal{T} \in \mathfrak{L}(\mathcal{C}_{\mathfrak{V}}(X,E),F)$ and G is a subspace of E, then the $\mathfrak{L}(G,F^{**})$-valued representation measure $\theta\|_G$ for the restriction of $\mathcal{T}$ to $\mathcal{C}_{\mathfrak{V}}(X,G)$ is the restriction of θ to G, that is, it coincides with θ on all sets $A \in \mathfrak{R}$ and $a \in G$.*

Proof. Let $\mathcal{T},\theta$ and $\theta\|_G$ be as stated. For $A \in \mathfrak{R}$ and the net $(\varphi_i)_{i\in\mathcal{I}}$ in $\mathcal{C}_{\mathcal{K}}(X)$ from Lemma 2.5.6, we observe that all the functions $\varphi_i \cdot a$ are contained in $\mathcal{C}_{\mathfrak{V}}(X,G)$. Thus, again using Lemma 2.5.6 for both θ and $\mathcal{T}$ as well as $\theta\|_G$ and the restriction of $\mathcal{T}$ to $\mathcal{C}_{\mathfrak{V}}(X,G)$, we infer that

$$\theta_A(a) = \lim_{i\in\mathcal{I}} \mathcal{T}(\varphi_i \cdot a) = (\theta\|_G)_A(a)$$

as claimed. □

Point evaluation operators

For a point $x \in X$ and a linear operator T from E into F, the formula $f \mapsto T(f(x))$ defines a linear operator $\mathcal{D}_x^T$ from $\mathcal{C}_{\mathfrak{V}}(X,E)$ to F. Since $\{f(x) \mid f \in \mathcal{C}_{\mathfrak{V}}(X,E)\} = E$, this operator is not zero if and only if $T \neq 0$. For $\mathcal{D}_x^T$ to be continuous, that is, an element of $\mathfrak{L}(\mathcal{C}_{\mathfrak{V}}(X,E),F)$, it is necessary that $T \in \mathfrak{L}(E,F)$. Indeed, if $\mathcal{D}_x^T \in \mathfrak{L}(\mathcal{C}_{\mathfrak{V}}(X,E),F)$, given $W \in \mathcal{W}$ there is $\mathfrak{v} \in \mathfrak{V}$ such that $\mathcal{D}_x^T(\mathfrak{v}) \subset W$. According to the definition of a function space neighborhood, there is $V \in \mathcal{V}$ and $\varphi \in \mathcal{C}_{\mathcal{K}}(X)$ such that $\varphi(x) = 1$ and $\varphi \cdot a \leqslant \mathfrak{v}$ for all $a \in V$. Thus $T(a) = \mathcal{D}_x^T(\varphi \cdot a) \in W$. Hence our claim. However, depending on the neighborhood systems $\mathfrak{V}$ for $\mathcal{C}_{\mathfrak{V}}(X,E)$ and $\mathcal{W}$ for F not all choices for $x \in X$ and $T \in \mathfrak{L}(E,F)$ render operators $\mathcal{D}_x^T \in \mathfrak{L}(\mathcal{C}_{\mathfrak{V}}(X,E),F)$. The point evaluation measure δ_x^T does nevertheless represent this operator.

2.6.9 Lemma. *If $x \in X$ and $T \in \mathfrak{L}(E,F)$, then $\mathcal{D}_x^T \in \mathfrak{L}(\mathcal{C}_{\mathfrak{V}}(X,E),F)$ if and only if $\mathfrak{V}$ is finer than $\mathfrak{V}_{\delta_x^T}$, where δ_x^T is the representation measure for $\mathcal{D}_x^T$. This condition is satisfied if there is $\mathfrak{v} \in \mathfrak{V}$ such that the set $\{f(x) \mid f \leqslant \mathfrak{v}\}$ is bounded in E.*

Proof. Let $x \in X$ and $T \in \mathfrak{L}(E,F)$. If $\mathcal{D}_x^T \in \mathfrak{L}(\mathcal{C}_{\mathfrak{V}}(X,E),F)$, then according to Theorem 2.6.1 the neighborhood system $\mathfrak{V}_{\delta_x^T}$ defined by its representation measure δ_x^T is coarser than $\mathfrak{V}$. Conversely, if $\mathfrak{V}$ is finer than $\mathfrak{V}_{\delta_x^T}$, then for $W \in \mathcal{W}$ the neighborhood $\mathfrak{v}_{(\delta_x^T,W)}$ is defined by the family $\mathcal{N}_{(\delta_x^T,W)}$ consisting of all neighborhood functions n such that

$$\int_X n \, d\delta_x^T = \sup\{T(a) \mid a \in n(x)\} \leqslant \sigma_W,$$

that is, $T(n(x)) \subset W$. There is $\mathfrak{v} \in \mathfrak{V}$ such that $\mathfrak{v} \leqslant \mathfrak{v}_{(\delta_x^T,W)}$. Hence $f \leqslant \mathfrak{v}$ for $f \in \mathcal{C}_{\mathfrak{V}}(X,E)$ implies that $f \leqslant n$ for some $n \in \mathcal{N}_{(\delta_x^T,W)}$ and, therefore, $\mathcal{D}_x^T(f) = T(f(x)) \in W$. Thus $\mathcal{D}_x^T \in \mathfrak{L}(\mathcal{C}_{\mathfrak{V}}(X,E),F)$, as claimed. Now suppose that there is $\mathfrak{v} \in \mathfrak{V}$ such that the set $A = \{f(x) \mid f \leqslant \mathfrak{v}\}$ is bounded in E. Given $W \in \mathcal{W}$, there is $V \in \mathcal{V}$ such that $T(V) \subset W$. There is $\lambda > 0$ such that $A \subset \lambda V$. Thus $\mathcal{D}_x^T((1/\lambda)\mathfrak{v}) \subset W$. □

A nonzero operator, $\mathcal{D}_x^T \in \mathfrak{L}(\mathcal{C}_{\mathfrak{V}}(X,E),F)$, is called a *point evaluation operator*. It is represented by the point evaluation measure δ_x^T. Point evaluations will play an important part in the development of Choquet theory for linear operators in the subsequent sections.

2.6.10 Lemma. *If $\mathcal{D}_x^T$ and $\mathcal{D}_y^S$ are point evaluation operators in $\mathfrak{L}(\mathcal{C}_{\mathfrak{V}}(X,E),F)$, then $\mathcal{D}_x^T = \mathcal{D}_y^S$ if and only if $x = y$ and $T = S$.*

Proof. Let $\mathcal{D}_x^T, \mathcal{D}_y^S \in \mathfrak{L}(\mathcal{C}_{\mathfrak{V}}(X,E),F)$. If $x \neq y$, there is $\varphi \in \mathcal{C}_{\mathcal{K}}(X)$ such that $\varphi(x) = 1$ and $\varphi(y) = 0$. Since $T \neq 0$, there is $a \in E$ such that $T(a) \neq 0$ and since $\varphi \cdot a \in \mathcal{C}_{\mathfrak{V}}(X,E)$, we infer that $\mathcal{D}_x^T(\varphi \cdot a) = T(a)$, whereas $\mathcal{D}_y^S(\varphi \cdot a) = 0$. Thus $\mathcal{D}_x^T \neq \mathcal{D}_y^S$. If

$x = y$ but $T \neq S$, there is $a \in E$ such that $T(a) \neq S(a)$. Choosing $\varphi \in \mathcal{C}_{\mathcal{K}}(X)$ such that $\varphi(x) = 1$, this yields $\mathcal{D}_x^T(\varphi \cdot a) \neq \mathcal{D}_x^S(\varphi \cdot a)$. Hence $\mathcal{D}_x^T \neq \mathcal{D}_x^S$. □

For the following, we recall the function space topology of pointwise convergence from Example 1.2.7(b).

2.6.11 Lemma. *Let $\mathcal{D}_x^T \in \mathfrak{L}(\mathcal{C}_{\mathfrak{V}}(X,E),F)$ be a point evaluation operator. If $\mathcal{D}_x^T$ is compact (or weakly compact), then the operator $T \in \mathfrak{L}(E,F)$ is compact (or weakly compact). Conversely, if $T \in \mathfrak{L}(E,F)$ is compact (or weakly compact) and if $\mathfrak{V}$ is finer than the topology of pointwise convergence, then $\mathcal{D}_x^T$ is compact (or weakly compact).*

Proof. Suppose that the operator $\mathcal{D}_x^T \in \mathfrak{L}(\mathcal{C}_{\mathfrak{V}}(X,E),F)$ is compact (or weakly compact) and let C be a bounded subset of E. We choose a function $\varphi \in \mathcal{C}_{\mathcal{K}}(X)$ such that $0 \leqslant \varphi \leqslant 1$ and $\varphi(x) = 1$. The set $\mathcal{G} = \{\varphi \cdot c \mid c \in C\}$ is bounded in $\mathcal{C}_{\mathfrak{V}}(X,E)$. Indeed, given $\mathfrak{v} \in \mathfrak{V}$ there is $V \in \mathcal{V}$ such that $\chi_K \cdot V \leqslant \mathfrak{v}$, where K denotes the compact support of φ. We have $C \subset \lambda V$ with some $\lambda \geqslant 0$, and thus $\mathcal{G} \leqslant \lambda\mathfrak{v}$. Hence the set $\mathcal{D}_x^T(\mathcal{G}) = T(C)$ is relatively compact (or relatively weakly compact) in F. Conversely, suppose that $\mathfrak{V}$ is finer than the topology of pointwise convergence and that the operator $T \in \mathfrak{L}(E,F)$ is compact (or weakly compact). Let $\mathcal{G}$ be a bounded subset of $\mathcal{C}_{\mathfrak{V}}(X,E)$. Then the set $C = \{f(x) \mid f \in \mathcal{G}\}$ is bounded in E. Indeed, given $V \in \mathcal{V}$, by our assumption there is $\mathfrak{v} \in \mathfrak{V}$ such that $\mathfrak{v} \leqslant \mathfrak{v}_{(\{x\},V)} = \{f \in \mathcal{F}(X,E) \mid f(x) \in V\}$. We have $\mathcal{G} \subset \lambda\mathfrak{v} \leqslant \lambda\mathfrak{v}_{(\{x\},V)}$ with some $\lambda \geqslant 0$, and therefore $C \subset \lambda V$. Hence the set $\mathcal{D}_x^T(\mathcal{G}) = T(C)$ is relatively compact (or relatively weakly compact) in F, our claim. □

2.6.12 Lemma. *Let $\mathcal{D}_x^T \in \mathfrak{L}(\mathcal{C}_{\mathfrak{V}}(X,E),F)$ be a weakly compact point evaluation operator. If $\mathcal{D}_x^T = \mathcal{S} + \mathcal{U}$ for weakly compact operators $\mathcal{S},\mathcal{U} \in \mathfrak{L}(\mathcal{C}_{\mathfrak{V}}(X,E),F)$, then there is a weakly compact operator $\mathcal{R} \in \mathfrak{L}(\mathcal{C}_{\mathfrak{V}}(X,E),F)$ whose representation measure is supported by $X\backslash\{x\}$ and such that both $\mathcal{S}-\mathcal{R}$ and $\mathcal{U}+\mathcal{R}$ are either zero or weakly compact point evaluation operators at x.*

Proof. Let $\mathcal{D}_x^T, \mathcal{S}$ and $\mathcal{U}$ be as stated. According to Proposition 2.6.7, the restriction of $\mathcal{S}$ to $\{x\}$ is a weakly compact operator in $\mathfrak{L}(\mathcal{C}_{\mathfrak{V}}(X,E),F)$, hence equals $\mathcal{D}_x^S$ for some $S \in \mathfrak{L}(E,F)$. The operator $\mathcal{R} = \mathcal{S} - \mathcal{D}_x^S$ is also weakly compact, and we have

$$\mathcal{U} = \mathcal{D}_x^T - \mathcal{S} = \mathcal{D}_x^T - (\mathcal{D}_x^S + \mathcal{R}) = \mathcal{D}_x^{(T-S)} - \mathcal{R}.$$ □

2.6.13 Examples and Remarks. (a) If E is indeed a function space $\mathcal{C}_{\mathfrak{U}}(Y,G)$, where G is a normed space, Y is compact and $\mathfrak{U}$ renders the topology of uniform convergence on Y, then according to Proposition 1.2.9 the function space $\mathcal{C}_{\mathfrak{V}}(X,\mathcal{C}_{\mathfrak{U}}(Y,G))$ is isomorphic to a function space $\mathcal{C}_{\widetilde{\mathfrak{V}}}(X \times Y,G)$ via the bijection $f \mapsto \tilde{f}$ (see 1.2.9), that is, $\tilde{f}(x,y) = f(x)(y)$ for $f \in \mathcal{C}_{\mathfrak{V}}(X,\mathcal{C}_{\mathfrak{U}}(Y,G))$. Thus an operator $\mathcal{T} \in \mathfrak{L}(\mathcal{C}_{\mathfrak{V}}(X,\mathcal{C}_{\mathfrak{U}}(Y,G)),F)$ corresponds to an operator

$$\widetilde{\mathcal{T}} \in \mathfrak{L}(\mathcal{C}_{\widetilde{\mathfrak{V}}}(X \times Y,G),F)$$

defined as $\widetilde{\mathcal{T}}(\widetilde{f}) = \mathcal{T}(f)$. The representation measure θ for $\mathcal{T}$ is defined on subsets of X and $\mathfrak{L}(\mathcal{C}_{\mathfrak{U}}(Y,G),F^{**})$-valued, whereas the representation measure $\widetilde{\theta}$ for $\widetilde{\mathcal{T}}$ is defined on subsets of $X \times Y$ and $\mathfrak{L}(G,F^{**})$-valued. We shall indicate how to establish a relation between the measures θ and $\widetilde{\theta}$. For compact subsets A of X and B of Y, let $(\varphi_i)_{i\in\mathcal{I}}$ and $(\psi_j)_{j\in\mathcal{J}}$ be nets in $\mathcal{C}_{\mathcal{K}}(X)$ and $\mathcal{C}_{\mathcal{K}}(Y) = \mathcal{C}(Y)$ converging to χ_A and χ_B as in Lemma 2.5.6, respectively. A straightforward compactness argument shows that for every open set O in $X \times Y$ containing $A \times B$ there are open sets U in X and V in Y such that $A \times B \subset U \times V \subset O$. Thus the net $(\varphi_i \cdot \psi_j)_{(i,j)\in\mathcal{I}\times\mathcal{J}}$ in $\mathcal{C}_{\mathcal{K}}(X \times Y)$ (the index set $\mathcal{I} \times \mathcal{J}$ is ordered componentwise) satisfies the assumptions of Lemma 2.5.6 for the set $A \times B$ and we have

$$\widetilde{\theta}_{A\times B}(c) = \lim_{(i,j)\in\mathcal{I}\times\mathcal{J}} \widetilde{\mathcal{T}}(\varphi_i\psi_j \cdot c),$$

for every $c \in G$. On the other hand, for every $j \in \mathcal{J}$ we have

$$\theta_A(\psi_j \cdot c) = \lim_{i\in\mathcal{I}} \mathcal{T}(\varphi_i \cdot \psi_j \cdot c),$$

also by Lemma 2.5.6. This renders

$$\lim_{j\in\mathcal{J}} \theta_A(\psi_j \cdot c) = \lim_{j\in\mathcal{J}} \lim_{i\in\mathcal{I}} \mathcal{T}(\varphi_i \cdot \psi_j \cdot c) = \lim_{(i,j)\in\mathcal{I}\times\mathcal{J}} \widetilde{\mathcal{T}}(\varphi_i\psi_j \cdot c) = \widetilde{\theta}_{A\times B}(c)$$

for all $c \in G$. All limits are taken pointwise on F^*, that is, in the weak* topology of F^{**}. If in particular $\mathcal{T} = \mathcal{D}_x^T$ is a point evaluation and $A = \{x\}$, then

$$\widetilde{\theta}_{\{x\}\times B}(c) = \lim_{j\in\mathcal{J}} T(\psi_j \cdot c),$$

that is, the $\mathfrak{L}(G,F^{**})$-valued set function $B \mapsto \widetilde{\theta}_{\{x\}\times B}$ on Y determines the representation measure on Y for the operator $T \in \mathfrak{L}(\mathcal{C}_{\mathfrak{U}}(Y,G),F)$. The case that $G = \mathbb{R}$ is of particular interest. The measures θ and $\widetilde{\theta}$ then are $\mathfrak{L}(\mathcal{C}_{\mathfrak{U}}(Y),F^{**})$- and F^{**}-valued, respectively. The above reads as

$$\widetilde{\theta}_{A\times B} = \lim_{j\in\mathcal{J}} \theta_A(\psi_j)$$

for compact subsets A of X and B of Y in this case. The case that $G = \mathbb{C}$ is similar.

(b) An $\mathfrak{L}(E,F)$-valued measure θ can also be considered to be $\mathfrak{L}(\mathbb{R},\mathfrak{L}(E,F))$- or $\mathfrak{L}(\mathbb{C},\mathfrak{L}(E,F))$-valued, depending on the scalar field for both E and F (see Example 2.1.8(e)), if we interpret the element $\theta_A \in \mathfrak{L}(E,F)$ as the operator $z \mapsto z\,\theta_A$, where $z \in \mathbb{R}$ or $z \in \mathbb{C}$. For a closer look at this perspective, we consider the complex case (the real one is similar) and endow $\mathfrak{L}(E,F)$ with its strong operator topology. According to Proposition 1.3.1, the dual $\mathfrak{L}(E,F)^*$ of $\mathfrak{L}(E,F)$ is the tensor product $\widetilde{E} \otimes F^*$, where $\widetilde{E}$ denotes the quotient space E/E_0 and $E_0 = \bigcap_{V\in\mathcal{V}} V$. If E is Hausdorff, then $E = \widetilde{E}$. The $\mathfrak{L}(E,F)$-valued measure θ transfers to an $\mathfrak{L}(\mathbb{C},\mathcal{Q}_{\mathfrak{L}(E,F)})$-valued measure $\widetilde{\theta}$ via

$$\widetilde{\theta}_A(z)(\omega) = \mathfrak{Re}\,\omega(z\theta_A) = \theta_A(za_1)(\mu_1) + \cdots + \theta_A(za_n)(\mu_n)$$

for $A \in \mathfrak{R}$, $z \in \mathbb{C}$, $\omega = \widetilde{a}_1 \otimes \mu_1 + \cdots + \widetilde{a}_n \otimes \mu_n \in \mathfrak{L}(E,F)^*$ and any $a_i \in \widetilde{a}_i$. Then $\widetilde{\theta}_A(z) \in \mathfrak{L}(E,F) \subset \mathcal{Q}_{\mathfrak{L}(E,F)}$, and the mapping $A \mapsto \widetilde{\theta}_A$ defines an $\mathfrak{L}(E,F)$-valued measure. Its countable additivity follows directly from the countable additivity of θ. For the unit ball $\mathbb{B}$ of $\mathbb{C}$, we evaluate the modulus of $\widetilde{\theta}$ as

$$\begin{aligned}|\widetilde{\theta}|_A(\mathbb{B})(\omega) &= \sup\left\{\sum_{i=1}^{n}\widetilde{\theta}_A(z_i)(\omega)\;\middle|\;|z_i| \leqslant 1,\ A_i \in \mathfrak{R} \text{ disjoint},\ A_i \subset A\right\}\\ &= \sup\left\{\sum_{i=1}^{n}|\omega(\theta_{A_i})|\;\middle|\;A_i \in \mathfrak{R} \text{ disjoint},\ A_i \subset A\right\}\end{aligned}$$

for $\omega \in \mathfrak{L}(E,F)^*$. For boundedness of the measure $\widetilde{\theta}$, let $A \in \mathfrak{R}$ and let

$$\mathcal{U}_B^W = \{T \in \mathfrak{L}(E,F) \mid T(B) \subset W\}$$

be a neighborhood in the strong operator topology of $\mathfrak{L}(E,F)$, whereby is $W \in \mathcal{W}$ and B is a finite subset of E. Since $\mathcal{U}_B^W$ is the polar in $\mathfrak{L}(E,F)$ of the set

$$\Omega = \{\widetilde{a} \otimes \mu \mid a \in B,\ \mu \in W^\circ\} \subset \mathfrak{L}(E,F)^*,$$

the polar of $\mathcal{U}_B^W$ in $\mathfrak{L}(E,F)^*$ is the bipolar, that is, the weak* closed convex hull of Ω. There is $V \in \mathcal{V}$ such that $|\theta|_A(V) \leqslant \sigma_W$ and there is $\lambda \geqslant 0$ such that $B \subset \lambda V$. Then for every $\widetilde{a} \otimes \mu \in \Omega$ we have

$$|\widetilde{\theta}|_A(\mathbb{B})(\widetilde{a} \otimes \mu) = \sup\left\{\sum_{i=1}^{n}\theta_{A_i}(z_i a)(\mu)\;\middle|\;\begin{array}{l}|z_i| \leqslant 1,\ a_i \in \widetilde{a}_i,\\ A_i \in \mathfrak{R} \text{ disjoint},\ A_i \subset A\end{array}\right\}.$$

Thus $|\widetilde{\theta}|_A(\mathbb{B})(\omega) \leqslant |\theta|_A(\lambda V)(\mu) \leqslant \lambda$ holds for all $\omega \in \Omega$ and, therefore, for all ω in the weak* closed convex hull of Ω, that is, the polar of $\mathcal{U}_B^W$. The latter follows since $|\widetilde{\theta}|_A(\mathbb{B})$ is a weak* lower semicontinuous sublinear functional on $\mathfrak{L}(E,F)^*$. We infer that the $\mathfrak{L}(\mathbb{C}, \mathcal{Q}_{\mathfrak{L}(E,F)})$-valued measure $\widetilde{\theta}$ is indeed bounded.

If $\mathfrak{V}$ is a measurable weakly lower continuous function space neighborhood system and θ is the representation measure for an operator $\mathcal{T} \in \mathfrak{L}(\mathcal{C}_{\mathfrak{V}}(X,E),F)$, then θ is regular, and so is $\widetilde{\theta}$, as can immediately be understood from its prior definition. The $\mathfrak{L}(\mathbb{C}, \mathcal{Q}_{\mathfrak{L}(E,F)})$-valued measure $\widetilde{\theta}$ induces a function space neighborhood system $\mathfrak{V}_{\widetilde{\theta}}$ for complex-valued functions as elaborated in Section 2.5 and according to Proposition 2.5.4(b) defines the continuous linear operator

$$\mathcal{T}_{(\widetilde{\theta},X)} : \mathcal{C}_{\mathfrak{V}_{\widetilde{\theta}}}(X,\mathbb{C}) \to \widehat{\mathfrak{L}(E,F)}$$

by $\mathcal{T}_{(\widetilde{\theta},X)}(\varphi) = \int_X \varphi\, d\widetilde{\theta}$ for $\varphi \in \mathcal{C}_{\mathfrak{V}_{\widetilde{\theta}}}(X,\mathbb{C})$. We shall apply Theorem 2.3.28 in order to directly relate the operators $\mathcal{T}$ and $\mathcal{T}_{(\widetilde{\theta},B)}$. We proceed to establish that

$$\left(\int_X \varphi\, d\tilde\theta\right)(\tilde a \otimes \mu) = \left(\int_X \varphi \cdot a\, d\theta\right)(\mu)$$

holds for all $\varphi \in \mathcal{C}_{\mathcal{K}}(X,\mathbb{C})$, all $\tilde a \otimes \mu \in \mathfrak{L}(E,F)^*$ and $a \in \tilde a$. Integrability of these functions with respect the measures $\tilde\theta$ and θ is of course guaranteed due to the fact that φ has a compact support. First, we use Corollary 2.3.29 for

$$\left(\int_X \varphi\, d\tilde\theta\right)(\tilde a \otimes \mu) = \left(\int_X f\, d((\tilde a \otimes \mu) \circ \tilde\theta)\right)(+1).$$

Next, we use Theorem 2.3.28 with $\mathbb{C}$ for both G and H and the operators $S \in \mathfrak{L}(\mathbb{C},E)$ such that $z \mapsto za$ and $T \in \mathfrak{L}(F,\mathbb{C})$ such that $c \mapsto \mu(c)$. Then $S \circ \varphi = \varphi \cdot a$ and Theorem 2.3.28 states that

$$\left(\int_X \varphi \cdot a\, d\theta\right)(\mu) = \left(\int_X f\, d(\mu \circ \theta \circ S)\right)(+1).$$

Both measures $(\tilde a \otimes \mu) \circ \tilde\theta$ and $\mu \circ \theta \circ S$ are complex-valued, that is, $\mathfrak{L}(\mathbb{C}, Q_{\mathbb{C}})$-valued, and indeed equal, as for every $A \in \mathfrak{R}$ and $z \in \mathbb{C}$ we calculate that

$$((\tilde a \otimes \mu) \circ \tilde\theta)_A(z)(+1) = \theta_A(za)(\mu) = (\mu \circ \theta \circ S)(z)(+1).$$

We infer that

$$\mathcal{T}_{(\tilde\theta,X)}(\varphi)(\tilde a \otimes \mu) = \mathcal{T}(\varphi \cdot a)(\mu)$$

holds for all $\varphi \in \mathcal{C}_{\mathcal{K}}(X,\mathbb{C})$ and $\tilde a \otimes \mu \in \mathfrak{L}(E,F)^*$. Moreover, this argument also implies that $\mathcal{T}_{(\tilde\theta,X)}(\varphi)$ is an element of the space $\mathfrak{L}(E,F)$ rather than its completion. Indeed, let $\varphi \in \mathcal{C}_{\mathcal{K}}(X,\mathbb{C})$ and let $A \in \mathfrak{R}$ be the support of φ. We examine the operator $L : E \to F$ such that $L(a) = \mathcal{T}(\varphi \cdot a)$ for all $a \in E$. The linearity of L is obvious, and for continuity let $W \in \mathcal{W}$. There is $\mathfrak{v} \in \mathfrak{V}$ such $\mathcal{T}(\mathfrak{v}) \subset W$, and in turn there is $V \in \mathcal{V}$ such that $\chi_A \cdot V \leqslant \mathfrak{v}$. Then $\varphi \cdot a \leqslant \mathfrak{v}$ for all $a \in V$ and, therefore, $L(V) \subset W$. Hence $L \in \mathfrak{L}(E,F)$. But we have

$$L(\tilde a \otimes \mu) = \mathcal{T}(\varphi \cdot a)(\mu) = \mathcal{T}_{(\tilde\theta,X)}(\varphi)(\tilde a \otimes \mu)$$

for all $\tilde a \otimes \mu \in \mathfrak{L}(E,F)^*$ and, therefore,

$$\mathcal{T}_{(\tilde\theta,X)}(\varphi) = L \in \mathfrak{L}(E,F),$$

our claim.

3 Choquet theory

In this final chapter, we shall develop an extension of classical Choquet theory to spaces of linear operators on function spaces utilizing the results of Chapters 1 and 2. The generality of our setting requires substantial adaptations to the technology that is used in the original situation (see [1] or [41]), which deals with continuous linear functionals on spaces $\mathcal{C}(X)$ of continuous real-valued functions on a compact set X, endowed with the topology of uniform convergence. There, given a subspace G of $\mathcal{C}(X)$, an order relation is established in the dual of $\mathcal{C}(X)$, that is, the space of all regular Borel measures on X. This order allows for maximal elements that deliver integral representations for continuous linear functionals on the subspace G with particular properties. The support of any such maximal representation measure is localized on the Choquet boundary, a subset of X characterized by G. The transfer of this technique to our situation is far from trivial. We deal with locally compact rather than compact domains X, with spaces $\mathcal{C}_{\mathfrak{V}}(X,E)$ of vector-valued rather than real-valued functions, and with linear operators into a second-vector space F rather than linear functionals. In the classical situation, the order on its dual space $\mathcal{C}(X)^*$ is established using a suitable subcone of $\mathcal{C}(X)$, and the choice of this subcone guarantees that the set of functionals that dominate a given one in this order is weak* compact. This assures the existence of maximal elements. All these features turn out to be problematic in our more general environment. We shall establish an order on the space $\mathfrak{L}(\mathcal{C}_{\mathfrak{V}}(X,E),F)$ of continuous linear operators from $\mathcal{C}_{\mathfrak{V}}(X,E)$ into F, but the use of a subcone of $\mathcal{C}_{\mathfrak{V}}(X,E)$ is no longer suitable for this purpose. We shall use sub and superlinear functionals on $\mathfrak{L}(\mathcal{C}_{\mathfrak{V}}(X,E),F)$ generated by certain subsets of the tensor product $\mathcal{C}_{\mathfrak{V}}(X,E)\otimes F^*$ instead. Section 3.1 is concerned with the study of this type of functionals and their properties. We shall however strive for minimal rather than maximal elements in this order, since this formulation fits more naturally into our general settings. This will be elaborated on in Section 3.2, which contains our main results. Some of the concepts and techniques in therein lean on ideas developed by the author in [47] and [52] and Batty in [9] for linear functionals on complex- and vector-valued function spaces. Section 3.3 is concerned with some special cases and examples.

The classical theory is based on the fundamental work by Choquet [14, 15, 18], Choquet and Deny [16], Choquet and Meyer [17], Mokobotzki [37, 38] and Bishop and de Leeuw [11]. Later material can be found in [9, 13, 28, 32, 34, 47, 52] and [59].

3.1 $C(X)$-convex sets and functionals

We shall use the previously established notation and results and universally assume that the function space $\mathcal{C}_{\mathfrak{V}}(X,E)$ satisfies the assumptions of Theorem 2.6.1. That is, the space X is locally compact and Hausdorff, and the σ-ring $\mathfrak{R}$ consists of all relatively compact Borel subsets of X. The function space neighborhood system $\mathfrak{V}$ is

https://doi.org/10.1515/9783111315478-003

measurable and weakly lower continuous. The measures under consideration are supposed to be representation measures for some operator in $\mathfrak{L}(\mathcal{C}_{\mathfrak{V}}(X,E),F)$, hence are $\mathfrak{L}(E,F^{**})$-valued and regular.

We generally employ the weak operator topology on the operator space $\mathfrak{L}(\mathcal{C}_{\mathfrak{V}}(X,E),F)$, which was seen to be Hausdorff (see Section 1.3). Then according to Proposition 1.3.1 its dual is the tensor product $\widetilde{\mathcal{C}_{\mathfrak{V}}(X,E)} \otimes F^*$, where $\widetilde{\mathcal{C}_{\mathfrak{V}}(X,E)}$ is the quotient space $\mathcal{C}_{\mathfrak{V}}(X,E)/\mathcal{C}_{\mathfrak{V}}(X,E)_0$ and

$$\mathcal{C}_{\mathfrak{V}}(X,E)_0 = \{f \in \mathcal{C}_{\mathfrak{V}}(X,E) \mid f \leq \mathfrak{v} \text{ for all } \mathfrak{v} \in \mathfrak{V}\}$$

(see Section 1.3). If $\mathcal{C}_{\mathfrak{V}}(X,E)$ is Hausdorff, then $\widetilde{\mathcal{C}_{\mathfrak{V}}(X,E)} = \mathcal{C}_{\mathfrak{V}}(X,E)$. This is a requirement, which would limit our choice of function spaces (see Examples 1.2.7 and Proposition 1.2.6) and which we therefore prefer not to impose. Some of the subsequent deliberations would however be simplified in the Hausdorff case. If $\mathcal{C}_{\mathfrak{V}}(X,E)$ is not Hausdorff, then the bilinear form

$$\left(T, \sum_{i=1}^{n} f_i \otimes \mu_i\right) \mapsto \sum_{i=1}^{n} \mu_i(T(f_i))$$

on

$$\mathfrak{L}(\mathcal{C}_{\mathfrak{V}}(X,E),F) \times \mathcal{C}_{\mathfrak{V}}(X,E) \otimes F^*$$

is degenerate. Nevertheless, every linear functional in $\mathfrak{L}(\mathcal{C}_{\mathfrak{V}}(X,E),F)^*$ is represented by at least one element of $\mathcal{C}_{\mathfrak{V}}(X,E) \otimes F^*$ (Corollary 1.3.2), and every element of $\mathcal{C}_{\mathfrak{V}}(X,E) \otimes F^*$ defines a linear functional in $\mathfrak{L}(\mathcal{C}_{\mathfrak{V}}(X,E),F)^*$. Polars of sets and polar topologies are defined in the same way as in the nondegenerate case, and most results for the latter can be transferred using suitable quotient spaces. The weak operator topology of $\mathfrak{L}(\mathcal{C}_{\mathfrak{V}}(X,E),F)$ coincides with the weak topology induced by $\mathcal{C}_{\mathfrak{V}}(X,E) \otimes F^*$, that is, the Hausdorff topology generated by the neighborhoods

$$V_{\{\omega_1,\ldots,\omega_n\}} = \left\{ T \in \mathfrak{L}(\mathcal{C}_{\mathfrak{V}}(X,E),F) \;\middle|\; \begin{array}{l} \mathfrak{Re}\langle T, \omega_i\rangle \leq 1 \\ \text{for all } i = 1,\ldots,n \end{array} \right\}$$

corresponding to finitely many elements $\omega_1,\ldots,\omega_n$ of $\mathcal{C}_{\mathfrak{V}}(X,E) \otimes F^*$. It concurs with the weak topology of the duality of $\mathfrak{L}(\mathcal{C}_{\mathfrak{V}}(X,E),F)$ with its dual space $\widetilde{\mathcal{C}_{\mathfrak{V}}(X,E)} \otimes F^*$ (Proposition 1.3.1). Similarly, we endow $\mathcal{C}_{\mathfrak{V}}(X,E) \otimes F^*$ with the weak topology induced by $\mathfrak{L}(\mathcal{C}_{\mathfrak{V}}(X,E),F)$ and generated by the neighborhoods

$$U_{\{T_1,\ldots,T_n\}} = \left\{ \omega \in \mathcal{C}_{\mathfrak{V}}(X,E) \otimes F^* \;\middle|\; \begin{array}{l} \mathfrak{Re}\langle T_i, \omega\rangle \leq 1 \\ \text{for all } i = 1,\ldots,n \end{array} \right\}$$

corresponding to finitely many operators $T_1,\ldots,T_n$ in $\mathfrak{L}(\mathcal{C}_{\mathfrak{V}}(X,E),F)$. This topology is Hausdorff if and only if $\mathcal{C}_{\mathfrak{V}}(X,E)$ is Hausdorff. The dual of $\mathcal{C}_{\mathfrak{V}}(X,E) \otimes F^*$ under this topology is $\mathfrak{L}(\mathcal{C}_{\mathfrak{V}}(X,E),F)$.

We consider the pointwise order for $\overline{\mathbb{R}}$- or $\underline{\mathbb{R}}$-valued functions on the operator space $\mathfrak{L}(\mathcal{C}_{\mathfrak{V}}(X,E),F)$. The cone of all lower semicontinuous (with respect to the weak operator topology) $\overline{\mathbb{R}}$-valued sublinear functionals on $\mathfrak{L}(\mathcal{C}_{\mathfrak{V}}(X,E),F)$ is denoted by $\mathfrak{P}$, and the cone of all upper semicontinuous $\underline{\mathbb{R}}$-valued superlinear functionals by $\mathfrak{Q}$, that is, $\mathfrak{Q} = -\mathfrak{P}$. Their intersection $\mathfrak{P} \cap \mathfrak{Q}$ is the space of all real-valued, real-linear continuous functionals on $\mathfrak{L}(\mathcal{C}_{\mathfrak{V}}(X,E),F)$. Corresponding to a nonempty convex subset Ω of $\mathcal{C}_{\mathfrak{V}}(X,E) \otimes F^*$ we define sub and superlinear functionals $\mathfrak{p}_\Omega \in \mathfrak{P}$ and $\mathfrak{q}_\Omega \in \mathfrak{Q}$ by

$$\mathfrak{p}_\Omega(\mathcal{T}) = \sup\{\mathfrak{Re}\langle \mathcal{T},\omega\rangle \mid \omega \in \Omega\}$$

and

$$\mathfrak{q}_\Omega(\mathcal{T}) = \inf\{\mathfrak{Re}\langle \mathcal{T},\omega\rangle \mid \omega \in \Omega\}$$

for $\mathcal{T} \in \mathfrak{L}(\mathcal{C}_{\mathfrak{V}}(X,E),F)$. We observe that $-\mathfrak{p}_\Omega = \mathfrak{q}_{(-\Omega)}$. Following the above remarks, $\mathfrak{P} \cap \mathfrak{Q}$ consists of the functionals $\mathfrak{p}_{\{\omega\}} = \mathfrak{q}_{\{\omega\}}$ for elements $\omega \in \mathcal{C}_{\mathfrak{V}}(X,E) \otimes F^*$, that is, the mappings

$$\mathcal{T} \mapsto \mathfrak{Re}\langle \mathcal{T},\omega\rangle,$$

for all $\mathcal{T} \in \mathfrak{L}(\mathcal{C}_{\mathfrak{V}}(X,E),F)$. Conversely, with a sublinear functional $\mathfrak{p} \in \mathcal{P}$ we associate the subset of $\mathcal{C}_{\mathfrak{V}}(X,E) \otimes F^*$,

$$\Omega_\mathfrak{p} = \left\{\omega \in \mathcal{C}_{\mathfrak{V}}(X,E) \otimes F^* \;\middle|\; \begin{array}{l} \mathfrak{Re}\langle \mathcal{T},\omega\rangle \le \mathfrak{p}(\mathcal{T}) \\ \text{for all } \mathcal{T} \in \mathfrak{L}(\mathcal{C}_{\mathfrak{V}}(X,E),F) \end{array}\right\}.$$

Proposition 1.4.3(a) implies that there are sufficiently many continuous linear functionals on $\mathfrak{L}(\mathcal{C}_{\mathfrak{V}}(X,E),F)$ such that every $\mathfrak{p} \in \mathcal{P}$ is the pointwise supremum of the real parts of the functionals that it dominates. Since every continuous linear functional on $\mathfrak{L}(\mathcal{C}_{\mathfrak{V}}(X,E),F)$ is represented by an element of $\mathcal{C}_{\mathfrak{V}}(X,E) \otimes F^*$, we infer that $\mathfrak{p} = \mathfrak{p}_{\Omega_\mathfrak{p}}$. That is, every element of $\mathfrak{P}$ is generated by a convex subset of $\mathcal{C}_{\mathfrak{V}}(X,E) \otimes F^*$. We have $\mathfrak{p}_{(\Omega+\Phi)} = \mathfrak{p}_\Omega + \mathfrak{p}_\Phi$ and $\mathfrak{p}_{(\alpha\Omega)} = \alpha\mathfrak{p}_\Omega$ for convex subsets Ω, Φ of $\mathcal{C}_{\mathfrak{V}}(X,E) \otimes F^*$ and $\alpha \ge 0$. Moreover, $\mathfrak{p}_\Omega \le \mathfrak{p}_\Phi$ if and only if $\Omega \subset \overline{\Phi}$, whereby the closure is taken in the weak topology of $\mathcal{C}_{\mathfrak{V}}(X,E) \otimes F^*$ (see Proposition 1.4.3(a) and the correspondence between the closures of sets in $\mathcal{C}_{\mathfrak{V}}(X,E) \otimes F^*$ and in $\widetilde{\mathcal{C}_{\mathfrak{V}}(X,E)} \otimes F^*$ as elaborated in Section 1.3). Thus, $\mathfrak{p}_\Omega = \mathfrak{p}_\Phi$ if and only if $\overline{\Omega} = \overline{\Phi}$. Dual statements apply to the superlinear functionals $\mathfrak{q}_\Omega$. Both $\mathfrak{p}_\Omega$ and $\mathfrak{q}_\Omega$ are real-valued if and only if Ω is bounded (Proposition 1.4.3(b)). We also notice that if $\Omega = \operatorname{conv}(\bigcup_{i \in \mathcal{I}} \Omega_i)$, that is, the convex hull of the union of the sets Ω_i, then $\mathfrak{p}_\Omega = \bigvee_{i \in \mathcal{I}} \mathfrak{p}_{\Omega_i}$ is the pointwise supremum of the functionals $\mathfrak{p}_{\Omega_i}$. If a functional $\mathfrak{p} \in \mathfrak{P}$ (or $\mathfrak{q} \in \mathfrak{Q}$) is upper (or lower) semicontinuous at $0 \in \mathfrak{L}(\mathcal{C}_{\mathfrak{V}}(X,E),F)$, then it is continuous and real-valued on all of $\mathfrak{L}(\mathcal{C}_{\mathfrak{V}}(X,E),F)$ (see Section 1.4).

For a nonempty convex subset Ω of $\mathcal{C}_{\mathfrak{V}}(X,E) \otimes F^*$, we denote

$$\mathfrak{H}_\Omega = \{\mathcal{T} \in \mathfrak{L}(\mathcal{C}_{\mathfrak{V}}(X,E),F) \mid \mathfrak{p}_\Omega(\mathcal{T}) \le 1\},$$

that is, the polar of Ω in $\mathfrak{L}(\mathcal{C}_{\mathfrak{V}}(X,E),F)$. This set is convex and closed in the weak operator topology of $\mathfrak{L}(\mathcal{C}_{\mathfrak{V}}(X,E),F)$. It is compact in the weak operator topology if and only if its polar in the dual $\widetilde{\mathcal{C}_{\mathfrak{V}}(X,E)} \otimes F^*$ of $\mathfrak{L}(\mathcal{C}_{\mathfrak{V}}(X,E),F)$ is a neighborhood in a topology consistent with this duality (see Theorem IV.3.2 in [56]).

3.1.1 Lemma. *Let Ω be a convex subset of $\mathcal{C}_{\mathfrak{V}}(X,E) \otimes F^*$ such that $\mathfrak{H}_\Omega$ is compact in the weak operator topology, and let Φ be the closed convex hull of $\Omega \cup \{0\}$.*
(a) *If $\mathfrak{p}_\Omega(\mathcal{T}) \leq 0$ for $\mathcal{T} \in \mathfrak{L}(\mathcal{C}_{\mathfrak{V}}(X,E),F)$, then $\mathcal{T} = 0$.*
(b) *Φ absorbs every bounded subset of $\mathcal{C}_{\mathfrak{V}}(X,E) \otimes F^*$.*
(c) *$\overline{\Lambda} \subset \Lambda + \Phi$ for every convex subset Λ of $\mathcal{C}_{\mathfrak{V}}(X,E) \otimes F^*$.*
(d) *Let ψ be a linear functional on $\mathcal{C}_{\mathfrak{V}}(X,E) \otimes F^*$ that vanishes at all elements $f \otimes \mu \in \mathcal{C}_{\mathfrak{V}}(X,E) \otimes F^*$ such that $f \in \mathcal{C}_{\mathfrak{V}}(X,E)_0$. If $\mathfrak{Re}\,\psi$ is bounded above on Φ, then there is $T \in \mathfrak{L}(\mathcal{C}_{\mathfrak{V}}(X,E),F)$ such that $\psi(\omega) = \langle \mathcal{T}, \omega \rangle$ for all $\omega \in \mathcal{C}_{\mathfrak{V}}(X,E) \otimes F^*$.*

Proof. For Part (a), let $\mathcal{T} \in \mathfrak{L}(\mathcal{C}_{\mathfrak{V}}(X,E),F)$. If $\mathfrak{p}_\Omega(\mathcal{T}) \leq 0$, then $\lambda\mathcal{T} \in \mathfrak{H}_\Omega$ for all $\lambda \geq 0$. Given $f \in \mathcal{C}_{\mathfrak{V}}(X,E)$ and $\mu \in F^*$, the mapping $\mathcal{S} \mapsto \mu(\mathcal{S}(f))$ is continuous, hence bounded on the compact set $\mathfrak{H}_\Omega$. Thus $\mu(\mathcal{T}(f)) = 0$ holds for all $f \in \mathcal{C}_{\mathfrak{V}}(X,E)$ and $\mu \in F^*$. This shows $\mathcal{T} = 0$.

The arguments for Parts (b), (c) and (d) demand a closer look at the involved quotient spaces and their projections. In keeping with the notation from Section 1.3, for a function $f \in \mathcal{C}_{\mathfrak{V}}(X,E)$, we denote by $\widetilde{f} = f + \mathcal{C}_{\mathfrak{V}}(X,E)_0$ its projection into $\mathcal{C}_{\mathfrak{V}}(X,E)/\mathcal{C}_{\mathfrak{V}}(X,E)_0$, and for

$$\omega = \sum_{i=1}^{n} f_i \otimes \mu_i \in \mathcal{C}_{\mathfrak{V}}(X,E) \otimes F^*$$

we set

$$\widetilde{\omega} = \sum_{i=1}^{n} \widetilde{f}_i \otimes \mu_i \in \widetilde{\mathcal{C}_{\mathfrak{V}}(X,E)} \otimes F^*.$$

This notation will also be used for subsets of $\mathcal{C}_{\mathfrak{V}}(X,E)$ and of $\mathcal{C}_{\mathfrak{V}}(X,E) \otimes F^*$, respectively. Recall that every operator $\mathcal{T} \in \mathfrak{L}(\mathcal{C}_{\mathfrak{V}}(X,E),F)$ vanishes on $\mathcal{C}_{\mathfrak{V}}(X,E)_0$, that is, we have $\langle \mathcal{T}, \widetilde{\omega} \rangle = \langle \mathcal{T}, \omega \rangle$ for all $\omega \in \mathcal{C}_{\mathfrak{V}}(X,E) \otimes F^*$. If Φ is the polar of $\mathfrak{H}_\Omega$ in $\mathcal{C}_{\mathfrak{V}}(X,E) \otimes F^*$, then $\widetilde{\Phi}$ is the polar of $\mathfrak{H}_\Omega$ in $\widetilde{\mathcal{C}_{\mathfrak{V}}(X,E)} \otimes F^*$, that is, the second polar of $\widetilde{\Omega}$, which is the closed convex hull of $\widetilde{\Omega} \cup \{0\}$ in its weak topology. Thus Φ is the closed convex hull of $\Omega \cup \{0\}$ in $\mathcal{C}_{\mathfrak{V}}(X,E) \otimes F^*$. We observe that $\eta \in \Phi$ whenever $\omega \in \Phi$ and $\widetilde{\eta} = \widetilde{\omega}$. If $\mathfrak{H}_\Omega$ is compact in the weak operator topology of $\mathfrak{L}(\mathcal{C}_{\mathfrak{V}}(X,E),F)$, then its polar $\widetilde{\Phi}$ is a neighborhood in $\widetilde{\mathcal{C}_{\mathfrak{V}}(X,E)} \otimes F^*$ that is consistent in its duality with $\mathfrak{L}(\mathcal{C}_{\mathfrak{V}}(X,E),F)$. The projection $\widetilde{\Theta}$ of a bounded subset Θ of $\mathcal{C}_{\mathfrak{V}}(X,E) \otimes F^*$ is bounded in $\widetilde{\mathcal{C}_{\mathfrak{V}}(X,E)} \otimes F^*$ and, therefore, absorbed by $\widetilde{\Phi}$. Hence Θ is absorbed by Φ, that is, our claim in Part (b).

If ω is an element of the closure of a convex subset Λ of $\mathcal{C}_{\mathfrak{V}}(X,E) \otimes F^*$, then $\widetilde{\omega}$ is in the closure of $\widetilde{\Lambda}$. Hence $\widetilde{\omega} \in \widetilde{\Lambda} + \widetilde{\Phi} = \widetilde{\Lambda + \Phi}$. The latter follows, because the closure of $\widetilde{\Lambda}$ coincides in all topologies of $\widetilde{\mathcal{C}_{\mathfrak{V}}(X,E)} \otimes F^*$ that are consistent in its duality with

$\mathfrak{L}(\mathcal{C}_{\mathfrak{V}}(X,E),F)$, and because $\widetilde{\Phi}$ is a neighborhood in the Mackey topology. Thus we find $\eta \in \Lambda+\Phi$ such that $\widetilde{\eta} = \widetilde{\omega}$. That is, $\eta-\lambda \in \Phi$ with some $\lambda \in \Lambda$, and since $\widetilde{\omega - \lambda} = \widetilde{\eta - \lambda}$, also $\omega - \lambda \in \Phi$. We conclude that $\omega \in \Lambda + \Phi$, our claim in Part (c).

A linear functional ψ on $\mathcal{C}_{\mathfrak{V}}(X,E) \otimes F^*$ that vanishes at all elements $f \otimes \mu \in \mathcal{C}_{\mathfrak{V}}(X,E) \otimes F^*$ such that $f \in \mathcal{C}_{\mathfrak{V}}(X,E)_0$, canonically corresponds to a linear functional $\widetilde{\psi}$ on $\widetilde{\mathcal{C}_{\mathfrak{V}}(X,E)} \otimes F^*$ setting

$$\widetilde{\psi}\left(\sum_{i=1}^{n} \widetilde{f}_i \otimes \mu_i\right) = \psi\left(\sum_{i=1}^{n} f_i \otimes \mu_i\right)$$

for any choice of $f_i \in \widetilde{f}_i$. If $\mathfrak{Re}\,\psi$ is bounded above on Φ, then $\mathfrak{Re}\,\widetilde{\psi}$ is bounded above on $\widetilde{\Phi}$, which was seen to be a neighborhood consistent with the duality. We infer that the linear functional ψ is continuous in the Mackey topology of $\widetilde{\mathcal{C}_{\mathfrak{V}}(X,E)} \otimes F^*$, hence the evaluation of an operator $\mathcal{T} \in \mathfrak{L}(\mathcal{C}_{\mathfrak{V}}(X,E),F)$ on $\widetilde{\mathcal{C}_{\mathfrak{V}}(X,E)} \otimes F^*$. That is, we have

$$\psi(\omega) = \widetilde{\psi}(\widetilde{\omega}) = \langle \mathcal{T}, \widetilde{\omega} \rangle = \langle \mathcal{T}, \omega \rangle$$

for all $\omega \in \mathcal{C}_{\mathfrak{V}}(X,E) \otimes F^*$. Hence our claim in Part (d). □

Fortunately, we shall not be concerned with similar constructions that arise from the non-Hausdorff case in the sequel. We shall look at the elements of $\mathcal{C}_{\mathfrak{V}}(X,E) \otimes F^*$ as continuous linear functionals on $\mathfrak{L}(\mathcal{C}_{\mathfrak{V}}(X,E),F)$ and write

$$\omega(\mathcal{T}) \quad \text{for } \langle \mathcal{T}, \omega \rangle,$$

whenever $\omega \in \mathcal{C}_{\mathfrak{V}}(X,E) \otimes F^*$ and $\mathcal{T} \in \mathfrak{L}(\mathcal{C}_{\mathfrak{V}}(X,E),F)$.

$\mathcal{C}(X)$-convex sets

In Section 3.2 below, we shall use a suitable family of functionals $\mathfrak{p} \in \mathfrak{P}$ to define an order on $\mathfrak{L}(\mathcal{C}_{\mathfrak{V}}(X,E),F)$, which will yield minimal elements. Our purposes require these functionals to be additive for operators with mutually singular representation measures. We proceed as follows. For an element $\omega = \sum_{i=1}^{n} f_i \otimes \mu_i$ of $\mathcal{C}_{\mathfrak{V}}(X,E) \otimes F^*$ and a bounded function $\varphi \in \mathcal{C}(X)$, we recall from Lemma 1.2.1(a) that $\varphi \cdot f_i \in \mathcal{C}_{\mathfrak{V}}(X,E)$ and define

$$\varphi \cdot \omega = \sum_{i=1}^{n} (\varphi \cdot f_i) \otimes \mu_i \in \mathcal{C}_{\mathfrak{V}}(X,E) \otimes F^*.$$

We have $(\varphi \cdot \omega)(\mathcal{T}) = \omega(\mathcal{T}_\varphi)$ for all $\mathcal{T} \in \mathfrak{L}(\mathcal{C}_{\mathfrak{V}}(X,E),F)$. Similarly, for a subset Ω of $\mathcal{C}_{\mathfrak{V}}(X,E) \otimes F^*$, we write $\varphi \cdot \Omega = \{\varphi \cdot \omega \mid \omega \in \Omega\}$. Clearly, $\varphi \cdot \Omega$ is convex whenever Ω is convex. The usual associative and distributive properties hold for this operation. We have

$$\varphi \cdot (\psi \cdot \Omega) = (\varphi\psi) \cdot \Omega \quad \text{and} \quad (\varphi + \psi) \cdot \Omega \subset \varphi \cdot \Omega + \psi \cdot \Omega$$

for all bounded functions $\varphi, \psi \in C(X)$. A nonempty subset Ω of $C_{\mathfrak{V}}(X,E) \otimes F^*$ is said to be *$C(X)$-convex* if

$$\varphi \cdot \omega_1 + (1-\varphi) \cdot \omega_2 \in \Omega$$

whenever $\omega_1, \omega_2 \in \Omega$ and $0 \le \varphi \le 1$ for $\varphi \in C(X)$. This requirement implies of course convexity in the usual sense. Clearly, sums, scalar multiples and intersections of families of $C(X)$-convex sets are again $C(X)$-convex. If Ω is $C(X)$-convex, so is $\varphi \cdot \Omega$ for all bounded functions $\varphi \in C(X)$. Every subset of $C_{\mathfrak{V}}(X,E) \otimes F^*$ possesses a $C(X)$-convex hull. Correspondingly, a sublinear functional $\mathfrak{p} \in \mathfrak{P}$ is said to be *$C(X)$-convex* if there is a $C(X)$-convex subset Ω of $C_{\mathfrak{V}}(X,E) \otimes F^*$ such that $\mathfrak{p} = \mathfrak{p}_\Omega$. These functionals form a subcone of $\mathfrak{P}$. A dual notion applies to the superlinear functionals in $\mathfrak{Q}$.

For the sake of simplicity, we shall formulate the following results only in terms of the sublinear functionals $\mathfrak{p}_\Omega$, but since $-\Omega$ is $C(X)$-convex if and only Ω is $C(X)$-convex, similar results apply to the corresponding superlinear functionals $\mathfrak{q}_\Omega$. We recall that $C_{\mathfrak{V}}(X,E) \otimes F^*$ is endowed with the weak topology induced by $\mathfrak{L}(C_{\mathfrak{V}}(X,E), F)$. The closure of a subset of $C_{\mathfrak{V}}(X,E) \otimes F^*$ refers to this topology.

3.1.2 Proposition. *If Ω is a convex subset of $C_{\mathfrak{V}}(X,E) \otimes F^*$ such that $\mathfrak{p}_\Omega(\mathcal{T}) = \mathfrak{p}_\Omega(\mathcal{T}_\varphi) + \mathfrak{p}_\Omega(\mathcal{T}_{1-\varphi})$ for all $\mathcal{T} \in \mathfrak{L}(C_{\mathfrak{V}}(X,E), F)$ and $0 \le \varphi \le 1$ for $\varphi \in C(X)$, then $\overline{\Omega}$ is $C(X)$-convex.*

Proof. Let $\omega = \varphi \cdot \omega_1 + (1-\varphi) \cdot \omega_2$ for $\omega_1, \omega_2 \in \overline{\Omega}$ and $0 \le \varphi \le 1$ for $\varphi \in C(X)$. Then both $\mathfrak{Re}\, \omega_1(\mathcal{T}) \le \mathfrak{p}_\Omega(\mathcal{T})$ and $\mathfrak{Re}\, \omega_2(\mathcal{T}) \le \mathfrak{p}_\Omega(\mathcal{T})$ holds for all $\mathcal{T} \in \mathfrak{L}(C_{\mathfrak{V}}(X,E), F)$. Thus

$$\mathfrak{Re}\, \omega(\mathcal{T}) = \mathfrak{Re}\, \omega_1(\mathcal{T}_\varphi) + \mathfrak{Re}\, \omega_2(\mathcal{T}_{1-\varphi}) \le \mathfrak{p}_\Omega(\mathcal{T}_\varphi) + \mathfrak{p}_\Omega(\mathcal{T}_{1-\varphi}) = \mathfrak{p}(\mathcal{T}).$$

Hence $\omega \in \overline{\Omega}$ by the prior remarks. □

3.1.3 Corollary. *The closure of a $C(X)$-convex subset of $C_{\mathfrak{V}}(X,E) \otimes F^*$ is also $C(X)$-convex.*

Proof. Suppose that $\Omega \subset C_{\mathfrak{V}}(X,E) \otimes F^*$ is $C(X)$-convex, let $\mathcal{T} \in \mathfrak{L}(C_{\mathfrak{V}}(X,E), F)$ and $0 \le \varphi \le 1$ for $\varphi \in C(X)$. Then

$$\begin{aligned} \mathfrak{p}_\Omega(\mathcal{T}_\varphi) + \mathfrak{p}_\Omega(\mathcal{T}_{1-\varphi}) &= \sup_{\omega_1, \omega_2 \in \Omega} \{(\varphi \cdot \omega_1)(\mathcal{T}) + (1-\varphi) \cdot \omega_2(\mathcal{T})\} \\ &= \sup_{\omega \in \Omega} \omega(\mathcal{T}) = \mathfrak{p}_\Omega(\mathcal{T}) \end{aligned}$$

since Ω is $C(X)$-convex. Thus Proposition 3.1.2 applies. □

Consequently, a sublinear functional $\mathfrak{p} \in \mathfrak{P}$ is $C(X)$-convex if and only if the set $\Omega_{\mathfrak{p}} \subset C_{\mathfrak{V}}(X,E) \otimes F^*$ is $C(X)$-convex. The statement of Corollary 3.1.3 can be strengthened further as follows.

3.1.4 Proposition. *If Ω is a $C(X)$-convex subset of $C_{\mathfrak{V}}(X,E) \otimes F^*$ and if $\varphi_1, \ldots, \varphi_n$ are bounded nonnegative functions in $C(X)$ such that $\sum_{i=1}^n \varphi_i$ is strictly positive, then $(\sum_{i=1}^n \varphi_i) \cdot \overline{\Omega} = \sum_{i=1}^n \varphi_i \cdot \overline{\Omega}$.*

Proof. Clearly, $(\sum_{i=1}^n \varphi_i) \cdot \overline{\Omega} \subset \sum_{i=1}^n \varphi_i \cdot \overline{\Omega}$ holds in any case. We shall verify the reverse inclusion first for the case that $\sum_{i=1}^n \varphi_i = 1$. For $n = 2$, our claim follows from Corollary 3.1.3 in this case. Suppose it holds for some $n \geq 2$ and let $\omega = \sum_{i=1}^{n+1} \varphi_i \cdot \omega_i$ for $\omega_i \in \Omega$ and $0 \leq \varphi_i \in \mathcal{C}(X)$ such that $\sum_{i=1}^{n+1} \varphi_i = 1$. Given any neighborhood Δ in the weak* topology of $\mathcal{C}_{\mathfrak{V}}(X,E) \otimes F^*$, there is $0 < \varepsilon < 1$ such that $\varepsilon\varphi_{n+1} \cdot (\omega_{n+1} - \omega_n) \in \Delta$. Set $\psi_i = \varphi_i$ for $i = 1, \dots, n-1$, $\psi_n = \varphi_n + \varepsilon\varphi_{n+1}$ and $\psi_{n+1} = (1-\varepsilon)\varphi_{n+1}$. Then $0 \leq \psi_i$ for all i, $\sum_{i=1}^{n+1} \psi_i = 1$ and setting $\psi = \sum_{i=1}^n \psi_i$ we have $\psi(x) > 0$ for all $x \in X$. The latter follows since $\varphi_i(x) > 0$ for at least one $i \in \{1, \dots, n+1\}$. The functions ψ_i/ψ are therefore nonnegative, continuous and $\sum_{i=1}^n \psi_i/\psi = 1$. Thus according to our assumption $\eta = \sum_{i=1}^n (\psi_i/\psi) \cdot \omega_i$ is an element of $\overline{\Omega}$. Moreover, since $\psi + \psi_{n+1} = 1$, according to the first step of our argument, that is, the case $n = 2$, the element $\zeta = \psi \cdot \eta + \psi_{n+1} \cdot \omega_{n+1}$ is also contained in $\overline{\Omega}$. We have

$$\zeta = \sum_{i=1}^{n-1} \varphi_i \cdot \omega_i + (\varphi_n + \varepsilon\varphi_{n+1}) \cdot \omega_n + \big((1-\varepsilon)\varphi_{n+1}\big) \cdot \omega_{n+1},$$

hence

$$\omega - \zeta = \varepsilon\varphi_{n+1} \cdot (\omega_{n+1} - \omega_n) \in \Delta$$

by the above. Since the neighborhood Δ in $\mathcal{C}_{\mathfrak{V}}(X,E) \otimes F^*$ was arbitrarily chosen, this demonstrates that $\omega \in \overline{\Omega}$, as claimed.

Now for the general case suppose that $\omega = \sum_{i=1}^n \varphi_i \cdot \omega_i$ for $\omega_i \in \overline{\Omega}$ and $0 \leq \varphi_i \in \mathcal{C}(X)$ and that the function $\varphi = \sum_{i=1}^n \varphi_i$ is strictly positive. The functions φ_i/φ are nonnegative, continuous and $\sum_{i=1}^n (\varphi_i/\varphi) = 1$. Thus $\eta = \sum_{i=1}^n (\varphi_i/\varphi) \cdot \omega_i$ is an element of $\overline{\Omega}$ by our previous argument for the first case. Hence $\omega = \varphi \cdot \eta \leq \varphi \cdot \overline{\Omega}$, our claim. □

3.1.5 Proposition. *If Ω is a $\mathcal{C}(X)$-convex subset of $\mathcal{C}_{\mathfrak{V}}(X,E) \otimes F^*$, then*

(a) $\mathfrak{p}_\Omega(\mathcal{T}_{\sum_{i=1}^n \varphi_i}) = \sum_{i=1}^n \mathfrak{p}_\Omega(\mathcal{T}_{\varphi_i})$ *for all $\mathcal{T} \in \mathfrak{L}(\mathcal{C}_{\mathfrak{V}}(X,E), F)$ and bounded nonnegative functions $\varphi_i \in \mathcal{C}(X)$ such that $\sum_{i=1}^n \varphi_i$ is strictly positive.*

(b) $\mathfrak{p}_\Omega(\mathcal{T} + \mathcal{S}) = \mathfrak{p}_\Omega(\mathcal{T}) + \mathfrak{p}_\Omega(\mathcal{S})$ *for all operators $\mathcal{T}, \mathcal{S} \in \mathfrak{L}(\mathcal{C}_{\mathfrak{V}}(X,E), F)$ with mutually singular representation measures.*

Proof. For (a), let $\mathcal{T} \in \mathfrak{L}(\mathcal{C}_{\mathfrak{V}}(X,E), F)$ and $0 \leq \varphi_i \in \mathcal{C}(X)$ such that $\varphi = \sum_{i=1}^n \varphi_i$ is strictly positive. Using Proposition 3.1.4, we calculate

$$\sum_{i=1}^n \mathfrak{p}_\Omega(\mathcal{T}_{\varphi_i}) = \sup_{\omega_1,\dots,\omega_n \in \Omega} \sum_{i=1}^n (\varphi_i \cdot \omega_i)(\mathcal{T}) = \sup_{\omega \in \varphi \cdot \overline{\Omega}} \omega(\mathcal{T}) = \mathfrak{p}_{\varphi \cdot \overline{\Omega}}(\mathcal{T}) = \mathfrak{p}_{\overline{\Omega}}(\mathcal{T}_\varphi) = \mathfrak{p}_\Omega(\mathcal{T}_\varphi).$$

For Part (b), let $\mathcal{T}, \mathcal{S} \in \mathfrak{L}(\mathcal{C}_{\mathfrak{V}}(X,E), F)$ with mutually singular representation measures θ and ϑ, that is, there is $B \in \mathfrak{A}$ supporting θ and such that $X \setminus B$ supports ϑ. Then $\eta = \theta + \vartheta$ is the measure representing the operator $\mathcal{T} + \mathcal{S}$. We have

$$T(f) = \int_X f\, d\theta = \int_B f\, d\theta \quad \text{and} \quad S(f) = \int_X f\, d\vartheta = \int_{X \setminus B} f\, d\vartheta$$

for all $f \in \mathcal{C}_{\mathfrak{V}}(X,E)$. According to Lemma 2.5.7, there is a net $(\varphi)_{i\in\mathcal{I}}$ of functions $\varphi_i \in \mathcal{C}(X)$ such that $0 \leq \varphi_i \leq 1$ and

$$\lim_{i\in\mathcal{I}} (T+S)_{\varphi_i}(f) = \int_B f\,d\eta = T(f)$$

for all $f \in \mathcal{C}_{\mathfrak{V}}(X,E)$. Convergence is meant pointwise on F^*. This yields both

$$\lim_{i\in\mathcal{I}} (T+S)_{\varphi_i} = T \quad \text{and} \quad \lim_{i\in\mathcal{I}} (T+S)_{1-\varphi_i} = S,$$

in the weak operator topology of $\mathfrak{L}(\mathcal{C}_{\mathfrak{V}}(X,E),F)$. The functional $\mathfrak{p}_\Omega$ on $\mathfrak{L}(\mathcal{C}_{\mathfrak{V}}(X,E),F)$ is sublinear and lower semicontinuous. Sublinearity yields $\mathfrak{p}_\Omega(\mathcal{T}+\mathcal{S}) \leq \mathfrak{p}_\Omega(\mathcal{T}) + \mathfrak{p}_\Omega(\mathcal{S})$. Lower semicontinuity yields

$$\mathfrak{p}_\Omega(\mathcal{T}) \leq \liminf_{i\in\mathcal{I}} \mathfrak{p}_\Omega((\mathcal{T}+\mathcal{S})_{\varphi_i}) \quad \text{and} \quad \mathfrak{p}_\Omega(\mathcal{S}) \leq \liminf_{i\in\mathcal{I}} \mathfrak{p}_\Omega((\mathcal{T}+\mathcal{S})_{1-\varphi_i}).$$

Hence using the usual limit rules together with Part (a), we obtain the reverse inequality, that is,

$$\begin{aligned}
\mathfrak{p}_\Omega(\mathcal{T}) + \mathfrak{p}_\Omega(\mathcal{S}) &\leq \liminf_{i\in\mathcal{I}} (\mathfrak{p}_\Omega((\mathcal{T}+\mathcal{S})_{\varphi_i}) + \mathfrak{p}_\Omega((\mathcal{T}+\mathcal{S})_{1-\varphi_i})) \\
&= \liminf_{i\in\mathcal{I}} (\mathfrak{p}_\Omega((\mathcal{T}+\mathcal{S})_{\varphi_i} + (\mathcal{T}+\mathcal{S})_{1-\varphi_i})) \\
&= \mathfrak{p}_\Omega(\mathcal{T}+\mathcal{S}).
\end{aligned}$$
□

3.1.6 Corollary. *For every nonzero operator in $\mathfrak{L}(\mathcal{C}_{\mathfrak{V}}(X,E),F)$ that is not a point evaluation there are noncollinear operators $\mathcal{T}_1, \mathcal{T}_2 \in \mathfrak{L}(\mathcal{C}_{\mathfrak{V}}(X,E),F)$ such that $\mathcal{T} = \mathcal{T}_1 + \mathcal{T}_2$ and $\mathfrak{p}_\Omega(\mathcal{T}) = \mathfrak{p}_\Omega(\mathcal{T}_1) + \mathfrak{p}_\Omega(\mathcal{T}_2)$ for all $\mathcal{C}(X)$-convex subsets Ω of $\mathcal{C}_{\mathfrak{V}}(X,E) \otimes F^*$.*

Proof. If $\mathcal{T} \in \mathfrak{L}(\mathcal{C}_{\mathfrak{V}}(X,E),F)$ is neither zero nor a point evaluation, then the support of its representing measure θ contains two distinct points x and y. There is $\varphi \in \mathcal{C}(X)$ such that $0 \leq \varphi \leq 1$, $\varphi(x) = 0$ and $\varphi(y) = 1$. Then $\mathcal{T} = \mathcal{T}_\varphi + \mathcal{T}_{1-\varphi}$ and $\mathfrak{p}_\Omega(\mathcal{T}) = \mathfrak{p}_\Omega(\mathcal{T}_\varphi) + \mathfrak{p}_\Omega(\mathcal{T}_{1-\varphi})$ for all $\mathcal{C}(X)$-convex subsets Ω of $\mathcal{C}_{\mathfrak{V}}(X,E)\otimes F^*$ by Proposition 3.1.5(a). Because the functions φ and $1-\varphi$ are noncollinear, likewise are the operators $\mathcal{T}_\varphi$ and $\mathcal{T}_{1-\varphi}$ according to to Proposition 2.6.6. □

Ω-extremal point evaluations

Given a $\mathcal{C}(X)$-convex subset Ω of $\mathcal{C}_{\mathfrak{V}}(X,E) \otimes F^*$ such that $0 \in \Omega$, we shall say that a point evaluation $\mathcal{D}_x^T \in \mathfrak{L}(\mathcal{C}_{\mathfrak{V}}(X,E),F)$ is *Ω-extremal* if $\mathfrak{p}_\Omega(\mathcal{D}_x^T) < +\infty$ and if $T = T_1 + T_2$ such that

$$\mathfrak{p}_\Omega(\mathcal{D}_x^T) = \mathfrak{p}_\Omega(\mathcal{D}_x^{T_1}) + \mathfrak{p}_\Omega(\mathcal{D}_x^{T_2})$$

for $T_1, T_2 \in \mathfrak{L}(E,F)$ implies that

$$\mathfrak{p}_\Omega(\mathcal{D}_x^T)\,T_1 = \mathfrak{p}_\Omega(\mathcal{D}_x^{T_1})\,T \quad \text{and} \quad \mathfrak{p}_\Omega(\mathcal{D}_x^T)\,T_2 = \mathfrak{p}_\Omega(\mathcal{D}_x^{T_2})\,T.$$

3.1.7 Corollary. *If Ω is a $C(X)$-convex subset of $\mathcal{C}_{\mathfrak{V}}(X,E) \otimes F^*$ such that $0 \in \Omega$, then all nonzero extreme points of $\mathfrak{H}_\Omega$ are Ω-extremal point evaluations $\mathcal{D}_x^T$ such that $\mathfrak{p}_\Omega(\mathcal{D}_x^T) = 1$. If $\mathfrak{H}_\Omega$ is compact in the weak operator topology and if all operators in $\mathfrak{H}_\Omega$ are weakly compact, then all such point evaluations are extreme points of $\mathfrak{H}_\Omega$.*

Proof. We have $\mathfrak{p}_\Omega(\mathcal{T}) \geq 0$ for all $\mathcal{T} \in \mathfrak{L}(\mathcal{C}_{\mathfrak{V}}(X,E),F)$ since $0 \in \Omega$. Let $0 \neq \mathcal{T} \in \mathfrak{H}_\Omega$. That is, $p_\Omega(\mathcal{T}) \leq 1$. We go forward to list conditions that exclude $\mathcal{T}$ from being an extreme point of $\mathfrak{H}_\Omega$. (i) If $\mathfrak{p}_\Omega(\mathcal{T}) < 1$, we choose $\mathfrak{p}_\Omega(\mathcal{T}) < \lambda < 1$ and set $\mathcal{T}_1 = (1/\lambda)\mathcal{T}$ and $\mathcal{T}_2 = 0$. Then $\mathcal{T}_1, \mathcal{T}_2 \in \mathfrak{H}_\Omega$ and $\mathcal{T} = \lambda\mathcal{T}_1 + (1-\lambda)\mathcal{T}_2$, hence $\mathcal{T}$ is not an extreme point of $\mathfrak{H}_\Omega$ in this case. (ii) If $\mathfrak{p}_\Omega(\mathcal{T}) = 1$ and $\mathcal{T} = \mathcal{T}_1 + \mathcal{T}_2$ for noncollinear operators $\mathcal{T}_1, \mathcal{T}_2 \in \mathfrak{L}(\mathcal{C}_{\mathfrak{V}}(X,E),F)$ such that $\mathfrak{p}_\Phi(\mathcal{T}) = \mathfrak{p}_\Phi(\mathcal{T}_1) + \mathfrak{p}_\Phi(\mathcal{T}_2)$ for all $C(X)$-convex subsets Φ of $\mathcal{C}_{\mathfrak{V}}(X,E) \otimes F^*$, we argue as follows: We abbreviate $\alpha_1 = \mathfrak{p}_\Omega(\mathcal{T}_1)$ and $\alpha_2 = \mathfrak{p}_\Omega(\mathcal{T}_2)$, that is, $\alpha_1 + \alpha_2 = 1$. If both $\alpha_1, \alpha_2 > 0$, we set $\mathcal{S}_1 = (1/\alpha_1)\mathcal{T}_1$ and $\mathcal{S}_2 = (1/\alpha_2)\mathcal{T}_2$. Then $\mathcal{S}_1, \mathcal{S}_2 \in \mathfrak{H}_\Omega$ since $\mathfrak{p}_\Omega(\mathcal{S}_1) = \mathfrak{p}_\Omega(\mathcal{S}_2) = 1$ and $\mathcal{T} = \alpha_1\mathcal{S}_1 + \alpha_2\mathcal{S}_2$. Because $\mathcal{T}_1$ and $\mathcal{T}_2$ are noncollinear, the operators $\mathcal{S}_1$ and $\mathcal{S}_2$ are distinct. If $\alpha_1 = 1$ and $\alpha_2 = 0$, we set $\mathcal{S}_1 = \mathcal{T}_1$ and $\mathcal{S}_2 = \mathcal{T}_1 + 2\mathcal{T}_2$. Again we realize that $\mathcal{S}_1, \mathcal{S}_2 \in \mathfrak{H}_\Omega$ since

$$\mathfrak{p}_\Omega(\mathcal{S}_2) \leq \mathfrak{p}_\Omega(\mathcal{T}_1) + 2\mathfrak{p}_\Omega(\mathcal{T}_2) = 1 \quad \text{and} \quad \mathcal{T} = (1/2)\mathcal{S}_1 + (1/2)\mathcal{T}_2.$$

The case that $\alpha_1 = 0$ and $\alpha_2 = 1$ is similar. Thus in any case $\mathcal{T}$ is a convex combination of distinct elements of $\mathfrak{H}_\Omega$ and, therefore, not an extreme point of $\mathfrak{H}_\Omega$. Using this, we infer the following: (iii) If $\mathfrak{p}_\Omega(\mathcal{T}) = 1$ and $\mathcal{T}$ is not a point evaluation, we choose $\mathcal{T} = \mathcal{T}_1 + \mathcal{T}_2$ for noncollinear operators $T_1, T_2 \in \mathfrak{L}(\mathcal{C}_{\mathfrak{V}}(X,E),F)$ as in 3.1.6 and (ii) yields that $\mathcal{T}$ is not an extreme point of $\mathfrak{H}_\Omega$ in this case. (iv) If $\mathfrak{p}_\Omega(\mathcal{T}) = 1$, where $\mathcal{T} = \mathcal{D}_x^T$ is a point evaluation that is not Ω-extremal, then there are $T_1, T_2 \in \mathfrak{L}(E,F)$ such that $T = T_1 + T_2$ and $\alpha_1 + \alpha_2 = 1$, where $\alpha_1 = \mathfrak{p}_\Omega(\mathcal{D}_x^{T_1})$ and $\alpha_2 = \mathfrak{p}_\Omega(\mathcal{D}_x^{T_2})$, but $\alpha_1 T \neq T_1$. If both $\alpha_1, \alpha_2 > 0$, then both $T_1, T_2 \neq 0$ and we set $S_1 = (1/\alpha_1)T_1$ and $\mathcal{S}_2 = (1/\alpha_2)T_2$. Then $\mathcal{D}_x^{S_1}, \mathcal{D}_x^{S_2} \in \mathfrak{H}_\Omega$ and $\mathcal{D}_x^T = \alpha_1\mathcal{D}_x^{S_1} + \alpha_2\mathcal{D}_x^{S_2}$. Since $S_1 \neq T$ by the above, we have $\mathcal{D}_x^T \neq \mathcal{D}_x^{S_1}$ by Lemma 2.6.10. Hence $\mathcal{D}_x^T$ is not an extreme point of $\mathfrak{H}_\Omega$ in this case. If $\alpha_1 = 1$ and $\alpha_2 = 0$, we set $S_1 = T_1$ and $S_2 = T_1 + 2T_2$. Again we realize that $\mathcal{D}_x^{S_1}, \mathcal{D}_x^{S_2} \in \mathfrak{H}_\Omega$ since

$$\mathfrak{p}_\Omega(\mathcal{D}_x^{S_2}) \leq \mathfrak{p}_\Omega(\mathcal{D}_x^{S_1}) + 2\mathfrak{p}_\Omega(\mathcal{D}_x^{S_2}) = 1,$$

and $\mathcal{D}_x^T = (1/2)\mathcal{D}_x^{S_1} + (1/2)\mathcal{D}_x^{S_2}$. Moreover, since $S_1 \neq T$ by the above, we infer $\mathcal{D}_x^T$ is not an extreme point of $\mathfrak{H}_\Omega$ in this case. The case that $\alpha_1 = 0$ and $\alpha_2 = 1$ is similar. This completes our first line of arguments.

Now suppose that $\mathfrak{H}_\Omega$ is compact in the weak operator topology and that all its elements are weakly compact operators. Let $\mathcal{D}_x^T \in \mathfrak{H}_\Omega$ be an Ω-extremal point evaluation such that $\mathfrak{p}_\Omega(\mathcal{D}_x^T) = 1$. We shall verify that $\mathcal{D}_x^T$ is an extreme point of $\mathfrak{H}_\Omega$. For this, suppose that $\mathcal{D}_x^T = \alpha_1\mathcal{T}_1 + \alpha_2\mathcal{T}_2$ for $\mathcal{T}_1, \mathcal{T}_2 \in \mathfrak{H}_\Omega$ and $\alpha_1, \alpha_2 > 0$ such that $\alpha_1 + \alpha_2 = 1$. Then according to Lemma 2.6.12 there are $S, U \in \mathfrak{L}(E,F)$ and a weakly compact operator $\mathcal{R} \in \mathfrak{L}(\mathcal{C}_{\mathfrak{V}}(X,E),F)$ whose representation measure is supported by $X \setminus \{x\}$ and such that $\alpha_1\mathcal{T}_1 = \mathcal{D}_x^S + \mathcal{R}$ and $\alpha_2\mathcal{T}_2 = \mathcal{D}_x^U - \mathcal{R}$. We have

$$\mathfrak{p}_\Omega(\mathcal{D}_x^S + \mathcal{R}) = \mathfrak{p}_\Omega(\mathcal{D}_x^S) + \mathfrak{p}_\Omega(\mathcal{R})$$

by Proposition 3.1.5(b), so $\mathfrak{p}_\Omega(\mathcal{D}_x^S) \le \alpha_1$ and likewise, $\mathfrak{p}_\Omega(\mathcal{D}_x^U) \le \alpha_2$. We have $\mathcal{D}_x^T = \mathcal{D}_x^S + \mathcal{D}_x^U$, hence

$$\mathfrak{p}_\Omega(\mathcal{D}_x^T) \le \mathfrak{p}(\mathcal{D}_x^S) + \mathfrak{p}_\Omega(\mathcal{D}_x^U)$$

and, therefore, $\mathfrak{p}_\Omega(\mathcal{D}_x^S) = \alpha_1$ and $\mathfrak{p}_\Omega(\mathcal{D}_x^U) = \alpha_2$. This yields $S = \alpha_1 T$ and $U = \alpha_2 T$ since $\mathcal{D}_x^T$ is supposed to be Ω-extremal. The above also implies that $\mathfrak{p}_\Omega(\mathcal{R}) = 0$, hence $\mathcal{R} = 0$ by Lemma 3.1.1(a). Summarizing, we realize that $\alpha_1 \mathcal{T}_1 = \alpha_1 \mathcal{D}_x^T$ and $\alpha_2 \mathcal{T}_2 = \alpha_2 \mathcal{D}_x^T$, that is, $\mathcal{T}_1 = \mathcal{T}_2 = \mathcal{D}_x^T$. Our final claim follows. □

3.1.8 Corollary. *If $\Omega \subset \mathcal{C}_\mathfrak{V}(X,E) \otimes F^*$ is $C(X)$-convex, if $0 \in \Omega$ and if $\mathfrak{H}_\Omega$ is compact in the weak operator topology of $\mathfrak{L}(\mathcal{C}_\mathfrak{V}(X,E),F)$, then for every $x \in X$ the set*

$$\mathcal{H}_\Omega^x = \{S \in \mathfrak{L}(E,F) \mid \mathfrak{p}_\Omega(\mathcal{D}_x^S) \le 1\}$$

is convex and compact in the weak operator topology of $\mathfrak{L}(E,F)$. A point evaluation $\mathcal{D}_x^T \in \mathfrak{L}(\mathcal{C}_\mathfrak{V}(X,E),F)$ such that $\mathfrak{p}_\Omega(\mathcal{D}_x^T) < +\infty$ is Ω-extremal if and only if T is an extreme point of $\mathfrak{p}_\Omega(\mathcal{D}_x^T)\,\mathcal{H}_\Omega^x$.

Proof. Let $x \in X$. Obviously, $\mathcal{H}_\Omega^x$ is a convex subset of $\mathfrak{L}(E,F)$. Let $(T_i)_{i\in\mathcal{I}}$ be an ultranet in $\mathcal{H}_\Omega^x$. Then $(\mathcal{D}_x^{T_i})_{i\in\mathcal{I}}$ is an ultranet in $\mathfrak{H}_\Omega$, hence convergent towards an element $\mathcal{T} \in \mathfrak{H}_\Omega$ in the weak operator topology of $\mathfrak{L}(\mathcal{C}_\mathfrak{V}(X,E),F)$. We use Proposition 2.5.9 to argue that the representation measure for $\mathcal{T}$ is supported by $\{x\}$. Indeed, for all $\varphi \in \mathcal{C}_\mathcal{K}(X)$ and $a \in E$ such that φ is supported by $X\backslash\{x\}$, we have $\mathcal{D}_x^{T_i}(\varphi\cdot a) = 0$ for all $i \in \mathcal{I}$, hence $\mathcal{T}(\varphi \cdot a) = 0$. This shows that $\mathcal{T} = \mathcal{D}_x^T$ for some $T \in \mathfrak{L}(E,F)$. For every $a \in E$, there is $f \in \mathcal{C}_\mathfrak{V}(X,E)$ such that $f(x) = a$, hence $\mathcal{D}_x^{T_i}(f) = T_i(a)$ converges weakly to $\mathcal{D}_x^T(f) = T(a)$. The ultranet $(T_i)_{i\in\mathcal{I}}$ therefore converges to T in $\mathcal{H}_\Omega^x$, which is seen to be compact in the weak operator topology of $\mathfrak{L}(E,F)$. For our second claim, let $\mathcal{D}_x^T \in \mathfrak{L}(\mathcal{C}_\mathfrak{V}(X,E),F)$ be a point evaluation such that $\mathfrak{p}_\Omega(\mathcal{D}_x^T) < +\infty$. We have $\mathfrak{p}_\Omega(\mathcal{D}_x^T) > 0$ by Lemma 3.1.1(a). If $\mathcal{D}_x^T$ is Ω-extremal and if $T = \lambda R + (1-\lambda)S$ for $R, S \in \mathfrak{p}_\Omega(\mathcal{D}_x^T)\,\mathcal{H}_\Omega^x$ and $0 < \lambda < 1$, then $\mathcal{D}_x^T = \mathcal{D}_x^{\lambda R} + \mathcal{D}_x^{(1-\lambda)S}$ and

$$\begin{aligned}\mathfrak{p}_\Omega(\mathcal{D}_x^T) &\le \mathfrak{p}_\Omega(\mathcal{D}_x^{\lambda R}) + \mathfrak{p}_\Omega(\mathcal{D}_x^{(1-\lambda)S})\\ &= \lambda\mathfrak{p}_\Omega(\mathcal{D}_x^R) + (1-\lambda)\mathfrak{p}_\Omega(\mathcal{D}_x^S)\\ &\le \mathfrak{p}_\Omega(\mathcal{D}_x^T),\end{aligned}$$

that is,

$$\mathfrak{p}_\Omega(\mathcal{D}_x^T) = \mathfrak{p}_\Omega(\mathcal{D}_x^R) = \mathfrak{p}_\Omega(\mathcal{D}_x^S) \quad \text{and} \quad \mathfrak{p}_\Omega(\mathcal{D}_x^T) = \mathfrak{p}_\Omega(\mathcal{D}_x^{\lambda R}) + \mathfrak{p}_\Omega(\mathcal{D}_x^{(1-\lambda)S}).$$

Hence

$$\mathfrak{p}_\Omega(\mathcal{D}_x^T)(\lambda R) = \mathfrak{p}_\Omega(\mathcal{D}_x^{\lambda R})\,T = \mathfrak{p}_\Omega(\mathcal{D}_x^R)(\lambda T)$$

by the above definition of $\mathcal{D}_X^T$ being Ω-extremal and, therefore, $R = T$ and also $S = T$. Thus T is indeed an extreme point of $\mathfrak{p}_\Omega(\mathcal{D}_X^T)\,\mathcal{H}_\Omega^X$. Conversely, suppose that T is an extreme point of $\mathfrak{p}_\Omega(\mathcal{D}_X^T)\,\mathcal{H}_\Omega^X$ and that $T = T_1 + T_2$ such that

$$\mathfrak{p}_\Omega(\mathcal{D}_X^T) = \mathfrak{p}_\Omega(\mathcal{D}_X^{T_1}) + \mathfrak{p}_\Omega(\mathcal{D}_X^{T_2})$$

for $T_1, T_2 \in \mathfrak{L}(E,F)$. There is nothing to prove if either $T_1 = 0$ or $T_2 = 0$. Otherwise, we have both $\mathfrak{p}_\Omega(\mathcal{D}_X^{T_1}) > 0$ and $\mathfrak{p}_\Omega(\mathcal{D}_X^{T_2}) > 0$. Setting $\alpha_1 = \mathfrak{p}_\Omega(\mathcal{D}_X^{T_1})/\mathfrak{p}_\Omega(\mathcal{D}_X^T)$ and $\alpha_2 = \mathfrak{p}_\Omega(\mathcal{D}_X^{T_2})/\mathfrak{p}_\Omega(\mathcal{D}_X^T)$, with $S_1 = (1/\alpha_1)T_1$ and $S_2 = (1/\alpha_2)T_2$ we have $S_1, S_2 \in \mathfrak{p}_\Omega(\mathcal{D}_X^T)\,\mathcal{H}_\Omega^X$ and $T = \alpha_1 S_1 + \alpha_2 S_2$. Thus $S_1 = S_2 = T$ since T is supposed to be an extreme point of $\mathfrak{p}_\Omega(\mathcal{D}_X^T)\mathcal{H}_\Omega^X$. We conclude that both

$$\mathfrak{p}_\Omega(\mathcal{D}_X^T)T_1 = \mathfrak{p}_\Omega(\mathcal{D}_X^{T_1})T \quad \text{and} \quad \mathfrak{p}_\Omega(\mathcal{D}_X^T)T_2 = \mathfrak{p}_\Omega(\mathcal{D}_X^{T_2})T$$

as is required for $\mathcal{D}_X^T$ to be Ω-extremal. □

Corollary 3.1.7 establishes the main motivation for the use of $\mathcal{C}(X)$-convex subsets of $\mathcal{C}_{\mathfrak{V}}(X,E) \otimes F^*$ and their associated functionals to define a Choquet-type order structure on $\mathfrak{L}(\mathcal{C}_{\mathfrak{V}}(X,E),F)$. To further our preparations, we shall introduce some concepts and examples how to build up suitable collections of these sets. For subsets $\mathfrak{u}$ of $\mathcal{C}_{\mathfrak{V}}(X,E)$ and Π of F^*, we denote

$$\Omega_{\mathfrak{u}}^{\Pi} = \left\{ \sum_{i=1}^{n} (\varphi_i \cdot f_i) \otimes \mu_i \;\middle|\; \begin{array}{l} f_i \in \mathfrak{u}, \mu_i \in \Pi,\ 0 \le \varphi_i \in \mathcal{C}(X) \\ \text{such that } \sum_{i=1}^{n} \varphi_i = 1 \end{array} \right\}.$$

It is straightforward to verify that $\Omega_{\mathfrak{u}}^{\Pi}$ is a $\mathcal{C}(X)$-convex subset of $\mathcal{C}_{\mathfrak{V}}(X,E) \otimes F^*$. We note that $\lambda\Omega_{\mathfrak{u}}^{\Pi} = \Omega_{\mathfrak{u}}^{\lambda\Pi} = \Omega_{\lambda\mathfrak{u}}^{\Pi}$ for λ in $\mathbb{R}$ or in $\mathbb{C}$ and that $\Omega_{\mathfrak{u}}^{\Pi+\Gamma} \subset \Omega_{\mathfrak{u}}^{\Pi} + \Omega_{\mathfrak{u}}^{\Gamma}$ and $\Omega_{\mathfrak{u}+\mathfrak{w}}^{\Pi} \subset \Omega_{\mathfrak{u}}^{\Pi} + \Omega_{\mathfrak{w}}^{\Pi}$ for subsets $\mathfrak{u}, \mathfrak{w}$ of $\mathcal{C}_{\mathfrak{V}}(X,E)$ and Π, Γ of F^*. The $\mathcal{C}(X)$-convex set $\Omega_{\mathfrak{u}}^{\Pi}$ corresponds to the functional in $\mathfrak{p}_{\Omega_{\mathfrak{u}}^{\Pi}} \in \mathfrak{P}$. In order to compute the values of this functional, we return to the notation of the preceding sections and consider the embedding of F into $\mathcal{Q}_F$. That is, for an element $\omega = f \otimes \mu \in \mathcal{C}_{\mathfrak{V}}(X,E) \otimes F^*$ and an operator $\mathcal{T} \in \mathfrak{L}(\mathcal{C}_{\mathfrak{V}}(X,E),F)$ we write $\mathcal{T}(f)(\mu)$ for $\mathfrak{Re}\,\omega(\mathcal{T}) = \mathfrak{Re}\,\mu(\mathcal{T}(f))$. We also use the $\overline{\mathbb{R}}$-valued sublinear functional p_Π on F defined as

$$p_\Pi(c) = \sup_{\mu\in\Pi} \mathfrak{Re}\,\mu(c)$$

for all $c \in F$. Utilizing this, we calculate that

$$\begin{aligned} \mathfrak{p}_{\Omega_{\mathfrak{u}}^{\Pi}}(\mathcal{T}) &= \sup \left\{ \sum_{i=1}^{n} (\mathcal{T}(\varphi_i \cdot f_i)(\mu_i)) \;\middle|\; \begin{array}{l} f_i \in \mathfrak{u},\ \mu_i \in \Pi,\ 0 \le \varphi_i \in \mathcal{C}(X) \\ \text{such that } \sum_{i=1}^{n} \varphi_i = 1 \end{array} \right\} \\ &= \sup \left\{ \sum_{i=1}^{n} p_\Pi(\mathcal{T}(\varphi_i \cdot f_i)) \;\middle|\; \begin{array}{l} f_i \in \mathfrak{u},\ 0 \le \varphi_i \in \mathcal{C}(X) \\ \text{such that } \sum_{i=1}^{n} \varphi_i = 1 \end{array} \right\} \end{aligned}$$

for all $\mathcal{T} \in \mathfrak{L}(\mathcal{C}_{\mathfrak{V}}(X,E),F)$. We note that $\mathfrak{p}_{\Omega_{\mathfrak{u}}^{\Pi}}(\mathcal{T}) = \mathfrak{p}_{\Omega_{\overline{\mathfrak{u}}}^{\overline{\Pi}}}(\mathcal{T})$, where $\overline{\mathfrak{u}}$ and $\overline{\Pi}$ denote the respective closures of $\mathfrak{u}$ in $\mathcal{C}_{\mathfrak{V}}(X,E)$ in its given topology, and of Π in F^* in its weak* topology. The superlinear functional $\mathfrak{q}_{\Omega_{\mathfrak{u}}^{\Pi}}$ is similarly defined.

3.1.9 Proposition. *If $\mathcal{D}_x^T \in \mathfrak{L}(\mathcal{C}_{\mathfrak{V}}(X,E),F)$ is a point evaluation, then*

$$\mathfrak{p}_{\Omega_{\mathfrak{u}}^{\Pi}}(\mathcal{D}_x^T) = \sup\{p_\Pi(T(f(x))) \mid f \in \mathfrak{u}\}.$$

Proof. If $\mathcal{D}_x^T \in \mathfrak{L}(\mathcal{C}_{\mathfrak{V}}(X,E),F)$ is a point evaluation at $x \in X$, then $\mathcal{D}_x^T(f) = T(f(x))$ for all $f \in \mathcal{C}_{\mathfrak{V}}(X,E)$. For any choice of $f_i \in \mathfrak{u}$ and $0 \le \varphi_i \in C(X)$ such that $\sum_{i=1}^n \varphi_i = 1$, there is i_0 such that $p_\Pi(T(f_{i_0}(x))) = \max_{i=1}^n p_\Pi(T(f_i(x)))$. Then

$$\sum_{i=1}^n p_\Pi(T(\varphi_i(x)f_i(x))) = \sum_{i=1}^n \varphi_i(x)\, p_\Pi(T(f_i(x))) \le p_\Pi(T(f_{i_0}(x)))$$

since $\sum_{i=1}^n \varphi_i(x) = 1$. Thus $\mathfrak{p}_{\Omega_{\mathfrak{u}}^{\Pi}}(\mathcal{D}_x^T) \le \sup\{p_\Pi(T(f(x))) \mid f \in \mathfrak{u}\}$. The reverse inequality is evident. □

In keeping with the notation from Section 1.3, the symbol $\mathfrak{L}_{\mathfrak{u}}^{\Pi^\circ}(\mathcal{C}_{\mathfrak{V}}(X,E),F)$ will represent the set of all operators $\mathcal{T} \in \mathfrak{L}(\mathcal{C}_{\mathfrak{V}}(X,E),F)$ such that $\mathcal{T}(\mathfrak{u}) \subset \Pi^\circ$. If $\mathfrak{p}_{\Omega_{\mathfrak{u}}^{\Pi}}(\mathcal{T}) \le 1$, then $\mathfrak{Re}\,\mu(\mathcal{T}(f)) \le 1$ for all $\mu \in \Pi$. Hence $\mathcal{T}(f) \in \Pi^\circ$ for all $f \in \mathfrak{u}$, that is, $\mathcal{T} \in \mathfrak{L}_{\mathfrak{u}}^{\Pi^\circ}(\mathcal{C}_{\mathfrak{V}}(X,E),F)$. This renders

$$\mathfrak{H}_{\Omega_{\mathfrak{u}}^{\Pi}} \subset \mathfrak{L}_{\mathfrak{u}}^{\Pi^\circ}(\mathcal{C}_{\mathfrak{V}}(X,E),F).$$

Both sets $\mathfrak{H}_{\Omega_{\mathfrak{u}}^{\Pi}}$ and $\mathfrak{L}_{\mathfrak{u}}^{\Pi^\circ}(\mathcal{C}_{\mathfrak{V}}(X,E),F)$ were seen to be closed and convex in the weak operator topology of $\mathfrak{L}(\mathcal{C}_{\mathfrak{V}}(X,E),F)$. If the operator $\mathcal{D}_x^T \in \mathfrak{L}(\mathcal{C}_{\mathfrak{V}}(X,E),F)$ is a point evaluation, then

$$\mathfrak{p}_{\Omega_{\mathfrak{u}}^{\Pi}}(\mathcal{D}_x^T) = \sup\{p_\Pi(T(f(x))) \mid f \in \mathfrak{u}\}$$

by Proposition 3.1.9. Thus $\mathcal{D}_x^T \in \mathfrak{H}_{\Omega_{\mathfrak{u}}^{\Pi}}$, that is, $\mathfrak{p}_{\Omega_{\mathfrak{u}}^{\Pi}}(\mathcal{D}_x^T) \le 1$, if and only if $\mathcal{D}_x^T \in \mathfrak{L}_{\mathfrak{u}}^{\Pi^\circ}(\mathcal{C}_{\mathfrak{V}}(X,E),F)$.

Corollary 1.3.9 identifies cases when the set $\mathfrak{L}_{\mathfrak{u}}^{\Pi^\circ}(\mathcal{C}_{\mathfrak{V}}(X,E),F)$, hence also $\mathfrak{H}_{\Omega_{\mathfrak{u}}^{\Pi}}$ is compact in the weak operator topology. We summarize using Corollary 3.1.7 and the Krein–Milman theorem (Theorem VII.3.1.in [45]).

3.1.10 Proposition. *Let $\mathfrak{u}$ be a subset of $\mathcal{C}_{\mathfrak{V}}(X,E)$ and let $\Pi \subset F^*$ such that $\mathfrak{L}_{\mathfrak{u}}^{\Pi^\circ}(\mathcal{C}_{\mathfrak{V}}(X,E),F)$ is compact in the weak operator topology. Then $\mathfrak{H}_{\Omega_{\mathfrak{u}}^{\Pi}}$ is also compact. If $0 \in \Omega_{\mathfrak{u}}^{\Pi}$, then $\mathfrak{H}_{\Omega_{\mathfrak{u}}^{\Pi}}$ is the closed convex hull of the set of all $\Omega_{\mathfrak{u}}^{\Pi}$-extremal point evaluations together with* 0.

We shall at times require that a subset Π of F^* satisfies one of the following conditions:

(Π1) $p_\Pi(c+d) = p_\Pi(c) + p_\Pi(d)$ for all $c, d \in F$ such that $p_\Pi(c), p_\Pi(d) < +\infty$.
(Π2) $p_\Pi(-c) = -p_\Pi(c)$ for all $c \in F$ such that $p_\Pi(c) < +\infty$.

These conditions are certainly stringent and fall just short of requiring that Π is a singleton set. We realize that (Π2) implies (Π1). Indeed, if (Π2) holds and if both $p_\Pi(c), p_\Pi(d) < +\infty$, then $p_\Pi(c+d) < +\infty$ and, therefore,

$$p_\Pi(c) + p_\Pi(d) = -p_\Pi(-c) - p_\Pi(-d) \le -p_\Pi(-c-d) = p_\Pi(c+d).$$

Condition (Π1) requires that p_Π is additive and lower semicontinuous on

$$F_\Pi = \{c \in F \mid p_\Pi(c) < +\infty\},$$

that is, the conic hull of the polar of Π in F. Condition (Π2) requires that F_Π is indeed a real-linear subspace of F and that p_Π coincides on F_Π with the evaluation of the real part of a linear functional in F^*. Singleton sets Π do obviously satisfy these criteria. If Π is a subcone of F^*, then $p_\Pi(c) = 0$ if $c \in \Pi^\circ$ and $p_\Pi(c) = +\infty$ else. Thus Π satisfies (Π1), and indeed (Π2), provided it forms a real-linear subspace of F^*. For more elaborate settings, we refer to Example 3.1.16(c) below.

We shall use the notion of $\mathcal{C}(X)$-convexity also for subsets of $\mathcal{C}_{\mathfrak{V}}(X,E)$. A subset $\mathfrak{u}$ of $\mathcal{C}_{\mathfrak{V}}(X,E)$ is called *$\mathcal{C}(X)$-convex* (see [9]) if $\varphi \cdot f + (1-\varphi)\cdot g \in \mathfrak{u}$ whenever $f, g \in \mathfrak{u}$ and $0 \le \varphi \le 1$ for $\varphi \in \mathcal{C}(X)$. If $\mathfrak{u} \subset \mathcal{C}_{\mathfrak{V}}(X,E)$ is $\mathcal{C}(X)$-convex, then for every choice for F and $\mu \in F^*$ the set $\Omega = \{f \otimes \mu \mid f \in \mathfrak{u}\}$ is $\mathcal{C}(X)$-convex in $\mathcal{C}_{\mathfrak{V}}(X,E) \otimes F^*$ in the previously defined sense. If F equals $\mathbb{R}$ or $\mathbb{C}$, that is, if $\mathfrak{L}(\mathcal{C}_{\mathfrak{V}}(X,E), F) = \mathcal{C}_{\mathfrak{V}}(X,E)^*$ and $\Omega = \{f \otimes 1 \mid f \in \mathfrak{u}\}$, we observe that $\overline{\Omega} = \{f \otimes 1 \mid f \in \overline{\mathfrak{u}}\}$. Thus Proposition 3.1.4 applies with $\overline{\mathfrak{u}}$, that is, the closure of $\mathfrak{u}$ in $\mathcal{C}_{\mathfrak{V}}(X,E)$, in place of $\overline{\Omega}$. We note that the function space neighborhoods in Examples 1.2.7(a) to (e) all give rise to $\mathcal{C}(X)$-convex neighborhoods in $\mathcal{C}_{\mathfrak{V}}(X,E)$.

3.1.11 Proposition. *Let $\mathfrak{u}$ be a $\mathcal{C}(X)$-convex subset of $\mathcal{C}_{\mathfrak{V}}(X,E)$ and suppose that $\Pi \subset F^*$ satisfies* (Π1). *If either $0 \in \mathfrak{u}$ or Π is a singleton set, then*

$$\mathfrak{p}_{\Omega_{\mathfrak{u}}^\Pi}(\mathcal{T}) = \sup\{p_\Pi(\mathcal{T}(f)) \mid f \in \mathfrak{u}\}$$

holds for all $\mathcal{T} \in \mathfrak{L}(\mathcal{C}_{\mathfrak{V}}(X,E), F)$. Consequently, $\Omega_{\mathfrak{u}}^\Pi$ is contained in the closed convex hull of the set $\{f \otimes \mu \mid f \in \mathfrak{u},\ \mu \in \Pi\}$ and $\mathfrak{H}_{\Omega_{\mathfrak{u}}^\Pi} = \mathfrak{L}_{\mathfrak{u}}^{\Pi^\circ}(\mathcal{C}_{\mathfrak{V}}(X,E), F)$.

Proof. Let $F_\Pi = \{c \in F \mid p_\Pi(c) < +\infty\}$. According to our assumption, p_Π is additive on F_Π. Let $\mathcal{T} \in \mathfrak{L}(\mathcal{C}_{\mathfrak{V}}(X,E), F)$. We first consider the case that $\mathcal{T}(\mathfrak{u}) \subset F_\Pi$. For any choice of $f_i \in \mathfrak{u}$ and $0 \le \varphi_i \in \mathcal{C}(X)$, such that $\sum_{i=1}^n \varphi_i = 1$ according to Proposition 3.1.4 the function $f = \sum_{i=1} \varphi_i \cdot f_i$ is contained in $\overline{\mathfrak{u}}$. If $0 \in \mathfrak{u}$, we have $\varphi_i \cdot f_i \in \mathfrak{u}$ as well, hence $T(\varphi_i \cdot f_i) \in F_\Pi$ for all $i = 1, \ldots, n$. The same holds true if $F_\Pi = F$, our second alternative. Our assumption on p_Π then renders

$$\sum_{i=1}^n p_\Pi(\mathcal{T}(\varphi_i \cdot f_i)) = p_\Pi\left(\sum_{i=1}^n \mathcal{T}(\varphi_i \cdot f_i)\right) = p_\Pi(\mathcal{T}(f)).$$

Hence

$$\mathfrak{p}_{\Omega^{\Pi}_{\mathfrak{u}}}(\mathcal{T}) = \mathfrak{p}_{\Omega^{\Pi}_{\overline{\mathfrak{u}}}}(\mathcal{T}) = \sup\{p_{\Pi}(\mathcal{T}(f)) \mid f \in \overline{\mathfrak{u}}\}$$

as claimed. In case that $\mathcal{T}(\mathfrak{u}) \not\subset F_{\Pi}$, there is $f \in \mathfrak{u}$ such that $\mathcal{T}(f) \notin F_{\Pi}$, that is, $p_{\Pi}(\mathcal{T}(f)) = +\infty$. This yields

$$\mathfrak{p}_{\Omega^{\Pi}_{\mathfrak{u}}}(\mathcal{T}) = \sup\{p_{\Pi}(\mathcal{T}(f)) \mid f \in \mathfrak{u}\} = +\infty$$

in this case as well. The last statement is an obvious consequence, since by the prior argument we have $\mathfrak{p}_{\Omega^{\Pi}_{\mathfrak{u}}} = \mathfrak{p}_{\Phi}$, where Φ denotes the convex hull of $\{f \otimes \mu \mid f \in \mathfrak{u},\ \mu \in \Pi\}$. This implies that $\Omega^{\Pi}_{\mathfrak{u}} \subset \overline{\Phi}$ following the observations at the top of this section. Moreover, we have

$$\mathfrak{H}_{\Omega^{\Pi}_{\mathfrak{u}}} = \overset{-1}{\mathfrak{p}}_{\Omega^{\Pi}_{\mathfrak{u}}}((-\infty, 1]) = \overset{-1}{\mathfrak{p}}_{\Phi}((-\infty, 1]) = \mathfrak{L}^{\Pi^{\circ}}_{\mathfrak{u}}(\mathcal{C}_{\mathfrak{V}}(X, E), F).$$

Hence our claim. □

If both $\mathfrak{u} = \{f\}$ and $\Pi = \{\mu\}$ are singleton sets, Proposition 3.1.11 states that

$$\mathfrak{p}_{\Omega^{\{\mu\}}_{\{f\}}}(\mathcal{T}) = \mathfrak{Re}\,\mu(\mathcal{T}(f))$$

holds for all $\mathcal{T} \in \mathfrak{L}(\mathcal{C}_{\mathfrak{V}}(X, E), F)$.

3.1.12 Corollary. *Let n be a weakly lower continuous neighborhood function that defines a function space neighborhood $\mathfrak{u} = \{f \in \mathcal{C}_{\mathfrak{V}}(X, E) \mid f \leq n\}$. Suppose that $\Pi \subset F^*$ satisfies (Π1). If θ is the representation measure for $\mathcal{T} \in \mathfrak{L}(\mathcal{C}_{\mathfrak{V}}(X, E), F)$, then*

$$\mathfrak{p}_{\Omega^{\Pi}_{\mathfrak{u}}}(\mathcal{T}) = \sup_{\mu \in \Pi}\Big(\int_X n\, d\theta\Big)(\mu).$$

Proof. According to Proposition 3.1.11, we have

$$\begin{aligned}\mathfrak{p}_{\Omega^{\Pi}_{\mathfrak{u}}}(\mathcal{T}) &= \sup\{p_{\Pi}(\mathcal{T}(f)) \mid f \in \mathfrak{u}\}\\ &= \sup\{(\mathcal{T}(f))(\mu) \mid f \in \mathfrak{u},\ \mu \in \Pi\}\\ &= \sup_{\mu \in \Pi} \mathfrak{p}_{\Omega^{\{\mu\}}_{\mathfrak{u}}}(\mathcal{T}).\end{aligned}$$

Our claim follows with Proposition 2.3.25. That is,

$$\Big(\int_X n\, d\theta\Big)(\mu) = \sup\{(\mathcal{T}(f))(\mu) \mid f \in \mathfrak{u}\} = \mathfrak{p}_{\Omega^{\{\mu\}}_{\mathfrak{u}}}(\mathcal{T})$$

for all $\mu \in F^*$. □

3.1.13 Corollary. *Let $\mathfrak{u}$ and $\mathfrak{w}$ be $C(X)$-convex subsets of $\mathcal{C}_{\mathfrak{V}}(X, E)$ and suppose that $\Pi \subset F^*$ satisfies (Π1). If either $0 \in \mathfrak{u} \cap \mathfrak{w}$ or Π is a singleton set, then $\mathfrak{p}_{\Omega^{\Pi}_{\mathfrak{u}+\mathfrak{w}}} = \mathfrak{p}_{\Omega^{\Pi}_{\mathfrak{u}}} + \mathfrak{p}_{\Omega^{\Pi}_{\mathfrak{w}}}$.*

Proof. We note that $\Omega^{\Pi}_{\mathfrak{u}+\mathfrak{w}} \subset \Omega^{\Pi}_{\mathfrak{u}} + \Omega^{\Pi}_{\mathfrak{w}}$ holds in general. Thus $\mathfrak{p}_{\Omega^{\Pi}_{\mathfrak{u}+\mathfrak{w}}} \leq \mathfrak{p}_{\Omega^{\Pi}_{\mathfrak{u}}} + \mathfrak{p}_{\Omega^{\Pi}_{\mathfrak{w}}}$. The reverse inequality is obvious for all operators $\mathcal{T} \in \mathfrak{L}(\mathcal{C}_{\mathfrak{V}}(X,E),F)$ such that $\mathfrak{p}_{\Omega^{\Pi}_{\mathfrak{u}+\mathfrak{w}}}(\mathcal{T}) = +\infty$. Otherwise, we have $p_{\Pi}(\mathcal{T}(f+g)) < +\infty$ for all $f \in \mathfrak{u}$ and $g \in \mathfrak{w}$. If $0 \in \mathfrak{u} \cap \mathfrak{w}$, this implies that $p_{\Pi}(\mathcal{T}(f)), p_{\Pi}(\mathcal{T}(g)) < +\infty$ holds for all $f \in \mathfrak{u}$ and $g \in \mathfrak{w}$. If Π is a singleton set, the latter holds trivially true. Therefore,

$$p_{\Pi}(\mathcal{T}(f+g)) = p_{\Pi}(\mathcal{T}(f)) + p_{\Pi}(\mathcal{T}(g))$$

holds by (Π1). The assumptions of Proposition 3.1.11 apply for all of the sets $\mathfrak{u}, \mathfrak{w}$ and $\mathfrak{u}+\mathfrak{w}$ and we calculate that

$$\begin{aligned}\mathfrak{p}_{\Omega^{\Pi}_{\mathfrak{u}+\mathfrak{w}}}(\mathcal{T}) &= \sup\{p_{\Pi}(\mathcal{T}(f+g)) \mid f \in \mathfrak{u},\ g \in \mathfrak{w}\}\\ &= \sup\{p_{\Pi}(\mathcal{T}(f)) + p_{\Pi}(\mathcal{T}(g)) \mid f \in \mathfrak{u},\ g \in \mathfrak{w}\}\\ &= \mathfrak{p}_{\Omega^{\Pi}_{\mathfrak{u}}}(\mathcal{T}) + \mathfrak{p}_{\Omega^{\Pi}_{\mathfrak{w}}}(\mathcal{T}),\end{aligned}$$

our claim. □

As agreed earlier, we use the same symbol for a function space neighborhood and the subset of $\mathcal{C}_{\mathfrak{V}}(X,E)$ determined by it. With this in mind, if $\mathfrak{u} \subset \mathcal{C}_{\mathfrak{V}}(X,E)$ is a weakly lower continuous function space neighborhood in $\mathcal{F}(X,E)$, we are able to calculate the value of $\mathfrak{p}_{\Omega^{\Pi}_{\mathfrak{u}}}(\mathcal{T})$ in terms of the variation of the representation measure θ for $\mathcal{T}$.

3.1.14 Proposition. *Let $\mathfrak{u}$ be a measurable weakly lower continuous function space neighborhood, which is a neighborhood in $\mathcal{C}_{\mathfrak{V}}(X,E)$ and let $\Pi \subset F^*$. If θ is the representation measure for an operator $\mathcal{T} \in \mathfrak{L}(\mathcal{C}_{\mathfrak{V}}(X,E),F)$, then*

$$\sup_{A\in\mathfrak{R}} \|\theta\|^{(\mathfrak{u},\Pi)}_{A\cap O} = \sup\left\{\sum_{i=1}^{n} \mathcal{T}(\varphi_i \cdot f_i)(\mu_i) \;\middle|\; \begin{array}{l} f_i \in \mathfrak{u},\ \mu_i \in \Pi,\ 0 \leq \varphi_i \in \mathcal{C}(X) \\ \text{such that } \sum_{i=1}^{n} \varphi_i \leq \chi_O \end{array}\right\}$$

for all open subsets O of X. In particular, $\mathfrak{p}_{\Omega^{\Pi}_{\mathfrak{u}}}(\mathcal{T}) = \sup_{A\in\mathfrak{R}} \|\theta\|^{(\mathfrak{u},\Pi)}_{A}$.

Proof. Let O be an open subset of X. A brief inspection of our definition of the variation of a measure in Section 2.1 together with the observation for regular measures from Proposition 2.1.14 shows that

$$\begin{aligned}\sup_{A\in\mathfrak{R}} \|\theta\|^{(\mathfrak{u},\Pi)}_{A\cap O} &= \sup\left\{\sum_{i=1}^{n} \theta_{A_i}(a_i)(\mu_i) \;\middle|\; \begin{array}{l} A_i \in \mathfrak{R} \text{ disjoint, } A_i \subset O, \\ a_i \in E, \chi_{A_i} \cdot a_i \leq \mathfrak{u} \text{ and } \mu_i \in \Pi \end{array}\right\}\\ &= \sup\left\{\sum_{i=1}^{n} \theta_{K_i}(a_i)(\mu_i) \;\middle|\; \begin{array}{l} K_i \text{ disjoint compact, } K_i \subset O, \\ a_i \in E,\ \chi_{K_i} \cdot a_i \leq \mathfrak{u} \text{ and } \mu_i \in \Pi \end{array}\right\}.\end{aligned}$$

For $i = 1,\ldots,n$, let $f_i \in \mathfrak{u}$, let $0 \leq \varphi_i \in \mathcal{C}(X)$ such that $\sum_{i=1}^{n} \varphi_i \leq \chi_O$ and $\mu_i \in \Pi$. There is $W \in \mathcal{W}$ such that $\mu_i \in W^{\circ}$, and because $\mathfrak{V}$ is finer that $\mathfrak{V}_{\theta}$ by the integral representation theorem 2.6.1, there is $\mathfrak{v} \in \mathfrak{V}$ such that $\mathfrak{v} \leq \mathfrak{v}_{(\theta,W)}$. Hence $|(\int_X h\, d\theta)(\mu_i)| \leq 1$

holds for all integrable functions $h \le \mathfrak{v}$ and all μ_i. Since $\mathfrak{u}$ is supposed to be a neighborhood in $\mathcal{C}_{\mathfrak{V}}(X,E)$, we may assume that $\mathfrak{v} \subset \mathfrak{u}$. Let $\varepsilon > 0$. There is a compact subset K of X such that $\chi_{X\setminus K} \cdot f_i \in \varepsilon\mathfrak{v}$ for all f_i. In turn, there is $V \in \mathcal{V}$ such that $\chi_K \cdot V \le \varepsilon\mathfrak{v}$. We find a disjoint partition of $K \cap O$ by subsets $A_1, \dots, A_m \in \mathfrak{R}$ such that both

$$f_i(x) - f_i(y) \in V \quad \text{and} \quad \varphi_i(x)f_i(x) - \varphi_i(y)f_i(y) \in V$$

holds for all $i = 1, \dots, n$ whenever $x, y \in A_k$ for the same k. We fix elements $x_k \in A_k$ and define step functions

$$s_i = \sum_{k=1}^{m} \chi_{A_k} \cdot (\varphi_i \cdot f_i)(x_k)$$

for $i = 1, \dots, n$. We claim that $\varphi_i \cdot f_i - s_i \in 2\varepsilon\mathfrak{v}$. Indeed, for $x \in K \cap O$ we have $x \in A_k$ for some $k = 1, \dots, m$, hence

$$(\varphi_i \cdot f_i)(x) - s_i(x) = (\varphi_i \cdot f_i)(x) - (\varphi_i \cdot f_i)(x_k) \in V$$

and, therefore, $\chi_{K\cap O} \cdot (\varphi_i \cdot f_i - s_i) \le \varepsilon\mathfrak{v}$ by the above. Since the functions $\varphi_i \cdot f_i$ vanish outside of O and the s_i vanish outside of $K \cap O$, this yields

$$\begin{aligned} \varphi_i \cdot f_i - s_i &= \chi_{K\cap O} \cdot (\varphi_i \cdot f_i - s_i) + \chi_{O\setminus K} \cdot (\varphi_i \cdot f_i - s_i) \\ &= \chi_{K\cap O} \cdot (\varphi_i \cdot f_i - s_i) + \chi_{O\setminus K} \cdot (\varphi_i \cdot f_i) \\ &\le \varepsilon\mathfrak{v} + \varepsilon\mathfrak{v}. \end{aligned}$$

Thus, since $\mathfrak{v} \le \mathfrak{v}_{(\theta,W)}$,

$$\begin{aligned} \Big(\int_X \varphi_i \cdot f_i \, d\theta\Big)(\mu_i) &\le \Big(\int_X s_i\Big)(\mu_i) + 2\varepsilon \\ &= \sum_{k=1}^{m} \varphi_i(x_k)\theta_{A_k}(f_i(x_k))(\mu_i) + 2\varepsilon. \end{aligned}$$

We note that $\int_X \varphi_i \cdot f_i \, d\theta = \mathcal{T}(\varphi_i \cdot f_i) \in F$. Now summing up the above inequality over i returns

$$\sum_{i=1}^{n} (\mathcal{T}(\varphi_i \cdot f_i))(\mu_i) \le \sum_{k=1}^{m} \sum_{i=1}^{n} \varphi_i(x_k)\theta_{A_k}(f_i(x_k))(\mu_i) + 2n\varepsilon.$$

By the above, for every $i = 1, \dots, n$ and $k = 1, \dots, m$, we have $f_i(x) - f_i(x_k) \in V$ for all $x \in A_k$, hence $\chi_{A_k} \cdot (f_i - f_i(x_k)) \le \varepsilon\mathfrak{v} \subset \varepsilon\mathfrak{u}$ and, therefore, $\chi_{A_k} \cdot f_i(x_k) \in (1+\varepsilon)\mathfrak{u}$. Hence from the definition of the variation $\|\theta\|^{(\mathfrak{u},\Pi)}$ we infer that

$$\theta_{A_k}(f_i(x_k))(\mu_i) \le (1+\varepsilon)\|\theta\|^{(\mathfrak{u},\Pi)}_{A_k}.$$

Now continuing with the preceding inequalities and using that $\sum_{i=1}^{n} \varphi_i(x_k) \leq 1$ holds for all k, we realize that

$$\begin{aligned}\sum_{i=1}^{n} (\mathcal{T}(\varphi_i \cdot f_i))(\mu_i) &\leq (1+\varepsilon) \sum_{k=1}^{m} \sum_{i=1}^{n} \varphi_i(x_k) \|\theta\|_{A_k}^{(\mathfrak{u},\Pi)} + 2n\varepsilon \\ &\leq (1+\varepsilon) \sum_{k=1}^{m} \|\theta\|_{A_k}^{(\mathfrak{u},\Pi)} + 2n\varepsilon \\ &\leq (1+\varepsilon) \|\theta\|_{O}^{(\mathfrak{u},\Pi)} + 2n\varepsilon\end{aligned}$$

by the countable additivity of $\|\theta\|^{(\mathfrak{u},\Pi)}$. As the above inequality is valid for all $\varepsilon > 0$, we infer that

$$\sum_{i=1}^{n} (\mathcal{T}(\varphi_i \cdot f_i))(\mu_i) \leq \sup_{A \in \mathfrak{R}} \|\theta\|_{A \cap O}^{(\mathfrak{u},\Pi)}$$

holds for any choice of functions $f_i \leq \mathfrak{u}$, $0 \leq \varphi_i \in \mathcal{C}(X)$ such that $\sum_{i=1}^{n} \varphi_i \leq \chi_O$ and $\mu_i \in \Pi$. Thus

$$\mathfrak{p}_{\Omega_{\mathfrak{u}}^{\Pi}}(\mathcal{T}) \leq \sup_{A \in \mathfrak{R}} \|\theta\|_{A \cap O}^{(\mathfrak{u},\Pi)}.$$

For the reverse inequality, let K_i be disjoint compact subsets sets of O, let $\chi_{K_i} \cdot a_i \in \mathfrak{u}$ and $\mu_i \in \Pi$ for $i = 1, \ldots, n$. Let $\varepsilon > 0$. According to Lemma 1.2.4(a), since $\mathfrak{u}$ is supposed to be weakly lower continuous there are disjoint open sets $K_i \subset O_i \in \mathfrak{O}$ such that $\chi_{O_i} \cdot a_i \in (1+\varepsilon)\mathfrak{u}$. We can therefore construct functions $f_i \in (1+\varepsilon)\mathfrak{u}$ such that $f_i(x) = a_i$ for all $x \in K_i$. Next, for each $i = 1, \ldots, n$, we use Lemma 2.5.6 with the functions f_i for f, the compact sets K_i for K and O_i for U to find functions $\varphi_i \in \mathcal{C}_K(X)$ such that $\chi_{K_i} \leq \varphi_i \leq \chi_{O_i}$ and $|(\mathcal{T}(\varphi_i \cdot f_i) - \theta_{A_i}(a_i))(\mu_i)| \leq \varepsilon$. We have $\sum_{i=1}^{n} \varphi_i \leq \chi_O$ since the sets O_i are disjoint subsets of O. Summing up over i then yields

$$\sum_{i=1}^{n} \theta_{A_i}(a_i)(\mu_i) \leq \sum_{i=0}^{n} (\mathcal{T}(\varphi_i \cdot f_i))(\mu_i) + n\varepsilon \leq (1+\varepsilon)\, \mathfrak{p}_{\Omega_{\mathfrak{u}}^{\Pi}}(\mathcal{T}) + n\varepsilon.$$

We infer that $\|\theta\|_{O}^{(\mathfrak{u},\Pi)} \leq \sum_{i=1}^{n} (\mathcal{T}(\varphi_i \cdot f_i))(\mu_i)$, thus completing our argument. Finally, for the special case that $O = X$, we consider an expression

$$\sum_{i=1}^{n} (\mathcal{T}(\varphi_i \cdot f_i))(\mu_i)$$

for $f_i \in \mathfrak{u}$, $0 \leq \varphi_i \in \mathcal{C}(X)$ such that $\sum_{i=1}^{n} \varphi_i \leq 1$ and $\mu_i \in \Pi$, and set $\varphi_0 = 1 - \sum_{i=1}^{n} \varphi_i \in \mathcal{C}(X)$, set $f_0 = 0 \in \mathfrak{u}$ and choose any $\mu_0 \in \Pi$. Then $\sum_{i=0}^{n} \varphi_i = 1$ and

$$\sum_{i=0}^{n} \mu_i (\mathcal{T}(\varphi_i \cdot f_i))(\mu_i) = \sum_{i=1}^{n} (\mathcal{T}(\varphi_i \cdot f_i))(\mu_i).$$

Hence the expression used in the proposition coincides with the one used in the definition of $\mathfrak{p}_{\Omega_{\mathfrak{u}}^{\Pi}}$. Our final claim follows. □

Elementary operators

For fixed elements $\nu \in \mathcal{C}_{\mathfrak{V}}(X,E)^*$ and $e \in F$, we define the *elementary operator*

$$\mathcal{L}_\nu^e \in \mathfrak{L}(\mathcal{C}_{\mathfrak{V}}(X,E),F)$$

as $\mathcal{L}_\nu^e(f) = \nu(f)e$ for all $f \in \mathcal{C}_{\mathfrak{V}}(X,E)$. Finite rank operators are sums of elementary ones.

In line with the conventions introduced in Section 1.3, the symbol $\mathrm{bconv}(A)$ stands for the balanced convex hull of a subset A of a locally convex topological vector space. Clearly, $\mathrm{bconv}(A)$ is bounded, whenever A is bounded. We augment this notation for subsets of $\mathcal{C}_{\mathfrak{V}}(X,E)$, denoting by $\mathrm{b}cx(\mathfrak{u})$ the balanced $C(X)$-convex hull of a subset $\mathfrak{u}$ of $\mathcal{C}_{\mathfrak{V}}(X,E)$, that is, the intersection of all balanced $C(X)$-convex subsets of $\mathcal{C}_{\mathfrak{V}}(X,E)$ that contain $\mathfrak{u}$. If all function space neighborhoods in $\mathfrak{V}$ create $C(X)$-convex neighborhoods in $\mathcal{C}_{\mathfrak{V}}(X,E)$, then $\mathrm{b}cx(\mathfrak{u})$ is bounded whenever $\mathfrak{u}$ is bounded.

3.1.15 Proposition. *Let $\mathfrak{u}$ be a $C(X)$-convex subset of $\mathcal{C}_{\mathfrak{V}}(X,E)$ and let $\Pi \subset F^*$.*

(a) *If $\nu \in \mathcal{C}_{\mathfrak{V}}(X,E)^*$ and $e \in F$, then*

$$\mathfrak{p}_{\Omega_{\mathfrak{u}}^{\Pi}}(\mathcal{L}_\nu^e) \le p_{\mathrm{bconv}(\Pi)}(e)\ \sup\{|\nu(f)| \mid f \in \mathrm{b}cx(\mathfrak{u})\}.$$

(b) *If $\nu \in \mathcal{C}_{\mathfrak{V}}(X,E)^*$, if $e \in F$ is such that $0 \le \mu(e) \in \mathbb{R}$ for all $\mu \in \Pi$, and if either $0 \in \mathfrak{u}$ or Π is a singleton set, then*

$$\mathfrak{p}_{\Omega_{\mathfrak{u}}^{\Pi}}(\mathcal{L}_\nu^e) = p_\Pi(e)\ \sup\{\mathfrak{Re}\,\nu(f) \mid f \in \mathfrak{u}\}.$$

(c) *If $\mathrm{b}cx(\mathfrak{u})$ is bounded in $\mathcal{C}_{\mathfrak{V}}(X,E)$ and Π is weak* bounded in F^*, then $\mathfrak{p}_{\Omega_{\mathfrak{u}}^{\Pi}}(\mathcal{T}) < +\infty$ for every finite rank operator $\mathcal{T} = \sum_{i=1}^n \mathcal{L}_{\nu_i}^{e_i}$, where $\nu_i \in \mathcal{C}_{\mathfrak{V}}(X,E)^*$ and $e_i \in F$.*

(d) *If $\mathfrak{L}_{\mathfrak{u}}^{\Pi^\circ}(\mathcal{C}_{\mathfrak{V}}(X,E),F)$ is compact in the weak operator topology, then for any choice of $\nu_i \in \mathrm{b}cx(\mathfrak{u})^\circ$, $e_i \in \mathrm{bconv}(\Pi)^\circ$ and $\lambda_i \ge 0$ such that $\sum_{i=1}^\infty \lambda_i = 1$ we have $\mathcal{T} = \sum_{i=1}^\infty \lambda_i \mathcal{L}_{\nu_i}^{e_i} \in \mathfrak{H}_{\Omega_{\mathfrak{u}}^{\Pi}}$. The series converges in the weak operator topology.*

Proof. Suppose that $\mathfrak{u} \subset \mathcal{C}_{\mathfrak{V}}(X,E)$ is $C(X)$-convex and that $\Pi \subset F^*$. Let $\nu \in \mathcal{C}_{\mathfrak{V}}(X,E)^*$ and $e \in F$. We reformulate that

$$\mathfrak{p}_{\Omega_{\mathfrak{u}}^{\Pi}}(\mathcal{L}_\nu^e) = \sup\left\{\sum_{i=1}^n \mathfrak{Re}\,\mu_i(e)\nu(\varphi_i \cdot f_i)\ \middle|\ \begin{array}{l} f_i \in \mathfrak{u},\ \mu_i \in \Pi,\ 0 \le \varphi_i \in C(X) \\ \text{such that } \sum_{i=1}^n \varphi_i = 1 \end{array}\right\}.$$

For Part (a), we observe that $p_{\mathrm{bconv}(\Pi)}(e) = \sup\{|\mu(e)| \mid \mu \in \Pi\}$. Let us consider a fixed choice of $f_i \in \mathfrak{u}$, $\mu_i \in \Pi$ and $\varphi_i \in C(X)$ in the above expression. We set $g_i = \gamma_i f_i \in \mathrm{b}\mathcal{C}\mathcal{X}(\mathfrak{u})$, where $|\gamma_i| = 1$ such that $\mathfrak{Re}\, \nu(\varphi_i \cdot g_i) = |\nu(\varphi_i \cdot f_i)|$. Then

$$\sum_{i=1}^{n} \mathfrak{Re}\, \mu_i(e)\nu(\varphi_i \cdot f_i) \leq \sum_{i=1}^{n} |\mu_i(e)||\nu(\varphi_i \cdot f_i)| \leq p_{\mathrm{bconv}(\Pi)}(e)\, \mathfrak{Re}\, \nu\left(\sum_{i=1}^{n} \varphi_i \cdot g_i\right).$$

Since $\mathrm{b}\mathcal{C}\mathcal{X}(\mathfrak{u})$ is $C(X)$-convex, we have $\sum_{i=1}^{n} \varphi_i \cdot g_i \in \overline{\mathrm{b}\mathcal{C}\mathcal{X}(\mathfrak{u})}$ (Proposition 3.1.4). Thus

$$\begin{aligned}\sum_{i=1}^{n} \mathfrak{Re}\, \mu_i(e)\nu(\varphi_i \cdot f_i) &\leq p_{\mathrm{bconv}(\Pi)}(e)\, \sup\{\mathfrak{Re}\, \nu(f) \mid f \in \overline{\mathrm{b}\mathcal{C}\mathcal{X}(\mathfrak{u})}\} \\ &\leq p_{\mathrm{bconv}(\Pi)}(e)\, \sup\{|\nu(f)| \mid f \in \overline{\mathrm{b}\mathcal{C}\mathcal{X}(\mathfrak{u})}\}.\end{aligned}$$

Our claim in Part (a) follows since

$$\sup\{|\nu(f)| \mid f \in \overline{\mathrm{b}\mathcal{C}\mathcal{X}(\mathfrak{u})}\} = \sup\{|\nu(f)| \mid f \in \mathrm{b}\mathcal{C}\mathcal{X}(\mathfrak{u})\}.$$

Now arguing Part (b), we suppose that $0 \leq \mu(e) \in \mathbb{R}$ holds for all $\mu \in \Pi$ and first assume that $0 \in \mathfrak{u}$. For a fixed choice of $f_i \in \mathfrak{u}$, $\mu_i \in \Pi$ and $\varphi_i \in C(X)$ we set $g_i = f_i$ if $\mathfrak{Re}\, \nu(\varphi_i f_i) \geq 0$ and $g_i = 0$ else. Then $g_i \in \mathfrak{u}$ and

$$\begin{aligned}\sum_{i=1}^{n} \mathfrak{Re}\, \mu_i(e)\nu(\varphi_i \cdot f_i) &\leq \sum_{i=1}^{n} \mu_i(e)\, \mathfrak{Re}\, \nu(\varphi_i \cdot g_i) \\ &\leq p_{\Pi}(e) \sum_{i=1}^{n} \mathfrak{Re}\, \nu(\varphi_i \cdot g_i) \\ &= p_{\Pi}(e)\, \mathfrak{Re}\, \nu\left(\sum_{i=1}^{n} \varphi_i \cdot g_i\right).\end{aligned}$$

Since $f = \sum_{i=1}^{n} \varphi_i \cdot g_i \in \overline{\mathfrak{u}}$, this renders

$$\mathfrak{p}_{\Omega_{\mathfrak{u}}^{\Pi}}(\mathcal{L}_\nu^e) \leq p_{\Pi}(e)\, \sup\{\mathfrak{Re}\, \nu(f) \mid f \in \mathfrak{u}\}.$$

For the reverse inequality, we recall that

$$\sup\{ab \mid a \in A,\ b \in B\} = (\sup A)(\sup B)$$

holds for nonempty subsets A, B of $\mathbb{R}$ such that all elements of A are nonnegative and such that $\sup B \geq 0$. With $A = \{\mu(e) \mid \mu \in \Pi\}$, that is, $\sup A = p_{\Pi}(e)$ and $B = \{\mathfrak{Re}\, \nu(f) \mid f \in \mathfrak{u}\}$ we infer that

$$p_{\Pi}(e)\, \sup\{\mathfrak{Re}\, \nu(f) \mid f \in \mathfrak{u}\} = \sup\{\mu(e)\, \mathfrak{Re}\, \nu(f) \mid f \in \mathfrak{u},\ \mu \in \Pi\} \leq \mathfrak{p}_{\Omega_{\mathfrak{u}}^{\Pi}}(\mathcal{L}_\nu^e),$$

hence our claim. The argument for the second alternative in (b), that is, the case that $\Pi = \{\mu\}$ is a singleton set is straightforward, because Proposition 3.1.11 returns immediately that

$$\begin{aligned} \mathfrak{p}_{\Omega_{\mathfrak{u}}^{\Pi}}(\mathcal{L}_{\nu}^{e}) &= \sup\{p_{\Pi}(\nu(f)e) \mid f \in \mathfrak{u}\} \\ &= \mu(e) \sup\{\mathfrak{Re}\, \nu(f) \mid f \in \mathfrak{u}\} \\ &= p_{\Pi}(e) \sup\{\mathfrak{Re}\, \nu(f) \mid f \in \mathfrak{u}\} \end{aligned}$$

holds in this case as well. Part (a) implies in particular $\mathfrak{p}_{\Omega_{\mathfrak{u}}^{\Pi}}(\mathcal{L}_{\nu}^{e}) \le 1$ provided that $\nu \in \mathrm{b}\mathcal{C}\mathcal{X}(\mathfrak{u})^{\circ}$ and $e \in \mathrm{bconv}(\Pi)^{\circ}$. Let $\mathcal{T}(f) = \sum_{i=1}^{n} \mathcal{L}_{\nu_i}^{e_i}$ be a finite rank operator, where $\nu_i \in \mathcal{C}_{\mathfrak{V}}(X,E)^*$ and $e_i \in F$. Suppose that $\mathrm{b}\mathcal{C}(\mathfrak{u})$ is bounded in $\mathcal{C}_{\mathfrak{V}}(X,E)$ and that Π is weak* bounded in F^*. Since $\mathrm{bconv}(\Pi)$ is also weak* bounded, the polars $\mathrm{b}\mathcal{C}\mathcal{X}(\mathfrak{u})^{\circ}$ and $\mathrm{bconv}(\Pi)^{\circ}$ are absorbing in $\mathcal{C}_{\mathfrak{V}}(X,E)^*$ and F, respectively. Thus there is $\lambda \ge 0$ such that both $\nu_i \in \lambda\, \mathrm{b}\mathcal{C}\mathcal{X}(\mathfrak{u})^{\circ}$ and $e_i \in \lambda\, \mathrm{bconv}(\Pi)^{\circ}$ for all $i = 1,\dots,n$. The former implies that $\sup\{|\nu(f)| \mid f \in \mathfrak{u}\} \le \lambda$, the latter that $p_{\mathrm{bconv}(\Pi)}(e_i) \le \lambda$. Then $\mathfrak{p}_{\Omega_{\mathfrak{u}}^{\Pi}}(\mathcal{L}_{\nu_i}^{e_i}) \le \lambda^2$ for the elementary operators $\mathcal{L}_{\nu_i}^{e_i}$. Thus

$$\mathfrak{p}_{\Omega_{\mathfrak{u}}^{\Pi}}(\mathcal{T}) \le \sum_{i=1}^{n} \mathfrak{p}_{\Omega_{\mathfrak{u}}^{\Pi}}(\mathcal{L}_{\nu_i}^{e_i}) \le n\lambda^2,$$

our claim in Part (c). Finally, since $\mathcal{L}_{\nu_i}^{e_i} \in \mathfrak{H}_{\Omega_{\mathfrak{u}}^{\Pi}}$ whenever $\nu_i \in \mathrm{b}\mathcal{C}\mathcal{X}(\mathfrak{u})^{\circ}$ and $e_i \in \mathrm{bconv}(\Pi)^{\circ}$, Part (d) follows with Proposition 1.3.10. □

We shall illustrate these concepts with a few examples.

3.1.16 Examples. (a) If $F = \mathbb{R}$ or $\mathbb{C}$, that is, $\mathfrak{L}(\mathcal{C}_{\mathfrak{V}}(X,E),F) = \mathcal{C}_{\mathfrak{V}}(X,E)^*$ (see Example 1.3.11(a)), then Proposition 3.1.11 applies with any $\mathcal{C}(X)$-convex neighborhood $\mathfrak{u}$ of $0 \in \mathcal{C}_{\mathfrak{V}}(X,E)$ and $\Psi = \{1\}$. Then $\mathfrak{H}_{\Omega_{\mathfrak{u}}^{\Pi}} = \mathfrak{u}^{\circ}$. According to Corollary 1.3.9 and Proposition 3.1.10 the set $\mathfrak{H}_{\Omega_{\mathfrak{u}}^{\Pi}}$ is weak* compact and all of its extreme points are point evaluations.

(b) Let $\mathfrak{u}$ be a $\mathcal{C}(X)$-convex neighborhood of $0 \in \mathcal{C}_{\mathfrak{V}}(X,E)$ and let F be a semireflexive space. For Π, we choose a subset of F^* such that the closure of its balanced convex hull $\mathrm{bconv}(\Pi)$ is absorbing in F^*. Then $\mathrm{bcore}(\Pi^{\circ}) = \mathrm{bconv}(\Pi)^{\circ}$ is bounded, hence weakly compact in F, and according to Corollary 1.3.9 the set $\mathfrak{L}_{\mathfrak{u}}^{\Pi^{\circ}}(\mathcal{C}_{\mathfrak{V}}(X,E),F)$ is compact in the weak operator topology. Let $\nu_i \in \mathrm{b}\mathcal{C}\mathcal{X}(\mathfrak{u})^{\circ}$, $e_i \in \mathrm{bconv}(\Pi)^{\circ}$ and $\lambda_i \ge 0$ such that $\sum_{i=1}^{\infty} \lambda_i \le 1$. Then $\mathcal{T} = \sum_{i=1}^{\infty} \lambda_i \mathcal{L}_{\nu_i}^{e_i} \in \mathfrak{H}_{\Omega_{\mathfrak{u}}^{\Pi}}$ by Proposition 3.1.15(d). We have

$$\mathcal{T}(f) = \sum_{i=1}^{\infty} \lambda_i \nu_i(f) e_i$$

for every $f \in \mathcal{C}_{\mathfrak{V}}(X,E)$. Hence $\mathcal{T}$ is a nuclear operator (see III.7.1 in [56]). Moreover, every nuclear operator $\mathcal{T} \in \mathfrak{L}(\mathcal{C}_{\mathfrak{V}}(X,E),F)$ is contained in such a subset $\mathfrak{H}_{\Omega_{\mathfrak{u}}^{\Pi}}$ for a suitable choice of $\mathfrak{u}$ and Π. However, not all operators in $\mathfrak{H}_{\mathfrak{Q}_{\mathfrak{u}}^{\Pi}}$ need to be nuclear, even if this set is compact in the weak operator topology. Indeed, according to Proposition 3.1.10 $\mathfrak{H}_{\mathfrak{Q}_{\mathfrak{u}}^{\Pi}}$ contains all point evaluations in $\mathfrak{L}_{\mathfrak{u}}^{\Pi^\circ}(\mathcal{C}_{\mathfrak{V}}(X,E),F)$, which are generally not nuclear.

For a concrete example, let F be the sequence space ℓ^p for $1 < p < +\infty$, let $\Pi = \mathbb{B}_q$ be the unit ball of ℓ^q, where $1/p + 1/q = 1$, and let e_i be the i-th unit sequence in ℓ^p. For any choice of $\lambda_i \geq 0$, such that $\sum_{i=1}^{\infty} \lambda_i \leq 1$ and $\nu_i \in \mathbb{B}_q$ the prior construction renders a nuclear operator $\mathcal{T} \in \mathfrak{H}_{\mathfrak{Q}_{\mathfrak{u}}^{\Pi}}$ such that

$$\mathcal{T}(f) = (\lambda_i \nu_i(f))_{i\in\mathbb{N}}$$

for all $f \in \mathcal{C}_{\mathfrak{V}}(X,E)$. However, not all operators in $\mathfrak{L}(\mathcal{C}_{\mathfrak{V}}(X,E),\ell^p)$ are of this type. For an example of an operator that is not absorbed by $\mathfrak{H}_{\mathfrak{Q}_{\mathfrak{u}}^{\Pi}}$ let $X = [0,1]$, let $E = \mathbb{R}$ and let $\mathcal{C}(X)$ be endowed with its usual supremum norm topology. Let $\mathfrak{u}$ be the unit ball of $\mathcal{C}(X)$. Define the operator $\mathcal{T} \in \mathfrak{L}(\mathcal{C}_{\mathfrak{V}}(X,\mathbb{R}),\ell^p)$ as $\mathcal{T}(f) = ((1/i)f(1/i))_{i\in\mathbb{N}}$. For every $n \in \mathbb{N}$, there are $0 \leq \varphi_i \in \mathcal{C}(X)$ such that $\sum_{i=1}^{n} \varphi_i = 1$, $\varphi_i(1/i) = 1$ and $\varphi_i(1/k) = 0$ for all $i \neq k$. Then $\mathfrak{p}_{\mathfrak{Q}_{\mathfrak{v}}^{\mathbb{B}_p}}(\mathcal{T}) \geq \sum_{i=1}^{n} 1/i$ for all $n \in \mathbb{N}$ and, therefore, $\mathfrak{p}_{\mathfrak{Q}_{\mathfrak{u}}^{\Pi}}(\mathcal{T}) = +\infty$. Hence $\mathcal{T}$ is absorbed by $\mathfrak{L}_{\mathfrak{u}}^{\Pi^\circ}(\mathcal{C}_{\mathfrak{V}}(X,E),F)$, but not by $\mathfrak{H}_{\mathfrak{Q}_{\mathfrak{u}}^{\Pi}}$.

(c) Let E be an ordered topological vector space and assume that there is a neighborhood $\mathfrak{v} \in \mathfrak{V}$ such that for $\mathfrak{u} = \mathfrak{v} \cap \mathcal{C}_{\mathfrak{V}}(X,E)_+$ the set $\mathrm{bcx}(\mathfrak{u})$ is a neighborhood of $0 \in \mathcal{C}_{\mathfrak{V}}(X,E)$. Let Y be a compact set and let F be the space of all regular real-valued Borel measures on Y, endowed with a polar topology of the dual pair $(F,\mathcal{C}(Y))$, where $\mathcal{C}(Y)$ is the space of all continuous functions on Y. Then $F^* = \mathcal{C}(Y)$ and the polar of every neighborhood in $\mathcal{C}(Y)$ is weakly compact in F. Both F and F^* carry their canonical order. Let $\rho : Y \to \overline{\mathbb{R}}$ be a (strictly) positive-valued lower semicontinuous function and let $\Pi = \{\varphi \in \mathcal{C}(Y) \mid \varphi \leq \rho\} \subset F^*$. Then $p_\Pi(\eta) = \sup\{\eta(\varphi) \mid \varphi \in \Pi\}$ for all measures $\eta \in F$, that is, $p_\Pi(\eta) = \int_Y \rho\, d\eta$ for all $\eta \in F_+$ and $p_\Pi(\eta) = +\infty$, else. The functional p_Π is additive but not necessarily continuous on $F_+ = \{c \in F \mid p_\Pi(c) < +\infty\}$, hence Π satisfies (Π1), but generally not (Π2). Proposition 3.1.11 applies to all operators $\mathcal{T} \in \mathfrak{L}(\mathcal{C}_{\mathfrak{V}}(X,E),F)$. The polar Π° of Π in F consists of all $\eta \in F_+$ such that $\int_Y \rho\, d\eta \leq 1$. Furthermore, Π° is a closed subset of $(\varepsilon\mathbb{B})^\circ$, where $\mathbb{B}$ is the unit ball for the maximum norm in $\mathcal{C}(Y)$ and $\varepsilon = \min\{\rho(y) \mid y \in Y\} > 0$. Hence Π° is weakly compact in F and Corollary 1.3.9 applies. The set $\mathfrak{H}_{\mathfrak{Q}_{\mathfrak{u}}^{\Pi}} = \mathfrak{L}_{\mathfrak{u}}^{\Pi^\circ}(\mathcal{C}_{\mathfrak{V}}(X,E),F)$ is convex and compact in the weak operator topology; all of its elements are weakly compact operators and its extreme points are point evaluations.

For a concrete example, let $Y = \{1/i \mid i \in \mathbb{N}\} \cup \{0\}$ endowed with the trace topology from $\mathbb{R}$. Then $F = \ell^1$ and $F^* = c$, the space of all convergent sequences in the following sense: For $\mu = (\alpha_i)_{i=1}^{\infty} \in c$, set $\alpha_0 = \lim_{i\to\infty} \alpha_i$ and for $b = (\beta_i)_{i=1}^{\infty} \in \ell^1$ set $\mu(b) = \sum_{i=0}^{\infty} \alpha_i \beta_{i+1}$. A strictly positive-valued lower semicontinuous function ρ on Y is represented by a sequence $(\rho_i)_{i=0}^{\infty}$ such that all $\rho_i > 0$ and $\rho_0 \leq \liminf_{n\in\mathbb{N}} \rho_i$. Then $p_\Pi((\alpha_i)_{n\in\mathbb{N}}) = \sum_{i=0}^{\infty} \rho_i \alpha_i$ for all $(\alpha_i)_{i=0}^{\infty} \in \ell_+^1$ and $p_\Pi((\alpha_i)_{n\in\mathbb{N}}) = +\infty$, else.

(d) The choice of F in Example (c) is a special case of the following: Let Q be a closed and locally compact subcone of F and $Q \cap (-Q) = \{0\}$. Then Q has a compact base $B^{\mu} = \mu^{-1}(1) \cap Q$ for some $\mu \in F^*$. (see Theorem II.2.6 in [1]). The convex hull B_0^{μ} of B^{μ} and $\{0\}$ is also compact. Let Γ be a convex subset of F^* such that B^{μ} is a compact base for Q for every $\mu \in \Gamma$, and for all $\mu, \nu \in \Gamma$ there is $\kappa \in \Gamma$ such that $B_0^{\kappa} \subset B_0^{\mu} \cap B_0^{\nu}$. We set $\Pi = \Gamma + Q^{\circ}$. Then $p_{\Pi}(c) = +\infty$ for all $c \in F \setminus Q$ and $p_{\Pi}(c) = \sup\{\mu(c) \mid \mu \in \Gamma\}$ for all $c \in Q$. The above condition for Γ implies that $\mathfrak{p}_{\Pi}$ is additive on Q. The same choice for $\mathfrak{u}$ as in Part (c) then satisfies the requirements of Proposition 3.1.11.

We observe that the concept of the variation of a measure can also be applied to its restriction with respect to a subspace G of E (see Proposition 2.6.8). That is, only elements $a_i \in G$ are used in the definition of the variation of the restricted measure. Proposition 3.1.14 then applies with the restriction of the corresponding operator to the subspace $\mathcal{C}_{\mathfrak{V}}(X, G)$ of $\mathcal{C}_{\mathfrak{V}}(X, E)$.

3.2 A Choquet ordering for linear operators on function spaces

Throughout this section, we shall consider the weak operator topology on the space $\mathfrak{L}(\mathcal{C}_{\mathfrak{V}}(X, E), F)$, which coincides with the weak topology of its duality with $\mathcal{C}_{\mathfrak{V}}(X, E) \otimes F^*$ (see Proposition 1.3.1). Similarly, we endow $\mathcal{C}_{\mathfrak{V}}(X, E) \otimes F^*$ with the weak topology induced by $\mathfrak{L}(\mathcal{C}_{\mathfrak{V}}(X, E), F)$. Unfortunately, some of the concepts to be introduced in the beginning of this section will be less than intuitive and reveal their usefulness only later on. For a convex subset Ω of $\mathcal{C}_{\mathfrak{V}}(X, E) \otimes F^*$, such that $0 \in \Omega$ we say that a set $\Phi \subset \mathcal{C}_{\mathfrak{V}}(X, E) \otimes F^*$ is *Ω-bounded* if $\Phi \subset \lambda\Omega$, that is, $\mathfrak{p}_{\Phi} \leq \lambda \mathfrak{p}_{\Omega}$ for the associated lower semicontinuous sublinear functionals $\mathfrak{p}_{\Phi}, \mathfrak{p}_{\Omega} \in \mathfrak{P}$, with some $\lambda \geq 0$. Clearly, sums and positive multiples of Ω-bounded sets are again Ω-bounded. Similarly, Φ is $\mathfrak{D}$-bounded for a family $\mathfrak{D}$ of such sets if it is Ω-bounded for all $\Omega \in \mathfrak{D}$.

Choquet cone

A *Choquet core* is a collection $\mathfrak{K}$ of nonempty $\mathcal{C}(X)$-convex subsets of $\mathcal{C}_{\mathfrak{V}}(X, E) \otimes F^*$, such that $0 \in \Omega$ for every $\Omega \in \mathfrak{K}$ and such that the polar $\mathfrak{H}_{\Omega}$ of Ω is compact in the weak operator topology of $\mathfrak{L}(\mathcal{C}_{\mathfrak{V}}(X, E), F)$. We also require that, given $\Phi, \Omega \in \mathfrak{K}$, the set Φ is contained in the union of all Ω-bounded sets in $\mathfrak{K}$, which are subsets of $\overline{\Phi}$, the closure of Φ in the weak* topology of $\mathcal{C}_{\mathfrak{V}}(X, E) \otimes F^*$.

A *Choquet cone over* $\mathfrak{K}$ is a cone $\mathfrak{C}$ of nonempty $\mathcal{C}(X)$-convex subsets of $\mathcal{C}_{\mathfrak{V}}(X, E) \otimes F^*$ that is generated by $\mathfrak{K}$ together with a collection of $\mathcal{C}(X)$-convex $\mathfrak{K}$-bounded subsets of $\mathcal{C}_{\mathfrak{V}}(X, E) \otimes F^*$.

Every set $\Phi \in \mathfrak{C}$ can be expressed as $\Phi = \Psi + \sum_{i=1}^{n} \lambda_i \Omega_i$, where $\Psi \in \mathfrak{C}$ is $\mathfrak{K}$-bounded, where $\lambda_i \geq 0$ and $\Omega_i \in \mathfrak{K}$. The sets in $\mathfrak{K}$ themselves are generally not required to be $\mathfrak{K}$-bounded.

We take notice that as polars of compact convex subsets of $\mathfrak{L}(\mathcal{C}_{\mathfrak{V}}(X,E),F)$, according to Lemma 3.1.1(b), the closures of the sets in $\mathfrak{K}$ absorb all bounded subsets of $\mathcal{C}_{\mathfrak{V}}(X,E)\otimes F^*$, which are therefore $\mathfrak{K}$-bounded. We have $\mathfrak{p}_\Omega(\mathcal{T})\geq 0$ for all $\Omega\in\mathfrak{K}$ and $\mathcal{T}\in\mathfrak{L}(\mathcal{C}_{\mathfrak{V}}(X,E),F)$, and $\mathfrak{p}_\Omega(\mathcal{T})=0$ if and only if $\mathcal{T}=0$ by Lemma 3.1.1(a).

The boundedness requirement for the elements of a Choquet core is satisfied if the following more easily verifiable criterion holds: (i) The elements of sets in $\mathfrak{K}$ are absorbed by all sets in $\mathfrak{K}$. (ii) Given $\Phi,\Omega\in\mathfrak{K}$ and $\alpha\geq 0$ there is $\beta\geq\alpha$ such that $\Phi\cap(\beta\Omega)\in\mathfrak{K}$. Indeed, suppose that this criterion holds and let $\omega\in\Phi$. Since Ω absorbs ω by (i), by (ii) there is $\beta\geq 0$ such that $\omega\in\Phi\cap(\beta\Omega)$ and $\Phi\cap(\beta\Omega)\in\mathfrak{K}$. Obviously, the set $\Phi\cap(\beta\Omega)$ is Ω-bounded and contained in Φ. Our claim follows.

Given a Choquet core $\mathfrak{K}$ we select a corresponding subset of operators in $\mathfrak{L}(\mathcal{C}_{\mathfrak{V}}(X,E),F)$ setting

$$\mathfrak{L}_{\mathfrak{K}}(\mathcal{C}_{\mathfrak{V}}(X,E),F)=\{\mathcal{T}\in\mathfrak{L}(\mathcal{C}_{\mathfrak{V}}(X,E),F)\mid \mathfrak{p}_\Omega(\mathcal{T})<+\infty \text{ for some } \Omega\in\mathfrak{K}\}.$$

3.2.1 Lemma. *If $\mathfrak{K}$ is not empty, then $\mathfrak{L}_{\mathfrak{K}}(\mathcal{C}_{\mathfrak{V}}(X,E),F)$ is a subcone of $\mathfrak{L}(\mathcal{C}_{\mathfrak{V}}(X,E),F)$.*

Proof. Clearly, $\lambda\mathcal{T}\in\mathfrak{L}_{\mathfrak{K}}(\mathcal{C}_{\mathfrak{V}}(X,E),F)$ whenever $\mathcal{T}\in\mathfrak{L}_{\mathfrak{K}}(\mathcal{C}_{\mathfrak{V}}(X,E),F)$ and $\lambda\geq 0$. For $\mathcal{S},\mathcal{T}\in\mathfrak{L}_{\mathfrak{K}}(\mathcal{C}_{\mathfrak{V}}(X,E),F)$, there are $\Phi,\Omega\in\mathfrak{K}$ such that $\mathfrak{p}_\Phi(\mathcal{S}),\mathfrak{p}_\Omega(\mathcal{T})<+\infty$. By our assumption on $\mathfrak{K}$, there is an Ω-bounded set $\Psi\in\mathfrak{K}$ such that $\Psi\subset\overline{\Phi}$. Thus there is $\lambda\geq 0$ such that

$$\mathfrak{p}_\Psi(\mathcal{S}+\mathcal{T})\leq\mathfrak{p}_\Psi(\mathcal{S})+\mathfrak{p}_\Psi(\mathcal{T})\leq\mathfrak{p}_\Phi(\mathcal{S})+\lambda\mathfrak{p}_\Omega(\mathcal{T})<+\infty.$$

Hence $\mathcal{S}+\mathcal{T}\in\mathfrak{L}_{\mathfrak{K}}(\mathcal{C}_{\mathfrak{V}}(X,E),F)$. □

Choquet ordering

Corresponding to a Choquet cone $\mathfrak{C}$ over $\mathfrak{K}$, we define an *order relation* $\prec_{\mathfrak{C}}$ on $\mathfrak{L}(\mathcal{C}_{\mathfrak{V}}(X,E),F)$ setting

$$\mathcal{S}\prec_{\mathfrak{C}}\mathcal{T}\quad\text{if } \mathfrak{p}_\Omega(\mathcal{S})\leq\mathfrak{p}_\Omega(\mathcal{T})\quad(\text{or } \mathfrak{q}_{(-\Omega)}(\mathcal{T})\leq\mathfrak{q}_{(-\Omega)}(\mathcal{S}))\quad\text{for all}\quad\Omega\in\mathfrak{C}.$$

This order is reflexive and transitive, but not necessarily antisymmetric. It induces an equivalence relation on $\mathfrak{L}(\mathcal{C}_{\mathfrak{V}}(X,E),F)$, that is, $\mathcal{S}\sim\mathcal{T}$ if both $\mathcal{S}\prec_{\mathfrak{C}}\mathcal{T}$ and $\mathcal{T}\prec_{\mathfrak{C}}\mathcal{S}$. It is compatible with the multiplication by nonnegative scalars, but generally not with the addition. However, Proposition 3.1.5 yields the following.

3.2.2 Proposition.

(a) *If $\mathcal{S}_i\prec_{\mathfrak{C}}\mathcal{T}_{\varphi_i}$ for $\mathcal{S}_i,\mathcal{T}\in\mathfrak{L}(\mathcal{C}_{\mathfrak{V}}(X,E),F)$ and bounded nonnegative functions $\varphi_1,\ldots,\varphi_n\in\mathcal{C}(X)$ such that $\sum_{i=1}^n\varphi_i$ is strictly positive, then $\sum_{i=1}^n\mathcal{S}_i\prec_{\mathfrak{C}}\mathcal{T}_{\sum_{i=1}^n\varphi_i}$.*

(b) *If $\mathcal{S}_1\prec_{\mathfrak{C}}\mathcal{T}_1$ and $\mathcal{S}_2\prec_{\mathfrak{C}}\mathcal{T}_2$ for $\mathcal{S}_1,\mathcal{S}_2,\mathcal{T}_1,\mathcal{T}_2\in\mathfrak{L}(\mathcal{C}_{\mathfrak{V}}(X,E),F)$ and if the operators $\mathcal{T}_1$ and $\mathcal{T}_2$ have mutually singular representation measures, then $\mathcal{S}_1+\mathcal{S}_2\prec_{\mathfrak{C}}\mathcal{T}_1+\mathcal{T}_2$.*

Proof. (a) Let $\mathcal{S}_i, \mathcal{T}$ and φ_i be as stated. We have

$$\mathfrak{p}_\Omega\left(\sum_{i=1}^n \mathcal{S}_i\right) \leq \sum_{i=1}^n \mathfrak{p}_\Omega(\mathcal{S}_i) \leq \sum_{i=1}^n \mathfrak{p}_\Omega(\mathcal{T}_{\varphi_i}) = \mathfrak{p}_\Omega(\mathcal{T}_{\sum_{i=1}^n \varphi_i})$$

for all $\Omega \in \mathfrak{C}$ by Proposition 3.1.5(a). Part (b) follows in similar fashion from 3.1.5(b). □

Upper and lower envelopes

We proceed to characterize the order $\prec_{\mathfrak{C}}$ in terms of upper and lower envelopes of functions in $\mathfrak{Q}$ and $\mathfrak{P}$, respectively. For every $\mathfrak{q} \in \mathfrak{Q}$, we define its $\overline{\mathbb{R}}$-valued upper envelope on $\mathfrak{L}(\mathcal{C}_{\mathfrak{V}}(X,E),F)$ by

$$\hat{\mathfrak{q}}(\mathcal{T}) = \inf\{\mathfrak{p}_\Omega(\mathcal{T}) \mid \Omega \in \mathfrak{C},\ \mathfrak{q} \leq \mathfrak{p}_\Omega\}$$

for every $\mathcal{T} \in \mathfrak{L}(\mathcal{C}_{\mathfrak{V}}(X,E),F)$. Similarly, the $\overline{\mathbb{R}}$-valued lower envelope for $\mathfrak{p} \in \mathfrak{P}$ is

$$\check{\mathfrak{p}}(\mathcal{T}) = \sup\{\mathfrak{q}_{(-\Omega)}(\mathcal{T}) \mid \Omega \in \mathfrak{C},\ \mathfrak{q}_{(-\Omega)} \leq \mathfrak{p}\}.$$

We observe that

$$\begin{aligned} -\hat{\mathfrak{q}}(\mathcal{T}) &= -\inf\{\mathfrak{p}_\Omega(\mathcal{T}) \mid \Omega \in \mathfrak{C},\ \mathfrak{q} \leq \mathfrak{p}_\Omega\} \\ &= \sup\{\mathfrak{q}_{(-\Omega)}(\mathcal{T}) \mid \Omega \in \mathfrak{C},\ \mathfrak{q}_{(-\Omega)} \leq -\mathfrak{q}\} = \widecheck{(-\mathfrak{q})}(\mathcal{T}) \end{aligned}$$

holds for all $\mathcal{T} \in \mathfrak{L}(\mathcal{C}_{\mathfrak{V}}(X,E),F)$ and $\mathfrak{q} \in \mathfrak{Q}$. In this sense, both upper and lower envelopes can be equivalently used in many statements. In the sequel, we shall indicate these equivalences in brackets. Moreover, we have

$$\widehat{(\mathfrak{q} + \mathfrak{r})}(\mathcal{T}) \leq \hat{\mathfrak{q}}(\mathcal{T}) + \hat{\mathfrak{r}}(\mathcal{T}) \quad \text{and} \quad \widehat{(\alpha\mathfrak{q})}(\mathcal{T}) = \alpha\hat{\mathfrak{q}}(\mathcal{T}),$$

as well as

$$\widecheck{(\mathfrak{p} + \mathfrak{s})}(\mathcal{T}) \geq \check{\mathfrak{p}}(\mathcal{T}) + \check{\mathfrak{s}}(\mathcal{T}) \quad \text{and} \quad \widecheck{(\alpha\mathfrak{p})}(\mathcal{T}) = \alpha\check{\mathfrak{p}}(\mathcal{T}),$$

for $\mathcal{T} \in \mathfrak{L}(\mathcal{C}_{\mathfrak{V}}(X,E),F)$, $\alpha > 0$ and $\mathfrak{q}, \mathfrak{r} \in \mathfrak{Q}$ or $\mathfrak{p}, \mathfrak{s} \in \mathfrak{P}$, respectively, provided that the sums on the right-hand sides of the inequalities are defined in $\overline{\mathbb{R}}$. Hence for a fixed operator $\mathcal{T} \in \mathfrak{L}(\mathcal{C}_{\mathfrak{V}}(X,E),F)$ the mapping

$$\mathfrak{p} \mapsto \check{\mathfrak{p}}(\mathcal{T}) : \mathfrak{P} \to \overline{\mathbb{R}}$$

defines an extended superlinear functional on $\mathfrak{P}$ in the sense of Section 1.4. On the subcone $\mathfrak{Q}_0$ of $\mathfrak{Q}$ consisting of all finitely-valued elements of $\mathfrak{Q}$, that is, all lower semicontinuous superlinear functionals on $\mathfrak{L}(\mathcal{C}_{\mathfrak{V}}(X,E),F)$ defined by bounded convex subsets of $\mathcal{C}_{\mathfrak{V}}(X,E)\otimes F^*$, the mapping

$$\mathfrak{q}\mapsto\widehat{\mathfrak{q}}(\mathcal{T}):\mathfrak{Q}_0\to\overline{\mathbb{R}}$$

defines a sublinear functional.

3.2.3 Proposition. *For $\mathcal{S},\mathcal{T}\in\mathfrak{L}(\mathcal{C}_{\mathfrak{V}}(X,E),F)$, the following are equivalent:*
(i) $\mathcal{S}\prec_{\mathfrak{C}}\mathcal{T}$.
(ii) $\widehat{\mathfrak{q}}(\mathcal{S})\le\widehat{\mathfrak{q}}(\mathcal{T})$ *for all* $\mathfrak{q}\in\mathfrak{Q}$ *(or* $\check{\mathfrak{p}}(\mathcal{T})\le\check{\mathfrak{p}}(\mathcal{S})$ *for all* $\mathfrak{p}\in\mathfrak{P}$*).*
(iii) $\mathfrak{r}(\mathcal{S})\le\widehat{\mathfrak{r}}(\mathcal{T})$ *for all* $\mathfrak{r}\in\mathfrak{P}\cap\mathfrak{Q}$ *(or* $\check{\mathfrak{r}}(\mathcal{T})\le\mathfrak{r}(\mathcal{S})$ *for all* $\mathfrak{r}\in\mathfrak{P}\cap\mathfrak{Q}$*).*

Proof. (i) ⇒ (ii): If $\mathcal{S}\prec_{\mathfrak{C}}\mathcal{T}$, then

$$\widehat{\mathfrak{q}}(\mathcal{S})=\inf\{\mathfrak{p}_\Omega(\mathcal{S})\mid\Omega\in\mathfrak{C},\ \mathfrak{q}\le\mathfrak{p}_\Omega\}\le\inf\{\mathfrak{p}_\Omega(\mathcal{T})\mid\Omega\in\mathfrak{C},\ \mathfrak{q}\le\mathfrak{p}_\Omega\}=\widehat{\mathfrak{q}}(\mathcal{T})$$

for every $\mathfrak{q}\in\mathfrak{Q}$. (ii) ⇒ (iii) is obvious since $\mathfrak{r}(\mathcal{S})\le\widehat{\mathfrak{r}}(\mathcal{T})$. For (iii) ⇒ (i), suppose that $\mathfrak{r}(\mathcal{S})\le\widehat{\mathfrak{r}}(\mathcal{S})$ holds for all $\mathfrak{r}\in\mathfrak{P}\cap\mathfrak{Q}$ and let $\Omega\in\mathfrak{C}$. For every $\omega\in\Omega$, we have $\mathfrak{Re}\,\omega\in\mathfrak{P}\cap\mathfrak{Q}$ and $\mathfrak{Re}\,\omega\le\mathfrak{p}_\Omega$, hence

$$\mathfrak{Re}\,\omega(\mathcal{S})\le\widehat{\mathfrak{Re}\,\omega}(\mathcal{T})=\inf\{\mathfrak{p}_\Phi(\mathcal{T})\mid\Phi\in\mathfrak{C},\ \mathfrak{Re}\,\omega\le\mathfrak{p}_\Phi\}\le\mathfrak{p}_\Omega(\mathcal{T})$$

and, therefore, $\mathfrak{p}_\Omega(\mathcal{S})=\sup\{\mathfrak{Re}\,\omega(\mathcal{S})\mid\omega\in\Omega\}\le\mathfrak{p}_\Omega(\mathcal{T})$. □

3.2.4 Lemma. *Let $\mathfrak{q}\in\mathfrak{Q}$ (or $\mathfrak{p}\in\mathfrak{P}$). For every $\Omega\in\mathfrak{K}$, there is $\rho\ge0$ such that $\widehat{\mathfrak{q}}\le\rho\mathfrak{p}_\Omega$ (or $\check{\mathfrak{p}}\ge\rho\mathfrak{q}_{-\Omega}$). In particular, if $\mathcal{T}\in\mathfrak{L}_{\mathfrak{K}}(\mathcal{C}_{\mathfrak{V}}(X,E),F)$, then $\widehat{\mathfrak{q}}(\mathcal{T})\in\underline{\mathbb{R}}$ and $\check{\mathfrak{p}}(\mathcal{T})\in\overline{\mathbb{R}}$.*

Proof. We shall argue the case of the upper envelopes. Let $\mathfrak{q}\in\mathfrak{Q}$ and $\Omega\in\mathfrak{K}$. Since $\mathfrak{q}$ is $\underline{\mathbb{R}}$-valued and upper semicontinuous, it is bounded above by some $\rho\ge0$ on the compact set $\mathfrak{H}_\Omega$. This yields $\mathfrak{q}\le\rho\mathfrak{p}_\Omega$. If $\mathcal{T}\in\mathfrak{L}_{\mathfrak{K}}(\mathcal{C}_{\mathfrak{V}}(X,E),F)$, then $\mathfrak{p}_\Omega(\mathcal{T})<+\infty$ for some $\Omega\in\mathfrak{K}$ and, therefore, $\widehat{\mathfrak{q}}(\mathcal{T})\le\rho\mathfrak{p}_\Omega(\mathcal{T})<+\infty$. The argument for a functional $\mathfrak{p}\in\mathfrak{P}$ is similar. □

3.2.5 Lemma. *Let $\Omega\in\mathfrak{K}$, let $\mathfrak{q}\in\mathfrak{Q}$ (or $\mathfrak{p}\in\mathfrak{P}$) and $\mathcal{T}\in\mathfrak{L}(\mathcal{C}_{\mathfrak{V}}(X,E),F)$ such that $\mathfrak{p}_\Omega(\mathcal{T})<+\infty$. Then*

$$\widehat{\mathfrak{q}}(\mathcal{T})=\inf\{\mathfrak{p}_\Upsilon(\mathcal{T})\mid\Upsilon\in\mathfrak{C}\text{ is }\Omega\text{-bounded and }\mathfrak{q}\le\mathfrak{p}_\Upsilon\}$$
$$(\text{or }\check{\mathfrak{p}}(\mathcal{T})=\sup\{\mathfrak{q}_{(-\Upsilon)}(\mathcal{T})\mid\Upsilon\in\mathfrak{C}\text{ is }\Omega\text{-bounded and }\mathfrak{q}_{(-\Upsilon)}\le\mathfrak{p}\}).$$

Proof. We shall prove the case for the upper envelope. Let $\Omega\in\mathfrak{K}$, let $\mathfrak{q}\in\mathfrak{Q}$ and $\mathcal{T}\in\mathfrak{L}(\mathcal{C}_{\mathfrak{V}}(X,E),F)$ such that $\mathfrak{p}_\Omega(\mathcal{T})<+\infty$. Clearly,

$$\widehat{\mathfrak{q}}(\mathcal{T})\le\inf\{\mathfrak{p}_\Upsilon(\mathcal{T})\mid\Upsilon\in\mathfrak{C}\text{ is }\Omega\text{-bounded and }\mathfrak{q}\le\mathfrak{p}_\Upsilon\}$$

holds in any case. For the reverse inequality, let

$$\Lambda = \{\omega \in \mathcal{C}_{\mathfrak{V}}(X,E) \otimes F^* \mid \mathfrak{Re}\,\omega \geq \mathfrak{q}\}.$$

Then

$$\mathfrak{q} = \inf\{\mathfrak{Re}\,\omega \mid \omega \in \Lambda\} = \mathfrak{q}_\Lambda = -\mathfrak{p}_{(-\Lambda)}$$

(see the introductory remarks in Section 3.1). Now let $\Phi \in \mathfrak{C}$ such that $\mathfrak{q} \leq \mathfrak{p}_\Phi$, that is, $0 \leq \mathfrak{p}_\Phi + \mathfrak{p}_{(-\Lambda)} = \mathfrak{p}_{\Phi-\Lambda}$. Thus $\mathfrak{p}_{\{0\}} \leq \mathfrak{p}_{\Phi-\Lambda}$, which implies that $0 \in \overline{\Phi - \Lambda}$, the closure of the convex set $\Phi - \Lambda$ in $\mathcal{C}_{\mathfrak{V}}(X,E) \otimes F$. Thus, given $\varepsilon > 0$, Lemma 3.1.1(c) yields that

$$0 \in (\Phi - \Lambda) + \varepsilon\overline{\Omega} = \Phi - (\Lambda - \varepsilon\overline{\Omega}).$$

Hence we find an element $\omega \in \Phi \cap (\Lambda - \varepsilon\overline{\Omega})$. We have $\Phi = \Psi + \sum_{i=1}^n \lambda_i \Omega_i$, where $\Psi \in \mathfrak{C}$ is $\mathfrak{K}$-bounded, $\lambda_i \geq 0$ and $\Omega_i \in \mathfrak{K}$. Thus $\omega = \psi + \sum_{i=1}^n \lambda_i \omega_i$ with $\psi \in \Psi$ and $\omega_i \in \Omega_i$. By our assumption on $\mathfrak{K}$, there are Ω-bounded sets $\widetilde{\Omega}_i \in \mathfrak{K}$ such that $\omega_i \in \widetilde{\Omega}_i \subset \overline{\Omega}_i$. We set

$$\Upsilon = \Psi + \sum_{i=1}^n \lambda_i \widetilde{\Omega}_i \in \mathfrak{C}$$

and observe that Υ is Ω-bounded, that $\omega \in \Upsilon$ and $\Upsilon \subset \Psi + \sum_{i=1}^n \lambda_i \overline{\Omega}_i$. Thus

$$\mathfrak{p}_\Upsilon \leq \mathfrak{p}_\Psi + \sum_{i=1}^n \lambda_i \mathfrak{p}_{\overline{\Omega}_i} = \mathfrak{p}_\Psi + \sum_{i=1}^n \lambda_i \mathfrak{p}_{\Omega_i} = \mathfrak{p}_\Phi.$$

Moreover, $\omega \in \Upsilon \cap (\Lambda - \varepsilon\overline{\Omega})$ implies that $0 \in \Upsilon + (-\Lambda + \varepsilon\overline{\Omega})$ and, therefore,

$$\mathfrak{q} = -\mathfrak{p}_{(-\Lambda)} \leq \mathfrak{p}_{(\Upsilon + \varepsilon\overline{\Omega})} = \mathfrak{p}_{(\Upsilon + \varepsilon\Omega)}.$$

We set $\Gamma = \Upsilon + \varepsilon\Omega \in \mathfrak{C}$ and observe that Γ is also Ω-bounded and that $\mathfrak{q} \leq \mathfrak{p}_\Gamma$. Moreover,

$$\mathfrak{p}_\Gamma(\mathcal{T}) = \mathfrak{p}_\Upsilon(\mathcal{T}) + \varepsilon\mathfrak{p}_\Omega(\mathcal{T}) \leq \mathfrak{p}_\Phi(\mathcal{T}) + \varepsilon\mathfrak{p}_\Omega(\mathcal{T}).$$

Since $\mathfrak{p}_\Omega(\mathcal{T}) < +\infty$ and since $\varepsilon > 0$ was arbitrarily chosen, we infer that

$$\inf\{\mathfrak{p}_\Gamma(\mathcal{T}) \mid \Gamma \in \mathfrak{C} \text{ is } \Omega\text{-bounded and } \mathfrak{q} \leq \mathfrak{p}_\Upsilon\} \leq \mathfrak{p}_\Phi(\mathcal{T}).$$

The latter holds true for all $\Phi \in \mathfrak{C}$ such that $\mathfrak{q} \leq \mathfrak{p}_\Phi$. Our claim follows. □

3.2.6 Proposition. *For every $\mathfrak{q} \in \mathfrak{Q}$ (or $\mathfrak{p} \in \mathfrak{P}$) and $\mathcal{T} \in \mathfrak{L}_{\mathfrak{K}}(\mathcal{C}_{\mathfrak{V}}(X,E),F)$, we have*

$$\widehat{\mathfrak{q}}(\mathcal{T}) = \max\{\mathfrak{q}(\mathcal{S}) \mid \mathcal{S} \in \mathfrak{L}(\mathcal{C}_{\mathfrak{V}}(X,E),F),\ \mathcal{S} \prec_{\mathfrak{C}} \mathcal{T}\}$$
$$(\textit{or } \check{\mathfrak{p}}(\mathcal{T}) = \min\{\mathfrak{p}(\mathcal{S}) \mid \mathcal{S} \in \mathfrak{L}(\mathcal{C}_{\mathfrak{V}}(X,E),F),\ \mathcal{S} \prec_{\mathfrak{C}} \mathcal{T}\}).$$

Proof. We shall prove the case for the lower envelope $\check{\mathfrak{p}}$ of a function $\mathfrak{p} \in \mathfrak{P}$. First, we observe that $\mathfrak{P}$ is a locally convex cone (see Section 1.4, also [33] and [48]), endowed with its pointwise order and the neighborhoods $V_{\mathfrak{H}}$ for $\mathfrak{P}$, corresponding to compact (in the weak operator topology) convex subsets $\mathfrak{H}$ of $\mathfrak{L}(C_{\mathfrak{V}}(X,E),F)$ such that $0 \in \mathfrak{H}$. We set

$$\mathfrak{g} \leq \mathfrak{h} + V_{\mathfrak{H}} \text{ for } \mathfrak{g},\mathfrak{h} \in \mathfrak{P}, \text{ if } \mathfrak{g}(\mathcal{S}) \leq \mathfrak{h}(\mathcal{S}) + 1 \text{ for all } \mathcal{S} \in \mathfrak{H}.$$

Every functional $\mathfrak{g} \in \mathfrak{P}$ is bounded below on $\mathfrak{H}$, hence there is $\lambda \geq 0$ such that $0 \leq \mathfrak{g} + \lambda V_{\mathfrak{H}}$, and $\mathfrak{g}$ is seen to be bounded below as an element of $\mathfrak{P}$ as is required for locally convex cones. The polar $V_{\mathfrak{H}}^{\circ}$ of $V_{\mathfrak{H}}$ in the dual cone $\mathfrak{P}^*$ of $\mathfrak{P}$ consists of all linear functionals Θ on $\mathfrak{P}$ such that $\Theta(\mathfrak{g}) \leq \Theta(\mathfrak{h}) + 1$ whenever $\mathfrak{g} \leq \mathfrak{h} + V_{\mathfrak{H}}$. Now let $\mathfrak{p} \in \mathfrak{P}$ and $\mathcal{T} \in \mathfrak{L}_{\mathfrak{K}}(C_{\mathfrak{V}}(X,E),F)$. Since our statement holds for positive multiples of $\mathcal{T}$ whenever it holds for $\mathcal{T}$, we may assume that there is $\Omega \in \mathfrak{K}$ such that $\mathfrak{p}_{\Omega}(\mathcal{T}) \leq 1$. We define $\overline{\mathbb{R}}$-valued sub and extended superlinear functionals p and q on the cone $\mathfrak{P}$ setting

$$p(\mathfrak{g}) = \alpha\check{\mathfrak{p}}(\mathcal{T}) \text{ if } \mathfrak{g} = \alpha\mathfrak{p} \text{ for } \alpha \geq 0, \text{ and } p(\mathfrak{g}) = +\infty \text{ else,}$$

and $q(\mathfrak{g}) = \check{\mathfrak{g}}(\mathcal{T})$, respectively, for $\mathfrak{g} \in \mathfrak{P}$. We shall use the sandwich theorem 1.4.1 for locally convex cones with the functionals p and q and the neighborhood $V_{\mathfrak{H}_{\Omega}}$. That is, we need to verify that

$$\mathfrak{g} \leq \mathfrak{h} + v_{\mathfrak{H}} \text{ for } \mathfrak{g},\mathfrak{h} \in \mathfrak{P} \text{ implies that } q(\mathfrak{g}) \leq p(\mathfrak{h}) + 1.$$

For $\mathfrak{h} \neq \alpha\mathfrak{p}$, there is nothing to prove. Otherwise, $\mathfrak{g} \leq \alpha\mathfrak{p} + V_{\mathfrak{H}_{\Omega}}$ for some $\alpha \geq 0$ is equivalent to $\mathfrak{g} \leq \alpha\mathfrak{p} + \mathfrak{p}_{\Omega}$. Hence $\mathfrak{q}_{-\Phi} \leq \mathfrak{g}$ for $\Phi \in \mathfrak{C}$ implies that

$$\mathfrak{q}_{-\Phi} - \mathfrak{p}_{\Omega} = \mathfrak{q}_{-\Phi} + \mathfrak{q}_{-\Omega} = \mathfrak{q}_{-(\Phi+\Omega)} \leq \alpha\mathfrak{p}.$$

Thus $\mathfrak{q}_{-(\Phi+\Omega)} \leq \alpha\check{\mathfrak{p}}$ since $\Phi + \Omega \in \mathfrak{C}$. Hence $\mathfrak{q}_{-\Phi} \leq \alpha\check{\mathfrak{p}} + \mathfrak{p}_{\Omega}$. The last inequality applies to all $\mathfrak{q}_{-\Phi} \leq \mathfrak{g}$ for $\Phi \in \mathfrak{C}$ and, therefore, yields $\check{\mathfrak{g}} \leq \alpha\check{\mathfrak{p}} + \mathfrak{p}_{\Omega}$. Thus in particular

$$q(\mathfrak{g}) = \check{\mathfrak{g}}(\mathcal{T}) \leq \alpha\check{\mathfrak{p}}(\mathcal{T}) + \mathfrak{p}_{\Omega}(\mathcal{T}) \leq p(\alpha\mathfrak{p}) + 1,$$

as required. We infer that there is a linear functional Θ on $\mathfrak{P}$ such that $\Theta \in V_{\mathfrak{H}_{\Omega}}^{\circ}$ and $q \leq \Theta \leq p$. The formula

$$\omega \mapsto \Theta(\mathfrak{p}_{\{\omega\}})$$

defines a real-linear functional on $C_{\mathfrak{V}}(X,E) \otimes F^*$, that is, in the complex case, the real part of a linear functional ψ on $C_{\mathfrak{V}}(X,E) \otimes F^*$. For $f \otimes \mu \in C_{\mathfrak{V}}(X,E) \otimes F^*$ such that $f \in C_{\mathfrak{V}}(X,E)_0$, we have $\mathfrak{p}_{\{f\otimes\mu\}} = 0$, because $\mathcal{T}(f) = 0$ for all $\mathcal{T} \in \mathfrak{L}(C_{\mathfrak{V}}(X,E),F)$. Thus $\Theta(\mathfrak{p}_{\{f\otimes\mu\}}) = 0$ and, therefore, $\psi(f \otimes \mu) = 0$. Moreover, $\Theta \in V_{\mathfrak{H}_{\Omega}}^{\circ}$ implies that $\Theta(\mathfrak{p}_{\{\omega\}}) \leq 1$, that is, $\mathfrak{Re}\,\psi(\omega) \leq 1$, for all $\mathfrak{p}_{\{\omega\}} \leq V_{\mathfrak{H}_\Omega}$. The latter means that $\mathfrak{p}_{\{\omega\}}(\mathcal{R}) = \mathfrak{Re}\,\omega(\mathcal{R}) \leq 1$ holds for all $\mathcal{R} \in V_{\mathfrak{H}_\Omega}$; or equivalently, that $\omega \in \overline{\Omega}$, the second polar of Ω. Therefore, following Lemma 3.1.1(d) there is $\mathcal{S} \in \mathfrak{L}(C_{\mathfrak{V}}(X,E),F)$ such that $\varphi(\omega) = \omega(\mathcal{S})$ for all $\omega \in C_{\mathfrak{V}}(X,E) \otimes F^*$. For every $\mathfrak{r} \in \mathfrak{P} \cap \mathfrak{Q}$, there is $\omega \in C_{\mathfrak{V}}(X,E) \otimes F^*$ such that $\mathfrak{r} = \mathfrak{p}_{\{\omega\}}$ and

$$\mathfrak{r}(\mathcal{S}) = \mathfrak{Re}\,\omega(\mathcal{S}) = \mathfrak{Re}\,\psi(\omega) = \Theta(\mathfrak{r}).$$

Since

$$\Theta(\mathfrak{r}) \geq q(\mathfrak{r}) = \breve{\mathfrak{r}}(\mathcal{T}),$$

we infer that $\mathfrak{r}(\mathcal{S}) \geq \breve{\mathfrak{r}}(\mathcal{T})$ holds for all $\mathfrak{r} \in \mathfrak{P} \cap \mathfrak{Q}$ and, therefore, that $\mathcal{S} \prec_{\mathfrak{C}} \mathcal{T}$ according to Proposition 3.2.3. All that is left to show is that $\mathfrak{p}(\mathcal{S}) = \breve{\mathfrak{p}}(\mathcal{T})$ holds true for the given functional $\mathfrak{p} \in \mathfrak{P}$. For this, let

$$\Omega_{\mathfrak{p}} = \{\omega \in C_{\mathfrak{V}}(X,E) \otimes F^* \mid \mathfrak{Re}\,\omega \leq \mathfrak{p}\}.$$

Then

$$\begin{aligned} \mathfrak{p}(\mathcal{S}) &= \sup\{\mathfrak{Re}\,\omega(\mathcal{S}) \mid \omega \in \Omega_{\mathfrak{p}}\} \\ &= \sup\{\Theta(\mathfrak{Re}\,\omega) \mid \mathfrak{Re}\,\omega \leq \mathfrak{p}\} \leq \Theta(\mathfrak{p}) \leq \breve{\mathfrak{p}}(\mathcal{T}). \end{aligned}$$

On the other hand, $\breve{\mathfrak{p}}(\mathcal{T}) \leq \breve{\mathfrak{p}}(\mathcal{R}) \leq \mathfrak{p}(\mathcal{R})$ follows with Proposition 3.2.3(ii) for all $\mathcal{R} \prec_{\mathfrak{C}} \mathcal{T}$. This implies $\mathfrak{p}(\mathcal{S}) = \breve{\mathfrak{p}}(\mathcal{T})$ in particular and completes our argument. We note that $\mathfrak{p}(\mathcal{T})$ can take the value $+\infty$ in our line of reasoning. Replacing $\mathfrak{p} \in \mathfrak{P}$ by $-\mathfrak{q}$ for $\mathfrak{q} \in \mathfrak{Q}$ yields the statement for the upper envelopes. □

3.2.7 Lemma. *For every $\mathcal{T} \in \mathfrak{L}(C_{\mathfrak{V}}(X,E),F)$, the set*

$$\mathcal{T}\downarrow^{\mathfrak{C}} = \{\mathcal{S} \in \mathfrak{L}(C_{\mathfrak{V}}(X,E),F) \mid \mathcal{S} \prec_{\mathfrak{C}} \mathcal{T}\}$$

is closed and convex. If $\mathcal{T} \in \mathfrak{L}_{\mathfrak{K}}(C_{\mathfrak{V}}(X,E),F)$, then $\mathcal{T}\downarrow^{\mathfrak{C}}$ is compact in the weak operator topology.

Proof. If $\mathfrak{R}, \mathcal{S} \prec_{\mathfrak{C}} \mathcal{T}$ and $0 \leq \lambda \leq 1$, then

$$\mathfrak{p}_\Omega(\lambda\mathfrak{R} + (1-\lambda)\mathcal{S}) \leq \lambda\mathfrak{p}_\Omega(\mathfrak{R}) + (1-\lambda)\mathfrak{p}_\Omega(\mathcal{S}) \leq \mathfrak{p}_\Omega(\mathcal{T}) \quad \text{for all} \quad \Omega \in \mathfrak{C}$$

by the sublinearity of the functionals $\mathfrak{p}_\Omega$. Hence $\lambda\mathfrak{R}+(1-\lambda)\mathcal{S} \prec_{\mathfrak{C}} \mathcal{T}$, and the set $\mathcal{T}\downarrow^{\mathfrak{C}}$ is seen to be convex. $\mathcal{T}\downarrow^{\mathfrak{C}}$ is closed, since it is the intersection of the inverse images of the intervals $(-\infty, \mathfrak{p}_{\mathfrak{u}}(\Omega)]$ under the lower semicontinuous mappings $\mathfrak{p}_\Omega$, for all $\Omega \in \mathfrak{C}$. If $\mathcal{T} \in \mathfrak{L}_{\mathfrak{K}}(C_{\mathfrak{V}}(X,E),F)$, then there is $\Omega \in \mathfrak{K}$ such $\mathfrak{p}_\Omega(\mathcal{T}) \leq \lambda$ for some $\lambda \geq 0$. Since $\mathcal{T}\downarrow^{\mathfrak{C}} \subset \lambda\mathfrak{H}_\Omega$, it is compact. □

Minimal elements

An element $\mathcal{T} \in \mathfrak{L}(\mathcal{C}_{\mathfrak{V}}(X,E),F)$ is *minimal* if $\mathcal{S} \prec_{\mathfrak{C}} \mathcal{T}$ for $\mathcal{S} \in \mathfrak{L}(\mathcal{C}_{\mathfrak{V}}(X,E),F)$ implies that $\mathcal{S} \sim \mathcal{T}$. Since $\mathcal{S} \prec_{\mathfrak{C}} \mathcal{T}$ if and only if $\lambda\mathcal{S} \prec_{\mathfrak{C}} \lambda\mathcal{T}$ for all $\lambda > 0$, minimality of an element $\mathcal{T} \in \mathfrak{L}(\mathcal{C}_{\mathfrak{V}}(X,E),F)$ implies the minimality of $\lambda\mathcal{T}$ for all $\lambda > 0$. If $\mathfrak{K}$ is not empty, then the zero operator $0 \in \mathfrak{L}(\mathcal{C}_{\mathfrak{V}}(X,E),F)$ is minimal, since $\mathcal{S} \prec_{\mathfrak{C}} 0$ implies that $\mathfrak{p}_{\Omega}(\mathcal{S}) = 0$ for all $\Omega \in \mathfrak{K}$, hence $\mathcal{S} = 0$ by Lemma 3.1.1(a).

3.2.8 Proposition. *Suppose that $\mathcal{T} \in \mathfrak{L}(\mathcal{C}_{\mathfrak{V}}(X,E),F)$ is minimal and that $\mathfrak{p}_{\Omega}(\mathcal{T}) < +\infty$ for all $\Omega \in \mathfrak{C}$.*
(a) *If φ is a bounded nonnegative function in $\mathcal{C}(X)$, then $\mathcal{T}_{\varphi}$ is also minimal.*
(b) *If $\mathcal{T} = \mathcal{T}_1 + \mathcal{T}_2$ such that the representation measures for $\mathcal{T}_1, \mathcal{T}_2 \in \mathfrak{L}(\mathcal{C}_{\mathfrak{V}}(X,E),F)$ are mutually singular, then both $\mathcal{T}_1$ and $\mathcal{T}_2$ are also minimal.*

Proof. (a) Let $\mathcal{T}$ and φ be as stated and let $\rho > 0$ be an upper bound for φ. We have $\mathcal{T}_{\rho} = \mathcal{T}_{\varphi} + \mathcal{T}_{\rho-\varphi}$ and

$$\rho\, \mathfrak{p}_{\Omega}(\mathcal{T}) = \mathfrak{p}_{\Omega}(\mathcal{T}_{\rho}) = \mathfrak{p}_{\Omega}(\mathcal{T}_{\varphi}) + \mathfrak{p}_{\Omega}(\mathcal{T}_{\rho-\varphi})$$

for all $\Omega \in \mathfrak{C}$ by Proposition 3.1.5(a). In particular, $\mathfrak{p}_{\Omega}(\mathcal{T}_{\varphi}), \mathfrak{p}_{\Omega}(\mathcal{T}_{\rho-\varphi}) < +\infty$. We continue to verify that $\mathcal{T}_{\varphi}$ is minimal. Let $\mathcal{S} \prec_{\mathfrak{C}} \mathcal{T}_{\varphi}$. Then

$$\mathcal{S} + \mathcal{T}_{\rho-\varphi} \prec_{\mathfrak{C}} \mathcal{T}_{\varphi} + \mathcal{T}_{\rho-\varphi} = \mathcal{T}_{\rho}$$

by Proposition 3.2.2(a). The latter implies that $\mathcal{T}_{\rho} \prec_{\mathfrak{C}} \mathcal{S} + \mathcal{T}_{\rho-\varphi}$ since $\mathcal{T}_{\rho} = \rho\mathcal{T}$ is minimal. Therefore,

$$\mathfrak{p}_{\Omega}(\mathcal{T}_{\varphi}) + \mathfrak{p}_{\Omega}(\mathcal{T}_{\rho-\varphi}) = \mathfrak{p}_{\Omega}(\mathcal{T}_{\rho}) \le \mathfrak{p}_{\Omega}(\mathcal{S} + \mathcal{T}_{\rho-\varphi}) \le \mathfrak{p}_{\Omega}(\mathcal{S}) + \mathfrak{p}_{\Omega}(\mathcal{T}_{\rho-\varphi})$$

holds for all $\Omega \in \mathfrak{C}$. Because $\mathfrak{p}_{\Omega}(\mathcal{T}_{\rho-\varphi}) < +\infty$, we infer that $\mathfrak{p}_{\Omega}(\mathcal{T}_{\varphi}) \le \mathfrak{p}_{\Omega}(\mathcal{S})$, hence $\mathcal{T}_{\varphi} \prec_{\mathfrak{C}} \mathcal{S}$. Our claim follows. Part (b) is argued in a similar fashion using Propositions 3.1.5(b) and 3.2.2(b). □

3.2.9 Proposition. *For every operator $\mathcal{T} \in \mathfrak{L}_{\mathfrak{K}}(\mathcal{C}_{\mathfrak{V}}(X,E),F)$ there is a minimal element $\overline{\mathcal{T}} \in \mathfrak{L}_{\mathfrak{K}}(\mathcal{C}_{\mathfrak{V}}(X,E),F)$ such that $\overline{\mathcal{T}} \prec_{\mathfrak{C}} \mathcal{T}$.*

Proof. Let $\mathcal{T} \in \mathfrak{L}_{\mathfrak{K}}(\mathcal{C}_{\mathfrak{V}}(X,E),F)$. Given a totally ordered subset Λ of the compact convex set $\mathcal{T}\downarrow^{\mathfrak{C}}$, the intersection of the sets $\mathcal{S}\downarrow^{\mathfrak{C}}$, for all $\mathcal{S} \in \Lambda$, cannot be empty, hence contains a lower bound for Λ in $\mathcal{T}\downarrow^{\mathfrak{C}}$. By Zorn's lemma, there is a minimal element in $\mathcal{T}\downarrow^{\mathfrak{C}}$. □

3.2.10 Proposition. *For $\mathcal{T} \in \mathfrak{L}_{\mathfrak{K}}(\mathcal{C}_{\mathfrak{V}}(X,E),F)$, the following are equivalent:*
(i) *$\mathcal{T}$ is minimal.*
(ii) $\hat{\mathfrak{q}}(\mathcal{T}) = \max\{\mathfrak{q}(\mathcal{S}) \mid \mathcal{S} \sim \mathcal{T}\}$ *for all* $\mathfrak{q} \in \mathfrak{Q}$
(or $\check{\mathfrak{p}}(\mathcal{T}) = \min\{\mathfrak{p}(\mathcal{S}) \mid \mathcal{S} \sim \mathcal{T}\}$ *for all* $\mathfrak{p} \in \mathfrak{P}$*).*

(iii) $\widehat{\mathfrak{q}_{-\Omega}}(\mathcal{T}) = \mathfrak{q}_{-\Omega}(\mathcal{T})$ *for all* $\Omega \in \mathfrak{C}$
(or $\widecheck{\mathfrak{p}_{\Omega}}(\mathcal{T}) = \mathfrak{p}_{\Omega}(\mathcal{T})$ *for all* $\Omega \in \mathfrak{C}$*).*
(iv) *For every* $\Omega \in \mathfrak{C}$ *and* $\varepsilon > 0$, *there is* $\Phi \in \mathfrak{C}$ *such that* $0 \in \overline{\Omega + \Phi}$ *and* $\mathfrak{p}_{(\Omega+\Phi)}(\mathcal{T}) \leq \varepsilon$.

Proof. Clearly, (ii) implies (iii), since $\mathfrak{q}_{-\Omega} \in \mathfrak{Q}$ for every $\Omega \in \mathfrak{C}$, and $\mathcal{S} \sim \mathcal{T}$ implies that $\mathfrak{q}_{-\Omega}(\mathcal{S}) = \mathfrak{q}_{-\Omega}(\mathcal{T})$. Now suppose that (iii) holds and let $\mathcal{S} \prec_{\mathfrak{C}} \mathcal{T}$. Then for every $\Omega \in \mathfrak{C}$ we have

$$\mathfrak{q}_{-\Omega}(\mathcal{S}) \leq \widehat{\mathfrak{q}_{-\Omega}}(\mathcal{S}) \leq \widehat{\mathfrak{q}_{-\Omega}}(\mathcal{T}) = \mathfrak{q}_{-\Omega}(\mathcal{T})$$

by Proposition 3.2.3(ii). This yields $\mathcal{T} \prec_{\mathfrak{C}} \mathcal{S}$, hence the minimality of $\mathcal{T}$ and, therefore, (i). If $\mathcal{T}$ is minimal and $\mathfrak{q} \in \mathfrak{Q}$, then

$$\widehat{\mathfrak{q}}(\mathcal{T}) = \max\{\mathfrak{q}(\mathcal{S}) \mid \mathcal{S} \prec_{\mathfrak{C}} \mathcal{T}\} = \max\{\mathfrak{q}(\mathcal{S}) \mid \mathcal{S} \sim \mathcal{T}\}$$

by Proposition 3.2.6. Therefore, (i) implies (ii) as well. Finally, (iv) is just a reformulation of (iii), since $\mathfrak{q}_{-\Omega} \leq \mathfrak{p}_{\Phi}$ for $\Omega, \Phi \in \mathfrak{C}$ is equivalent to $0 \leq \mathfrak{p}_{\Omega} + \mathfrak{p}_{\Phi} = \mathfrak{p}_{(\Omega+\Phi)}$, that is, $0 \in \overline{\Omega + \Phi}$, and $\mathfrak{p}_{\Phi}(\mathcal{T}) \leq \mathfrak{q}_{-\Omega}(\mathcal{T}) + \varepsilon$ is equivalent to $\mathfrak{p}_{(\Omega+\Phi)}(\mathcal{T}) \leq \varepsilon$. □

The Choquet boundary

As in classical Choquet theory, we try to characterize minimal operators in $\mathfrak{L}(C_{\mathfrak{V}}(X,E),F)$ and their representation measures in terms of supporting subsets of X. For this, we introduce an analogue to the classical *Choquet boundary*. For a set $\Omega \in \mathfrak{K}$, we define

$$\Delta_{\mathfrak{C}}^{\Omega} = \left\{ \mathcal{D}_x^T \in \mathfrak{L}(C_{\mathfrak{V}}(X,E),F) \;\middle|\; \begin{array}{l} x \in X,\ T \in \mathfrak{L}(E,F),\ \mathcal{D}_x^T \text{ is nonzero,} \\ \Omega\text{-extremal and minimal} \end{array} \right\}$$

and

$$\partial_{\mathfrak{C}}^{\Omega} = \{x \in X \mid \mathcal{D}_x^T \in \Delta_{\mathfrak{C}}^{\Omega} \text{ for some } T \in \mathfrak{L}(E,F)\}.$$

Furthermore, we denote

$$\Delta_{\mathfrak{C}} = \bigcup\{\Delta_{\mathfrak{C}}^{\Omega} \mid \Omega \in \mathfrak{K}\} \quad \text{and} \quad \partial_{\mathfrak{C}} = \bigcup\{\partial_{\mathfrak{C}}^{\Omega} \mid \Omega \in \mathfrak{K}\}.$$

We realize that Corollary 3.1.8 for the characterization of Ω-extremal point evaluations applies. We shall say that a point evaluation $\mathcal{D}_x^T$ is *$\mathfrak{K}$-extremal* if it is Ω-extremal for some $\Omega \in \mathfrak{K}$. Thus $\Delta_{\mathfrak{C}}$ consists of all $\mathfrak{K}$-extremal point evaluations in $\mathfrak{L}(C_{\mathfrak{V}}(X,E),F)$. Note that our general settings allow that $\Delta_{\mathfrak{C}}$ is empty. Indeed, choose $\mathfrak{C} = \mathfrak{K} = \{C_{\mathfrak{V}}(X,E) \otimes F^*\}$. Then $\mathfrak{p}_{\Omega}(\mathcal{T}) = +\infty$ for all $\mathcal{T} \neq 0$ and $\Omega \in \mathfrak{K}$. Hence there are no nonzero point evaluations in $\mathfrak{L}_{\mathfrak{K}}(C_{\mathfrak{V}}(X,E),F)$. We shall further elaborate on this in Corollary 3.2.12 below.

The main tool for advancing the theory is an analogue to the classical case (Proposition I.4.10 in [1]). However, the presence of unbounded and indeed infinity-valued functions defining the Choquet ordering will complicate matters. Let $\Omega \in \mathfrak{K}$. We shall say that a function $\mathfrak{p} \in \mathfrak{P}$ is Ω-bounded above if $\mathfrak{p} \leq \lambda \mathfrak{p}_\Omega$, that is, if

$$\Omega_\mathfrak{p} = \{\omega \in \mathcal{C}_\mathfrak{V}(X, E) \otimes F^* \mid \mathfrak{Re}\, \omega \leq \mathfrak{p}\} \subset \lambda\Omega$$

for some $\lambda \geq 0$, that is, if the set $\Omega_\mathfrak{p}$ is Ω-bounded. Similarly, $\mathfrak{q} \in \mathfrak{Q}$ is Ω-bounded below on $\mathfrak{H}_\Omega$ if $\mathfrak{q} \geq \lambda \mathfrak{q}_{-\Omega}$, that is, if $\{\omega \in \mathcal{C}_\mathfrak{V}(X, E) \otimes F^* \mid \mathfrak{Re}\, \omega \geq \mathfrak{q}\} \subset -\lambda\Omega$ for some $\lambda \geq 0$. A sequence $(\mathfrak{p}_n)_{n\in\mathbb{N}}$ in $\mathfrak{P}$ or $(\mathfrak{q}_n)_{n\in\mathbb{N}}$ in $\mathfrak{Q}$ is Ω-bounded above or below if the respective property holds with the same λ for all $n \in \mathbb{N}$.

3.2.11 Theorem. *Let $\Omega \in \mathfrak{K}$ and let $(\mathfrak{q}_n)_{n\in\mathbb{N}}$ be an Ω-bounded below sequence in $\mathfrak{Q}$ (or $(\mathfrak{p}_n)_{n\in\mathbb{N}}$ an Ω-bounded above sequence in $\mathfrak{P}$). Let $a \geq 0$. If*

$$\liminf_{n\in\mathbb{N}} \widehat{\mathfrak{q}}_n(\mathcal{D}_x^T) \geq a\mathfrak{q}_{-\Omega}(\mathcal{D}_x^T) \quad \Big(or\ \limsup_{n\in\mathbb{N}} \check{\mathfrak{p}}_n(\mathcal{D}_x^T) \leq a\mathfrak{p}_\Omega(\mathcal{D}_x^T)\Big)$$

for all $\mathcal{D}_x^T \in \Delta_\mathfrak{C}^\Omega$, then

$$\liminf_{n\in\mathbb{N}} \widehat{\mathfrak{q}}_n(\mathcal{T}) \geq a\mathfrak{q}_{-\Omega}(\mathcal{T}) \quad \Big(or\ \limsup_{n\in\mathbb{N}} \check{\mathfrak{p}}_n(\mathcal{T}) \leq a\mathfrak{p}_\Omega(\mathcal{T})\Big)$$

for all $\mathcal{T} \in \mathfrak{L}(\mathcal{C}_\mathfrak{V}(X, E), F)$ such that $\mathfrak{q}_{-\Omega}(\mathcal{T}) > -\infty$ (or $\mathfrak{p}_\Omega(\mathcal{T}) < +\infty$).

Proof. We will to some extent follow the outlines of the proof of Theorem 2.5 in [47]. Our more general approach will however require some substantial adaptations. We shall verify the case of the lower envelopes. Let $(\mathfrak{p}_n)_{n\in\mathbb{N}}$ in $\mathfrak{P}$, $\mathcal{T} \in \mathfrak{L}(\mathcal{C}_\mathfrak{V}(X, E), F)$, $\Omega \in \mathfrak{K}$ and $a \geq 0$ be as stated. If $\mathfrak{p}_\Omega(\mathcal{T}) = 0$, then $\mathcal{T} = 0$ by Lemma 3.1.1(a). But $\check{\mathfrak{p}}_n(0) \leq \mathfrak{p}_n(0) = 0$ for all $n \in \mathbb{N}$, hence our claim is true in this case. We may therefore assume that $\mathfrak{p}_\Omega(\mathcal{T}) = 1$, hence $\mathcal{T} \in \mathfrak{H}_\Omega$. The sequence $(\mathfrak{p}_n)_{n\in\mathbb{N}}$ is supposed to be Ω-bounded above, that is, $\mathfrak{p}_n \leq \lambda \mathfrak{p}_\Omega$ holds with some $\lambda \geq 0$ for all $n \in \mathbb{N}$. This implies that $\check{\mathfrak{p}}_n \leq \lambda \mathfrak{p}_\Omega$ holds as well and, in particular, $\check{\mathfrak{p}}_n(\mathcal{T}) \leq \lambda$. Using Lemma 3.2.5, we construct a corresponding sequence $(\Phi_n)_{n\in\mathbb{N}}$ of sets in $\mathfrak{C}$ such that Ω absorbs Φ_n and such that

$$\mathfrak{q}_{-\Phi_n} \leq \mathfrak{p}_n \quad \text{and} \quad \mathfrak{q}_{-\Phi_n}(\mathcal{T}) \geq \check{\mathfrak{p}}_n(\mathcal{T}) - 1/n.$$

We note that the sequence $(\mathfrak{q}_{-\Phi_n})_{n\in\mathbb{N}}$ is bounded above by λ on $\mathfrak{H}_\Omega$ and that each of the functions $\mathfrak{q}_{-\Phi_n}$ is bounded below on $\mathfrak{H}_\Omega$ since $\Phi_n \subset \lambda_n\Omega$ for some $\lambda_n \geq 0$ implies that $\mathfrak{q}_{-\Phi_n} = -\mathfrak{p}_{\Phi_n} \geq -\lambda_n \mathfrak{q}_\Omega$. Hence $\mathfrak{q}_{-\Phi_n} \geq -\lambda_n$ on $\mathfrak{H}_\Omega$. As in the proof for the classical case, we shall use a representation of $\mathfrak{H}_\Omega$ in the space $\mathbb{R}^\mathbb{N}$ and apply Choquet's theorem for metrizable compact sets. For this, define $\Theta : \mathfrak{H}_\Omega \to \mathbb{R}^\mathbb{N}$ by

$$\Theta(\mathcal{S}) = \big(\mathfrak{q}_{-\Phi_n}(\mathcal{S})\big)_{n\in\mathbb{N}} \quad \text{for} \quad \mathcal{S} \in \mathfrak{H}_\Omega.$$

Since all of the functions $\mathfrak{q}_{-\Phi_n}$ are bounded on $\mathfrak{H}_\Omega$, the sets $\mathfrak{q}_{-\Phi_n}(\mathfrak{H}_\Omega)$ are all bounded in $\mathbb{R}$. The Cartesian product of their compact closures is therefore compact in the metric space $\mathbb{R}^{\mathbb{N}}$, and so is the set $K = \overline{\Theta(\mathfrak{H}_\Omega)}$, as it is a closed subset of this product. Now let C be the sup-stable point-separating cone of continuous real-valued functions on K, generated by the constants and the projections $\pi_m : \mathbb{R}^{\mathbb{N}} \to \mathbb{R}$, for all $m \in \mathbb{N}$. Let $\partial_C \subset K$ denote the classical Choquet boundary of C in K (see I.5 in [1]). Recall that ∂_C consists of those points in K whose point evaluation measure is maximal among all probability measures on K with respect to the classical Choquet ordering defined by the functions in C. We shall first establish that for every nonzero $(\alpha_n)_{n\in\mathbb{N}} \in \partial_C$ there is $\mathcal{D}_x^T \in \Delta_{\mathfrak{C}}^{\Omega}$ such that $\Theta(\mathcal{D}_x^T) = (\alpha_n)_{n\in\mathbb{N}}$. Indeed, for $0 \neq (\alpha_n)_{n\in\mathbb{N}} \in \partial_C$ set

$$Z = \{\mathcal{S} \in \mathfrak{H}_\Omega \mid \mathfrak{q}_{-\Phi_n}(\mathcal{S}) \geq \alpha_n \text{ for all } n \in \mathbb{N}\}.$$

Because the functions $\mathfrak{q}_{-\Phi_n}$ are superlinear and upper semicontinuous on the space $\mathfrak{L}(C_{\mathfrak{V}}(X,E),F)$, Z is a compact convex subset of $\mathfrak{H}_\Omega$. Let us verify that it is not empty. There is a net $(\mathcal{S}_i)_{i\in\mathcal{I}}$ in $\mathfrak{H}_\Omega$ such that both $\mathcal{S}_i \to \mathcal{S}$ in $\mathfrak{H}_\Omega$ and $\Theta(\mathcal{S}_i) \to (\alpha_n)_{n\in\mathbb{N}}$ in $\mathbb{R}^{\mathbb{N}}$. From the upper semicontinuity of the functions $\mathfrak{q}_{-\Phi_n}$, we infer that

$$\alpha_n = \lim_{i\in\mathcal{I}} \mathfrak{q}_{-\Phi_n}(\mathcal{S}_i) \leq \mathfrak{q}_{-\Phi_n}\Big(\lim_{i\in\mathcal{I}} \mathcal{S}_i\Big) = \mathfrak{q}_{-\Phi_n}(\mathcal{S})$$

holds for all $n \in \mathbb{N}$. Hence $\mathcal{S} \in Z$. Let $\mathcal{L}$ be the collection of all nonempty compact convex subsets L of Z with the following property:

If $\mathfrak{q}_{-\Phi}(\mathcal{S}) \leq \lambda\mathfrak{q}_{-\Phi}(\mathcal{S}_1) + (1-\lambda)\mathfrak{q}_{-\Phi}(\mathcal{S}_2)$ for all $\Phi \in \mathfrak{C}$,
some $\mathcal{S} \in L, \mathcal{S}_1, \mathcal{S}_2 \in \mathfrak{H}_\Omega$ and $0 \leq \lambda \leq 1$, then $\mathcal{S}_1, \mathcal{S}_2 \in L$.

Let us verify that $Z \in \mathcal{L}$. Let $\mathfrak{q}_{-\Phi}(\mathcal{S}) \leq \lambda\mathfrak{q}_{-\Phi}(\mathcal{S}_1) + (1-\lambda)\mathfrak{q}_{-\Phi}(\mathcal{S}_2)$ for all $\Phi \in \mathfrak{C}$, some $\mathcal{S} \in Z$, and $\mathcal{S}_1, \mathcal{S}_2 \in \mathfrak{H}_\Omega$. Set $\beta_n = \mathfrak{q}_{-\Phi_n}(\mathcal{S}_1)$ and $\gamma_n = \mathfrak{q}_{-\Phi_n}(\mathcal{S}_2)$. Then

$$\alpha_n \leq \mathfrak{q}_{-\Phi_n}(\mathcal{S}) \leq \lambda\mathfrak{q}_{-\Phi_n}(\mathcal{S}_1) + (1-\lambda)\mathfrak{q}_{-\Phi_n}(\mathcal{S}_2) = \lambda\beta_n + (1-\lambda)\gamma_n$$

for all $n \in \mathbb{N}$. If $\delta_{(\alpha_n)}, \delta_{(\beta_n)}$ and $\delta_{(\gamma_n)}$ denote the point evaluations on K at the points $(\alpha_n)_{n\in\mathbb{N}}$, $(\beta_n)_{n\in\mathbb{N}}$ and $(\gamma_n)_{n\in\mathbb{N}}$, respectively, then the above shows that the measure $\lambda\delta_{(\beta_n)} + (1-\lambda)\delta_{(\gamma_n)}$ dominates the measure $\delta_{(\alpha_n)}$ in the classical Choquet ordering generated by C on K. Hence $\delta_{(\alpha_n)} = \delta_{(\beta_n)} = \delta_{(\gamma_n)}$, and $\alpha_n = \beta_n = \gamma_n$, since $(\alpha_n)_{n\in\mathbb{N}} \in \partial_C$ is maximal. Thus $\mathcal{S}_1, \mathcal{S}_2 \in Z$. This shows in particular that the collection $\mathcal{L}$ is not empty. Next, we observe that every set $L \in \mathcal{L}$ is decreasing with respect to the order $\prec_{\mathfrak{C}}$. Indeed, suppose that $\mathcal{R} \prec_{\mathfrak{C}} \mathcal{S}$ for $\mathcal{S} \in L$ and $\mathcal{R} \in \mathfrak{L}(C_{\mathfrak{V}}(X,E),F)$. Then $\mathcal{R} \in \mathfrak{H}_\Omega$ since $\mathfrak{p}_\Omega(\mathcal{R}) \leq \mathfrak{p}_\Omega(\mathcal{S}) \leq 1$, and $\mathfrak{q}_{-\Phi}(\mathcal{S}) \leq \mathfrak{q}_{-\Phi}(\mathcal{R})$ for all $\Phi \in \mathfrak{C}$, implies that $\mathcal{R} \in L$. Furthermore, every extreme point $\mathcal{S}$ of L is some Ω-extremal point evaluation $\mathcal{D}_x^T$, because otherwise, according to Corollary 3.1.7, $\mathcal{S}$ is not an extreme point of $\mathfrak{H}_\Omega$ and can be expressed as $\mathcal{S} = \lambda\mathcal{S}_1 + (1-\lambda)\mathcal{S}_2$ with $\mathcal{S}_1, \mathcal{S}_2 \in \mathfrak{H}_\Omega$ distinct from $\mathcal{S}$ and $0 \le \lambda \le 1$ and such that

$$\mathfrak{q}_{-\Phi}(\mathcal{S}) = \lambda\mathfrak{q}_{-\Phi}(\mathcal{S}_1) + (1-\lambda)\mathfrak{q}_{-\Phi}(\mathcal{S}_2)$$

holds for all $\Phi \in \mathfrak{C}$. Thus $\mathcal{S}_1, \mathcal{S}_2 \in F$ by the definition of $\mathcal{L}$, a contradiction.

The family $\mathcal{L}$ is ordered by set inclusion, the intersection of any subfamily of its elements is again in $\mathfrak{L}$ and, therefore, every chain has a lower bound. Thus $\mathcal{L}$ contains a minimal element L_0. We shall verify that all operators in L_0 are equivalent and minimal with respect to the order $\prec_{\mathfrak{C}}$ in $\mathfrak{L}(C_{\mathfrak{V}}(X,E),F)$. Indeed, let $\Phi_0 \in \mathfrak{C}$ and

$$G_0 = \{\mathcal{S} \in L_0 \mid \mathfrak{q}_{-\Phi_0}(\mathcal{S}) = \max\{\mathfrak{p}_{-\Phi_0}(L_0)\}\}.$$

Because the functional $\mathfrak{q}_{-\Phi_0}$ is upper semicontinuous and superlinear, G_0 is a nonempty compact convex subset of L_0 and

$$\mathfrak{q}_{-\Phi}(\mathcal{S}) \le \lambda\mathfrak{q}_{-\Phi}(\mathcal{S}_1) + (1-\lambda)\mathfrak{q}_{-\Phi}(\mathcal{S}_2)$$

for all $\Phi \in \mathfrak{C}$, some $\mathcal{S} \in G_0$, $\mathcal{S}_1, \mathcal{S}_2 \in \mathfrak{H}_\Omega$ and $0 \le \lambda \le 1$, implies that $\mathcal{S}_1, \mathcal{S}_2 \in L_0$, and consequently,

$$\mathfrak{p}_{-\Phi_0}(\mathcal{S}_1) = \mathfrak{p}_{-\Phi_0}(\mathcal{S}_2) = \max\{\mathfrak{p}_{-\Phi_0}(L_0)\}.$$

We infer that $\mathcal{S}_1, \mathcal{S}_2 \in G_0$, and that G_0 is also an element of $\mathcal{L}$. Now the minimality of L_0 implies that $G_0 = L_0$ and, therefore, that all the functionals $\mathfrak{p}_{-\Phi}$, for $\Phi \in \mathfrak{C}$, are constant on L_0, our claim. Moreover, all elements of L_0 are minimal because L_0 is decreasing. Since $(\alpha_n)_{n\in\mathbb{N}} \neq 0$, since $L_0 \neq \{0\}$, and every nonzero extreme point of L_0 is a minimal Ω-extremal point evaluation by the above and, therefore, an element of $\Delta_{\mathfrak{C}}^{\Omega}$, we conclude that

$$\Theta^{-1}((\alpha_n)_{n\in\mathbb{N}}) \cap \Delta_{\mathfrak{C}}^{\Omega} \neq \varnothing.$$

Next, we choose any $\mathcal{D}_x^T \in \Theta^{-1}((\alpha_n)_{n\in\mathbb{N}}) \cap \Delta_{\mathfrak{C}}^{\Omega}$ and observe that $\mathfrak{q}_{-\Phi_m} \le \mathfrak{p}_m$ implies that $\mathfrak{q}_{-\Phi_m} \le \breve{\mathfrak{p}}_m$ by the definition of the lower envelope. Thus

$$\begin{aligned}\limsup_{m\in\mathbb{N}} \pi_m((\alpha_n)_{n\in\mathbb{N}}) &= \limsup_{m\in\mathbb{N}} \alpha_m\\ &= \limsup_{m\in\mathbb{N}} \mathfrak{q}_{-\Phi_m}(\mathcal{D}_x^T)\\ &\le \limsup_{m\in\mathbb{N}} \breve{\mathfrak{p}}_m(\mathcal{D}_x^T)\\ &\le \alpha\mathfrak{p}_\Omega(\mathcal{D}_x^T) \le \alpha\end{aligned}$$

by our assumption. This argument applies to all elements $(\alpha_n)_{n\in\mathbb{N}}$ of the Choquet boundary $\partial C \subset K$, that is, $\limsup_{m\in\mathbb{N}} \pi_m \le \alpha$ holds pointwise on ∂C. Choquet's theorem for compact convex metric spaces guarantees the existence of a probability measure θ on K, which is supported by ∂C and on the cone C of continuous real-valued functions on K dominates the point evaluation at $\Theta(\mathcal{T}) \in K$. That is to say,

$$\int_K \pi_m \, d\theta = \int_{\partial C} \pi_m \, d\theta \geq \pi_m(\Phi(\mathcal{T})) = \mathfrak{q}_{-\Phi_m}(\mathcal{T})$$

holds for all $m \in \mathbb{N}$. The sequence $(\pi_m)_{m\in\mathbb{N}}$ in C is uniformly bounded above on K, because this holds for the sequence $(\mathfrak{q}_{-\Phi_m})_{m\in\mathbb{N}}$ on $\mathfrak{H}_\Omega$. Thus we may use Fatou's lemma for

$$\limsup_{m\in\mathbb{N}} \int_{\partial C} \pi_m \, d\theta \leq \int_{\partial C} \limsup_{m\in\mathbb{N}} \pi_m \, d\theta \leq \alpha.$$

Moreover, our selection of the functionals $\mathfrak{q}_{-\Phi_m}$ renders that

$$\limsup_m \breve{\mathfrak{p}}_m(\mathcal{T}) = \limsup_{m\in\mathbb{N}} \mathfrak{q}_{-\Phi_m}(\mathcal{T}).$$

Now combining all of the above, we have

$$\limsup_m \breve{\mathfrak{p}}_m(\mathcal{T}) \leq \alpha = \alpha \mathfrak{p}_\Omega(\mathcal{T}).$$

This completes our argument. □

3.2.12 Corollary. *Let $\Omega \in \mathfrak{K}$. The Choquet boundary $\Delta_{\mathfrak{C}}^{\Omega}$ is nonempty if and only if there is an operator $\mathcal{T} \in \mathfrak{L}(C_{\mathfrak{V}}(X,E),F)$, such that $\mathfrak{p}_\Omega(\mathcal{T}) < +\infty$ and $\mathfrak{p}_\Phi(\mathcal{T}) < 0$ for some $\Phi \in \mathfrak{C}$.*

Proof. If $\mathcal{D}_x^T \in \Delta_x^\Omega$, then $\mathfrak{p}_\Omega(\mathcal{D}_x^T) < +\infty$. Since $\mathcal{D}_x^T \neq 0$, we have $\mathfrak{p}_\Omega(\mathcal{D}_x^T) > 0$ by Lemma 3.1.1(a). Thus $\mathcal{D}_x^T \not\prec 0$ since $\mathfrak{p}_\Omega(0) = 0$ and, therefore, $0 \not\prec_{\mathfrak{C}} \mathcal{D}_x^T$ since $\mathcal{D}_x^T$ is minimal. The latter means that $\mathfrak{p}_\Phi(\mathcal{D}_x^T) < \mathfrak{p}_\Phi(0) = 0$ for some $\Phi \in \mathfrak{C}$, our claim. For the converse statement, let $\mathcal{T} \in \mathfrak{L}(C_{\mathfrak{V}}(X,E),F)$ such that $\mathfrak{p}_\Omega(\mathcal{T}) < +\infty$ and $\mathfrak{p}_\Phi(\mathcal{T}) < 0$ for $\Phi \in \mathfrak{C}$ and assume that $\Delta_{\mathfrak{C}}^\Omega = \emptyset$. We apply the bracketed case of Theorem 3.2.11 with the constant sequence $\mathfrak{p}_n = \mathfrak{p}_\Omega$ and $\alpha = 0$. Since the condition involving $\Delta_{\mathfrak{C}}^\Omega$ is void, the theorem renders that $\breve{\mathfrak{p}}_\Omega(\mathcal{T}) \leq 0$. On the other hand, $\mathfrak{q}_{\{0\}} \leq \mathfrak{p}_\Omega$ implies that $\breve{\mathfrak{p}}_\Omega(\mathcal{T}) \geq 0$ and, therefore, $\breve{\mathfrak{p}}_\Omega(\mathcal{T}) = 0$. Thus according to Proposition 3.2.6 there is $S \in \mathfrak{L}(C_{\mathfrak{V}}(X,E),F)$ such that $S \prec_{\mathfrak{C}} T$ and $\mathfrak{p}_\Omega(S) = 0$. By Lemma 3.1.1(a), the latter implies that $S = 0$. Hence $\mathfrak{p}_\Phi(0) \leq \mathfrak{p}_\Phi(T) < 0$, a contradiction since $\mathfrak{p}_\Phi(0) = 0$. □

In special cases, the expressions in Theorem 3.2.11 involving limits and upper (or lower) envelopes of functions can be reformulated using the statements of Proposition 3.2.6.

3.2.13 Lemma. *Let $\mathcal{T} \in \mathfrak{L}_{\mathfrak{K}}(C_{\mathfrak{V}}(X,E),F)$ and let $(\mathfrak{q}_n)_{n\in\mathbb{N}}$ be a sequence in $\mathfrak{Q}$ that is increasing (or $(\mathfrak{p}_n)_{n\in\mathbb{N}}$ a sequence in $\mathfrak{P}$ that is decreasing) on $\mathcal{T}{\downarrow}^{\mathfrak{C}}$. Then*

$$\liminf_{n\in\mathbb{N}} \widehat{\mathfrak{q}}_n(\mathcal{T}) = \sup_{\mathcal{S}\prec_{\mathfrak{C}}\mathcal{T}} \left\{\sup_{n\in\mathbb{N}} \mathfrak{q}_n(\mathcal{S})\right\}$$

$$\left(or \ \limsup_{n\in\mathbb{N}} \check{\mathfrak{p}}_n(\mathcal{T}) = \inf_{\mathcal{S}\prec_{\mathfrak{C}}\mathcal{T}} \left\{\inf_{n\in\mathbb{N}} \mathfrak{p}_n(\mathcal{S})\right\}\right).$$

If the sequence $(\mathfrak{q}_n)_{n\in\mathbb{N}}$ *is increasing (or the sequence* $(\mathfrak{p}_n)_{n\in\mathbb{N}}$ *is decreasing) on all of* $\mathfrak{L}(\mathcal{C}_{\mathfrak{V}}(X,E),F)$ *and if one of the functions* $\mathfrak{q}_n$ *(or* $\mathfrak{p}_n$*) is continuous at* $0 \in \mathfrak{L}_{\mathfrak{K}}(\mathcal{C}_{\mathfrak{V}}(X,E),F)$*, then the suprema (or infima) with respect to* $\mathcal{S} \prec_{\mathfrak{C}} \mathcal{T}$ *in the above expressions are in fact maxima (or minima).*

Proof. We shall prove the case of the upper envelopes. Let $\mathcal{T} \in \mathfrak{L}_{\mathfrak{K}}(\mathcal{C}_{\mathfrak{V}}(X,E),F)$ and $(\mathfrak{q}_n)_{n\in\mathbb{N}}$ be as stated. The sequence $(\widehat{\mathfrak{q}}_n(\mathcal{T}))_{n\in\mathbb{N}}$ is increasing in $\underline{\mathbb{R}}$ and we have

$$\liminf_{n\in\mathbb{N}} \widehat{\mathfrak{q}}_n(\mathcal{T}) = \sup_{n\in\mathbb{N}} \widehat{\mathfrak{q}}_n(\mathcal{T}) = \sup_{n\in\mathbb{N}}\left\{\max_{\mathcal{S}\prec_{\mathfrak{C}}\mathcal{T}} \mathfrak{q}_n(\mathcal{S})\right\} = \sup_{\mathcal{S}\prec_{\mathfrak{C}}\mathcal{T}} \left\{\sup_{n\in\mathbb{N}} \mathfrak{q}_n(\mathcal{S})\right\}$$

by Proposition 3.2.6. If the sequence $(\mathfrak{q}_n)_{n\in\mathbb{N}}$ is increasing on the whole space $\mathfrak{L}(\mathcal{C}_{\mathfrak{V}}(X,E),F)$ and if the function $\mathfrak{q}_{n_0}$ is continuous at $0 \in \mathfrak{L}_{\mathfrak{K}}(\mathcal{C}_{\mathfrak{V}}(X,E),F)$, then it is real-valued and so are all $\mathfrak{q}_n$ for $n \geq n_0$. Under these assumptions, we continue our argument as follows: There are $\mathcal{S}_n \prec_{\mathfrak{C}} \mathcal{T}$ such that

$$\mathfrak{q}_n(\mathcal{S}_n) = \max_{\mathcal{S}\prec_{\mathfrak{C}}\mathcal{T}} \mathfrak{q}_n(\mathcal{S}) = \widehat{\mathfrak{q}}_n(\mathcal{T}),$$

and using the compactness of $\mathcal{T}\downarrow^{\mathfrak{C}}$ we find an accumulation point $\mathcal{S} \prec_{\mathfrak{C}} \mathcal{T}$ of the sequence $(\mathcal{S}_n)_{n\in\mathbb{N}}$. We claim that

$$\sup_{n\in\mathbb{N}} \mathfrak{q}_n(\mathcal{S}) = \sup_{n\in\mathbb{N}} \mathfrak{q}_n(\mathcal{S}_n).$$

Indeed, since $\mathfrak{q}_n(\mathcal{S}) \leq \mathfrak{q}_n(\mathcal{S}_n)$ for all $n \in \mathbb{N}$, we have

$$\sup_{n\in\mathbb{N}} \mathfrak{q}_n(\mathcal{S}) \leq \sup_{n\in\mathbb{N}} \mathfrak{q}_n(\mathcal{S}_n).$$

On the other hand, given $\varepsilon > 0$, by the continuity of the function $\mathfrak{q}_{n_0}$ at $0 \in \mathfrak{L}(\mathcal{C}_{\mathfrak{V}}(X,E),F)$ there is $n_1 \geq n_0$ such that

$$\mathfrak{q}_n(\mathcal{S} - \mathcal{S}_n) \geq \mathfrak{q}_{n_0}(\mathcal{S} - \mathcal{S}_n) \geq -\varepsilon$$

for all $n \geq n_1$. Now using the superlinearity of the functionals $\mathfrak{q}_n$, we infer that

$$\mathfrak{q}_n(\mathcal{S}) \geq \mathfrak{q}_n(\mathcal{S}_n) + \mathfrak{q}_n(\mathcal{S} - \mathcal{S}_n),$$

hence

$$\mathfrak{q}_n(\mathcal{S}_n) \leq \mathfrak{q}_n(\mathcal{S}) - \mathfrak{q}_n(\mathcal{S} - \mathcal{S}_n) \leq \mathfrak{q}_n(\mathcal{S}) + \varepsilon$$

for all $n \geq n_1$. As $\mathfrak{q}_n(\mathcal{S}_n) = \hat{\mathfrak{q}}_n(\mathcal{T}) \leq \hat{\mathfrak{q}}_m(\mathcal{T}) = \mathfrak{q}_m(\mathcal{S}_m)$ whenever $n \leq m$, the sequence $(\mathfrak{q}_n(\mathcal{S}_n))_{n\in\mathbb{N}}$ is increasing in $\overline{\mathbb{R}}$ and we have

$$\sup_{n\in\mathbb{N}} \mathfrak{q}_n(\mathcal{S}_n) = \sup_{n\geq n_1} \mathfrak{q}_n(\mathcal{S}_n) \leq \sup_{n\in\mathbb{N}} \mathfrak{q}_n(\mathcal{S}) + \varepsilon.$$

The latter holds for all $\varepsilon > 0$, and since

$$\sup_{n\in\mathbb{N}} \mathfrak{q}_n(\mathcal{S}_n) = \sup_{n\in\mathbb{N}} \hat{\mathfrak{q}}_n(\mathcal{T}) = \liminf_{n\in\mathbb{N}} \hat{\mathfrak{q}}_n(\mathcal{T}),$$

our claim follows. □

The next corollary relates to the classical Choquet–Bishop–DeLeeuw theorem (see Corollary I.4.12 in [1]). It establishes a localization for the support of the representation measures of minimal operators. Recall that a G_δ-subset of X is defined to be a countable intersection of open sets. An F_σ-set is a countable union of closed sets. All G_δ- and F_σ-sets are obviously contained in the σ-field $\mathfrak{A}$.

3.2.14 Corollary. *Let* $\mathfrak{u}$ *be a function space neighborhood, which is a neighborhood in* $\mathcal{C}_{\mathfrak{V}}(X,E)$ *and let* $\Pi \subset F^*$ *such that* $\Omega_{\mathfrak{u}}^{\Pi} \in \mathfrak{K}$. *Let* $\mathcal{T} \in \mathfrak{L}_{\mathfrak{K}}(\mathcal{C}_{\mathfrak{V}}(X,E),F)$ *be minimal and* $\mathfrak{p}_{\Omega_{\mathfrak{u}}^{\Pi}}(\mathcal{T}) < +\infty$. *If* B *is a* G_δ-*subset of* X *disjoint from* $\partial_{\mathfrak{C}}^{\Omega_{\mathfrak{u}}^{\Pi}}$, *then*

$$\inf_{\mathcal{S}\sim\mathcal{T}} \sup_{A\in\mathfrak{R}} \{\|\theta\|_{A\cap B}^{(\mathfrak{u},\Pi)} \mid \theta \text{ represents } \mathcal{S}\} = 0.$$

Proof. Let O_n be open subsets of X such that $B = \bigcap_{n\in\mathbb{N}} O_n \in \mathfrak{A}$ is disjoint from $\partial_{\mathfrak{C}}^{\Omega_{\mathfrak{u}}^{\Pi}}$. We may assume that the sequence $(O_n)_{n\in\mathbb{N}}$ is decreasing. The subsets

$$\Omega_n = \left\{ \sum_{i=1}^{n} (\varphi_i \cdot f_i) \otimes \mu_i \;\middle|\; \begin{array}{l} f_i \in \mathfrak{u},\ \mu_i \in \Pi,\ 0 \leq \varphi_i \in C(X) \\ \text{such that } \sum_{i=1}^{n} \varphi_i \leq \chi_{O_n} \end{array} \right\}$$

of $\mathcal{C}_{\mathfrak{V}}(X,E) \otimes F^*$ are $C(X)$-convex and the functionals $\mathfrak{p}_n = \mathfrak{p}_{\Omega_n}$ are contained in $\mathfrak{P}$. We have

$$\mathfrak{p}_n(\mathcal{T}) = \sup \left\{ \sum_{i=1}^{n} (\mathcal{T}(\varphi_i \cdot f_i))(\mu_i) \;\middle|\; \begin{array}{l} f_i \in \mathfrak{u},\ \mu_i \in \Pi,\ 0 \leq \varphi_i \in C(X) \\ \text{such that } \sum_{i=1}^{n} \varphi_i \leq \chi_{O_n} \end{array} \right\}$$

for all $\mathcal{T} \in \mathfrak{L}_{\mathfrak{K}}(\mathcal{C}_{\mathfrak{V}}(X,E),F)$. Proposition 3.1.14 states that

$$\sup_{A\in\mathfrak{R}} \|\theta\|_{A\cap B}^{(\mathfrak{u},\Pi)} \leq \sup_{A\in\mathfrak{R}} \|\theta\|_{A\cap O_n}^{(\mathfrak{u},\Pi)} = \mathfrak{p}_n(\mathcal{T}) \leq \mathfrak{p}_{\Omega_{\mathfrak{u}}^{\Pi}}(\mathcal{T}),$$

where θ denotes the representation measure for $\mathcal{T}$. The sequence $(\mathfrak{p}_n)_{n\in\mathbb{N}}$ is decreasing and since $(\bigcap_{n\in\mathbb{N}} O_n) \cap \partial_{\mathfrak{C}}^{\Omega_{\mathfrak{u}}^{\Pi}} = \emptyset$, we infer that

$$\limsup_{n\in\mathbb{N}} \breve{\mathfrak{p}}_n(\mathcal{D}_X^\eta) \le \limsup_{n\in\mathbb{N}} \mathfrak{p}_n(\mathcal{D}_X^\eta) = 0$$

for all $\mathcal{D}_X^\eta \in \Delta_{\mathfrak{C}}^{\Omega_{\mathfrak{u}}^\Pi}$. If $\mathfrak{p}_{\Omega_{\mathfrak{u}}^\Pi}(\mathcal{T}) < +\infty$ for $\mathcal{T} \in \mathfrak{L}(C_{\mathfrak{V}}(X,E),F)$, we may therefore apply the bracketed case in Theorem 3.2.11 with $\alpha = 0$ and obtain

$$\limsup_{n\in\mathbb{N}} \breve{\mathfrak{p}}_n(\mathcal{T}) = \inf_{n\in\mathbb{N}} \breve{\mathfrak{p}}_n(\mathcal{T}) = 0.$$

If $\mathcal{T}$ is indeed minimal, then the above together with Proposition 3.2.10 shows that

$$\inf_{\mathcal{S}\sim\mathcal{T}} \{\|\theta\|_{A\cap B}^{(\mathfrak{u},\Pi)} \mid \theta \text{ represents } \mathcal{S}\} \le \inf_{\mathcal{S}\sim\mathcal{T}} \mathfrak{p}_n(\mathcal{S}) = \breve{\mathfrak{p}}_n(\mathcal{T})$$

holds for all $n \in \mathbb{N}$. Our claim follows. □

For the next corollary, we recall from Section 2.1 that a subset Π of F^* is called θ-dense if for all $A \in \mathfrak{R}$ and $a \in E$ we have $\theta_A(a) = 0$ whenever $\theta_A(a)(\mu) = 0$ for all $\mu \in \Pi$. For this condition to hold for an $\mathfrak{L}(E,F^{**})$-valued representation measure θ, it suffices that the linear span of Π is dense in of F^* in its weak topology. If θ is indeed $\mathfrak{L}(E,F)$-valued, then density in the weak* topology is enough.

3.2.15 Corollary. *Let $\mathfrak{u}$ be a function space neighborhood, which is a neighborhood in $C_{\mathfrak{V}}(X,E)$, and let $\Pi \subset F^*$ such that $\Omega_{\mathfrak{u}}^\Pi \in \mathfrak{K}$ and that $\|\theta\|^{(\mathfrak{u},\Pi)} = \|\vartheta\|^{(\mathfrak{u},\Pi)}$ whenever θ and ϑ are the representation measures for equivalent operators. If $\mathcal{T} \in \mathfrak{L}_{\mathfrak{K}}(C_{\mathfrak{V}}(X,E),F)$ is minimal and $\mathfrak{p}_{\Omega_{\mathfrak{u}}^\Pi}(\mathcal{T}) < +\infty$, and if Π is θ-dense in F^* for the representation measure θ for $\mathcal{T}$, then θ is supported by every F_σ-subset of X that contains $\partial_{\mathfrak{C}}^{\Omega_{\mathfrak{u}}^\Pi}$.*

Proof. Suppose that the operator $\mathcal{T} \in \mathfrak{L}(C_{\mathfrak{V}}(X,E),F)$ is minimal and that $\mathfrak{p}_{\Omega_{\mathfrak{u}}^\Pi}(\mathcal{T}) < +\infty$. Let θ be the representation measure for $\mathcal{T}$. The complement of an F_σ-subset of X that contains $\partial_{\mathfrak{C}}^{\Omega_{\mathfrak{u}}^\Pi}$ is a G_δ-subset $B \in \mathfrak{A}$ disjoint from $\partial_{\mathfrak{C}}^{\Omega_{\mathfrak{u}}^\Pi}$. Under our assumptions, Corollary 3.2.14 yields that $\|\theta\|_A^{(\mathfrak{u},\Pi)} = 0$ for all subsets $A \in \mathfrak{R}$ of B. Our claim follows with Proposition 2.1.11(b). □

$\mathfrak{C}$-superharmonic sets

Theorem 3.2.11 and its corollaries suggest the benefit of having small or indeed singleton equivalence classes in $\mathfrak{L}_{\mathfrak{K}}(C_{\mathfrak{V}}(X,E),F)$ while preserving the given Choquet boundary. We shall say that a $C(X)$-convex subset Φ of $C_{\mathfrak{V}}(X,E) \otimes F^*$ is *$\mathfrak{C}$-superharmonic* if $\mathfrak{p}_\Phi(\mathcal{T}) \le \mathfrak{p}_\Phi(\mathcal{D}_X^T)$ whenever $\mathcal{T} \prec_{\mathfrak{C}} \mathcal{D}_X^T$ for $\mathcal{T}, \mathcal{D}_X^T \in \mathfrak{L}_{\mathfrak{K}}(C_{\mathfrak{V}}(X,E),F)$ such that $\mathcal{D}_X^T$ is a $\mathfrak{K}$-extremal point evaluation. Clearly, sums and positive multiples of $\mathfrak{C}$-superharmonic sets are again $\mathfrak{C}$-superharmonic.

We define the *completion* $\widetilde{\mathfrak{C}}$ of a Choquet cone $\mathfrak{C}$ over $\mathfrak{K}$ as the cone $C(X)$-convex subsets of $C_{\mathfrak{V}}(X,E) \otimes F^*$, which is generated by $\mathfrak{K}$ and the collection of all nonempty

$\mathfrak{K}$-bounded $\mathfrak{C}$-superharmonic subsets Φ of $C_{\mathfrak{V}}(X,E)\otimes F^*$. Every set $\Phi \in \widetilde{\mathfrak{C}}$ can be expressed as $\Phi = \Psi + \sum_{i=1}^{n} \lambda_i \Omega_i$, where $\Psi \in \mathfrak{C}$ is $\mathfrak{K}$-bounded and $\mathfrak{C}$-superharmonic, where $\lambda_i \geq 0$ and $\Omega_i \in \mathfrak{K}$. All elements of $\widetilde{\mathfrak{C}}$ are $\mathfrak{C}$-superharmonic.

3.2.16 Proposition. *The completion $\widetilde{\mathfrak{C}}$ of a Choquet cone $\mathfrak{C}$ is itself a Choquet cone over $\mathfrak{K}$ that contains $\mathfrak{C}$ and $\Delta_{\widetilde{\mathfrak{C}}} \subset \Delta_{\mathfrak{C}}$.*

Proof. Obviously, $\mathfrak{C}$ is a Choquet cone over $\mathfrak{K}$ and $\mathfrak{C} \subset \widetilde{\mathfrak{C}}$. Let us verify that $\Delta_{\widetilde{\mathfrak{C}}} \subset \Delta_{\mathfrak{C}}$. If $\mathcal{D}_x^T \in \Delta_{\widetilde{\mathfrak{C}}}$ and $\mathcal{T} \prec_{\mathfrak{C}} \mathcal{D}_x^T$ for $\mathcal{T}, \mathcal{D}_x^T \in \mathfrak{L}_{\mathfrak{K}}(C_{\mathfrak{V}}(X,E),F)$, then $\mathcal{T} \prec_{\widetilde{\mathfrak{C}}} \mathcal{D}_x^T$ holds as well, since all elements of $\widetilde{\mathfrak{C}}$ are $\mathfrak{C}$-superharmonic. Thus $\mathcal{D}_x^T \prec_{\widetilde{\mathfrak{C}}} \mathcal{T}$ and, therefore, $\mathcal{D}_x^T \prec_{\mathfrak{C}} \mathcal{T}$. Hence $\mathcal{D}_x^T$ is $\prec_{\mathfrak{C}}$-minimal, and consequently, an element of $\Delta_{\mathfrak{C}}$. □

Our construction reveals that the completion $\widetilde{\mathfrak{C}}$ of a Choquet cone $\mathfrak{C}$ is not necessarily the largest Choquet cone over $\mathfrak{K}$ containing $\mathfrak{C}$ without increasing its Choquet boundary. But it serves our purpose since the preceding results are applicable with the completion $\widetilde{\mathfrak{C}}$ of $\mathfrak{C}$ in place of $\mathfrak{C}$. We proceed to recognize some $\mathfrak{C}$-superharmonic subsets of $C_{\mathfrak{V}}(X,E)\otimes F^*$.

If $\mathfrak{K} \neq \emptyset$, we consider the closed $C(X)$-convex subcone

$$\mathfrak{K}_0 = \bigcap \{\varepsilon\Omega \mid \Omega \in \mathfrak{K},\ \varepsilon > 0\}$$

of $C_{\mathfrak{V}}(X,E)\otimes F^*$. For every $\mathcal{T} \in \mathfrak{L}_{\mathfrak{K}}(C_{\mathfrak{V}}(X,E),F)$, there is $\Omega \in \mathfrak{K}$ such that $\mathfrak{p}_\Omega(\mathcal{T}) < +\infty$. Thus $\mathfrak{p}_\Phi(\mathcal{T}) \leq 0$ for every $C(X)$-convex subset Φ of $\mathfrak{K}_0$, and $\mathfrak{p}_{\mathfrak{K}_0}(\mathcal{T}) = 0$ in particular, since $0 \in \mathfrak{K}_0$. Moreover, if $\omega \in \mathfrak{K}_0$ and if $0 \leq \varphi \in C(X)$ is nonnegative and bounded, then $\varphi \cdot \omega \in \mathfrak{K}_0$.

3.2.17 Proposition. *Let $\mathfrak{C}$ be a Choquet cone over $\mathfrak{K} \neq \emptyset$.*
(a) *If $\mathfrak{K} \neq \emptyset$, then $\mathfrak{K}_0 \in \widetilde{\mathfrak{C}}$.*
(b) *If $\omega \in \mathfrak{K}_0$ and if $\{\varphi_i \cdot \omega\} \in \widetilde{\mathfrak{C}}$ for $i = 1, \ldots, n$ with bounded functions $\varphi_i \in C(X)$, then $\{\varphi \cdot \omega\} \in \widetilde{\mathfrak{C}}$, where $\varphi = \bigvee_{i=1}^{n} \varphi_i$.*

Proof. (a) Clearly, $\mathfrak{K}_0$ is $\mathfrak{C}$-superharmonic since $\mathfrak{p}_{\mathfrak{K}_0}(\mathcal{T}) = 0$ for all operators $\mathcal{T} \in \mathfrak{L}_{\mathfrak{K}}(C_{\mathfrak{V}}(X,E),F)$, and $\mathfrak{K}_0$ is $\mathfrak{K}$-bounded. For (b), let ω, φ_i and φ be as stated. The singleton sets $\{\varphi_i \cdot \omega\}$ and $\{\varphi \cdot \omega\}$ are obviously $C(X)$-convex. All the elements $(\varphi - \varphi_i)\cdot\omega$ are contained in $\mathfrak{K}_0$ and, therefore, $\mathfrak{p}_{\{(\varphi-\varphi_i)\cdot\omega\}}(\mathcal{T}) \leq 0$, that is, $\mathfrak{p}_{\{\varphi\cdot\omega\}}(\mathcal{T}) \leq \mathfrak{p}_{\{\varphi_i\cdot\omega\}}(\mathcal{T})$ holds for all $\mathcal{T} \in \mathfrak{L}_{\mathfrak{K}}(C_{\mathfrak{V}}(X,E),F)$ by the above. For a point evaluation $\mathcal{D}_x^T$ and a bounded function $\psi \in C(X)$, we evaluate

$$\mathfrak{p}_{\{\psi\cdot\omega\}}(\mathcal{D}_x^T) = \mathfrak{Re}(\psi\cdot\omega)(\mathcal{D}_x^T) = \psi(x)\mathfrak{Re}\,\omega(\mathcal{D}_x^T) = \psi(x)\mathfrak{p}_{\{\omega\}}(\mathcal{D}_x^T).$$

Thus $\mathcal{T} \prec_{\mathfrak{C}} \mathcal{D}_x^T$ for $\mathcal{T}, \mathcal{D}_x^T \in \mathfrak{L}_{\mathfrak{K}}(C_{\mathfrak{V}}(X,E),F)$ such that $\mathcal{D}_x^T$ is a $\mathfrak{K}$-extremal point evaluation implies that

$$\mathfrak{p}_{\{\varphi_i\cdot\omega\}}(\mathcal{T}) \leq \varphi_i(x)\mathfrak{Re}\,\omega(\mathcal{D}_x^T)$$

for all $i = 1, \dots, n$. Now using that $\mathfrak{Re}\,\omega(\mathcal{D}_x^T) \le 0$ we calculate that

$$\begin{aligned} \mathfrak{p}_{\{\varphi\cdot\omega\}}(\mathcal{T}) &\le \inf_{i=1}^{n}\{\varphi_i(x)\mathfrak{Re}\,\omega(\mathcal{D}_x^T)\} \\ &= \mathfrak{Re}\,\omega(\mathcal{D}_x^T)\sup_{i=1}^{n}\{\varphi_i(x)\} \\ &= \varphi(x)\mathfrak{Re}\,\omega(\mathcal{D}_x^T) \\ &= \mathfrak{p}_{\{\varphi\cdot\omega\}}(\mathcal{D}_x^T). \end{aligned}$$

The singleton set $\{\varphi\cdot\omega\}$ is therefore indeed $\mathfrak{C}$-superharmonic and contained in $\widetilde{\mathfrak{C}}$. □

Proposition 3.2.17 is of course meaningful only if $\mathfrak{K}_0$ contains nonzero elements. This is not always the case. We shall therefore provide some further criteria, which do not rely on this assumption.

3.2.18 Proposition. *Let $\mathfrak{C}$ be a Choquet cone and let $\mathfrak{D}$ be a family of $\mathfrak{C}$-superharmonic subsets of $\mathcal{C}_{\mathfrak{V}}(X,E)\otimes F^*$. If $\Psi = \bigcap_{\Phi\in\mathfrak{D}} \Phi$ is not empty and if $\mathfrak{p}_\Psi(\mathcal{D}_x^T) = \inf_{\Phi\in\mathfrak{D}} \mathfrak{p}_\Phi(\mathcal{D}_x^T)$ for all $\mathfrak{K}$-extremal point evaluations $\mathcal{D}_x^T \in \mathfrak{L}_{\mathfrak{K}}(\mathcal{C}_{\mathfrak{V}}(X,E),F)$, then Ψ is also $\mathfrak{C}$-superharmonic.*

Proof. Let $\mathfrak{D}$ be as stated and let Ψ denote the nonempty intersection of its elements. Clearly, $\mathfrak{p}_\Psi(\mathcal{T}) \le \inf_{\Phi\in\mathfrak{D}} \mathfrak{p}_\Phi(\mathcal{T})$ for every $\mathcal{T} \in \mathfrak{L}(\mathcal{C}_{\mathfrak{V}}(X,E),F)$. Let $\mathcal{T} \prec_{\mathfrak{C}} \mathcal{D}_x^T$ for $\mathcal{T}, \mathcal{D}_x^T \in \mathfrak{L}_{\mathfrak{K}}(\mathcal{C}_{\mathfrak{V}}(X,E),F)$ such that $\mathcal{D}_x^T$ is a $\mathfrak{K}$-extremal point evaluation. Then

$$\mathfrak{p}_\Psi(\mathcal{T}) \le \inf_{\Phi\in\mathfrak{D}} \mathfrak{p}_\Phi(\mathcal{T}) \le \inf_{\mathfrak{u}\in\mathfrak{D}} \mathfrak{p}_\Phi(\mathcal{D}_x^T) = \mathfrak{p}_\Psi(\mathcal{D}_x^T)$$

by our assumption, hence our claim. □

As an immediate consequence of Corollary 3.1.13, we notice the following.

3.2.19 Proposition. *Let $\mathfrak{C}$ be a Choquet cone, let $\mathfrak{u}$ and $\mathfrak{w}$ be $C(X)$-convex subsets of $\mathcal{C}_{\mathfrak{V}}(X,E)$, suppose that $\Pi \subset F^*$ satisfies (Π1) and that both $\Omega_{\mathfrak{u}}^{\Pi}$ and $\Omega_{\mathfrak{w}}^{\Pi}$ are $\mathfrak{C}$-superharmonic. If either $0 \in \mathfrak{u} \cap \mathfrak{w}$ or Π is a singleton set, then $\Omega_{\mathfrak{u}+\mathfrak{w}}^{\Pi}$ is also $\mathfrak{C}$-superharmonic.*

The following criteria are rather technical, yet useful in many cases. For a convex subset U of E, such that $0 \in U$ and a lower semicontinuous $\overline{\mathbb{R}}$+-valued function $\rho : X \to \overline{\mathbb{R}}$ we set

$$\mathfrak{u}_U^\rho = \{f \in \mathcal{C}_{\mathfrak{V}}(X,E) \mid f(x) \in \rho(x)U \text{ for all } x \in X\}.$$

Recall from Section 1.1 that $(+\infty)U$ is defined to be the conic hull of U in E. Clearly, $\mathfrak{u}_U^\rho$ is a $C(X)$-convex subset of $\mathcal{C}_{\mathfrak{V}}(X,E)$. For the following lemma, also recall that we generally set $0(+\infty) = 0$ and $(+\infty)(+\infty) = +\infty$ (see Section 1.1).

3.2.20 Lemma. *Let U be a convex subset of E such that $0 \in U$, let $\rho : X \to \overline{\mathbb{R}}$ be a lower semicontinuous $\overline{\mathbb{R}}_+$-valued function and let $\Pi \subset F^*$. Then*

$$\mathfrak{p}_{\Omega^{\Pi}_{\mathfrak{u}^{\rho}_{U}}}(\mathcal{D}^T_x) = \rho(x) \sup_{a \in U} p_{\Pi}(T(a))$$

for every point evaluation $\mathcal{D}^T_x \in \mathfrak{L}(\mathcal{C}_{\mathfrak{V}}(X,E),F)$.

Proof. Let U, ρ and Π be as stated and let $\mathcal{D}^T_x \in \mathfrak{L}(\mathcal{C}_{\mathfrak{V}}(X,E),F)$ be a point evaluation. We abbreviate $\Phi = \Omega^{\Pi}_{\mathfrak{u}^{\rho}_{U}}$ and $\beta = \sup_{a \in U} p_{\Pi}(T(a))$. According to Proposition 3.1.9, we have

$$\mathfrak{p}_{\Phi}(\mathcal{D}^T_x) = \sup\{p_{\Pi}(T(f(x))) \mid f \in \mathfrak{u}^{\rho}_{U}\}$$

If $\rho(x) = 0$, there is nothing to prove since $f(x) = 0$ for every $f \in \mathfrak{u}^{\rho}_{U}$, and both sides in the equation in our lemma equal zero. If $\beta = 0$, then $p_{\Pi}(T(a)) \leq 0$ holds for all $a \in U$, and indeed for all elements of the conic hull of U. Since $T(f(x))$ is an element of this subcone for any $f \in \mathfrak{u}^{\rho}_{U}$, we conclude again that both sides in the above equation equal zero. We may therefore assume that both $\rho(x) > 0$ and $\beta > 0$ and continue as follows. We have $f(x) \in \rho(x)U$ for all $f \in \mathfrak{u}^{\rho}_{U}$. Thus $\mathfrak{p}_{\Phi}(\mathcal{D}^T_x) \leq \rho(x)\beta$. For the reverse inequality, let $0 < \gamma < \rho(x)$. There is a neighborhood $\mathcal{N} \in \mathfrak{N}$ of x such that $\rho(y) > \gamma$ for all $y \in \mathcal{N}$. There is $\psi \in \mathcal{C}(X)$, such that $0 \leq \psi \leq 1$, $\psi(x) = 1$ and $\psi(y) = 0$ for all $y \in X \setminus \mathcal{N}$. Then for any choice of $a \in U$ the function $\gamma(\psi \cdot a)$ is contained in $\mathfrak{u}^{\rho}_{U}$, hence $\mathfrak{p}_{\Phi}(\mathcal{D}^T_x) \geq p_{\Pi}(T(\gamma a))$ for all $a \in U$, hence $\mathfrak{p}_{\Phi}(\mathcal{D}^T_x) \geq \gamma\beta$. The latter holds true for all choices of $0 < \gamma < \rho(x)$ and yields $\mathfrak{p}_{\Phi}(\mathcal{D}^T_x) \geq \rho(x)\beta$. □

Using the notation from above, we formulate our next criterion.

3.2.21 Proposition. *Let $\mathfrak{C}$ be a Choquet cone. Let U be a closed convex subset of E such that $0 \in U$ and let $\rho : X \to \overline{\mathbb{R}}$ be a strictly positive lower semicontinuous function. Let Γ be a multiplicative subgroup of the unit sphere of $\mathbb{R}$ or $\mathbb{C}$ and suppose that $\gamma U \subset U$ for all $\gamma \in \Gamma$. For $i = 1, \ldots, n$, let φ_i be continuous real-or complex-valued functions such that*

$$|\varphi_i|(x) < \rho(x) \quad \textit{and} \quad \varphi_i(x)/|\varphi_i(x)| \in \Gamma$$

for all $x \in X$ such that $\varphi_i(x) \neq 0$. Suppose that $\Pi \subset F^*$ satisfies (Π2). If for all $a \in U$ and $i = 1, \ldots, n$ the sets

$$\Omega^{\Pi}_{\mathfrak{w}^a_{\varphi_i}}, \quad \textit{where} \quad \mathfrak{w}^a_{\varphi_i} = \mathfrak{u}^{\rho}_{U} - \varphi_i \cdot a,$$

are $\mathfrak{C}$-superharmonic, then the set

$$\Omega^{\Pi}_{\mathfrak{u}^{\sigma}_{U}}, \quad \textit{where} \quad \sigma = \rho - \vee_{i=1}^{n} |\varphi_i|,$$

is also $\mathfrak{C}$-superharmonic.

Proof. Let U, ρ, φ_i and Γ be as stated. For $a \in U$, we denote $\mathfrak{w}^a_{\varphi_i} = \mathfrak{u}^\rho_U - \varphi_i \cdot a$ and

$$\mathfrak{w} = \bigcap \{\mathfrak{w}^a_{\varphi_i} \mid a \in U,\ i = 1, \ldots, n\}.$$

We claim that $\mathfrak{u}^\sigma_U \subset \mathfrak{w}$, where $\sigma = \rho - \vee^n_{i=1} |\varphi_i|$. We note that the function $\sigma : X \to \overline{\mathbb{R}}$ is also strictly positive and lower semicontinuous. To verify our claim, let $f \in \mathfrak{u}^\sigma_U$, $a \in U$, $i \in \{1, \ldots, n\}$ and $x \in X$. We shall establish that $f(x) + \varphi_i(x)a \in \rho(x)U$. We have $f(x) \in (\rho(x) - |\varphi_i(x)|)\, U$. Our assumptions on U and φ_i guarantee that there is $\gamma_i \in \Gamma$ such that $\varphi_i(x) = \gamma_i |\varphi_i(x)|$ and $\gamma_i a \in U$. Thus

$$\varphi_i(x)a = |\varphi_i(x)|(\gamma_i a) \in |\varphi_i(x)| U$$

and

$$f(x) + \varphi_i(x)a \in (\rho(x) - |\varphi_i(x)|)\, U + |\varphi_i(x)| U = \rho(x)U.$$

This demonstrates that $f + \varphi_i \cdot a \in \mathfrak{u}^\rho_U$ holds for all $a \in U$ and $i = 1, \ldots, n$ and, therefore, that $f \in \mathfrak{w}$. For the second step of our argument, we consider the family

$$\mathfrak{D} = \{\Omega^\Pi_{\mathfrak{w}^a_{\varphi_i}} \mid a \in U,\ i = 1, \ldots, n\}$$

of subsets of $\mathcal{C}(X, E) \otimes F^*$. By our assumption, all sets in $\mathfrak{D}$ are $\mathfrak{C}$-superharmonic. Using the first step of our argument, we realize that

$$\Psi = \bigcap \{\Omega^\Pi_{\mathfrak{w}^a_{\varphi_i}} \in \mathfrak{D}\} \supset \Omega^\Pi_{\mathfrak{u}^\sigma_U}.$$

Thus

$$\mathfrak{p}_{\Omega^\Pi_{\mathfrak{u}^\sigma_U}} \le \mathfrak{p}_\Psi \le \inf\{\mathfrak{p}_{\Omega^\Pi_{\mathfrak{w}^a_{\varphi_i}}} \mid \Omega^\Pi_{\mathfrak{w}^a_{\varphi_i}} \in \mathfrak{D}\}$$

holds pointwise on $\mathfrak{L}(\mathcal{C}_{\mathfrak{V}}(X, E), F)$. We proceed to verify that these expressions coincide at all point evaluations $\mathcal{D}^T_x \in \mathfrak{L}(\mathcal{C}_{\mathfrak{V}}(X, E), F)$. For this, we abbreviate $\beta = \sup\{p_\Pi(T(a)) \mid a \in U\}$ and recall from Lemma 3.2.20 that

$$\mathfrak{p}_{\Omega^\Pi_{\mathfrak{u}^\sigma_U}}(\mathcal{D}^T_x) = \sigma(x)\beta.$$

If $\beta = +\infty$, this shows $\mathfrak{p}_\Psi(\mathcal{D}^T_x) = +\infty$ since $\sigma(x) > 0$ and proves the reverse of the above inequality. If $\beta < +\infty$, we continue as follows. Since $\Omega^\Pi_{\mathfrak{w}^a_{\varphi_i}} \subset \Omega^\Pi_{\mathfrak{u}^\rho_U} + \Omega^\Pi_{\mathfrak{u}^\rho_{\{-\varphi_i \cdot a\}}}$, we have

$$\mathfrak{p}_{\Omega^\Pi_{\mathfrak{w}^a_{\varphi_i}}}(\mathcal{D}^T_x) \le \mathfrak{p}_{\Omega^\Pi_{\mathfrak{u}^\rho_U}}(\mathcal{D}^T_x) + \mathfrak{p}_{\Omega^\Pi_{\{-\varphi_i \cdot a\}}}(\mathcal{D}^T_x) = \rho(x)\beta + \mathfrak{p}_{\Omega^\Pi_{\{-\varphi_i \cdot a\}}}(\mathcal{D}^T_x)$$

for all $i = 1, \ldots, n$ and use Proposition 3.1.9 for

$$\mathfrak{p}_{\Omega^{\Pi}_{\{-\varphi_i \cdot a\}}}(\mathcal{D}_x^T) = \mathfrak{p}_{\Pi}(\varphi_i(x)T(-a)).$$

For every i, there is $\gamma_i \in \Gamma$ such that $\varphi_i(x) = \gamma_i|\varphi_i(x)|$. Hence

$$p_{\Pi}(\varphi_i(x)T(-a)) = |\varphi_i(x)|p_{\Pi}(T(-\gamma_i a)).$$

As $\gamma_i a \in U$ if and only if $a \in U$, for every i we calculate

$$\inf_{a \in U} \mathfrak{p}_{\Omega^{\Pi}_{\mathfrak{w}^a_{\varphi_i}}}(\mathcal{D}_x^T) = \rho(x)\beta + |\varphi_i(x)| \inf\{\mathfrak{p}_{\Pi}(T(-a)) \mid a \in U\}.$$

Because $p_{\Pi}(T(a)) \le \beta < +\infty$ for all $a \in U$, our assumption on the functional p_{Π} yields that $p_{\Pi}(T(-a)) = -p_{\Pi}(T(a))$ and, therefore,

$$\inf\{\mathfrak{p}_{\Pi}(T(-a)) \mid a \in U\} = -\sup\{\mathfrak{p}_{\Pi}(T(a)) \mid a \in U\} = -\beta.$$

Now taking the infimum over all the elements of $\mathfrak{D}$, that is,

$$\inf_{i=1}^{n} \inf_{a \in U} \mathfrak{p}_{\Omega^{\Pi}_{\mathfrak{w}^a_{\varphi_i}}}(\mathcal{D}_x^T)$$

we obtain

$$\inf_{\Omega^{\Pi}_{\mathfrak{w}^a_{\varphi_i}} \in \mathfrak{D}} \mathfrak{p}_{\Omega^{\Pi}_{\mathfrak{w}^a_{\varphi_i}}}(\mathcal{D}_x^T) = \rho(x)\beta - \vee_{i=1}^{n}|\varphi_i(x)|\beta = \sigma(x)\beta = \mathfrak{p}_{\Omega^{\Pi}_{\mathfrak{u}^{\sigma}_U}}(\mathcal{D}_x^T),$$

thus demonstrating that

$$\mathfrak{p}_{\Psi}(\mathcal{D}_x^T) = \inf\{\mathfrak{p}_{\Omega^{\Pi}_{\mathfrak{w}^a_{\varphi_i}}}(\mathcal{D}_x^T) \mid \Omega^{\Pi}_{\mathfrak{w}^a_{\varphi_i}} \in \mathfrak{D}\} = \mathfrak{p}_{\Omega^{\Pi}_{\mathfrak{u}^{\sigma}_U}}(\mathcal{D}_x^T).$$

Now according to Proposition 3.2.18 the set Ψ is $\mathfrak{C}$-superharmonic. So is $\Omega^{\Pi}_{\mathfrak{u}^{\sigma}_U}$, because $\mathcal{T} \prec_{\mathfrak{C}} \mathcal{D}_x^T$ for $\mathcal{T}, \mathcal{D}_x^T \in \mathfrak{L}_{\mathfrak{K}}(C_{\mathfrak{V}}(X,E),F)$ such that $\mathcal{D}_x^T$ is a $\mathfrak{K}$-extremal point evaluation implies that

$$\mathfrak{p}_{\Omega^{\Pi}_{\mathfrak{u}^{\sigma}_U}}(\mathcal{T}) \le \mathfrak{p}_{\Phi}(\mathcal{T}) \le \mathfrak{p}_{\Phi}(\mathcal{D}_x^T) = \mathfrak{p}_{\Omega^{\Pi}_{\mathfrak{u}^{\sigma}_U}}(\mathcal{D}_x^T). \qquad \square$$

We shall formulate the next criterion for the special case that E is a Hilbert space over $\mathbb{R}$ or $\mathbb{C}$ with the inner product $\langle , \rangle$ and unit ball $\mathbb{B}$. For functions $f, g \in \mathcal{F}(X,E)$, we denote the real- or complex-valued function $x \mapsto \langle f(x), g(x)\rangle$ on X by $\langle f, g\rangle$.

3.2.22 Proposition. *Let $\mathfrak{C}$ be a Choquet cone. Let E be a Hilbert space with unit ball $\mathbb{B}$ and let $\rho : X \to \overline{\mathbb{R}}$ be a strictly positive lower semicontinuous function such that $\mathfrak{u}^{\rho}_{\mathbb{B}}$ is a neighborhood in $C_{\mathfrak{V}}(X,E)$. Let $\mu \in F^*$ such that $\Omega^{\{\mu\}}_{u^{\rho}_{\mathbb{B}}} \in \mathfrak{C}$. If for $\mathcal{T} \in \mathfrak{L}(C_{\mathfrak{V}}(X,E),F)$ there is a function $g \in \mathfrak{u}^{\rho}_{\mathbb{B}}$ such that*

$$\Omega^{\{\mu\}}_{\{-g\}} \in \mathfrak{C} \quad \textit{and} \quad \mathfrak{Re}\,\mu(\mathcal{T}(g)) = \mathfrak{p}_{\Omega^{\{\mu\}}_{\mathfrak{u}^{\rho}_{\mathbb{B}}}}(\mathcal{T}) < +\infty,$$

then for every $\mathcal{S} \in \mathfrak{L}(\mathcal{C}_{\mathfrak{V}}(X,E),F)$ *such that* $\mathcal{S} \prec_{\mathfrak{C}} \mathcal{T}$ *we have*

$$\mathfrak{Re}\,\mu(\mathcal{S}(f)) \leq 0$$

for all $f \in \mathcal{C}_{\mathfrak{V}}(X,E)$ *such that* $\mathfrak{Re}\langle f(x), g(x)\rangle \leq 0$ *for all* $x \in Y$, *where* Y *is the compact set* $\{x \in X \mid \|g(x)\| = \rho(x)\}$.

Proof. Unfortunately, the verification of our claim will be rather laborious. We shall formulate our argument for the complex case. Let ρ, g, μ and $\mathcal{S} \prec_{\mathfrak{C}} \mathcal{T}$ be as stated. We shall begin with a few preliminary observations. For every $f \in \mathcal{C}_{\mathfrak{V}}(X,E)$, we have

$$\begin{aligned} -\mathfrak{Re}\,\mu(\mathcal{S}(g)) &= \mathfrak{Re}\,\mu(\mathcal{S}(-g)) = \mathfrak{p}_{\Omega^{\{\mu\}}_{\{-g\}}}(\mathcal{S}) \\ &\leq \mathfrak{p}_{\Omega^{\{\mu\}}_{\{-g\}}}(\mathcal{T}) = \mathfrak{Re}\,\mu(\mathcal{T}(-g)) = -\mathfrak{Re}\,\mu(\mathcal{T}(g)). \end{aligned}$$

That is, $\mathfrak{Re}\,\mu(\mathcal{T}(g)) \leq \mathfrak{Re}\,\mu(\mathcal{S}(g))$. We shall make use of the following facts:

(i) *If for* $f \in \mathcal{C}_{\mathfrak{V}}(X,E)$, *there is* $\alpha \geq 1$ *such that* $f + g \in \alpha\mathfrak{u}^{\rho}_{\mathbb{B}}$, *then* $\mathfrak{Re}\,\mu(\mathcal{S}(f)) \leq (\alpha - 1)\mathfrak{p}_{\Omega^{\Pi}_{\mathfrak{u}^{\rho}_{\mathbb{B}}}}(\mathcal{T})$.

Indeed, if $f + g \in \alpha\mathfrak{u}^{\rho}_{\mathbb{B}}$, then

$$\mathfrak{p}_{\Omega^{\{\mu\}}_{\{f+g\}}}(\mathcal{S}) \leq \alpha\mathfrak{p}_{\Omega^{\{\mu\}}_{\mathfrak{u}^{\rho}_{\mathbb{B}}}}(\mathcal{S}) \leq \alpha\mathfrak{p}_{\Omega^{\{\mu\}}_{\mathfrak{u}^{\rho}_{\mathbb{B}}}}(\mathcal{T}) = \alpha\mathfrak{Re}\,\mu(\mathcal{T}(g)),$$

that is,

$$\mathfrak{Re}\,\mu(\mathcal{S}(f)) + \mathfrak{Re}\,\mu(\mathcal{S}(g)) \leq \alpha\mathfrak{Re}\,\mu(\mathcal{T}(g)) \leq \alpha\mathfrak{Re}\,\mu(\mathcal{S}(g)).$$

Hence our claim. We note that $g \in \mathfrak{u}^{\rho}_{\mathbb{B}}$ implies that $\|g(x)\| \leq \rho(x)$ for all $x \in X$. We define the nonnegative lower semicontinuous function $\sigma : X \to \overline{\mathbb{R}}$ setting $\sigma(x) = \rho(x) - \|g(x)\|$ and observe that

$$Y = \{x \in X \mid \sigma(x) = 0\}$$

is a closed subset of X. Next we shall establish the following:

(ii) *If for* $f \in \mathcal{C}_{\mathfrak{V}}(X,E)$, *there are* $\alpha, \beta \geq 0$ *such that*

$$\|f(x)\| \leq \alpha\sigma(x) + \beta\rho(x)$$

for all $x \in X$, *then* $|\mathfrak{Re}\,\mu(\mathcal{S}(f))| \leq \beta\,\mathfrak{p}_{\Omega^{\Pi}_{\mathfrak{u}^{\rho}_{\mathbb{B}}}}(\mathcal{T})$.

Indeed, we have $\|f(x)\| \leq \alpha(\rho(x) - \|g(x)\|) + \beta\rho(x)$, that is, $\|(f + \alpha g)(x)\| \leq (\alpha + \beta)\rho(x)$ for all $x \in X$. If $\alpha = 0$, this shows $f \in \beta \mathfrak{u}_{\mathbb{B}}^{\rho}$. Hence

$$\mathfrak{Re}\,\mu(\mathcal{S}(f)) \leq \beta\, \mathfrak{p}_{\Omega^{\{\mu\}}_{\mathfrak{u}_{\mathbb{B}}^{\rho}}}(\mathcal{S}) \leq \beta\, \mathfrak{p}_{\Omega^{\{\mu\}}_{\mathfrak{u}_{\mathbb{B}}^{\rho}}}(\mathcal{T}).$$

If $\alpha > 0$, the above shows that

$$\frac{1}{\alpha} f + g \in \left(\frac{\alpha + \beta}{\alpha}\right) \mathfrak{u}_{\mathbb{B}}^{\rho},$$

hence

$$\frac{1}{\alpha}\, \mathfrak{Re}\,\mu(\mathcal{S}(f)) \leq \left(\frac{\alpha + \beta}{\alpha} - 1\right) \mathfrak{p}_{\Omega^{\{\mu\}}_{\mathfrak{u}_{\mathbb{B}}^{\rho}}}(\mathcal{T})$$

and, therefore, $\mathfrak{Re}\,\mu(\mathcal{S}(f)) \leq \beta\, \mathfrak{p}_{\Omega^{\Pi}_{\mathfrak{u}_{\mathbb{B}}^{\rho}}}(\mathcal{T})$ by (i). These arguments also apply to the function $-f$ and, therefore, yield (ii).

Let $f \in \mathcal{C}_{\mathfrak{V}}(X, E)$. We initially assume that $\mathfrak{Re}\langle f(x), g(x)\rangle \leq 0$ holds for all $x \in X$ and shall demonstrate that $\mathfrak{Re}\,\mu(\mathcal{S}(f)) \leq 0$. Indeed, since $\mathfrak{u}_{\mathbb{B}}^{\rho}$ is supposed to be a neighborhood in $\mathcal{C}_{\mathfrak{V}}(X, E)$, hence absorbing, we may assume that $f \in \mathfrak{u}_{\mathbb{B}}^{\rho}$. For any $\lambda \geq 1$, we claim that

$$h = (\lambda - 1)g + \sqrt{\lambda - (1/2)}\, f \in \lambda \mathfrak{u}_{\mathbb{B}}^{\rho}.$$

For this, we shall verify that $\|h(x)\| \leq \lambda\rho(x)$ holds for all $x \in X$. If $g(x) = 0$ or $\lambda = 1$, this is evident since $\|h(x)\| = \sqrt{\lambda - (1/2)}\, \|f\| \leq \lambda\rho(x)$ in this case. If $g(x) \neq 0$ and $\lambda > 1$, we set $m(x) = g(x)/\|g(x)\|$ and

$$\begin{aligned} a(x) &= \langle h(x), m(x)\rangle\, m(x) \\ &= (\lambda - 1)g(x) + (\sqrt{\lambda - (1/2)}\, \langle f(x), m(x)\rangle) m(x), \end{aligned}$$

that is, the orthogonal projection of $h(x)$ onto the linear span of $g(x)$ in E, and

$$b(x) = h(x) - a(x) = \sqrt{\lambda - (1/2)}(f(x) - \langle f(x), m(x)\rangle\, m(x)).$$

We have $\|h(x)\|^2 = \|a(x)\|^2 + \|b(x)\|^2$. It is now straightforward to establish that

$$\|b(x)\|^2 \leq (\lambda - (1/2))\|f(x)\|^2 \leq (\lambda - (1/2))\rho(x)^2.$$

Furthermore, since $g(x) = \|g(x)\|\, m(x)$ and since $\mathfrak{Re}\langle f(x), m(x)\rangle \leq 0$, we calculate

$$\begin{aligned} \|a(x)\|^2 &= |(\lambda - 1)\|g(x)\| + \sqrt{\lambda - (1/2)}\langle f(x), m(x)\rangle|^2 \\ &= (\lambda - 1)^2\|g(x)\|^2 + 2(\lambda - 1)\|g(x)\|\sqrt{\lambda - (1/2)}\, \mathfrak{Re}\langle f(x), m(x)\rangle \end{aligned}$$

$$
\begin{aligned}
&+ (\lambda - (1/2))|\langle f(x), m(x)\rangle|^2 \\
&\le (\lambda-1)^2\|g(x)\|^2 + (\lambda - (1/2))\|f(x)\|^2 \\
&\le (\lambda-1)^2\rho(x)^2 + (\lambda - (1/2))\rho(x)^2 \\
&\le (\lambda^2 - \lambda + (1/2))\rho(x)^2.
\end{aligned}
$$

Thus indeed, $\|h(x)\|^2 \le \lambda^2\rho(x)^2$. This verifies that $h \in \lambda\mathfrak{u}_{\mathbb{B}}^{\rho}$ and

$$(\sqrt{\lambda - (1/2)}/(\lambda-1))f + g \in (\lambda/(\lambda-1))\mathfrak{u}_{\mathbb{B}}^{\rho}.$$

Using statement (i), we infer that

$$\frac{\sqrt{\lambda-(1/2)}}{(\lambda-1)}\mathfrak{Re}\,\mu(\mathcal{S}(f)) \le \left(\frac{\lambda}{\lambda-1} - 1\right)\mathfrak{p}_{\mathfrak{Q}^{\Pi}_{\mathfrak{u}_{\mathbb{B}}^{\rho}}}(\mathcal{T}),$$

that is,

$$\sqrt{\lambda-(1/2)}\,\mathfrak{Re}\,\mu(\mathcal{S}(f)) \le \mathfrak{p}_{\mathfrak{Q}^{\Pi}_{\mathfrak{u}_{\mathbb{B}}^{\rho}}}(\mathcal{T}).$$

The latter holds for all $\lambda \ge 1$ and, therefore, renders $\mathfrak{Re}\,\mu(\mathcal{S}(f)) \le 0$. This completes the initial step in the argument for our claim.

Now in the second step of the argument we suppose that $f \in \mathfrak{u}_{\mathbb{B}}^{\rho}$ and that

$$\mathfrak{Re}\langle f(x), g(x)\rangle \le 0$$

holds for all $x \in Y$. Both functions f and g vanish at infinity relative to the neighborhood $\mathfrak{u}_{\mathbb{B}}^{\rho}$. Thus there is a compact subset K of X such that both $\|f(x)\| \le (1/2)\rho(x)$ and $\|g(x)\| \le (1/2)\rho(x)$ for all $x \in X \setminus K$. Let Z be another compact subset of X whose topological interior U contains K, and let $\alpha > 0$ be the minimum value of the lower semicontinuous function ρ, and β the maximum value of the continuous function $x \mapsto \|f(x)\|$ on Z. Let

$$B = \{x \in Z \mid \sigma(x) \le \alpha/2\} \quad \text{and} \quad O = \{x \in U \mid \|g(x)\| > \alpha/4\}.$$

Clearly, B is compact, O is open and $Y \subset B \subset O \subset Z$. The latter follows since for any $x \in B$ we have $\rho(x) - \|g(x)\| \le \alpha/2$, hence $\|g(x)\| \ge \rho(x) - \alpha/2 \ge \alpha/2$. The above implies in particular that Y is compact, one of our assertions. There is a function $\varphi \in \mathcal{C}_{\mathcal{K}}(X)$ such that $\chi_B \le \varphi \le \chi_O$. Setting $h = \varphi \cdot f$ and $k = (1-\varphi)\cdot f$, both of these functions are contained in $\mathcal{C}_{\mathfrak{V}}(X, E)$, we have

$$\mathfrak{Re}\,\mu(S(f)) = \mathfrak{Re}\,\mu(S(h)) + \mathfrak{Re}\,\mu(S(k))$$

and claim that $\mathfrak{Re}\,\mu(S(k)) = 0$. Indeed, $k(x) = 0$ if $x \in B$,

$$\|k(x)\| \leq \|f(x)\| \leq (1/2)\rho(x) \leq \rho(x) - \|g(x)\| = \sigma(x)$$

if $x \in X \setminus K$ and

$$\|k(x)\| \leq \|f(x)\| \leq \beta = (2\beta/\alpha)(\alpha/2) \leq (2\beta/\alpha)\sigma(x)$$

if $x \in K \setminus B$. Setting $\lambda = \max\{1, (2\beta/\alpha)\}$, we have $\|k(x)\| \leq \lambda\sigma(x)$ for all $x \in X$, and our claim follows from (ii). We infer that $\mathfrak{Re}\, \mu(S(f)) = \mathfrak{Re}\, \mu(S(h))$ and observe that $\chi_{X \setminus O} \cdot h = 0$. Next, we define the function l setting

$$l(x) = \frac{\mathfrak{Re}\langle h(x), g(x)\rangle \vee 0}{\|g(x)\|^2} g(x) \quad \text{for } x \in O, \text{ and } l(x) = 0 \text{ else.}$$

Since $\|l(x)\| \leq \|h(x)\|$ holds for all $x \in X$, the function l is continuous at all points $x \in X$ where $h(x) = 0$, in particular, at all $x \in X \setminus O$. Moreover, since $\|g(x)\| > 0$ holds for are $x \in O$, the function l is also continuous at the points in O. Its support is contained in the compact set Z, and l is seen to be an element of $\mathcal{C}_{\mathfrak{V}}(X, E)$. Because f and h coincide on Y, we note that $\mathfrak{Re}\langle h(x), g(x)\rangle \leq 0$, hence $l(x) = 0$ holds for all $x \in Y$. For the function $h - l$, we realize that

$$\mathfrak{Re}\big\langle (h - l)(x), g(x)\big\rangle = \mathfrak{Re}\big\langle h(x), g(x)\big\rangle - \big(\mathfrak{Re}\big\langle h(x), g(x)\big\rangle \vee 0\big) \leq 0$$

for all $x \in X$. This implies $\mathfrak{Re}\, \mu(S(h - l)) \leq 0$ by the first step of our argument. Hence

$$\mathfrak{Re}\, \mu\big(S(f)\big) = \mathfrak{Re}\, \mu\big(S(h)\big) \leq \mathfrak{Re}\, \mu\big(S(l)\big).$$

All left to show is that $\mathfrak{Re}\, \mu(S(l)) \leq 0$. Given $\varepsilon > 0$, we claim that for every $x \in Z$ there is $\lambda_x \geq 0$ such that $\|l(y)\| < \lambda_x \sigma(y) + \varepsilon\rho(y)$ holds for all y in an open neighborhood U_x of x. We note that the mapping $x \mapsto \|l(x)\|$ is continuous. Our claim is obvious for $x \in Y$ since $\|l(x)\| = \sigma(x) = 0$ and $\rho(x) > 0$ in this case. If $x \in Z \setminus Y$, then $\sigma(x) > 0$, and we find $\lambda_x \geq 0$ such that $\|l(x)\| < \lambda_x \sigma(x)$. Since σ is lower semicontinuous, this relation holds for all elements in an open neighborhood U_x of x as stated. The compact set Z is covered by a finite number U_{x_i}, for $i = 1, \ldots, n$, and if we set $\lambda = \bigvee_{i=1}^{n} \lambda_{x_i}$, then $\|l(x)\| \leq \lambda\sigma(x) + \varepsilon\rho(x)$ holds for all $x \in Z$, and indeed for all $x \in X$ since l is supported by Z. Now statement (ii) from above renders that

$$\big|\mathfrak{Re}\, \mu\big(\mathcal{S}(l)\big)\big| \leq \varepsilon\, \mathfrak{p}_{\Omega^{\{\mu\}}_{u^{\rho}_{\mathbb{B}}}}(\mathcal{T}).$$

The latter holds true for all $\varepsilon > 0$ and shows that $\mathfrak{Re}\, \mu(\mathcal{S}(f)) \leq \mathfrak{Re}\, \mu(\mathcal{S}(l)) = 0$. Hence our claim. □

Let Π be a subset of F^* such that its linear span is weak*-dense in F^*. This condition implies that $c = 0$ for $c \in F$ whenever $\mathfrak{Re}\, \mu(\gamma c) \leq 0$ for all $\mu \in \Pi$ and $\gamma \in \mathbb{C}$, because otherwise we could find $\mu \in \Pi$ such that $\mu(c) \neq 0$ and $\gamma \in \mathbb{C}$ such that

$\mathfrak{Re}\,\mu(\gamma c) > 0$. Suppose that for every $\mu \in \Pi$ there is a function $g_\mu \in \mathfrak{u}_{\mathbb{B}}^{\rho}$ as in the assumptions of Proposition 3.2.22, that is, $\Omega_{\{-g_\mu\}}^{\{\mu\}} \in \mathfrak{C}$ and

$$\mathfrak{Re}\,\mu(\mathcal{T}(g_\mu)) = \mathfrak{p}_{\Omega_{\mathfrak{u}_{\mathbb{B}}^{\rho}}^{\{\mu\}}}(\mathcal{T}) < +\infty,$$

and let Y be the closure in X of the set

$$\left\{x \in X \mid \|g_\mu(x)\| = \rho(x) \text{ for some } \mu \in \Pi\right\}.$$

Let $\mathcal{S} \prec_{\mathfrak{C}} \mathcal{T}$ and let θ be the representation measure for $\mathcal{S}$. If $f \in C_{\mathfrak{V}}(X,E)$ is supported by $X \setminus Y$, then $\langle f(x), g(x)\rangle = 0$ holds for all $x \in Y$, and we have

$$\mathfrak{Re}\,\mu(\gamma\mathcal{S}(f)) = \mathfrak{Re}\,\mu(\mathcal{S}(\gamma f)) \leq 0$$

for all $\mu \in \Pi$ and all $\gamma \in \mathbb{C}$ by Proposition 3.2.22. Hence $\mathcal{S}(f) = 0$ by the above. Now using Proposition 2.5.9 we conclude that θ is supported by Y. We formulate this observation as a corollary.

3.2.23 Corollary. *Let $\mathfrak{C}$ be a Choquet cone. Let E be a Hilbert space with unit ball $\mathbb{B}$ and let $\rho : X \to \overline{\mathbb{R}}$ be a strictly positive lower semicontinuous function such that $\mathfrak{u}_{\mathbb{B}}^{\rho}$ is a neighborhood in $C_{\mathfrak{V}}(X,E)$. Let $\Pi \subset F^*$ such that its linear span is weak*-dense in F^* and $\Omega_{u_{\mathbb{B}}^{\rho}}^{\{\mu\}} \in \mathfrak{C}$ for all $\mu \in \Pi$ and let $\mathcal{T} \in \mathfrak{L}(C_{\mathfrak{V}}(X,E),F)$. If for every $\mu \in \Pi$ there is a function $g_\mu \in \mathfrak{u}_{\mathbb{B}}^{\rho}$ such that*

$$\Omega_{\{-g_\mu\}}^{\{\mu\}} \in \mathfrak{C} \quad \textit{and} \quad \mathfrak{Re}\,\mu(\mathcal{T}(g_\mu)) = \mathfrak{p}_{\Omega_{\mathfrak{u}_{\mathbb{B}}^{\rho}}^{\{\mu\}}}(\mathcal{T}) < +\infty,$$

then the representation measure for any operator $\mathcal{S} \in \mathfrak{L}(C_{\mathfrak{V}}(X,E),F)$ such that $\mathcal{S} \prec_{\mathfrak{C}} \mathcal{T}$ is supported by the closure of the set

$$\left\{x \in X \mid \|g_\mu(x)\| = \rho(x) \text{ for some } \mu \in \Pi\right\}.$$

3.3 Special cases and examples

This final section provides a range of examples and applications of the concepts developed in the preceding sections. We begin with the classical origins of Choquet theory (see [1] and [41]).

The case that X is compact and that $E = F = \mathbb{R}$

Let X be a compact Hausdorff space and let $E = F = \mathbb{R}$. We endow $C(X)$ with the topology of uniform convergence with the unit ball $\mathbb{B}$. With these insertions, we have

$\mathfrak{L}(\mathcal{C}_{\mathfrak{V}}(X,E),F) = \mathcal{C}(X)^*$ endowed with its weak* topology, and $\mathcal{C}_{\mathfrak{V}}(X,E) \otimes F^* = \mathcal{C}(X)$ endowed with its weak topology. In order to model classical Choquet theory, we use the Choquet core $\mathfrak{K} = \{\mathfrak{u}\}$, where

$$\mathfrak{u} = \{f \in \mathcal{C}(X) \mid f \leq 1\}.$$

The polar of $\mathfrak{u}$ is weak*-compact (see Corollary 1.3.9 with $S = [-\infty, 1)$). Let H be a subspace of $\mathcal{C}_{\mathfrak{V}}(X,E)$ that contains the constant functions and separates the points of X. The suitable choice for our Choquet cone is

$$\mathfrak{C} = \{\alpha\mathfrak{u} + f \mid \alpha \geq 0,\, f \in H\}.$$

The operators in $\mathfrak{L}_{\mathfrak{K}}(\mathcal{C}_{\mathfrak{V}}(X,E),F)$ are positive linear functionals in this case, and $\mathcal{S} \prec_{\mathfrak{C}} \mathcal{T}$ for $\mathcal{S},\mathcal{T} \in \mathfrak{L}_{\mathfrak{K}}(\mathcal{C}_{\mathfrak{V}}(X,E),F)$ means that $\mathcal{S}(f) = \mathcal{T}(f)$ for all $f \in H$. Let

$$C = \left\{\bigwedge_{i=1}^{n} f_i \mid f_i \in L\right\}.$$

It is straightforward to check that all the sets $\{f\}$ for $f \in C$ are $\mathfrak{C}$-superharmonic. Indeed, let $f = \wedge_{i=1}^{n} f_i \in C$ and let $\mathcal{T}, \mathcal{D}_x \in \mathfrak{L}_{\mathfrak{K}}(\mathcal{C}_{\mathfrak{V}}(X,E),F)$ such that $\mathcal{T} \prec_{\mathfrak{C}} \mathcal{D}_x$. Then $\mathcal{T}(f_i) = f_i(x)$ for all $i = 1, \dots, n$ hence

$$\mathcal{T}(f) \leq \bigwedge_{i=1}^{n} T(f_i) \leq \bigwedge_{i=1}^{n} f_i(x) = f(x),$$

that is, $\mathfrak{p}_{\{f\}}(\mathcal{T}) \leq \mathfrak{p}_{\{f\}}(\mathcal{D}_x)$. The order $\prec_{\tilde{C}}$ then leads to the classical approach to Choquet theory.

Sample settings in the general case

In this sample setting, the Choquet core $\mathfrak{K}$ is constructed in the following way. Let $\mathfrak{U}$ be a nonempty family of $\mathcal{C}(X)$-convex, but not necessarily balanced neighborhoods of 0 in $\mathcal{C}_{\mathfrak{V}}(X,E)$, such that for all $\mathfrak{u}, \mathfrak{w} \in \mathfrak{U}$ and $\alpha \geq 0$ there is $\beta \geq \alpha$ such that $\mathfrak{u} \cap (\beta\mathfrak{w}) \in \mathfrak{U}$, and let Π be a subset of F^* such that the balanced core $\operatorname{bcore}(\Pi^\circ)$ of its polar in F is weakly compact. We set

$$\mathfrak{K} = \{\Omega_{\mathfrak{u}}^{\Pi} \mid \mathfrak{u} \in \mathfrak{U}\}$$

and verify the conditions for a Choquet core from the beginning of Section 3.2. First, for every $\mathfrak{u} \in \mathfrak{U}$ according to Corollary 1.3.9 the set $\mathfrak{L}_{\mathfrak{u}}^{\Pi^\circ}(\mathcal{C}_{\mathfrak{V}}(X,E),F)$ is compact in the weak operator topology of $\mathfrak{L}(\mathcal{C}_{\mathfrak{V}}(X,E),F)$. Hence, keeping with Proposition 3.1.10 the set $\mathfrak{H}_{\Omega_{\mathfrak{u}}^{\Pi}}$, that is, the polar of $\Omega_{\mathfrak{u}}^{\Pi}$ in $\mathfrak{L}(\mathcal{C}_{\mathfrak{V}}(X,E),F)$, is also compact. Next, for the boundedness condition let $\Omega_{\mathfrak{u}}^{\Pi}, \Omega_{\mathfrak{w}}^{\Pi} \in \mathfrak{K}$ and let $\omega \in \Omega_{\mathfrak{u}}^{\Pi}$, that is, $\omega = \sum_{i=1}^{n}(\varphi_i \cdot f_i) \otimes \mu_i$

for $f_i \in \mathfrak{u}$, $\mu_i \in \Pi$ and $0 \le \varphi_i \in \mathcal{C}(X)$ such that $\sum_{i=1}^n \varphi_i = 1$. There is $\alpha \ge 0$ such that $f_i \in \alpha\mathfrak{w}$ for all $i = 1, \dots, n$. Hence by our condition on $\mathfrak{U}$ there is $\beta \ge \alpha$ such that $\mathfrak{s} = \mathfrak{u} \cap (\beta\mathfrak{w}) \in \mathfrak{U}$. Then

$$\Omega_{\mathfrak{s}}^{\Pi} \in \mathfrak{K}, \quad \omega \in \Omega_{\mathfrak{s}}^{\Pi}, \quad \Omega_{\mathfrak{s}}^{\Pi} \subset \Omega_{\mathfrak{u}}^{\Pi} \quad \text{and} \quad \Omega_{\mathfrak{s}}^{\Pi} \subset \beta\Omega_{\mathfrak{w}}^{\Pi}.$$

This proves our claim. We remark that the weak*-closure of the balanced convex hull $\mathrm{bconv}(\Pi)$ of Π is the polar of the balanced core $\mathrm{bcore}(\Pi^\circ)$ of Π° in F, which is supposed to be weakly compact. But the polar of a weakly compact set is known to be absorbing. This implies in particular that the linear span of Π is weak*-dense in F^* as required in some of the results of the preceding sections.

Let $\mathfrak{C}_0$ be a collection of nonempty subsets $\Omega_{\mathfrak{w}}^{\Gamma}$ of $\mathcal{C}_{\mathfrak{V}}(X,E) \otimes F^*$ such that $\Gamma \subset \Pi$ and $\mathfrak{w} \subset \mathcal{C}_{\mathfrak{V}}(X,E)$ is $\mathcal{C}(X)$-convex and $\mathfrak{U}$-bounded, that is, for every $\mathfrak{u} \in \mathfrak{U}$ there is $\lambda \ge 0$ such that $\mathfrak{w} \subset \lambda\mathfrak{u}$. Every such set $\Omega_{\mathfrak{w}}^{\Gamma}$ is obviously $\mathfrak{K}$-bounded. For the Choquet cone $\mathfrak{C}$, we choose the cone of subsets of $\mathcal{C}_{\mathfrak{V}}(X,E) \otimes F^*$ generated by $\mathfrak{K}$ and $\mathfrak{C}_0$.

We notice that all operators $\mathcal{T} \in \mathfrak{L}_{\mathfrak{K}}(\mathcal{C}_{\mathfrak{V}}(X,E),F)$ are weakly compact in this setting. Hence their representation measures are $\mathfrak{L}(E,F)$-valued (Theorem 2.6.1). Indeed, if $\mathfrak{p}_{\Omega_{\mathfrak{u}}^{\Pi}}(\mathcal{T}) < +\infty$ for $\Omega_{\mathfrak{u}}^{\Pi} \in \mathfrak{K}$, then

$$\mathcal{T} \in \lambda\mathfrak{H}_{\Omega_{\mathfrak{u}}^{\Pi}} \subset \lambda\mathfrak{L}_{\mathfrak{u}}^{\Pi^\circ}(\mathcal{C}_{\mathfrak{V}}(X,E),F)$$

for some $\lambda \ge 0$, and Corollary 1.3.9 applies. As a consequence, all representation measures of operators in $\mathfrak{L}_{\mathfrak{K}}(\mathcal{C}_{\mathfrak{V}}(X,E),F)$ are $\mathfrak{L}(E,F)$-valued and the set Π is θ-dense in F^* with respect to any such measure θ.

According to Corollary 3.1.8, a point evaluation $\mathcal{D}_x^T \in \mathfrak{L}(\mathcal{C}_{\mathfrak{V}}(X,E),F)$ is an element of the Choquet boundary

$$\Delta_{\mathfrak{C}}^{\Omega_{\mathfrak{u}}^{\Pi}} \quad \text{for} \quad \Omega_{\mathfrak{u}}^{\Pi} \in \mathfrak{K}$$

if and only if it is minimal, if $\mathfrak{p}_{\Omega_{\mathfrak{u}}^{\Pi}}(\mathcal{D}_x^T) < +\infty$ and if T is an extreme point of the set

$$\{S \in \mathfrak{L}(E,F) \mid \mathfrak{p}_{\Omega_{\mathfrak{u}}^{\Pi}}(\mathcal{D}_x^S) \le \mathfrak{p}_{\Omega_{\mathfrak{u}}^{\Pi}}(\mathcal{D}_x^T)\}.$$

We continue to formulate our observations for the case that both E and F are vector spaces over $\mathbb{C}$. The real case is similar.

3.3.1 Proposition. *Let $\mathfrak{K}$ and $\mathfrak{C}$ be as specified and suppose in addition that all sets in $\mathfrak{U}$ are balanced and that $\Omega_{\gamma\mathfrak{w}}^{\Gamma} \in \mathfrak{C}_0$ for all $|\gamma| = 1$ whenever $\Omega_{\mathfrak{w}}^{\Gamma} \in \mathfrak{C}_0$. Then the following hold:*

(a) *If $\Omega \in \mathfrak{C}$, then $\alpha\Omega \in \mathfrak{C}$ for all $\alpha \in \mathbb{C}$.*

(b) *If $\mathcal{S} \prec_{\mathfrak{C}} \mathcal{T}$ for $\mathcal{S}, \mathcal{T} \in \mathfrak{L}(\mathcal{C}_{\mathfrak{V}}(X,E),F)$, then $\alpha\mathcal{S} \prec_{\mathfrak{C}} \alpha\mathcal{T}$ for all $\alpha \in \mathbb{C}$.*

(c) *If $\Phi \subset \mathcal{C}_{\mathfrak{V}}(X,E) \otimes F^*$ is $\mathfrak{C}$-superharmonic, then $\alpha\Phi$ is $\mathfrak{C}$-superharmonic for all $\alpha \in \mathbb{C}$.*

(d) *$\mathfrak{L}_{\mathfrak{K}}(\mathcal{C}_{\mathfrak{V}}(X,E),F)$ is a subspace of $\mathfrak{L}(\mathcal{C}_{\mathfrak{V}}(X,E),F)$.*

Proof. According to Lemma 3.2.1, $\mathfrak{L}_{\mathfrak{K}}(C_{\mathfrak{V}}(X,E),F)$ is a subcone of the operator space $\mathfrak{L}(C_{\mathfrak{V}}(X,E),F)$. Thus for Part (d), we only have to validate that $\alpha\mathcal{T} \in \mathfrak{L}_{\mathfrak{K}}(C_{\mathfrak{V}}(X,E),F)$ whenever $\mathcal{T} \in \mathfrak{L}_{\mathfrak{K}}(C_{\mathfrak{V}}(X,E),F)$ and $\alpha \in \mathbb{C}$. This, as well as the statements in (a), (b) and (c), is obvious for $\alpha = 0$. If $\alpha \neq 0$, we set $\gamma = \alpha/|\alpha|$ and argue as follows. For Part (a), let $\Omega \in \mathfrak{C}$, that is,

$$\Omega = \sum_{i=1}^{n} \kappa_i \Omega^{\Pi}_{\mathfrak{u}_i} + \sum_{k=1}^{m} \lambda_k \Omega^{\Gamma_k}_{\mathfrak{w}_k}$$

for $\Omega^{\Pi}_{\mathfrak{u}_i} \in \mathfrak{K}$, $\Omega^{\Gamma_k}_{\mathfrak{w}_k} \in \mathfrak{C}_0$ and $\kappa_i, \lambda_k \geq 0$. We observe that

$$\alpha\Omega^{\Pi}_{\mathfrak{u}_i} = |\alpha|\Omega^{\Pi}_{\mathfrak{u}_i} \quad \text{and} \quad \alpha\Omega^{\Gamma_k}_{\mathfrak{w}_k} = \Omega^{\Gamma_k}_{\alpha\mathfrak{w}_k} = |\alpha|\Omega^{\Gamma_k}_{\gamma\mathfrak{w}_k}$$

since the sets $\mathfrak{u}_i$ are supposed to be balanced. Thus

$$\alpha\Omega = |\alpha|\left(\sum_{i=1}^{n} \kappa_i \Omega^{\Pi}_{\mathfrak{u}_i} + \sum_{k=1}^{m} \lambda_k \Omega^{\Gamma_k}_{\gamma\mathfrak{w}_k}\right) \in \mathfrak{C}$$

by our assumption on $\mathfrak{C}_0$. For Part (b), suppose that $\mathcal{S} \prec_{\mathfrak{C}} \mathcal{T}$ for $\mathcal{S},\mathcal{T} \in \mathfrak{L}(C_{\mathfrak{V}}(X,E),F)$ and let $\alpha \in \mathbb{C}$. Then for every $\Omega \in \mathfrak{C}$ we have

$$\mathfrak{p}_{\Omega}(\alpha\mathcal{S}) = \mathfrak{p}_{\alpha\Omega}(\mathcal{S}) \leq \mathfrak{p}_{\alpha\Omega}(\mathcal{T}) = \mathfrak{p}_{\Omega}(\alpha\mathcal{T}).$$

Thus $\alpha\mathcal{S} \prec_{\mathfrak{C}} \alpha\mathcal{T}$. For Part (d), let $\mathcal{T} \in \mathfrak{L}_{\mathfrak{K}}(C_{\mathfrak{V}}(X,E),F)$. There is $\Omega^{\Pi}_{\mathfrak{u}} \in \mathfrak{K}$ such that $\mathfrak{p}_{\Omega^{\Pi}_{\mathfrak{u}}}(\mathcal{T}) < +\infty$. Since

$$\mathfrak{p}_{\bar{\gamma}\Omega^{\Pi}_{\mathfrak{u}}}(\alpha\mathcal{T}) = |\alpha|\mathfrak{p}_{\bar{\gamma}\Omega^{\Pi}_{\mathfrak{u}}}(\gamma\mathcal{T}) = |\alpha|\mathfrak{p}_{\Omega^{\Pi}_{\mathfrak{u}}}(\mathcal{T}) < +\infty,$$

we infer that $\alpha\mathcal{T} \in \mathfrak{L}_{\mathfrak{K}}(C_{\mathfrak{V}}(X,E),F)$. For Part (c), suppose that $\Phi \subset C_{\mathfrak{V}}(X,E) \otimes F^*$ is $\mathfrak{C}$-superharmonic and let $\mathcal{T} \prec_{\mathfrak{C}} \mathcal{D}^T_x$ for $\mathcal{T},\mathcal{D}^T_x \in \mathfrak{L}_{\mathfrak{K}}(C_{\mathfrak{V}}(X,E),F)$ such that $\mathcal{D}^T_x$ is a $\mathfrak{K}$-extremal point evaluation. Then $\alpha\mathcal{T},\alpha\mathcal{D}^T_x \in \mathfrak{L}_{\mathfrak{K}}(C_{\mathfrak{V}}(X,E),F)$ by Part (d) and $\alpha\mathcal{D}^T_x$ is also a $\mathfrak{K}$-extremal point evaluation. We have $\alpha\mathcal{T} \prec_{\mathfrak{C}} \alpha\mathcal{D}^T_x$ by Part (b) and, therefore,

$$\mathfrak{p}_{\alpha\Phi}(\mathcal{T}) = \mathfrak{p}_{\Phi}(\alpha\mathcal{T}) \leq \mathfrak{p}_{\Phi}(\alpha\mathcal{D}^T_x) = \mathfrak{p}_{\alpha\Phi}(\mathcal{D}^T_x)$$

since Φ is $\mathfrak{C}$-superharmonic. Hence our claim. □

We prepare to further specify our settings for $\mathfrak{C}_0$, hence for $\mathfrak{C}$. Let $\mathcal{H}$ be a subcone of $C_{\mathfrak{V}}(X,E)$. Let $V \in \mathcal{V}$ and let ρ be a strictly positive $\overline{\mathbb{R}}_+$-valued lower semicontinuous function on X, such that $\mathfrak{u}^{\rho}_V$ is a $\mathfrak{U}$-bounded neighborhood in $C_{\mathfrak{V}}(X,E)$. For a subspace G of E, we denote

$$H_G = \{\varphi \in C(X,\mathbb{C}) \mid \varphi \cdot a \in \mathcal{H} \text{ for all } a \in G\}$$

and

$$C_G = \left\{ \bigvee_{i=1}^{n} |\varphi_i| \mid \varphi_i \in H_G \right\}.$$

Clearly, H_G is a subspace of $\mathcal{C}(X,\mathbb{C})$, C_G is a subcone and $C_G - C_G$, that is, the subspace spanned by C_G, is a sublattice of $\mathcal{C}(X)$. Let $\mu \in \Pi$. We consider the $\mathcal{C}(X)$-convex subsets $\Omega^{\{\mu\}}_{\mathfrak{u}^{\rho}_{V\cap G}}$ of $C_{\mathfrak{V}}(X,E) \otimes F^*$, which are $\mathfrak{K}$-bounded, and the functionals on $\mathfrak{L}(C_{\mathfrak{V}}(X,E),F)$ that are defined by these sets. If θ is the representation measure for an operator $\mathcal{T} \in \mathfrak{L}(C_{\mathfrak{V}}(X,E),F)$, we recall from Proposition 2.6.8 that its $\mathfrak{L}(G,F^{**})$-valued restriction $\theta\|_G$ to G represents the restriction of $\mathcal{T}$ to $C_{\mathfrak{V}}(X,G)$. The latter coincides with θ on all sets $A \in \mathfrak{R}$ and all elements of the subspace G of E. (See Proposition 2.6.8 and the remarks at the end of Section 3.1.) Using Propositions 2.3.11 and 3.1.14 and Corollary 3.1.12, we calculate that

$$\mathfrak{p}_{\Omega^{\{\mu\}}_{\mathfrak{u}^{\rho}_{V\cap G}}}(\mathcal{T}) = \sup\{\mathfrak{Re}\,\mu(\mathcal{T}(f)) \mid f \in \mathfrak{u}^{\rho}_{V\cap G}\} = \int_X \rho\, d[|\theta\|_G|(V\cap G)(\mu)].$$

Since the integral is additive, the above implies in particular that

$$\mathfrak{p}_{\Omega^{\{\mu\}}_{\mathfrak{u}^{\rho+\sigma}_{V\cap G}}}(\mathcal{T}) = \mathfrak{p}_{\Omega^{\{\mu\}}_{\mathfrak{u}^{\rho}_{V\cap G}}}(\mathcal{T}) + \mathfrak{p}_{\Omega^{\{\mu\}}_{\mathfrak{u}^{\sigma}_{V\cap G}}}(\mathcal{T})$$

holds for strictly positive $\overline{\mathbb{R}}_+$-valued lower semicontinuous functions ρ and σ.

Employing these concepts, we shall specify our choice for the family $\mathfrak{C}_0$ of subsets of $C_{\mathfrak{V}}(X,E) \otimes F^*$ to be used in our arrangement. Let $\mathcal{G}$ be a family of subspaces of E and for each $G \in \mathcal{G}$ we choose a neighborhood $V_G \in \mathcal{V}$ and a function ρ_G as above. For the sake of readability, we shall omit the space G in the notation for V and ρ in the sequel. This dependency will be evident from the context. We are now ready for the completion of our setup. For $\mathfrak{C}_0$, we choose the following subsets of $C_{\mathfrak{V}}(X,E) \otimes F^*$:

(i) all sets $\Omega^{\{\mu\}}_{\{f\}}$ for $f \in \mathcal{H}$ and $\mu \in \Pi$ and

(ii) all sets $\Omega^{\{\mu\}}_{\mathfrak{u}^{\rho}_{V\cap G}}$ for $G \in \mathcal{G}$ and $\mu \in \Pi$.

We take note that

$$\mathfrak{p}_{\Omega^{\{\mu\}}_{\{f\}}}(\mathcal{T}) = \mathfrak{Re}\,\mu(\mathcal{T}(f)) = (\mathcal{T}(f))(\mu)$$

holds for all $\mathcal{T} \in \mathfrak{L}(C_{\mathfrak{V}}(X,E),F)$ (Proposition 3.1.11). We shall use these settings throughout the remainder of this subsection.

3.3.2 Proposition. *Suppose that $\mathcal{S} \prec_{\mathfrak{C}} \mathcal{T}$ for $\mathcal{S},\mathcal{T} \in \mathfrak{L}(C_{\mathfrak{V}}(X,E),F)$.*

(a) *If $\mathcal{H}$ is a subspace of $C_{\mathfrak{V}}(X,E)$, then $\mathcal{S}(f) = \mathcal{T}(f)$ for all $f \in \mathcal{H}$.*

(b) *If Π absorbs all elements of F^*, then $\mathcal{S}(\mathfrak{u}^{\rho}_{V\cap G}) \subset \mathcal{T}(\mathfrak{u}^{\rho}_{V\cap G})$ for all $G \in \mathcal{G}$.*

Proof. Suppose that $\mathcal{S} \prec_{\mathfrak{C}} \mathcal{T}$ for $\mathcal{S},\mathcal{T} \in \mathfrak{L}(\mathcal{C}_{\mathfrak{V}}(X,E),F)$. (a) For $f \in \mathcal{H}$, we have

$$\mathfrak{Re}\, \alpha\mu(\mathcal{S}(f)) = \mathfrak{p}_{\Omega^{\{\mu\}}_{\{\alpha f\}}}(\mathcal{S}) \le \mathfrak{p}_{\Omega^{\{\mu\}}_{\alpha\{\alpha f\}}}(\mathcal{T}) = \mathfrak{Re}\, \alpha\mu(\mathcal{T}(f))$$

for all $\alpha \in \mathbb{C}$ and $\mu \in \Pi$, that is, $\mu(\mathcal{S}(f)) = \mu(\mathcal{T}(f))$ for all $\mu \in \Pi$ and, therefore, $\mathcal{S}(f) = \mathcal{T}(f)$ since the linear span of Π is weak*-dense in F^*. (b) If $\mathcal{S}(\mathfrak{u}^{\rho}_{V\cap G}) \not\subset \overline{\mathcal{T}(\mathfrak{u}^{\rho}_{V\cap G})}$ for $G \in \mathcal{G}$, then there are $f \in \mathfrak{u}^{\rho}_{V\cap G}$ and $\nu \in F^*$ such that

$$\mathfrak{Re}\, \nu(\mathcal{S}(f)) > \sup\{\mathfrak{Re}\, \nu(\mathcal{T}(g)) \mid g \in \mathfrak{u}^{\rho}_{V\cap G}\} = \mathfrak{p}_{\Omega^{\{\mu\}}_{\mathfrak{u}^{\rho}_{V\cap G}}}(\mathcal{T}).$$

By our assumption, there are $\mu \in \Pi$ and $\lambda \ge 0$ such that $\nu = \lambda\mu$. Then

$$\mathfrak{p}_{\Omega^{\{\mu\}}_{\mathfrak{u}^{\rho}_{V\cap G}}}(\mathcal{S}) \le \mathfrak{p}_{\Omega^{\{\mu\}}_{\mathfrak{u}^{\rho}_{V\cap G}}}(\mathcal{T}) \quad \text{implies that} \quad \mathfrak{p}_{\Omega^{\{\nu\}}_{\mathfrak{u}^{\rho}_{V\cap G}}}(\mathcal{S}) \le \mathfrak{p}_{\Omega^{\{\nu\}}_{\mathfrak{u}^{\rho}_{V\cap G}}}(\mathcal{T}),$$

contradicting the above. □

Propositions 3.2.21 and 3.2.22 permit to identify certain $\mathfrak{C}$-superharmonic subsets of $\mathcal{C}_{\mathfrak{V}}(X,E)$. We begin with a technical lemma.

3.3.3 Lemma. *Let θ be the representation measure for $\mathcal{T} \in \mathfrak{L}(\mathcal{C}_{\mathfrak{V}}(X,E),F)$. Let $G \in \mathcal{G}$, let $\mu \in \Pi$ and suppose that $\mathfrak{p}_{\Omega^{\{\mu\}}_{\mathfrak{u}^{\rho}_{V\cap G}}}(\mathcal{T}) < +\infty$.*

(a) *If $V \cap G = G$, then the $\overline{\mathbb{R}}_+$-valued measure $[|\theta\|_G|(V \cap G)(\mu)]$ equals 0.*
(b) *If $V \cap G \ne G$, then for every $\varphi \in H_G$ and $\varepsilon > 0$ there is $A \in \mathfrak{R}$ and $\lambda \ge 0$ such that $|\varphi(x)| \le \lambda\rho(x)$ for all $x \in X$ and $|\varphi(x)| \le \varepsilon\rho(x)$ for all $x \in X \setminus A$.*

Proof. (a) If $V \cap G = G$, then $\mathfrak{u}^{\rho}_{V\cap G} = \mathcal{C}_{\mathfrak{V}}(X,G)$, and $\mathfrak{p}_{\Omega^{\{\mu\}}_{\mathfrak{u}^{\rho}_{V\cap G}}}(\mathcal{T}) < +\infty$ implies that $\mu(\mathcal{T}(f)) = 0$ for all $f \in \mathcal{C}_{\mathfrak{V}}(X,G)$. Hence

$$\begin{aligned}\mathfrak{p}_{\Omega^{\{\mu\}}_{\mathfrak{u}^{\kappa}_{V\cap G}}}(\mathcal{T}) &= \sup\{\mathfrak{Re}\, \mu(\mathcal{T}(f)) \mid f \in \mathcal{C}_{\mathfrak{V}}(X,G)\} \\ &= \int_X \kappa\, d[|\theta\|_G|(V \cap G)(\mu)] = 0\end{aligned}$$

for every strictly positive lower semicontinuous $\overline{\mathbb{R}}_+$-valued function κ. As for every $A \in \mathfrak{R}$, there is such a function such that $\chi_A \le \kappa$, we have

$$0 \le [|\theta\|_G|(V \cap G)(\mu)]_A \le \int_X \kappa\, d[|\theta\|_G|(V \cap G)(\mu)] = 0$$

as claimed. (b) If $V \cap G \ne G$, then there is $a \in G \setminus V$. Consequentially, $\alpha a \in \beta V$ for $\alpha \in \mathbb{C}$ and $\beta \ge 0$ implies that either $\alpha = 0$ or $a \in (\beta/\alpha)V$ and, therefore, $|\beta/\alpha| > 1$. Thus $|\alpha| < \beta$ holds in any case. Let $\varphi \in H_G$. Since $\varphi \cdot a \in \mathcal{C}_{\mathfrak{V}}(X,E)$ and since $\mathfrak{u}^{\rho}_V$ is supposed to be a neighborhood in $\mathcal{C}_{\mathfrak{V}}(X,E)$, there is $\lambda \geq 0$ such that $\varphi \cdot a \in \lambda \mathfrak{u}_V^\rho = \mathfrak{u}_V^{\lambda\rho}$. By the above, this implies that $|\varphi(x)| < \lambda\rho(x)$ for all $x \in X$. Now, given $\varepsilon > 0$, since $\varphi{\cdot}a$ vanishes at infinity, there is $A \in \mathfrak{R}$ such that $\chi_{(X\setminus A)} \cdot (\varphi \cdot a) \in \varepsilon \mathfrak{u}_V^\rho$. Thus $|\varphi(x)| \leq \varepsilon\rho(x)$ for all $x \in X \setminus A$, our claim. □

3.3.4 Proposition. *Let $G \in \mathcal{G}$ and $\tau \in C_G$.*

(a) *If $\tau(x) < \rho(x)$ for all $x \in X$, then for all $\mu \in \Pi$ the sets $\Omega^{\{\mu\}}_{\mathfrak{u}^\sigma_{V\cap G}}$, where $\sigma = \rho - \tau$, are $\mathfrak{C}$-superharmonic and elements of $\widetilde{\mathfrak{C}}$.*

(b) *If $\mathcal{S} \prec_{\widetilde{\mathfrak{C}}} \mathcal{T}$ for $\mathcal{S}, \mathcal{T} \in \mathfrak{L}(\mathcal{C}_{\mathfrak{V}}(X,E),F)$ and ϑ and θ are the representation measures for $\mathcal{S}$ and $\mathcal{T}$, respectively, and if $\mathfrak{p}_{\Omega^{\{\mu\}}_{\mathfrak{u}^\rho_{V\cap G}}}(\mathcal{S}) = \mathfrak{p}_{\Omega^{\{\mu\}}_{\mathfrak{u}^\rho_{V\cap G}}}(\mathcal{T}) < +\infty$ for $\mu \in \Pi$, then*

$$\int_X \tau \, d[|\vartheta\|_G|(V\cap G)(\mu)] \geq \int_X \tau \, d[|\theta\|_G|(V\cap G)(\mu)].$$

Proof. Let $\tau = \bigvee_{i=1}^n \varphi_i \in C_G$ for $\varphi_i \in C(X,\mathbb{C})$ be as stated. For Part (a), in Proposition 3.2.21 we set $U = V_G$, Γ the full unit sphere of $\mathbb{C}$ and $\Pi = \{\mu\}$. Since the sets $\Omega^{\{\mu\}}_{\mathfrak{u}^\rho_{V\cap G}}$ and $\Omega^{\{\mu\}}_{\{\varphi_i\cdot a\}}$ for all i and $a \in G$ are contained in $\mathfrak{C}$, Proposition 3.2.19 applies and states that the sets

$$\Omega^{\{\mu\}}_{\mathfrak{w}^a_{\varphi_i}}, \quad \text{where} \quad \mathfrak{w}^a_{\varphi_i} = \mathfrak{u}^\rho_{V\cap G} - \varphi_i.a,$$

are $\mathfrak{C}$-superharmonic. Proposition 3.2.21 now yields that the set $\Omega^{\{\mu\}}_{\mathfrak{u}^\sigma_{V\cap G}}$ is also $\mathfrak{C}$-superharmonic. As a subset of $\Omega^{\{\mu\}}_{\mathfrak{u}^\rho_{V\cap G}}$ is $\mathfrak{K}$-bounded, and thus an element of $\widetilde{\mathfrak{C}}$. For Part (b), suppose that $\mathcal{S} \prec_{\widetilde{\mathfrak{C}}} \mathcal{T}$ for $\mathcal{S}, \mathcal{T} \in \mathfrak{L}(\mathcal{C}_{\mathfrak{V}}(X,E),F)$. We distinguish two cases. If $V\cap G = G$, then

$$[|\theta\|_G|(V\cap G)(\mu)] = [|\vartheta\|_G|(V\cap G)(\mu)] = 0$$

by Lemma 3.3.3(a). Our claim holds trivially true in this case. In the second case, if $V\cap G \neq G$ following Lemma 3.3.3(b) there is $\lambda \geq 0$ such that $|\varphi_i(x)| < \lambda\rho(x)$ for all $x \in X$, and we may apply Part (a) with $\lambda\rho$ in place of ρ. We abbreviate $\sigma = \lambda\rho - \tau$ and infer that

$$\mathfrak{p}_{\Omega^{\{\mu\}}_{\mathfrak{u}^\sigma_{V\cap G}}}(\mathcal{S}) \leq \mathfrak{p}_{\Omega^{\{\mu\}}_{\mathfrak{u}^\sigma_{V\cap G}}}(\mathcal{T})$$

holds by Part (a). Using this, we calculate

$$\begin{aligned}\mathfrak{p}_{\Omega^{\{\mu\}}_{\mathfrak{u}^{\lambda\rho}_{V\cap G}}}(\mathcal{S}) - \mathfrak{p}_{\Omega^{\{\mu\}}_{\mathfrak{u}^\tau_{V\cap G}}}(\mathcal{S}) &= \mathfrak{p}_{\Omega^{\{\mu\}}_{\mathfrak{u}^\sigma_{V\cap G}}}(\mathcal{S}) \\ &\leq \mathfrak{p}_{\Omega^{\{\mu\}}_{\mathfrak{u}^\sigma_{V\cap G}}}(\mathcal{T}) = \mathfrak{p}_{\Omega^{\{\mu\}}_{\mathfrak{u}^{\lambda\rho}_{V\cap G}}}(\mathcal{T}) - \mathfrak{p}_{\Omega^{\{\mu\}}_{\mathfrak{u}^\tau_{V\cap G}}}(\mathcal{T}),\end{aligned}$$

hence

$$\int_X \tau\, d[|\vartheta\|_G|(V\cap G)(\mu)] = \mathfrak{p}_{\Omega^{\{\mu\}}_{\mathfrak{u}^{\tau}_{V\cap G}}}(\mathcal{S})$$

$$\geq \mathfrak{p}_{\Omega^{\{\mu\}}_{\mathfrak{u}^{\tau}_{V\cap G}}}(\mathcal{T}) = \int_X \tau\, d[|\theta\|_G|(V\cap G)(\mu)]$$

since $\mathfrak{p}_{\Omega^{\{\mu\}}_{\mathfrak{u}^{\lambda\rho}_{V\cap G}}}(\mathcal{S}) = \mathfrak{p}_{\Omega^{\{\mu\}}_{\mathfrak{u}^{\lambda\rho}_{V\cap G}}}(\mathcal{T}) < +\infty$. Hence our claim. □

3.3.5 Corollary. *Let $G \in \mathcal{G}$ and suppose that for every $x \in X$ there is $\varphi \in H_G$ such that $\varphi(x) \neq 0$, and for distinct points $x, y \in X$ there is $\varphi \in H_G$ such that $\varphi(x) = 0 \neq \varphi(y)$. If ϑ and θ are the representation measures for $\mathcal{S}, \mathcal{T} \in \mathfrak{L}(\mathcal{C}_{\mathfrak{V}}(X,E),F)$, respectively, if $\mathcal{S} \sim \mathcal{T}$ with respect to the order $\prec_{\widetilde{\mathfrak{C}}}$ and if $\mathfrak{p}_{\Omega^{\{\mu\}}_{\mathfrak{u}^{\rho}_{V\cap G}}}(\mathcal{T}) < +\infty$ for $\mu \in \Pi$, then*

$$[|\vartheta\|_G|(V\cap G)(\mu)] = [|\theta\|_G|(V\cap G)(\mu)].$$

If $\mathcal{T}$ is minimal in this order and if there is a set $\mathfrak{u} \in \mathfrak{U}$ that is defined by a function space neighborhood and such that $\mathfrak{u}^{\rho}_{V\cap G} \leq \mathfrak{u}$ and $\mathfrak{p}_{\Omega^{\Pi}_{\mathfrak{u}}}(\mathcal{T}) < +\infty$, then $\theta\|_G$ is supported by any F_σ-subset of X that contains $\partial^{\Omega^{\Pi}_{\mathfrak{u}}}_{\mathfrak{C}}$.

Proof. Let $G \in \mathcal{G}$ be as stated. If $V \cap G = G$, then

$$[|\theta\|_G|(V\cap G)(\mu)] = [|\vartheta\|_G|(V\cap G)(\mu)] = 0$$

holds for all $\mu \in \Pi$ by Lemma 3.3.3(a). Otherwise, we shall use a suitable space $\mathcal{C}_{\mathfrak{V}_{\mathbb{R}}}(X)$ of real-valued functions and an appropriate version of the Stone–Weierstrass theorem to establish our argument. For this, we define the function space neighborhood w in $\mathcal{F}(X)$ by the single neighborhood function $x \mapsto [-\rho(x), +\rho(x)]$, that is,

$$w = \{\varphi \in \mathcal{F}(X) \mid |\varphi(x)| \leq \rho(x) \text{ for all } x \in X\},$$

and the function space neighborhood system $\mathfrak{V}_{\mathbb{R}} = \{aw \mid a > 0\}$ for $\mathcal{F}(X)$. Lemma 3.3.3(b) then shows that all functions in H_G, hence all functions in C_G vanish at infinity and are therefore contained in $\mathcal{C}_{\mathfrak{V}_{\mathbb{R}}}(X)$. The linear span of C_G, that is, $D = C_G - C_G$, forms a sublattice of $\mathcal{C}_{\mathfrak{V}_{\mathbb{R}}}(X)$. Applying a version of the Stone–Weierstrass theorem from [49] we can demonstrate that D is dense $\mathcal{C}_{\mathfrak{V}}(X)$. We use Corollaries 2.2 and 2.3 in [49], which require that (i) for every $x \in X$ there is $\varphi \in D$ such that $\varphi(x) \neq 0$ and (ii) for all $x \neq y \in X$ there is $\varphi \in D$ such that $\varphi \geq 0$, $\varphi(x) = 0$ and $\varphi(y) > 0$. Both conditions are immediately implied by our assumptions on H_G. Now let $\mathcal{S}, \mathcal{T} \in \mathfrak{L}(\mathcal{C}_{\mathfrak{V}}(X,E),F)$ such that $\mathcal{S} \sim \mathcal{T}$ with respect to the order $\prec_{\widetilde{\mathfrak{C}}}$ and let $\mu \in \Pi$ such that

$$\mathfrak{p}_{\Omega^{\{\mu\}}_{\mathfrak{u}^{\rho}_{V\cap G}}}(\mathcal{S}) = \mathfrak{p}_{\Omega^{\{\mu\}}_{\mathfrak{u}^{\rho}_{V\cap G}}}(\mathcal{T}) < +\infty.$$

Then Proposition 3.3.4(b) applies. We have

$$\int_X \tau\, d[|\vartheta\|_G|(V \cap G)(\mu)] = \int_X \tau\, d[|\theta\|_G|(V \cap G)(\mu)]$$

for all $\tau \in C_G$. For $\varphi \in \mathcal{C}_{\mathfrak{V}_\mathbb{R}}(X)$, such that $|\varphi| \leq \rho$ and $\mu \in \Pi$ we realize that

$$\left|\int_X \varphi\, d[|\theta\|_G|(V \cap G)(\mu)]\right| \leq \int_X \rho\, d[|\theta\|_G|(V \cap G)(\mu)] = \mathfrak{p}_{\Omega^{\{\mu\}}_{\mathfrak{u}^\rho_{V\cap G}}}(\mathcal{T}) < +\infty.$$

Thus the mapping

$$\varphi \mapsto \int_X \varphi\, d[|\theta\|_G|(V \cap G)(\mu)] \;:\; \mathcal{C}_{\mathfrak{V}_\mathbb{R}}(X) \to \mathbb{R}$$

is a continuous linear functional on $\mathcal{C}_{\mathfrak{V}_\mathbb{R}}(X)$. In particular, the measure

$$[|\theta\|_G|(V \cap G)(\mu)]$$

is finitely valued on $\mathfrak{R}$ and, therefore, regular by Proposition 2.1.15. The same considerations apply with ϑ in place of θ. Both functionals coincide on the dense subspace D and, therefore, on all of $\mathcal{C}_{\mathfrak{V}_\mathbb{R}}(X)$. Now using Proposition 2.5.8, we infer that

$$[|\vartheta\|_G|(V \cap G)(\mu)] = [|\theta\|_G|(V \cap G)(\mu)],$$

that is, the first part of our claim. Now suppose that $\mathcal{T} \in \mathfrak{L}(\mathcal{C}_{\mathfrak{V}}(X,E),F)$ is minimal in the order $\prec_{\widetilde{\mathfrak{C}}}$ and that there is a function space neighborhood $\mathfrak{u} \in \mathfrak{U}$, that is, $\Omega^\Pi_\mathfrak{u} \in \mathfrak{K}$, such that $\mathfrak{u}_{V\cap G} \leq \mathfrak{u}$ and $\mathfrak{p}_{\Omega^\Pi_\mathfrak{u}}(\mathcal{T}) < +\infty$. (Recall that we agreed to use the same symbols for function space neighborhoods and their intersections with $\mathcal{C}_{\mathfrak{V}}(X,E)$.) Since

$$\mathfrak{p}_{\Omega^{\{\mu\}}_{\mathfrak{u}^\rho_{V\cap G}}}(\mathcal{T}) \leq \mathfrak{p}_{\Omega^\Pi_\mathfrak{u}}(\mathcal{T})$$

for all $\mu \in \Pi$, our first claim applies to all such μ. Let B be a G_δ-subset of X disjoint from $\partial^{\Omega^\Pi_\mathfrak{u}}_{\mathfrak{C}}$ and let $A \in \mathfrak{R}$ be a subset of B. Since A is relatively compact and ρ is strictly positive and lower semicontinuous, there is $\lambda \geq 0$ such that $\chi_A \cdot V \leq \lambda\mathfrak{u}$. Hence using Lemma 2.1.5(a), we infer that

$$|\theta\|_G|_A(V \cap G)(\mu) \leq \|\theta\|_G\|_A^{(\lambda\mathfrak{u},\{\mu\})} \leq \lambda \sup_{A\in\mathfrak{R}} \|\theta\|_G\|_{A\cap B}^{(\mathfrak{u},\Pi)}$$

holds for all $\mu \in \Pi$. The same holds true with ϑ in place of θ. Next, we use Corollary 3.2.14, the first part of the current corollary and the prior observation for

$$\begin{aligned}|\theta\|_G|_A(V \cap G)(\mu) &= \inf_{\mathcal{S}\sim\mathcal{T}}\{|\vartheta\|_G|_A(V \cap G)(\mu) \mid \vartheta \text{ represents } \mathcal{S}\}\\ &\leq \inf_{\mathcal{S}\sim\mathcal{T}}\{\|\vartheta\|_A^{(\mathfrak{u},\Pi)} \mid \vartheta \text{ represents } \mathcal{S}\}\end{aligned}$$

$$\leq \inf_{\mathcal{S}\sim\mathcal{T}} \sup_{A\in\mathfrak{R}} \{\|\vartheta\|_{A\cap B}^{(\mathfrak{u},\Pi)} \mid \vartheta \text{ represents } \mathcal{S}\} = 0.$$

Thus $|\theta\|_G|_A(V\cap G)(\mu) = 0$ for all $\mu \in \Pi$. Finally, since the operator $\mathcal{T}$ is weakly compact, according to Theorem 2.6.1 its representation measure θ is $\mathfrak{L}(E,F)$-valued. Hence for a fixed $V \in \mathcal{V}$ and $A \in \mathfrak{R}$ the mapping

$$\mu \mapsto |\theta\|_G|_A(V\cap G)(\mu)$$

is sublinear and lower semicontinuous with respect to the weak* topology of F^* (see the definition of the modulus of a measure in Section 2.1). Thus we have

$$|\theta\|_G|_A(V\cap G)(\mu) = 0$$

for all μ in the weak*-closure of the balanced convex hull $\operatorname{bconv}(\Pi)$ of Π, which was understood to be absorbing. This demonstrates that $|\theta\|_G|_A(V\cap G)(\mu) = 0$ holds for all $\mu \in F^*$, our claim. □

The application of Proposition 3.2.22 in this standard setting requires that the family $\mathcal{G}$ of subspaces of E contains E itself.

3.3.6 Proposition. *Let E be a Hilbert space with unit ball $\mathbb{B}$ and suppose that $E \in \mathcal{G}$, that $V_E = \mathbb{B}$ and $\rho = \rho_E$. Let $\mathcal{S},\mathcal{T} \in \mathfrak{L}(\mathcal{C}_{\mathfrak{V}}(X,E),F)$ such that $\mathcal{S} \prec_{\mathfrak{C}} \mathcal{T}$. If for $\mu \in \Pi$ there is $g_\mu \in (-\mathcal{H}) \cap \mathfrak{u}_{\mathbb{B}}^{\rho}$ such that $\mathfrak{Re}\,\mu(\mathcal{T}(g_\mu)) = \mathfrak{p}_{\Omega^{\{\mu\}}_{\mathfrak{u}^{\rho}_{\mathbb{B}}}}(\mathcal{T})$, then*

$$\mathfrak{Re}\,\mu(\mathcal{S}(f)) \leq 0$$

for all $f \in \mathcal{C}_{\mathfrak{V}}(X,E)$ such that $\mathfrak{Re}\langle f(x), g_\mu(x)\rangle \leq 0$ for all $x \in Y_\mu$, where Y_μ is the compact set $\{x \in X \mid \|g_\mu(x)\| = \rho(x)\}$. If for every $\mu \in \Pi$ there is such a function $g_\mu \in \mathcal{H} \cap \mathfrak{u}$, then the representation measure for S is supported by the closure of the set $\bigcup_{\mu\in\Pi} Y_\mu$.

Proof. Since $\Omega^{\{\mu\}}_{\mathfrak{u}^{\rho}_{\mathbb{B}}}, \Omega^{\{\mu\}}_{\{-g\}} \in \mathfrak{C}$ in our sample settings, the first part of our claim follows directly from Proposition 3.2.22. The second part is a reformulation of Corollary 3.2.23. □

We go forward to consider a few special cases using the preceding sample settings.

The case that $F = \mathbb{R}$ or $F = \mathbb{C}$

In this setting, $\mathfrak{L}(E,F)$ is the dual E^* of E, endowed with its weak* topology, $\mathfrak{L}(\mathcal{C}_{\mathfrak{V}}(X,E),F) = \mathcal{C}_{\mathfrak{V}}(X,E)^*$, that is, the dual of $\mathcal{C}_{\mathfrak{V}}(X,E)$ endowed with its weak* topology, and $\mathcal{C}_{\mathfrak{V}}(X,E) \otimes F^* = \mathcal{C}_{\mathfrak{V}}(X,E)$ endowed with its weak topology. We shall use the previously established sample settings with $\Pi = \{1\}$. That is, $\mathfrak{K} = \mathfrak{U}$, where $\mathfrak{U}$ is a family of (not necessarily balanced) $\mathcal{C}(X)$-convex neighborhoods of 0 in $\mathcal{C}_{\mathfrak{V}}(X,E)$

with the required properties, $\mathcal{H}$ is a subcone of $\mathcal{C}_{\mathfrak{V}}(X,E)$ and $\mathcal{G}$ and the sets $\mathfrak{u}_V^\rho$ are as stated in the general setting. Then $\mathfrak{C}_0 \subset \mathcal{C}_{\mathfrak{V}}(X,E)$ consists of:
(i) all sets $\{f\}$ for $f \in \mathcal{H}$ and
(ii) all sets $\mathfrak{u}_{V\cap G}^\rho$ for $G \in \mathcal{G}$.

We notice that according to Proposition 3.1.11 we have

$$\mathfrak{p}_{\mathfrak{w}}(\mathcal{T}) = \sup\{\mathfrak{Re}\, \mathcal{T}(f) \mid f \in \mathfrak{w}\}$$

for every $C(X)$-convex subset $\mathfrak{w}$ of $\mathcal{C}_{\mathfrak{V}}(X,E)$ containing 0 and $\mathcal{T} \in \mathcal{C}_{\mathfrak{V}}(X,E)^*$. The measures $|\theta\|_G|(V_G)$ are $\overline{\mathbb{R}}_+$-valued. The formulations in Propositions 3.3.1 and 3.3.4, Lemma 3.3.3 and Corollary 3.3.5 simplify accordingly: There is no mention of Π or the functional $\mu \in \Pi$ in the statements, and the sets

$$\Omega^{\Pi}_{\mathfrak{u}^\rho_{V\cap G}}, \quad \Omega^{\{\mu\}}_{\mathfrak{u}^\rho_{V\cap G}} \quad \text{and} \quad \Omega^{\{\mu\}}_{\{f\}}$$

read as $\mathfrak{u}_V^\rho$, $\mathfrak{u}_{V\cap G}^\rho$ and $\{f\}$, respectively. Proposition 3.3.6 requires that $E \in \mathcal{H}$ and simplifies accordingly.

The case that $E = \mathbb{R}$ or $E = \mathbb{C}$

In this setting, we have $\mathfrak{L}(E,F) = F$, endowed with its weak topology, and $\mathcal{C}_{\mathfrak{V}}(X,E)\otimes F^*$ equals $\mathcal{C}_{\mathfrak{V}}(X)\otimes F^*$ or $\mathcal{C}_{\mathfrak{V}}(X,\mathbb{C})\otimes F^*$. We shall formulate the case that $E = \mathbb{C}$ and use the previously established sample settings. The real case is similar. That is, we have

$$\mathfrak{K} = \{\Omega^{\Pi}_{\mathfrak{u}} \mid \mathfrak{u} \in \mathfrak{U}\},$$

where $\mathfrak{U}$ is a family of (not necessarily balanced) $C(X)$-convex neighborhoods of 0 in $\mathcal{C}_{\mathfrak{V}}(X,\mathbb{C})$ with the required properties, and $\Pi \subset F^*$ is such that $\operatorname{bcore}(\Pi^\circ)$ is weakly compact in F, $\mathcal{H}$ is a subcone of $\mathcal{C}_{\mathfrak{V}}(X,\mathbb{C})$ and $\{0\}$ and $\mathbb{C}$ are the only possible elements of $\mathcal{G}$. If $G = \mathbb{C}$ is in $\mathcal{H}$, then it is associated with a set $\mathfrak{u}_{\mathbb{B}}^\rho$ in $\mathcal{C}_{\mathfrak{V}}(X,\mathbb{C})$, whereby $\mathbb{B}$ denotes the unit ball of $\mathbb{C}$. The corresponding subspace H_G of $C(X,\mathbb{C})$ and subcone C_G of $C(X)$ are

$$H_G = \operatorname{bcore}(\mathcal{H}) = \bigcap_{|\gamma|=1} \gamma\mathcal{H} \quad \text{and} \quad C_G = \left\{\bigvee_{i=1}^{n} |\varphi_i| \mid \varphi_i \in \mathcal{H}\right\}.$$

Then $\mathfrak{C}_0 \subset \mathcal{C}_{\mathfrak{V}}(X,\mathbb{C})\otimes F^*$ consists of
(i) all sets $\Omega^{\{\mu\}}_{\{f\}}$ for $f \in \mathcal{H}$ and $\mu \in \Pi$ and
(ii) if $\mathcal{G} = \{\mathbb{C}\}$: all sets $\Omega^{\{\mu\}}_{\mathfrak{u}^\rho_{\mathbb{B}}}$ for $\mu \in \Pi$.

The measures $|\theta|(\mathbb{B})(\mu)$ are $\mathbb{R}_+$-valued. The formulations in Propositions 3.3.4 and Corollary 3.3.5 simplify only in the sense that they do not involve any subspaces G of E. Proposition 3.3.6 requires that $\mathbb{C} \in \mathcal{H}$ and simplifies accordingly.

We continue to investigate this case and relate the given Choquet ordering of $\mathfrak{L}(\mathcal{C}_{\mathfrak{V}}(X,\mathbb{C}),F)$ to a corresponding Choquet ordering in $\mathfrak{L}(\mathcal{C}_{\mathfrak{V}}(X,\mathbb{C}),\mathbb{C})$ in the following way. We employ a parallel Choquet cone $\mathfrak{C}_{\mathbb{C}}$ for $\mathfrak{L}(\mathcal{C}_{\mathfrak{V}}(X,\mathbb{C}),\mathbb{C})$ is spanned by the Choquet kernel $\mathfrak{K}_{\mathbb{C}} = \mathfrak{U}$ and the collection

$$(\mathfrak{C}_{\mathbb{C}})_0 = \{\mathfrak{w} \subset \mathcal{C}_{\mathfrak{V}}(X,\mathbb{C}) \mid \Omega_{\mathfrak{w}}^{\{\mu\}} \in \mathfrak{C}_0 \text{ for some } \mu \in \Pi\}.$$

That is to say, $(\mathfrak{C}_{\mathbb{C}})_0 = \{\{f\} \mid f \in \mathcal{H}\}$ in case that $\mathcal{G} = \varnothing$, and

$$(\mathfrak{C}_{\mathbb{C}})_0 = \{\{f\} \mid f \in \mathcal{H}\} \cup \{\mathfrak{u}_{\mathbb{B}}^{\rho}\}$$

in case that $\mathcal{G} = \{\mathbb{C}\}$. We choose $\Pi_{\mathbb{C}} = \{1\}$. With an operator $\mathcal{T} \in \mathfrak{L}(\mathcal{C}_{\mathfrak{V}}(X,\mathbb{C}),\mathbb{C})$, that is, $\mathcal{T} \in \mathcal{C}_{\mathfrak{V}}(X,\mathbb{C})^*$, and an element $e \in F$ we associate the elementary operator $\mathcal{L}_{\mathcal{T}}^{e} \in \mathfrak{L}(\mathcal{C}_{\mathfrak{V}}(X,\mathbb{C}),F)$ (see Section 3.1) defined by $\mathcal{L}_{\mathcal{T}}^{e}(f) = \mathcal{T}(f)\,e$ for $f \in \mathcal{C}_{\mathfrak{V}}(X,\mathbb{C})$. We observe that all point evaluations in $\mathfrak{L}(\mathcal{C}_{\mathfrak{V}}(X,\mathbb{C}),F)$ are elementary operators $\mathcal{D}_x^e = \mathcal{L}_{\mathcal{D}_x}^{e}$ in this sense, where $x \in X$ and $\mathcal{D}_x = \mathcal{D}_x^1$ is a point evaluation in $\mathfrak{L}(\mathcal{C}_{\mathfrak{V}}(X,\mathbb{C}),\mathbb{C})$. We shall say that an element $e \in F$ is *Π-positive* if $0 \le \mu(e) \in \mathbb{R}$ holds for all $\mu \in \Pi$. Note that $p_\Pi(e) = 0$ for a Π-positive element $e \in F$ implies that $e = 0$.

3.3.7 Proposition. *Let $\mathcal{S},\mathcal{T} \in \mathfrak{L}(\mathcal{C}_{\mathfrak{V}}(X,\mathbb{C}),\mathbb{C})$ and let $e \in F$ be Π-positive.*
(a) *If $\mathfrak{w}$ is a $\mathcal{C}(X)$-convex subset of $\mathcal{C}_{\mathfrak{V}}(X,\mathbb{C})$ and $\Gamma \subset \Pi$ such that either $0 \in \mathfrak{w}$ or that $\Gamma = \{\mu\}$ is a singleton set, then $\mathfrak{p}_{\Omega_{\mathfrak{w}}^{\Gamma}}(\mathcal{L}_{\mathcal{T}}^{e}) = p_\Gamma(e)\,\mathfrak{p}_{\mathfrak{w}}(\mathcal{T})$.*
(b) *If $\mathcal{S} \prec_{\mathfrak{C}_{\mathbb{C}}} \mathcal{T}$, then $\mathcal{L}_{\mathcal{S}}^{e} \prec_{\mathfrak{C}} \mathcal{L}_{\mathcal{T}}^{e}$.*
(c) *If $\mathcal{L}_{\mathcal{S}}^{e} \prec_{\mathfrak{C}} \mathcal{L}_{\mathcal{T}}^{e}$, if $e \ne 0$ and $p_\Pi(e) < +\infty$, then $\mathcal{S} \prec_{\mathfrak{C}_{\mathbb{C}}} \mathcal{T}$.*
(d) *If $\mathcal{L}_{\mathcal{T}}^{e} \ne 0$, then $\mathcal{L}_{\mathcal{T}}^{e} \in \mathfrak{L}_{\mathfrak{K}}(\mathcal{C}_{\mathfrak{V}}(X,\mathbb{C}),F)$ if and only if $\mathcal{T} \in \mathfrak{L}_{\mathfrak{K}_{\mathbb{C}}}(\mathcal{C}_{\mathfrak{V}}(X,\mathbb{C}),\mathbb{C})$ and $p_\Pi(e) < +\infty$.*

Proof. Recall that $\mathfrak{p}_{\mathfrak{w}}(\mathcal{T}) = \sup\{\mathfrak{Re}\,\mathcal{T}(f) \mid f \in \mathfrak{w}\}$ for a $\mathcal{C}(X)$-convex subset $\mathfrak{w}$ of $\mathcal{C}_{\mathfrak{V}}(X,\mathbb{C})$ and $\mathcal{T} \in \mathfrak{L}(\mathcal{C}_{\mathfrak{V}}(X,\mathbb{C}),\mathbb{C})$. Thus our claim in (a) follows from Proposition 3.1.15(b). For (b), in order to establish that $\mathcal{S} \prec_{\mathfrak{C}_{\mathbb{C}}} \mathcal{T}$ we only have to verify that $\mathfrak{p}_{\mathfrak{w}}(\mathcal{S}) \le \mathfrak{p}_{\mathfrak{w}}(\mathcal{T})$ holds for all $\mathfrak{w} \in \mathfrak{K}_{\mathbb{C}} \cup (\mathfrak{C}_{\mathbb{C}})_0$. A similar assertion applies to the operators in $\mathfrak{L}(\mathcal{C}_{\mathfrak{V}}(X,\mathbb{C}),F)$. Thus if $\mathcal{S} \prec_{\mathfrak{C}_{\mathbb{C}}} \mathcal{T}$, since all elements of $\mathfrak{K} \cup \mathfrak{C}_0$ satisfy one of its alternative requirements, Part (a) yields that $\mathcal{L}_{\mathcal{S}}^{e} \prec_{\mathfrak{C}} \mathcal{L}_{\mathcal{T}}^{e}$ also holds. For (c), suppose that $p_\Pi(e) < +\infty$ for $e \ne 0$ and that $\mathcal{L}_{\mathcal{T}}^{e} \prec_{\mathfrak{C}} \mathcal{L}_{\mathcal{T}}^{e}$. Since the linear span of Π is weak*-dense in F^*, there is $\mu \in \Pi$ such that $\mu(e) > 0$ and, therefore, $p_\Pi(e) > 0$. Thus Part (a) yields that $\mathfrak{p}_{\mathfrak{u}}(\mathcal{S}) \le \mathfrak{p}_{\mathfrak{u}}(\mathcal{T})$ holds for all $\mathfrak{u} \in \mathfrak{U}$. Furthermore, for every $\mathfrak{w} \in (\mathfrak{C}_{\mathbb{C}})_0$ and $\mu \in \Pi$ the set $\Omega_{\mathfrak{w}}^{\{\mu\}}$ is contained in $\mathfrak{C}_0$, and Part (a) with $\Gamma = \{\mu\}$ renders that $\mathfrak{p}_{\mathfrak{w}}(\mathcal{S}) \le \mathfrak{p}_{\mathfrak{w}}(\mathcal{T})$. We infer that $\mathcal{S} \prec_{\mathfrak{C}_{\mathbb{C}}} \mathcal{T}$ as claimed. For (d), first suppose that $\mathcal{T} \in \mathfrak{L}_{\mathfrak{K}_{\mathbb{C}}}(\mathcal{C}_{\mathfrak{V}}(X),\mathbb{C})$, that is, $\mathfrak{p}_{\mathfrak{u}}(\mathcal{T}) < +\infty$ for some $\mathfrak{u} \in \mathfrak{U}$, and that $p_\Pi(e) < +\infty$. Then $\mathfrak{p}_{\Omega_{\mathfrak{u}}^{\Pi}}(\mathcal{L}_{\mathcal{T}}^{e}) < +\infty$ by Part (a) and, therefore, $\mathcal{L}_{\mathcal{T}}^{e} \in \mathfrak{L}_{\mathfrak{K}}(\mathcal{C}_{\mathfrak{V}}(X,\mathbb{C}),F)$. Conversely, if $0 \neq \mathcal{L}_{\mathcal{T}}^{e} \in \mathfrak{L}_{\mathfrak{K}}(\mathcal{C}_{\mathfrak{V}}(X,\mathbb{C}),F)$, there is $\mathfrak{u} \in \mathfrak{U}$ such that $\mathfrak{p}_{\Omega_{\mathfrak{u}}^{\Pi}}(\mathcal{L}_{\mathcal{T}}^{e}) = p_{\Pi}(e)\,\mathfrak{p}_{\mathfrak{u}}(\mathcal{T}) < +\infty$. Since $e \neq 0$, we have $p_{\Pi}(e) > 0$ and $\mathfrak{p}_{\mathfrak{u}}(\mathcal{T}) > 0$ since $\mathfrak{u}$ is a neighborhood in $\mathcal{C}_{\mathfrak{V}}(X)$ and $\mathcal{T} \neq 0$. Thus both $p_{\Pi}(e) < +\infty$ and $\mathfrak{p}_{\mathfrak{u}}(\mathcal{T}) < +\infty$, that is, $\mathcal{T} \in \mathfrak{L}_{\mathfrak{K}_{\mathbb{C}}}(\mathcal{C}_{\mathfrak{V}}(X,\mathbb{C}),\mathbb{C})$, our claim. □

3.3.8 Proposition. *Suppose that for every* $\mathfrak{u} \in \mathfrak{U}$ *the linear span of the set* $\bigcap\{\gamma\mathfrak{u} \mid \gamma \in \mathbb{C},\ \mathfrak{Re}\,\gamma > 0\}$ *is dense in* $\mathcal{C}_{\mathfrak{V}}(X,\mathbb{C})$.

(a) *Let* $\mathcal{T} \in \mathfrak{L}(\mathcal{C}_{\mathfrak{V}}(X,\mathbb{C}),\mathbb{C})$ *and* $e \in F$. *If* $\mathcal{L}_{\mathcal{T}}^{e} \in \mathfrak{L}_{\mathfrak{K}}(\mathcal{C}_{\mathfrak{V}}(X,\mathbb{C}),F)$ *is not zero, then there is* $\alpha \neq 0$ *such that* αe *is* Π*-positive and* $p_{\Pi}(\alpha e) = 1$.

(b) *We have* $\partial_{\mathfrak{C}} \subset \partial_{\mathfrak{C}_{\mathbb{C}}}$.

Proof. For (a), let $0 \neq \mathcal{L}_{\mathcal{T}}^{e} \in \mathfrak{L}_{\mathfrak{K}}(\mathcal{C}_{\mathfrak{V}}(X,\mathbb{C}),F)$. There is $\mathfrak{u} \in \mathfrak{U}$ such that $\mathfrak{p}_{\Omega_{\mathfrak{u}}^{\Pi}}(\mathcal{L}_{\mathcal{T}}^{e}) < +\infty$. Since $\mathcal{T} \neq 0$, by our assumption there is $f \in \bigcap\{\gamma\mathfrak{u} \mid \gamma \in \mathbb{C},\ \mathfrak{Re}\,\gamma > 0\}$ such that $\beta = -\mathcal{T}(f) \neq 0$. For every $\mu \in \Pi$, we have

$$\mathfrak{Re}\,\mu(e)\mathcal{T}(g) \leq \mathfrak{p}_{\Omega_{\mathfrak{u}}^{\Pi}}(\mathcal{L}_{\mathcal{T}}^{e}) < +\infty$$

for all $g \in \mathfrak{u}$. Since $\gamma f \in \mathfrak{u}$ for all $\gamma \in \mathbb{C}$ such that $\mathfrak{Re}\,\gamma > 0$, this implies that $\mathfrak{Re}\,\gamma\mu(e)(-\beta) \leq 0$ for all such γ. We conclude that $0 \leq \mu(\beta e) \in \mathbb{R}$. The element $\beta e \in F$ is therefore Π-positive. Moreover, since

$$\mathcal{L}_{(1/\beta)\mathcal{T}}^{\beta e} = \mathcal{L}_{\mathcal{T}}^{e} \in \mathfrak{L}_{\mathfrak{K}}(\mathcal{C}_{\mathfrak{V}}(X,\mathbb{C}),F),$$

Proposition 3.3.7(d) renders that $0 < p_{\Pi}(\beta e) < +\infty$. Our claim follows with $\alpha = \beta/p_{\Pi}(\beta e)$. For (b), let $x \in \partial_{\mathfrak{C}}$. Then there is $\Omega_{\mathfrak{u}}^{\Pi} \in \mathfrak{K}$ and a point evaluation $\mathcal{D}_{x}^{e} \in \Delta_{\mathfrak{C}}^{\Omega_{\mathfrak{u}}^{\Pi}}$. We shall establish that $\mathcal{D}_{x}^{(1/\alpha)} \in \Delta_{\mathfrak{C}_{\mathbb{C}}}^{\mathfrak{u}}$ with some $\alpha \neq 0$. Since $\mathcal{D}_{x}^{e} = \mathcal{L}_{\mathcal{D}_{x}}^{e}$, Part (a) demonstrates that there is $\alpha \neq 0$ such that αe is Π-positive and $p_{\Pi}(\alpha e) = 1$. We write $\mathcal{D}_{x}^{e} = \mathcal{L}_{\mathcal{D}_{x}^{(1/\alpha)}}^{\alpha e}$ and infer using Proposition 3.3.7(d) that $\mathcal{D}_{x}^{(1/\alpha)} \in \mathfrak{L}_{\mathfrak{K}_{\mathbb{C}}}(\mathcal{C}_{\mathfrak{V}}(X,\mathbb{C}),\mathbb{C})$ is a point evaluation. Now applying 3.3.7(a) for every choice of $\beta \in \mathbb{C}$, we calculate

$$\mathfrak{p}_{\Omega_{\mathfrak{u}}^{\Pi}}(\mathcal{D}_{x}^{\beta e}) = \mathfrak{p}_{\Omega_{\mathfrak{u}}^{\Pi}}(\mathcal{L}_{\mathcal{D}_{x}^{(\beta/\alpha)}}^{\alpha e}) = \mathfrak{p}_{\mathfrak{u}}(\mathcal{D}_{x}^{(\beta/\alpha)}).$$

We proceed to verify that $\mathcal{D}_{x}^{(1/\alpha)}$ is an element of $\Delta_{\mathfrak{C}_{\mathbb{C}}}^{\mathfrak{u}}$. First, we establish that it is minimal in $\mathfrak{L}(\mathcal{C}_{\mathfrak{V}}(X,\mathbb{C}),\mathbb{C})$. For this, let $\mathcal{T} \in \mathfrak{L}(\mathcal{C}_{\mathfrak{V}}(X,\mathbb{C}),\mathbb{C})$ such that $\mathcal{T} \prec_{\mathfrak{C}_{\mathbb{C}}} \mathcal{D}_{x}^{(1/\alpha)}$. Then

$$\mathcal{L}_{\mathcal{T}}^{\alpha e} \prec_{\mathfrak{C}} \mathcal{L}_{\mathcal{D}_{x}^{(1/\alpha)}}^{\alpha e} = \mathcal{D}_{x}^{e}$$

by 3.3.7(b) and, therefore, $\mathcal{D}_{x}^{e} \prec_{\mathfrak{C}} \mathcal{L}_{\mathcal{T}}^{\alpha e}$ because $\mathcal{D}_{x}^{e}$ is minimal. Since $\mathcal{D}_{x}^{e} = \mathcal{L}_{\mathcal{D}_{x}^{(1/\alpha)}}^{\alpha e}$, Proposition 3.3.7(c) yields that $\mathcal{D}_{x}^{(1/\alpha)} \prec_{\mathfrak{C}_{\mathbb{C}}} \mathcal{T}$, and the point evaluation $\mathcal{D}_{x}^{(1/\alpha)}$ is seen to be minimal as claimed. All that is left to show is that $\mathcal{D}_{x}^{(1/\alpha)}$ is $\mathfrak{u}$-extremal. For this, let $1/\alpha = \alpha_1 + \alpha_2$ for $\alpha_1, \alpha_2 \in \mathbb{C}$ such that

$$\mathfrak{p}_{\mathfrak{u}}(\mathcal{D}_X^{(1/\alpha)}) = \mathfrak{p}_{\mathfrak{u}}(\mathcal{D}_X^{\alpha_1}) + \mathfrak{p}_{\mathfrak{u}}(\mathcal{D}_X^{\alpha_1}).$$

Then $e = \alpha_1 \alpha e + \alpha_2 \alpha e$, and we have

$$\mathfrak{p}_{\Omega_{\mathfrak{u}}^{\Pi}}(\mathcal{D}_X^e) = \mathfrak{p}_{\mathfrak{u}}(\mathcal{D}_X^{(1/\alpha)}), \quad \mathfrak{p}_{\Omega_{\mathfrak{u}}^{\Pi}}(\mathcal{D}_X^{\alpha_1 \alpha e}) = \mathfrak{p}_{\mathfrak{u}}(\mathcal{D}_X^{\alpha_1})$$

and

$$\mathfrak{p}_{\Omega_{\mathfrak{u}}^{\Pi}}(\mathcal{D}_X^{\alpha_2 \alpha e}) = \mathfrak{p}_{\mathfrak{u}}(\mathcal{D}_X^{\alpha_2})$$

by the above. Thus

$$\begin{aligned} \mathfrak{p}_{\Omega_{\mathfrak{u}}^{\Pi}}(\mathcal{D}_X^e) &= \mathfrak{p}_{\mathfrak{u}}(\mathcal{D}_X^{(1/\alpha)}) \\ &= \mathfrak{p}_{\mathfrak{u}}(\mathcal{D}_X^{\alpha_1}) + \mathfrak{p}_{\mathfrak{u}}(\mathcal{D}_X^{\alpha_1}) \\ &= \mathfrak{p}_{\Omega_{\mathfrak{u}}^{\Pi}}(\mathcal{D}_X^{\alpha_1 \alpha e}) + \mathfrak{p}_{\Omega_{\mathfrak{u}}^{\Pi}}(\mathcal{D}_X^{\alpha_2 \alpha e}). \end{aligned}$$

Because $\mathcal{D}_X^e$ is $\Omega_{\mathfrak{u}}^{\Pi}$-extremal, the latter implies that

$$\mathfrak{p}_{\Omega_{\mathfrak{u}}^{\Pi}}(\mathcal{D}_X^e)\, \alpha_1 e = \mathfrak{p}_{\Omega_{\mathfrak{u}}^{\Pi}}(\mathcal{D}_X^{\alpha_1 \alpha e})(1/\alpha)e$$

and

$$\mathfrak{p}_{\Omega_{\mathfrak{u}}^{\Pi}}(\mathcal{D}_X^e)\alpha_2 e = \mathfrak{p}_{\Omega_{\mathfrak{u}}^{\Pi}}(\mathcal{D}_X^{\alpha_2 \alpha e})(1/\alpha)e,$$

hence

$$\mathfrak{p}_{\mathfrak{u}}(\mathcal{D}_X^{(1/\alpha)})\alpha_1 = \mathfrak{p}_{\mathfrak{u}}(\mathcal{D}_X^{\alpha_1})(1/\alpha)$$

and

$$\mathfrak{p}_{\mathfrak{u}}(\mathcal{D}_X^{(1/\alpha)})\alpha_2 = \mathfrak{p}_{\mathfrak{u}}(\mathcal{D}_X^{\alpha_2})(1/\alpha).$$

Thus $\mathcal{D}_X^{(1/\alpha)}$ is indeed $\mathfrak{u}$-extremal, and we conclude that $\mathcal{D}_X^{(1/\alpha)} \in \Delta_{\mathfrak{C}_\mathbb{C}}^{\mathfrak{u}}$ and therefore that $x \in \partial_{\mathfrak{C}_\mathbb{C}}$. □

Part 3.3.8(b) is of course the most consequential of these statements. We shall illustrate it in the following example using the real case.

3.3.9 Example. Let X be a compact convex set and suppose that $C_{\mathfrak{V}}(X) = C(X)$ is endowed with the topology of uniform convergence. Let $\mathfrak{U} = \{\mathfrak{u}\}$, where $\mathfrak{u} = \{f \in C(X) \mid f \leq 1\}$. Then $\bigcap\{\gamma\mathfrak{u} \mid \mathfrak{Re}\,\gamma > 0\}$ is the cone of all negative functions, and the assumptions of Corollary 3.3.8 hold. For $\mathcal{H}$, we choose the cone of all concave functions in $C(X)$. We opt for $\mathcal{G} = \{\mathbb{R}\}$ and the neighborhood $\mathfrak{u}_{\mathbb{B}}^1 = \{f \in C(X) \mid |f| \leq 1\}$ corresponding to $\mathbb{R} \in \mathcal{G}$. Then $\partial_{\mathfrak{C}_\mathbb{R}}$ is the set of extreme points of X according to classical Choquet theory (see [1] or [41]). Moreover, $H_{\mathbb{R}}$ is the subspace of all affine functions in $\mathcal{C}(X)$ and Corollary 3.3.5 applies. For F, we choose a sequence space ℓ^p for $1 < p < +\infty$ furnished with its pointwise order and for Π the positive elements of the unit ball in $F^* = \ell^q$, where $1/p + 1/q = 1$. The Π-positive elements of ℓ^p are its positive sequences. The requirements of Propositions 3.3.7 and 3.3.8 are easily verified. Thus 3.3.8(b) applies, but the main results of Section 3.2 lend themselves only to operators $\mathcal{T} \in \mathfrak{L}(\mathcal{C}_{\mathfrak{V}}(X), \ell^p)$ such that $\mathfrak{p}_{\mathfrak{Q}^\Pi_\mathfrak{u}}(\mathcal{T}) < +\infty$. Since $\mathfrak{u}$ contains all negative functions in $\mathcal{C}_{\mathfrak{V}}(X)$, the latter requires that the operator $\mathcal{T}$ is positive. We observe that $p_\Pi(c) = \|c\|_p \le \|c\|_1$ holds for positive sequences $c \in \ell^p$ and that the norm $\| \, \|_1$ is monotone and additive on the positive cone of ℓ^p. Using this, for a positive operator $\mathcal{T} \in \mathfrak{L}(\mathcal{C}_{\mathfrak{V}}(X), \ell^p)$ we calculate

$$\begin{aligned}
\mathfrak{p}_{\mathfrak{Q}^\Pi_\mathfrak{u}}(\mathcal{T}) &= \sup\left\{ \sum_{i=1}^{n} \|\mathcal{T}(\varphi_i \cdot f_i)\|_p \;\middle|\; \begin{array}{l} f_i \in \mathfrak{u},\ 0 \le \varphi_i \in \mathcal{C}(X) \\ \text{such that } \sum_{i=1}^n \varphi_i = 1 \end{array} \right\} \\
&\le \sup\left\{ \sum_{i=1}^{n} \|\mathcal{T}(\varphi_i)\|_1 \;\middle|\; \begin{array}{l} 0 \le \varphi_i \in \mathcal{C}(X) \\ \text{such that } \sum_{i=1}^n \varphi_i = 1 \end{array} \right\} \\
&= \|\mathcal{T}(1)\|_1.
\end{aligned}$$

This provides a sufficient condition for $\mathfrak{p}_{\mathfrak{Q}^\Pi_\mathfrak{u}}(\mathcal{T})$ to be finite. It is however not necessary, as a simple example can show: Let $x \in X$ and define $\mathcal{T} \in \mathfrak{L}(\mathcal{C}_{\mathfrak{V}}(X), \ell^p)$ by $\mathcal{T}(f) = f(x)c$, where $c = (1/i)_{i \in \mathbb{N}} \in \ell^p$. That is, $\mathcal{T}$ is the elementary operator $\mathcal{L}^c_{\delta_x}$ (see Section 3.1). Then

$$\mathfrak{p}_{\mathfrak{Q}^\Pi_\mathfrak{u}}(\mathcal{T}) = p_\Pi(c) \sup\{f(x) \mid f \in \mathfrak{u}\} = \|c\|_p < +\infty$$

by Proposition 3.1.15(b), but $\|\mathcal{T}(1)\|_1 = +\infty$. If $\mathfrak{p}_{\mathfrak{Q}^\Pi_\mathfrak{u}}(\mathcal{T}) < +\infty$ for an operator $\mathcal{T} \in \mathfrak{L}(\mathcal{C}_{\mathfrak{V}}(X), \ell^p)$, then the results of Section 3.2 apply. The Choquet boundary $\partial_{\mathfrak{C}}$ is contained in the set of extreme points by Proposition 3.3.8(b) and Proposition 3.2.9 guarantees that there is an operator $\mathcal{S} \prec_{\tilde{\mathfrak{C}}} \mathcal{T}$, which is minimal with respect to the order $\prec_{\tilde{\mathfrak{C}}}$, and according to Corollary 3.3.5 the representation measure for $\mathcal{S}$ is supported by any F_σ-subset of X that contains all extreme points of X. (Recall that $\partial_{\tilde{\mathfrak{C}}} \subset \partial_{\mathfrak{C}}$.) For a concrete example of this setting, let $X = [0, 1]$ and consider the operator $\mathcal{T} \in \mathfrak{L}(\mathcal{C}_{\mathfrak{V}}(X), \ell^p)$ defined by

$$\mathcal{T}(f) = \big((1/i^2) f(1/i)\big)_{i \in \mathbb{N}}.$$

Then

$$\mathfrak{p}_{\mathfrak{Q}^\Pi_\mathfrak{u}}(\mathcal{T}) \le \|\mathcal{T}(1)\|_1 = \sum_{i=1}^{\infty} 1/i^2 < +\infty$$

by the above, and the results of Section 3.2 apply. The Choquet boundary $\partial_{\tilde{\mathfrak{C}}}$ is contained in the set $\{0, 1\}$, and in this example we can easily construct a minimal operator $\mathcal{S} \prec_{\tilde{\mathfrak{C}}} \mathcal{T}$, which is supported by this set. We have $\mathcal{S} = \mathcal{D}^c_0 + \mathcal{D}^d_1$, where $c = (\gamma_i)_{i \in \mathbb{N}}$ and $d = (\delta_i)_{i \in \mathbb{N}}$, both contained in ℓ^p and easily determined by the fact that $\mathcal{S}$ and $\mathcal{T}$

coincide on all affine functions on $[0,1]$. For the function $f(x) = x$, this requires that $\delta_i = 1/i^3$ for all $i \in \mathbb{N}$ and for the constant function $f(x) = 1$ that $\gamma_i + \delta_i = 1/i^2$. Thus $\gamma_i = (i-1)/i^3$.

We recall from Proposition 1.2.9 that the assumptions of this subsection also apply if E is a function space $\mathcal{C}_{\mathfrak{U}}(Y, G)$, where G is a normed space, Y is compact and $\mathfrak{U}$ renders the topology of uniform convergence on Y. The space $\mathcal{C}_{\mathfrak{V}}(X, \mathcal{C}_{\mathfrak{U}}(Y,G))$ was understood to be isomorphic to a space $\mathcal{C}_{\overline{\mathfrak{V}}}(X \times Y, G)$ in this case.

The case that both *E* and *F* are either $\mathbb{R}$ or $\mathbb{C}$

Again we formulate the complex case. In this setting, $\mathfrak{L}(E,F)$ equals $\mathbb{C}$, the operator space $\mathfrak{L}(\mathcal{C}_{\mathfrak{V}}(X,E),F)$ is the dual $\mathcal{C}_{\mathfrak{V}}(X,\mathbb{C})^*$ of $\mathcal{C}_{\mathfrak{V}}(X,\mathbb{C})$ endowed with its weak* topology, and the dual $\mathcal{C}_{\mathfrak{V}}(X,\mathbb{C}) \otimes F^*$ of $\mathfrak{L}(\mathcal{C}_{\mathfrak{V}}(X,E),F)$ equals $\mathcal{C}_{\mathfrak{V}}(X,\mathbb{C})$ endowed with its weak topology. We shall use the previously established sample settings with $\Pi = \{1\}$ and consider the balanced case. That is, $\mathfrak{K} = \mathfrak{U}$, where $\mathfrak{U}$ is a family of balanced $C(X)$-convex neighborhoods of 0 in $\mathcal{C}_{\mathfrak{V}}(X,E)$ with the required properties, $\mathcal{H}$ is a subspace of $\mathcal{C}_{\mathfrak{V}}(X,\mathbb{C})$ and either $\mathcal{G} = \{\mathbb{C}\}$ or $\mathcal{G} = \emptyset$. For $\mathfrak{u} = \Omega_{\mathfrak{u}}^{\{1\}} \in \mathfrak{C}$ and $\mathcal{T} \in \mathcal{C}_{\mathfrak{V}}(X,\mathbb{C})^*$, we have

$$\mathfrak{p}_{\mathfrak{u}}(\mathcal{T}) = \sup\{\mathfrak{Re}\,\mathcal{T}(f) \mid f \in \mathfrak{u}\}.$$

Proposition 3.3.1 applies and Proposition 3.3.4 and Corollary 3.3.5 simplify in obvious ways. This special case allows for the extension and further investigation of some of the concepts from the general case. The representation measure θ for an operator $\mathcal{T} \in \mathcal{C}_{\mathfrak{V}}(X,\mathbb{C})^*$ is complex-valued. We abbreviate $[\theta]$ for the $\overline{\mathbb{R}}_+$-valued measure $[|\theta|(\mathbb{B})(1)]$, which is the usual absolute value of a complex-valued measure. (See Example 2.1.8(f).) Boundedness means that $[\theta]_A < +\infty$ for all $A \in \mathfrak{R}$, that is, that $[\theta]$ is real-valued. Corollary 3.3.5 applies with $[\theta]$ in place of $[\|\theta\|_G|(V \cap G)(\mu)]$. Moreover, it is well known that there is a measurable function ω such that $|\omega(x)| = 1$ and $\theta = \omega[\theta]$ (see Theorem 6.12 in [55]), that is, $\theta_A = \int_A \omega\, d[\theta]$ for all $A \in \mathfrak{R}$. The measure $[\theta]$ defines and represents a functional $[\mathcal{T}] \in \mathcal{C}_{\mathfrak{V}}(X,\mathbb{C})^*$, called the *absolute value* of $\mathcal{T}$ and defined by $[\mathcal{T}](f) = \int_X f\, d[\theta]$ for $f \in \mathcal{C}_{\mathfrak{V}}(X,\mathbb{C})$. We have

$$[\mathcal{T}](f) = \sup\{\mathfrak{Re}\,\mathcal{T}(g) \mid g \in \mathcal{C}_{\mathfrak{V}}(X,\mathbb{C}),\ |g| \le f\} = \mathfrak{p}_{\mathfrak{u}^f_{\mathbb{B}}}(\mathcal{T})$$

for nonnegative functions $f \in \mathcal{C}_{\mathfrak{V}}(X)$ (see also Propositions 2.3.25 and 3.1.11).

The statement of Proposition 3.3.6 can be developed further in this particular situation. We extend the notation of the composition of an operator with a bounded continuous real- or complex-valued function from Section 2.6 in the following way: Let $\mathcal{T} \in \mathcal{C}_{\mathfrak{V}}(X,\mathbb{C})^*$ and let φ be a bounded measurable complex-valued function. It is straightforward to argue that for all $f \in \mathcal{C}_{\mathfrak{V}}(X,\mathbb{C})$ the function $\varphi \cdot f$ is integrable over all sets $A \in \mathfrak{R}$ with respect to the representation measure θ for $\mathcal{T}$. We shall argue that this function is also integrable over X. Indeed, the measure θ is continuous relative to some neighborhood $\mathfrak{v} \in \mathfrak{V}$ (Theorem 2.6.1), that is, $|\int_X g\, d\theta| \leq 1$ for all integrable functions $g \leq \mathfrak{v}$. Then, given $\varepsilon > 0$, since f vanishes at infinity, there is $A \in \mathfrak{R}$ such that $\chi_{X\setminus A} \cdot f \in \varepsilon\mathfrak{v}$. Thus for $A, B, C \in \mathfrak{R}$ such that $A \subset C \subset B$ we have $\chi_{B\setminus C} \cdot f \in \varepsilon\mathfrak{v}$ and, therefore,

$$\left| \int_B \varphi \cdot f\, d\theta - \int_C \varphi \cdot f\, d\theta \right| = \left| \int_{B\setminus C} \varphi \cdot f\, d\theta \right| < \lambda\varepsilon,$$

where $\lambda \geq 0$ is a bound for the function $|\varphi|$. Thus the net $(\int_A \varphi \cdot f)_{A\in\mathfrak{R}}$ is indeed convergent and its limit is $\int_X \varphi\cdot f$. Furthermore, since $\int_X \varphi\cdot f \leq \lambda$ for all $f \leq \mathfrak{v}$, the formula $\mathcal{T}_\varphi(f) = \int_X \varphi \cdot f\, d\theta$ defines a functional in $\mathcal{C}_{\mathfrak{V}}(X, \mathbb{C})^*$. Moreover, if θ is the complex-valued representation measure for $\mathcal{T}$, then $\varphi\theta$ is the representation measure for $\mathcal{T}_\varphi$.

A *unimodular function* is a measurable function $\omega \in \mathcal{F}(X, \mathbb{C})$ such that $|\omega|$ takes only the values 0 and 1 on X. We set $X_\omega = \{x \in X \mid |\omega(x)| = 1\}$. For a function $f \in \mathcal{F}(X, \mathbb{C})$, we define

$$\omega_f(x) = \begin{cases} f(x)/|f(x)|, & \text{if } f(x) \neq 0 \\ 0, & \text{if } f(x) = 0. \end{cases}$$

Hence $f = |f| \cdot \omega_f$. For the following, we assume that one of the neighborhoods used in the setup for $\mathfrak{U}$ equals $\mathfrak{u}^\rho_{\mathbb{B}}$, where ρ is a strictly positive lower semicontinuous $\overline{\mathbb{R}}$-valued function. Proposition 3.3.6 returns the following.

3.3.10 Proposition. *Suppose that $\mathfrak{u}^\rho_{\mathbb{B}} \in \mathfrak{U}$. If for $\mathcal{T} \in \mathcal{C}_{\mathfrak{V}}(X, \mathbb{C})^*$ there is $g \in \mathcal{H} \cap \mathfrak{u}^\rho_{\mathbb{B}}$ such that $\mathfrak{Re}\,\mathcal{T}(g) = \mathfrak{p}_{\mathfrak{u}^\rho_{\mathbb{B}}}(\mathcal{T})$, then the representation measure for every $\mathcal{S} \in \mathcal{C}_{\mathfrak{V}}(X, \mathbb{C})^*$ such that $\mathcal{S} \prec_{\mathfrak{C}} \mathcal{T}$ is supported by the compact set $Y = \{x \in X \mid |g(x)| = \rho(x)\}$. We have $[\mathcal{S}] = \mathcal{S}_{\omega_g}$ and $\mathcal{S} = [\mathcal{S}]_{\overline{\omega}_g}$ and $\mathfrak{p}_{\mathfrak{u}^\rho_{\mathbb{B}}}(\mathcal{S}) = \mathfrak{p}_{\mathfrak{u}^\rho_{\mathbb{B}}}(\mathcal{T})$.*

Proof. If the functions ρ and g and the linear functionals $\mathcal{S}, \mathcal{T} \in \mathcal{C}_{\mathfrak{V}}(X, \mathbb{C})^*$ are as stated, we may utilize Proposition 3.3.6 with $\mu = 1$, which states that the representation measure ϑ for $\mathcal{S}$ is supported by the compact set $Y = \{x \in X \mid |g(x)| = \rho(x)\}$. Because $g(x) \neq 0$ for all $x \in Y$, the function ω_g is continuous on Y and admits a continuous extension $\varphi \in \mathcal{C}_{\mathfrak{V}}(X, \mathbb{C})$ with compact support and such that $|\varphi| \leq 1$. For every $f \in \mathcal{C}_{\mathfrak{V}}(X, \mathbb{C})$ and $x \in Y$, we have $\mathfrak{Re}\,(\varphi \cdot f)(x)\overline{g(x)} = |g(x)|\mathfrak{Re} f(x)$. Thus according to Proposition 3.3.6, if $\mathfrak{Re} f(x) \leq 0$ for all $x \in Y$, then $\mathfrak{Re}\,\mathcal{S}(\varphi \cdot f) \leq 0$. We infer that $\mathfrak{Re}\,\mathcal{S}(\varphi \cdot f) = 0$ whenever $\mathfrak{Re} f = 0$ and $\mathfrak{Im}\,\mathcal{S}(\varphi \cdot f) = 0$ whenever $\mathfrak{Im} f = 0$. Hence $\mathcal{S}_\varphi(f)$ is real-valued for real-valued functions f and $\mathcal{S}_\varphi(f) \leq 0$ whenever $f \leq 0$. Since $\varphi \cdot f = \varphi \cdot \mathfrak{Re} f + i\varphi \cdot \mathfrak{Im} f$, this yields

$$\mathfrak{Re}\,\mathcal{S}_\varphi(f) = \mathcal{S}_\varphi(\mathfrak{Re} f) \quad \text{and} \quad \mathfrak{Im}\,\mathcal{S}_\varphi(f) = \mathcal{S}_\varphi(\mathfrak{Im} f)$$

for all $f \in \mathcal{C}_{\mathfrak{V}}(X,\mathbb{C})$. Moreover, because ϑ is supported by Y and $(\varphi \cdot \overline{\varphi})(x) = 1$ for all $x \in Y$, we observe that $\mathcal{S}(f) = \mathcal{S}_\varphi(f \cdot \overline{\varphi})$ holds for all $f \in \mathcal{C}_{\mathfrak{V}}(X,\mathbb{C})$. Using this, for a nonnegative real-valued function $f \in \mathcal{C}_{\mathfrak{V}}(X,\mathbb{C})$, we calculate

$$\begin{aligned}[\mathcal{S}](f) &= \sup\{\mathfrak{Re}\,\mathcal{S}(h) \mid h \in \mathcal{C}_{\mathfrak{V}}(X,\mathbb{C}),\ |h| \le f\}\\ &= \sup\{\mathfrak{Re}\,\mathcal{S}_\varphi(h \cdot \overline{\varphi}) \mid h \in \mathcal{C}_{\mathfrak{V}}(X,\mathbb{C}),\ |h| \le f\}\\ &= \sup\{\mathcal{S}_\varphi(\mathfrak{Re}(h \cdot \overline{\varphi})) \mid h \in \mathcal{C}_{\mathfrak{V}}(X,\mathbb{C}),\ |h| \le f\} = \mathcal{S}_\varphi(f),\end{aligned}$$

that is, $[\mathcal{S}] = \mathcal{S}_\varphi = \mathcal{S}_{\omega_g}$. The last equality arises since φ and ω_g coincide on Y. Conversely, since

$$\mathcal{S}(f) = \mathcal{S}_\varphi(f \cdot \overline{\varphi}) = [\mathcal{S}](f \cdot \overline{\varphi}) = [\mathcal{S}]_{\overline{\varphi}}(f)$$

holds for all $f \in \mathcal{C}_{\mathfrak{V}}(X,E)$, we have $\mathcal{S} = [\mathcal{S}]_{\overline{\varphi}} = [\mathcal{S}]_{\overline{\omega}_g}$ as well. Moreover, $\mathcal{S} \prec_{\mathfrak{C}} \mathcal{T}$ implies that $\mathfrak{p}_{\mathfrak{u}_{\mathbb{B}}^\rho}(\mathcal{S}) \le \mathfrak{p}_{\mathfrak{u}_{\mathbb{B}}^\rho}(\mathcal{T})$ and $\mathcal{S}(g) = \mathcal{T}(g)$. Thus

$$\mathfrak{p}_{\mathfrak{u}_{\mathbb{B}}^\rho}(\mathcal{S}) \ge \mathfrak{Re}\,\mathcal{S}(g) = \mathfrak{Re}\,\mathcal{T}(g) = \mathfrak{p}_{\mathfrak{u}_{\mathbb{B}}^\rho}(\mathcal{T}).$$

Hence our claim. □

Following Proposition 3.3.1, a point evaluation $\mathcal{D}_x^\eta$ for $x \in X$ and $0 \ne \eta \in \mathbb{C}$ is minimal if and only if $\mathcal{D}_x = \mathcal{D}_x^1$ is minimal. Moreover, if

$$0 < \mathfrak{p}_{\mathfrak{u}}(\mathcal{D}_x^\eta) = |\eta| \sup\{|f(x)| \mid f \in \mathfrak{u}\} < +\infty$$

for some $\mathfrak{u} \in \mathfrak{U}$ and if $\eta = \eta_1 + \eta_2$ and $\mathfrak{p}_{\mathfrak{u}}(\mathcal{D}_x^\eta) = \mathfrak{p}_{\mathfrak{u}}(\mathcal{D}_x^{\eta_1}) + \mathfrak{p}_{\mathfrak{u}}(\mathcal{D}_x^{\eta_2})$ holds for $0 \ne \eta_1, \eta_2 \in \mathbb{C}$, then $\eta/|\eta| = \eta_1/|\eta_1| = \eta_2/|\eta_2|$ and, therefore,

$$\mathfrak{p}_{\mathfrak{u}}(\mathcal{D}_x^\eta)\,\eta_1 = \mathfrak{p}_{\mathfrak{u}}(\mathcal{D}_x^{\eta_1})\,\eta \quad\text{and}\quad \mathfrak{p}_{\mathfrak{u}}(\mathcal{D}_x^\eta)\,\eta_2 = \mathfrak{p}_{\mathfrak{u}}(\mathcal{D}_x^{\eta_2})\,\eta.$$

This demonstrates that any point evaluation operator $\mathcal{D}_x^\eta \in \mathcal{C}_{\mathfrak{V}}(X,\mathbb{C})^*$ is $\mathfrak{K}$-extremal, provided that $0 < \mathfrak{p}_{\mathfrak{u}}(\mathcal{D}_x) < +\infty$ for some $\mathfrak{u} \in \mathfrak{U}$. We summarize with the following.

3.3.11 Lemma. *A point evaluation $\mathcal{D}_x^\eta$ for $x \in X$ and $0 \ne \eta \in \mathbb{C}$ is an element of $\Delta_{\mathfrak{C}}$ if and only if $\mathcal{D}_x$ is minimal and $0 < \mathfrak{p}_{\mathfrak{u}}(\mathcal{D}_x) < +\infty$ for some $\mathfrak{u} \in \mathfrak{U}$. A $C(X)$-convex subset $\mathfrak{w}$ of $\mathcal{C}_{\mathfrak{V}}(X,\mathbb{C})$ is $\mathfrak{C}$-superharmonic if and only if $\mathfrak{p}_{\mathfrak{u}}(\mathcal{T}) \le \mathfrak{p}_{\mathfrak{u}}(\mathcal{D}_x)$ whenever $\mathcal{T} \prec_{\mathfrak{C}} \mathcal{D}_x$ for $\mathcal{T}, \mathcal{D}_x \in \mathcal{C}_{\mathfrak{V}}(X,\mathbb{C})^*$, where $\mathcal{D}_x$ is a point evaluation such that $0 < \mathfrak{p}_{\mathfrak{u}}(\mathcal{D}_x) < +\infty$ for some $\mathfrak{u} \in \mathfrak{U}$.*

3.3.12 Proposition. *Let Y be a closed subset of X. Suppose that for every $y \in Y$ there is $\mathfrak{u}_{\mathbb{B}}^{\rho_y} \in \mathfrak{U}$ and $g_y \in \mathcal{H} \cap \mathfrak{u}_{\mathbb{B}}^{\rho_y}$ such that $|g_y(y)| = \rho_y(y)$ and $|g_y(x)| < \rho_y(x)$ for all $x \in X \setminus Y$. Then the following hold:*

(a) *If $\mathcal{T} \prec_{\mathfrak{C}} \mathcal{D}_y$ for $\mathcal{T}, \mathcal{D}_y \in \mathcal{C}_{\mathfrak{V}}(X,\mathbb{C})^*$, where $\mathcal{D}_y$ is a point evaluation such that $y \in Y$, then the representation measure for $\mathcal{T}$ is supported by Y and $\mathcal{T} = [\mathcal{T}]_{\gamma\overline{\omega}_{g_y}}$, where $\gamma = \overline{\omega}_{g_y}(y)$.*

(b) *If $\mathfrak{C}$ contains an absorbing $\mathfrak{U}$-bounded set, if $\mathcal{S} \prec_{\widetilde{\mathfrak{C}}} \mathcal{T}$ for $\mathcal{S}, \mathcal{T} \in \mathcal{C}_{\mathfrak{V}}(X, \mathbb{C})^*$, and if the representation measure for $\mathcal{T}$ is supported by Y, then the representation measure for $\mathcal{S}$ is also supported by Y.*

Proof. (a) Let $y \in Y$. We employ Proposition 3.3.10 with $\mathfrak{u}_{\mathbb{B}}^{\rho_y} \in \mathfrak{U}$, the function $g = \gamma g_y \in \mathfrak{u}_{\mathbb{B}}^{\rho_y} \cap L$, where $\gamma = \overline{\omega}_{g_y}(y) = \overline{g_y(y)/|g_y(y)|}$ and the point evaluation $\mathcal{D}_y$ in place of $\mathcal{T}$. We have

$$g(y) = |g(y)| = |g_y(y)| = \rho_y(y),$$

and

$$\mathfrak{p}_{\mathfrak{u}_{\mathbb{B}}^{\rho_y}}(\mathcal{D}_y) = \rho_y(x) = g(x) = \mathcal{D}_y(g),$$

as required. Thus, if $\mathcal{T} \prec_{\mathfrak{C}} \mathcal{D}_y$ for $\mathcal{T} \in \mathcal{C}_{\mathfrak{V}}(X, \mathbb{C})^*$, then according to 3.3.10 the representation measure for $\mathcal{T}$ is supported by Y and $\mathcal{T} = \mathcal{T}_{\overline{\omega}_g} = [\mathcal{T}]_{\gamma\overline{\omega}_{g_y}}$ For Part (b), suppose that $\mathfrak{C}$ contains an absorbing $\mathfrak{U}$-bounded set $\mathfrak{u}$. Then the set

$$\tilde{\mathfrak{u}} = \{f \in \mathfrak{w} \mid f|_Y = 0\}$$

is also $\mathfrak{U}$-bounded, and $C(X)$-superharmonic as we go on to demonstrate. For this, let $\mathcal{T} \prec_{\mathfrak{C}} \mathcal{D}_x$ for $x \in X$ such that $0 < \mathfrak{p}_{\mathfrak{w}}(\mathcal{D}_x) < +\infty$ for some $\mathfrak{w} \in \mathfrak{U}$. If $x \in Y$, then $\mathcal{T}$ is supported by Y by Part (a) and we have $\mathfrak{p}_{\tilde{\mathfrak{u}}}(\mathcal{T}) = \mathfrak{p}_{\tilde{\mathfrak{u}}}(\mathcal{D}_x) = 0$. If $x \in X \setminus Y$, we find a function $\varphi \in \mathcal{C}_{\mathcal{K}}(X)$ such that $0 \le \varphi \le \chi_{X \setminus Y}$ and $\varphi(x) = 1$. Then $\varphi \cdot f \in \tilde{\mathfrak{u}}$ whenever $f \in \mathfrak{u}$ and

$$\begin{aligned}
\mathfrak{p}_{\mathfrak{u}}(\mathcal{D}_x) &= \sup\{|f(x)| \mid f \in \mathfrak{u}\} \\
&= \sup\{|(\varphi \cdot f)(x)| \mid f \in \mathfrak{u}\} \\
&= \sup\{|f(x)| \mid f \in \tilde{\mathfrak{u}}\} \\
&= \mathfrak{p}_{\tilde{\mathfrak{u}}}(\mathcal{D}_x).
\end{aligned}$$

Thus

$$\mathfrak{p}_{\tilde{\mathfrak{u}}}(\mathcal{T}) \le \mathfrak{p}_{\mathfrak{u}}(\mathcal{T}) \le \mathfrak{p}_{\mathfrak{u}}(\mathcal{D}_x) = \mathfrak{p}_{\tilde{\mathfrak{u}}}(\mathcal{D}_x).$$

We infer that $\tilde{\mathfrak{u}}$ is indeed $C(X)$-superharmonic and, therefore, an element of $\widetilde{\mathfrak{C}}$. Hence we have $\mathfrak{p}_{\tilde{\mathfrak{u}}}(\mathcal{S}) \le \mathfrak{p}_{\tilde{\mathfrak{u}}}(\mathcal{T}) = 0$ whenever $\mathcal{S} \prec_{\widetilde{\mathfrak{C}}} \mathcal{T}$ for $\mathcal{S}, \mathcal{T} \in \mathcal{C}_{\mathfrak{V}}(X, \mathbb{C})^*$ such that the representation measure for $\mathcal{T}$ is supported by Y. That is, $\mathfrak{p}_{\tilde{\mathfrak{u}}}(\mathcal{S}) = 0$. Now Proposition 2.5.9 yields that the representation measure for $\mathcal{S}$ is also supported by Y. Indeed, let $\varphi \in \mathcal{C}_{\mathcal{K}}(X)$ and $a \in E$ such that φ is supported by $X \setminus Y$. Since $\mathfrak{u}$ is absorbing and balanced, there is $\varepsilon > 0$ such that $\gamma\varphi{\cdot}a \in \tilde{\mathfrak{u}}$ for all $|\gamma| \le \varepsilon$. Hence $|\mathcal{S}(\varphi{\cdot}a)| \le \mathfrak{p}_{\tilde{\mathfrak{u}}}(\mathcal{S}) = 0$. Our claim in Part (b) follows. □

In preparation for our next result, for a closed subset Y of X and $f \in \mathcal{F}(X,\mathbb{C})$, we abbreviate $f|_Y$ for $\chi_Y \cdot f$. For a $\mathcal{C}(X)$-convex subset $\mathfrak{u}$ of $\mathcal{C}_{\mathfrak{V}}(X,\mathbb{C})$ and $Y \subset X$ we set

$$\mathfrak{u}|Y = \{f \in \mathcal{C}_{\mathfrak{V}}(X,\mathbb{C}) \mid f|_Y = g|_Y \text{ for some } g \in \mathfrak{u}\}$$

and observe the following.

3.3.13 Lemma. *Let $\mathfrak{u}$ be a $\mathcal{C}(X)$-convex subset of $\mathcal{C}_{\mathfrak{V}}(X,\mathbb{C})$ and let Y be a closed subset of X. Let $\mathcal{S},\mathcal{T} \in \mathcal{C}_{\mathfrak{V}}(X,\mathbb{C})^*$, and let ϑ and θ be their respective representation measures.*

(a) *If $\mathfrak{p}_{\mathfrak{u}|Y}(\mathcal{T}) < +\infty$, then θ is supported by Y.*
(b) *If θ is supported by Y, then $\mathfrak{p}_{\mathfrak{u}|Y}(\mathcal{T}) = \mathfrak{p}_{\mathfrak{u}}(\mathcal{T})$.*
(c) *If $\mathfrak{u}|Y \in \widetilde{\mathfrak{C}}$, if $\mathfrak{p}_{\mathfrak{u}|Y}(\mathcal{T}) < +\infty$, and if $\mathcal{S} \prec_{\widetilde{\mathfrak{C}}} \mathcal{T}$, then ϑ is supported by Y.*
(d) *If Y is the union of a family $\{Y_i\}_{i\in\mathcal{I}}$ of closed subsets of X and if all the sets $\mathfrak{u}|Y_i$ are $\mathfrak{C}$-superharmonic, then $\mathfrak{u}|Y$ is also $\mathfrak{C}$-superharmonic.*

Proof. For (a), suppose that there is $x \in \operatorname{supp}\theta$ for $x \notin Y$. There is $g \in \mathcal{C}_{\mathfrak{V}}(X,\mathbb{C})$ such that $g|_Y = 0$ and $\mathcal{T}(g) = \int_X g\, d\theta \neq 0$. Thus $f + \alpha g \in \mathfrak{u}|Y$ whenever $f \in \mathfrak{u}$ and $\alpha \in \mathbb{C}$. Hence $\mathfrak{p}_{\mathfrak{u}|Y}(\mathcal{T}) = +\infty$, our claim. For (b), suppose that θ is supported by Y. Then

$$\begin{aligned}\mathfrak{p}_{\mathfrak{u}|Y}(\mathcal{T}) &= \sup\left\{\mathfrak{Re}\int_Y f\, d\theta \mid f \in \mathfrak{u}|Y\right\}\\ &= \sup\left\{\mathfrak{Re}\int_Y f\, d\theta \mid f \in \mathfrak{u}\right\} = \mathfrak{p}_{\mathfrak{u}}(\mathcal{T}).\end{aligned}$$

For (c), if $\mathfrak{u}|Y \in \widetilde{\mathfrak{C}}$, if $\mathcal{S} \prec_{\widetilde{\mathfrak{C}}} \mathcal{T}$ and if $\mathfrak{p}_{\mathfrak{u}|Y}(\mathcal{T}) < +\infty$, then $\mathfrak{p}_{\mathfrak{u}|Y}(\mathcal{S}) \le \mathfrak{p}_{\mathfrak{u}|Y}(\mathcal{T}) < +\infty$. Hence by (a) the representation measure for $\mathcal{S}$ is supported by Y. We use Lemma 3.3.11 for Part (d). Suppose that the sets $\mathfrak{u}|Y_i$ are $\mathfrak{C}$-superharmonic and let $\mathcal{S} \prec_{\mathfrak{C}} \mathcal{D}_X$ for $\mathcal{S} \in \mathcal{C}_{\mathfrak{V}}(X,\mathbb{C})^*$ and $x \in X$ such that $0 < \mathfrak{p}_{\mathfrak{w}}(\mathcal{D}_X) < +\infty$ for some $\mathfrak{w} \in \mathfrak{U}$. There is nothing to prove if $\mathfrak{p}_{\mathfrak{u}|Y}(\mathcal{D}_X) = +\infty$. Otherwise, we have $x \in Y$ by Part (a), that is, $x \in Y_i$ for some $i \in \mathcal{I}$. Then

$$\mathfrak{p}_{\mathfrak{u}|Y_i}(\mathcal{D}_X) = \mathfrak{p}_{\mathfrak{u}}(\mathcal{D}_X) = \mathfrak{p}_{\mathfrak{u}|Y}(\mathcal{D}_X) < +\infty$$

by Part (b). Now, since $\mathcal{S} \prec_{\mathfrak{C}} \mathcal{D}_X$ implies that $\mathfrak{p}_{\mathfrak{u}|Y_i}(\mathcal{S}) \le \mathfrak{p}_{\mathfrak{u}|Y_i}(\mathcal{D}_X) < +\infty$, Part (a) renders that the representation measure ϑ for $\mathcal{S}$ is supported by Y_i. Thus

$$\mathfrak{p}_{\mathfrak{u}|Y}(\mathcal{S}) = \mathfrak{p}_{\mathfrak{u}|Y_i}(\mathcal{S}) \le \mathfrak{p}_{\mathfrak{u}|Y_i}(\mathcal{D}_X) = \mathfrak{p}_{\mathfrak{u}|Y}(\mathcal{D}_X).$$ □

Given a unimodular function ω, for every $f \in \mathcal{F}(X,\mathbb{C})$ we define its ω-*real* and ω-*imaginary* parts as the real-valued functions $\overset{\omega}{\mathfrak{Re}} f = \mathfrak{Re}(f \cdot \overline{\omega})$ and $\overset{\omega}{\mathfrak{Im}} f = \mathfrak{Im}(f \cdot \overline{\omega})$. Then

$$\overset{\omega}{\mathfrak{Re}} f \cdot \omega + i \overset{\omega}{\mathfrak{Im}} f \cdot \omega = f|_{X_\omega},$$

where $X_\omega = \{x \in X \mid |\omega(x)| = 1\}$. If both f and ω are continuous on a subset Y of X, so are both $\overset{\omega}{\mathfrak{Re}} f$ and $\overset{\omega}{\mathfrak{Im}} f$. The following criterion is formulated in very general, therefore, somewhat complicated terms.

3.3.14 Proposition. *Let $\omega \in \mathcal{F}(X, \mathbb{C})$ be a unimodular function and let Y be a closed subset of X_ω. Suppose that for every $y \in Y$ there is $\mathfrak{u}_{\mathbb{B}}^{\rho_y} \in \mathfrak{U}$ and a continuous nonnegative real-valued function φ_y such that $\varphi_y \cdot \omega \in \mathcal{H} \cap \mathfrak{u}_{\mathbb{B}}^{\rho_y}$ and such that $\varphi_y(y) = \rho_y(y)$ and $\varphi_y(x) < \rho_y(x)$ for all $x \in X \setminus Y$. Then the following hold:*

(a) *If $\mathfrak{u} \subset \mathcal{C}_{\mathfrak{Y}}(X, \mathbb{C})$ is $\mathfrak{C}$-superharmonic, then $\mathfrak{u}|Y$ is $\mathfrak{C}$-superharmonic.*

(b) *If $\mathfrak{u} \subset \mathcal{C}_{\mathfrak{Y}}(X, \mathbb{C})$ and $\mathfrak{u}|Y$ is $\mathfrak{C}$-superharmonic, then $\{\overset{\omega}{\mathfrak{Re}} f \cdot \omega \mid f \in \mathfrak{u}\}|_Y$ is $\mathfrak{C}$-superharmonic.*

(c) *If $f_1, \dots, f_n$ and $g_1, \dots, g_m$ are functions in $\mathcal{H}$, then both sets*

$$\{h \cdot \omega \mid h \in C(X),\ \bigvee_{i=1}^n \overset{\omega}{\mathfrak{Re}} f_i|_Y \le h|_Y\}|_Y$$

and

$$\{h \cdot \omega \mid h \in C(X),\ h|_Y \le \bigwedge_{k=1}^m \overset{\omega}{\mathfrak{Re}}\, g_k|_Y\}|_Y$$

are $\mathfrak{C}$-superharmonic, so is their intersection, provided it is not empty.

(d) *If $\{0\}|_Y$ is $\mathfrak{U}$-bounded, if $\mathcal{S} \prec_{\tilde{\mathfrak{C}}} \mathcal{T}$ for $\mathcal{S}, \mathcal{T} \in \mathcal{C}_{\mathfrak{Y}}(X, \mathbb{C})^*$, and if the representation measure for $\mathcal{T}$ is supported by Y, then the representation measure for $\mathcal{S}$ is also supported by Y.*

Proof. Our claims in Parts (a) to (c) are concerned with $\mathfrak{C}$-superharmonicity of sets $\mathfrak{u}|Y$, where $\mathfrak{u}$ is a $C(X)$-convex subset of $\mathcal{C}_{\mathfrak{Y}}(X, \mathbb{C})$. For this, let $\mathcal{S} \prec_{\mathfrak{C}} \mathcal{D}_x$ for $x \in X$ such that $0 < \mathfrak{p}_{\mathfrak{w}}(\mathcal{D}_x) < +\infty$ for some $\mathfrak{w} \in \mathfrak{U}$. We need to establish that $\mathfrak{p}_{\mathfrak{u}|Y}(\mathcal{S}) \le \mathfrak{p}_{\mathfrak{u}|Y}(\mathcal{D}_x)$. If $x \notin Y$, then $\mathfrak{p}_{\mathfrak{u}|Y}(\mathcal{S}) \le \mathfrak{p}_{\mathfrak{u}|_Y}(\mathcal{D}_x) = +\infty$ by Lemma 3.3.13(a) for every $C(X)$-convex subset $\mathfrak{u}$ of $\mathcal{C}_{\mathfrak{Y}}(X, \mathbb{C})$. Thus if investigating $C(X)$-superharmonicity of the set $\mathfrak{u}|Y$ we only need to consider point evaluations supported by Y. If $x \in Y$, we employ Proposition 3.3.12(a) with the functions $g_y = \varphi_y \cdot \omega \in \mathfrak{u}_{\mathbb{B}}^{\rho_y} \cap L$ for all $y \in Y$. Then $|g_y(y)| = \varphi_y(y) = \rho_y(y)$ and $|g_y(x)| < \rho_y(x)$ for all $x \in X \setminus Y$. Consequently, if $\mathcal{S} \prec_{\mathfrak{C}} \mathcal{D}_x$ according to 3.3.12(a) the representation measure for $\mathcal{S}$ is supported by Y and we have $[\mathcal{S}] = \mathcal{S}_{\gamma\omega}$ and $\mathcal{S} = [\mathcal{S}]_{\overline{\gamma\omega}}$, where $\gamma = \overline{\omega}(x)$. The latter follows since $\omega_{g_y} = \omega$. From this, we infer that $\mathfrak{p}_{\mathfrak{u}|Y}(\mathcal{S}) = \mathfrak{p}_{\mathfrak{u}}(\mathcal{S})$ by Lemma 3.3.13(b). Therefore, if $\mathfrak{u} \subset \mathcal{C}_{\mathfrak{Y}}(X, \mathbb{C})$ is $\mathfrak{C}$-superharmonic as supposed in Part (a), we have $\mathfrak{p}_{\mathfrak{u}|Y}(\mathcal{S}) \le \mathfrak{p}_{\mathfrak{u}|Y}(\mathcal{D}_x)$ and the set $\mathfrak{u}|Y$ is seen to be also $\mathfrak{C}$-superharmonic.

In Parts (b) and (c), we claim $\mathfrak{C}$-superharmonicity for sets $\mathfrak{w}|Y$, where $f \cdot \overline{\omega}$ is a real-valued function for all $f \in \mathfrak{w}$. For function of this type, we observe that

$$\mathfrak{Re}\, \mathcal{S}(f) = \mathfrak{Re}\, [\mathcal{S}](\overline{\gamma} f \cdot \overline{\omega}) = \mathfrak{Re}\, \overline{\gamma}\, [\mathcal{S}](f \cdot \overline{\omega}) = (\mathfrak{Re}\, \gamma)[\mathcal{S}](f \cdot \overline{\omega})$$

Using this, we calculate

$$\mathfrak{p}_{\mathfrak{w}|_Y}(\mathcal{S}) = \mathfrak{p}_{\mathfrak{w}}(\mathcal{S}) = \sup\{\mathfrak{Re}\,\mathcal{S}(f) \mid f \in \mathfrak{w}\} = \sup\{(\mathfrak{Re}\,\gamma)\,[\mathcal{S}](f \cdot \overline{\omega}) \mid f \in \mathfrak{w}\}.$$

If $\mathfrak{Re}\,\gamma = 0$, this shows $\mathfrak{p}_{\mathfrak{w}|Y}(\mathcal{S}) = \mathfrak{p}_{\mathfrak{w}}(\mathcal{S}) = \mathfrak{p}_{\mathfrak{w}|Y}(\mathcal{D}_X) = 0$. Thus we may assume that $\mathfrak{Re}\,\gamma \neq 0$. Furthermore, we notice that if $f = \overset{\omega}{\mathfrak{Re}}\,h \cdot \omega$ for some $h \in C_{\mathfrak{V}}(X, E)$, then

$$\begin{aligned}[\mathcal{S}](f \cdot \overline{\omega}) &= [\mathcal{S}](\overset{\omega}{\mathfrak{Re}}\,h) = [\mathcal{S}](\mathfrak{Re}(h \cdot \overline{\omega})) \\ &= \mathfrak{Re}\,[\mathcal{S}](h \cdot \overline{\omega}) = \mathfrak{Re}\,[\mathcal{S}](\gamma h \cdot (\overline{\gamma\omega})) = \mathfrak{Re}\,\mathcal{S}(\gamma h).\end{aligned}$$

Now for Part (b) let $\mathfrak{u}$ be a $\mathfrak{C}$-superharmonic subset of $\mathfrak{C}$ and let

$$\mathfrak{w} = \{\overset{\omega}{\mathfrak{Re}}\,h \cdot \omega \mid h \in \mathfrak{u}\}.$$

Then for $f = \overset{\omega}{\mathfrak{Re}}\,h \cdot \omega \in \mathfrak{w}$ we have $(\mathfrak{Re}\,\gamma)[\mathcal{S}](f \cdot \overline{\omega}) = (\mathfrak{Re}\,\gamma)\mathfrak{Re}\,\mathcal{S}(\gamma h)$ by the preceding. We set $\zeta = \overline{\gamma}/\mathfrak{Re}\,\gamma$ and evaluate

$$\begin{aligned}\mathfrak{p}_{\mathfrak{w}|Y}(\mathcal{S}) &= \sup\{(\mathfrak{Re}\,\gamma)\,\mathfrak{Re}\,\mathcal{S}(\gamma h) \mid h \in \mathfrak{u}\} \\ &= \sup\{\mathfrak{Re}\,\mathcal{S}(h) \mid h \in \zeta\mathfrak{u}\} \\ &= \mathfrak{p}_{\zeta\mathfrak{u}}(\mathcal{S}).\end{aligned}$$

According to Proposition 3.3.1(c), the set $\zeta\mathfrak{u}$ is also $\mathfrak{C}$-superharmonic. Hence

$$\mathfrak{p}_{\mathfrak{w}|Y}(\mathcal{S}) = \mathfrak{p}_{\zeta\mathfrak{u}}(\mathcal{S}) \leq \mathfrak{p}_{\zeta\mathfrak{u}}(\mathcal{D}_X) = \mathfrak{p}_{\mathfrak{w}|Y}(\mathcal{D}_X).$$

This argument demonstrates that $\mathfrak{w}|Y$ is indeed $\mathfrak{C}$-superharmonic as claimed in Part (b).

For Part (c), let $f_1, \ldots, f_n \in L$ and let

$$\mathfrak{w} = \{h \cdot \omega \mid h \in C(X),\ \vee_{i=1}^{n} \overset{\omega}{\mathfrak{Re}}\,f_i|_Y \leq h|_Y\}|_Y.$$

We abbreviate $f = \vee_{i=1}^{n} \overset{\omega}{\mathfrak{Re}}\,f_i$. Let ϑ be the representation measure for $\mathcal{S}$, which is supported by Y. For $f = h \cdot \omega \in \mathfrak{w}$, we have $[\mathcal{S}](f \cdot \overline{\omega}) = [\mathcal{S}](h) = \int_Y h\,d[\vartheta]$ and

$$\mathfrak{p}_{\mathfrak{w}|Y}(\mathcal{S}) = \sup\left\{\mathfrak{Re}\,\gamma \int_Y h\,d[\vartheta] \mid h \in C(X),\ f \leq h|_Y\right\}$$

by the above. Thus

$$\mathfrak{p}_{\mathfrak{w}|Y}(\mathcal{D}_X) = \sup\{h(x) \mid h \in C(X),\ f \leq h|_Y\} = +\infty,$$

if $\mathfrak{Re}\,\gamma > 0$, and our requirement is satisfied. If $\mathfrak{Re}\,\gamma < 0$, then

$$\mathfrak{p}_{\mathfrak{w}|Y}(\mathcal{S}) = \mathfrak{Re}\,\gamma \inf\left\{\int_Y h\,d[\vartheta] \mid h \in C(X),\ f \leq h|_Y\right\} = \mathfrak{Re}\,\gamma \int_Y f\,d[\vartheta].$$

In particular, $\mathfrak{p}_{\mathfrak{w}|Y}(\mathcal{D}_x) = \mathfrak{Re}\,\gamma f(x)$. We remark that $\int_Y f\,d[\vartheta] \geq \int_Y \overset{\omega}{\mathfrak{Re}}\, f_i\,d[\vartheta]$, hence

$$\mathfrak{p}_{\mathfrak{w}|Y}(\mathcal{S}) \leq \mathfrak{Re}\,\gamma \int_Y \overset{\omega}{\mathfrak{Re}}\, f_i\,d[\vartheta] = \mathfrak{Re}\,\gamma\,[\mathcal{S}](\overset{\omega}{\mathfrak{Re}}\, f_i)$$

for all $i = 1, \dots, n$. For the functions $f_i \in L$, we have $\mathcal{S}(f_i) = \mathcal{D}_x(f_i)$. Thus

$$[\mathcal{S}](\overset{\omega}{\mathfrak{Re}}\, f_i) = \mathfrak{Re}\,\mathcal{S}(\gamma f_i) = \mathfrak{Re}\,\mathcal{D}_x(\gamma f_i) = \mathcal{D}_x(\overset{\omega}{\mathfrak{Re}}\, f_i) = \overset{\omega}{\mathfrak{Re}}\, f_i(x)$$

by the above and

$$\begin{aligned}\mathfrak{p}_{\mathfrak{w}|Y}(\mathcal{S}) &\leq \textstyle\bigwedge_{i=1}^n \mathfrak{Re}\,\gamma\,[\mathcal{S}](\overset{\omega}{\mathfrak{Re}}\, f_i) = \mathfrak{Re}\,\gamma \textstyle\bigvee_{i=1}^n (\overset{\omega}{\mathfrak{Re}}\, f_i)(x)\\ &= \mathfrak{Re}\,\gamma(f)(x) = \mathfrak{p}_{\mathfrak{w}|Y}(\mathcal{D}_x).\end{aligned}$$

Hence $\mathfrak{p}_{\gamma\mathfrak{u}}(\mathcal{S}) \leq \mathfrak{p}_{\gamma\mathfrak{u}}(\mathcal{D}_x)$ holds in any case, and the set $\mathfrak{w}|Y$ is seen to be $\mathfrak{C}$-superharmonic as claimed in Part (c). For the second statement in (c), we set

$$\mathfrak{u} = \{h \cdot \omega \mid h \in C(X) \text{ and } h|_Y \leq \textstyle\bigwedge_{k=1}^m \overset{\omega}{\mathfrak{Re}}\, g_k|_Y\}|Y$$

and observe that

$$\mathfrak{u} = -\{h \cdot \omega \mid h \in C(X) \text{ and } \textstyle\bigvee_{k=1}^m \overset{\omega}{\mathfrak{Re}}\,(-g_k)|_Y \leq h|_Y\}|Y.$$

Our claim follows from the first statement with Proposition 3.3.1(c). If the intersection of $\mathfrak{u}$ and $\mathfrak{w}$ is not empty, that is, if $\overset{\omega}{\mathfrak{Re}}\, f_i|_Y \leq \overset{\omega}{\mathfrak{Re}}\, g_k|_Y$ for all i, k, then Proposition 3.2.18 yields that the set

$$\mathfrak{u} \cap \mathfrak{w} = \{h \cdot \omega \mid h \in C(X),\ \textstyle\bigvee_{i=1}^n \overset{\omega}{\mathfrak{Re}}\, f_i|_Y \leq h|_Y \leq \textstyle\bigwedge_{k=1}^m \overset{\omega}{\mathfrak{Re}}\, g_k|_Y\}|Y$$

is also $\mathfrak{C}$-superharmonic. For this, we abbreviate $g = \bigwedge_{k=1}^m \overset{\omega}{\mathfrak{Re}}\, g_k|_Y$. For $x \in Y$, a point evaluation $\mathcal{D}_x$ and $\gamma = \overline{\omega}(x)$ using the above we have $\mathfrak{p}_{\mathfrak{u}}(\mathcal{D}_x) = \mathfrak{p}_{\mathfrak{w}}(\mathcal{D}_x) = \mathfrak{p}_{\mathfrak{u}\cap\mathfrak{w}}(\mathcal{D}_x) = 0$ if $\mathfrak{Re}\,\gamma = 0$, we have $\mathfrak{p}_{\mathfrak{w}}(\mathcal{D}_x) = \mathfrak{Re}\,\gamma f(x)$, $\mathfrak{p}_{\mathfrak{u}}(\mathcal{D}_x) = +\infty$ and $\mathfrak{p}_{\mathfrak{u}\cap\mathfrak{w}}(\mathcal{D}_x) = \mathfrak{Re}\,\gamma f(x)$ if $\mathfrak{Re}\,\gamma > 0$, and we have $\mathfrak{p}_{\mathfrak{w}}(\mathcal{D}_x) = +\infty$, $\mathfrak{p}_{\mathfrak{u}}(\mathcal{D}_x) = \mathfrak{Re}\,\gamma g(x)$ and $\mathfrak{p}_{\mathfrak{u}\cap\mathfrak{w}}(\mathcal{D}_x) = \mathfrak{Re}\,\gamma g(x)$ if $\mathfrak{Re}\,\gamma < 0$. Thus $\mathfrak{p}_{\mathfrak{u}\cap\mathfrak{w}}(\mathcal{D}_x) = \mathfrak{p}_{\mathfrak{u}}(\mathcal{D}_x) \wedge \mathfrak{p}_{\mathfrak{w}}(\mathcal{D}_x)$ holds in any case, as is required in 3.2.18.

Finally, for Part (d) suppose that $\mathcal{S} \prec_{\widetilde{\mathfrak{C}}} \mathcal{T}$ and that $\mathcal{T}$ is supported by Y. The set $\{0\}|_Y$ is $\mathfrak{C}$-superharmonic by Part (a), and if it is $\mathfrak{U}$-bounded, an element of $\widetilde{\mathfrak{C}}$. We have $\mathfrak{p}_{\{0\}|Y}(\mathcal{T}) = \mathfrak{p}_{\{0\}}(\mathcal{T}) = 0$ by Lemma 3.3.14(b). Thus the support of the representation measure for $\mathcal{S}$ is contained in Y by 3.3.14(d). □

Given a unimodular function $\omega \in \mathcal{F}(X, \mathbb{C})$, for an operator $\mathcal{T} \in C_{\mathfrak{V}}(X, \mathbb{C})^*$ we define its *ω-real* and *ω-imaginary* parts $\overset{\omega}{\mathfrak{Re}}\,\mathcal{T}$ and $\overset{\omega}{\mathfrak{Im}}\,\mathcal{T}$ as the operators

$$\overset{\omega}{\mathfrak{Re}}\,\mathcal{T}(f) = \mathfrak{Re}\,\mathcal{T}_\omega(\mathfrak{Re} f) + i\,\mathfrak{Re}\,\mathcal{T}_\omega(\mathfrak{Im} f)$$

and

$$\overset{\omega}{\mathfrak{Im}}\,\mathcal{T}(f) = \mathfrak{Im}\,\mathcal{T}_\omega(\overset{\omega}{\mathfrak{Re}}\,f) + i\,\mathfrak{Im}\,\mathcal{T}_\omega(\mathfrak{Im}\,f)$$

for $f \in \mathcal{C}_{\mathfrak{V}}(X, \mathbb{C})$. We observe that $\mathcal{T}_\omega = \overset{\omega}{\mathfrak{Re}}\,\mathcal{T} + i\,\overset{\omega}{\mathfrak{Im}}\,\mathcal{T}$, that is,

$$\mathcal{T}_{\chi_{X_\omega}} = (\overset{\omega}{\mathfrak{Re}}\,\mu)_{\overline{\omega}} + i\,\mu(\overset{\omega}{\mathfrak{Im}}\,\mu)_{\overline{\omega}},$$

where $X_\omega = \{x \in X \mid |\omega(x)| = 1\}$. Both operators $\overset{\omega}{\mathfrak{Re}}\,\mathcal{T}, \overset{\omega}{\mathfrak{Im}}\,\mathcal{T} \in \mathcal{C}_{\mathfrak{V}}(X, \mathbb{C})^*$ take real values at real-valued functions, hence their respective representation measures are real-valued (see Corollary 2.6.2 with the insertion of $\mathbb{R}$ for both E_+ and F_+). Operators $\mathcal{R} \in \mathcal{C}_{\mathfrak{V}}(X, \mathbb{C})^*$ of this type can be split into positive and negative parts that are represented by positive measures, that is, $\mathcal{R} = \mathcal{R}_+ - \mathcal{R}_-$, where

$$\mathcal{R}_+(f) = \sup\{\mathcal{R}(g) \mid 0 \le g \le f\} \quad \text{and} \quad \mathcal{R}_-(f) = \sup\{\mathcal{R}(g) \mid -f \le g \le 0\}$$

for nonnegative functions $f \in \mathcal{C}_{\mathfrak{V}}(X)$. The representation measures for $\mathcal{R}_+$ and $\mathcal{R}_-$ are mutually singular (see Proposition 11.22 in [53]). We use Proposition 3.3.14 for the following corollary.

3.3.15 Corollary. *Let $\omega \in \mathcal{F}(X, \mathbb{C})$ be a unimodular function and let Y be a closed subset of X_ω such that $\{0\}|_Y$ is $\mathfrak{U}$-bounded. Suppose that for every $y \in Y$ there is $\mathfrak{u}_{\mathbb{B}}^{\rho_y} \in \mathfrak{K}$ and a continuous nonnegative real-valued function φ_y such that $\varphi_y \cdot \omega \in \mathfrak{u}_{\mathbb{B}}^{\rho_y} \cap L$ and such that $\varphi_y(y) = \rho_y(y)$ and $\varphi_y(x) < \rho_y(x)$ for all $x \in X \backslash Y$. If $\mathcal{S} \prec_{\widetilde{\mathfrak{C}}} \mathcal{T}$ for $\mathcal{S}, \mathcal{T} \in \mathcal{C}_{\mathfrak{V}}(X, \mathbb{C})^*$ and if the representation measure for $\mathcal{T}$ is supported by Y, then*

$$(\overset{\omega}{\mathfrak{Re}}\,\mathcal{S})_{\overline{\omega}} \prec_{\widetilde{\mathfrak{C}}} (\overset{\omega}{\mathfrak{Re}}\,\mathcal{T})_{\overline{\omega}} \quad \textit{and} \quad (\overset{\omega}{\mathfrak{Im}}\,\mathcal{S})_{\overline{\omega}} \prec_{\widetilde{\mathfrak{C}}} (\overset{\omega}{\mathfrak{Im}}\,\mathcal{T})_{\overline{\omega}}.$$

If $f_1, \ldots, f_n$ are functions in $\mathcal{H}$ such that $f = \wedge_{i=1}^n \overset{\omega}{\mathfrak{Re}}\,f_i|_Y \ge 0$, then

$$(\overset{\omega}{\mathfrak{Re}}\,\mathcal{S})_+(f) \le (\overset{\omega}{\mathfrak{Re}}\,\mathcal{T})_+(f) \quad \textit{and} \quad (\overset{\omega}{\mathfrak{Re}}\,\mathcal{S})_-(f) \le (\overset{\omega}{\mathfrak{Re}}\,\mathcal{T})_-(f)$$

as well as

$$(\overset{\omega}{\mathfrak{Im}}\,\mathcal{S})_+(f) \le (\overset{\omega}{\mathfrak{Im}}\,\mathcal{T})_+(f) \quad \textit{and} \quad (\overset{\omega}{\mathfrak{Im}}\,\mathcal{S})_-(f) \le (\mathfrak{Im}\,\mathcal{T})_-(f).$$

Proof. Since $\mathfrak{u}|Y = \mathfrak{u} + \{0\}|_Y$ for every $C(X)$-convex subset $\mathfrak{u}$ of $\mathcal{C}_{\mathfrak{V}}(X, \mathbb{C})$, the set $\mathfrak{u}|Y$ is $\mathfrak{U}$-bounded whenever $\mathfrak{u}$ is $\mathfrak{U}$-bounded and an element of $\widetilde{\mathfrak{C}}$, provided that it is $\mathfrak{C}$-superharmonic in addition. If $\mathcal{S} \prec_{\widetilde{\mathfrak{C}}} \mathcal{T}$ and if the representation measure for $\mathcal{T}$ is supported by Y, following Proposition 3.3.14(d) the representation measure for $\mathcal{S}$ is also supported by Y. Hence also the representation measures for $\overset{\omega}{\mathfrak{Re}}\,\mathcal{S}$ and $\overset{\omega}{\mathfrak{Re}}\,\mathcal{T}$ and their respective positive and negative parts. The functionals $\mathfrak{p}_{\mathfrak{u}}$ and $\mathfrak{p}_{\mathfrak{u}|Y}$ therefore coincide for these (Lemma 3.3.13(b)). Let $\mathfrak{u} \in \widetilde{\mathfrak{C}}$ and set $\mathfrak{w} = \{\overset{\omega}{\mathfrak{Re}}\,f \cdot \omega \mid f \in \mathfrak{u}\}$. For a function $f \in \mathcal{C}_{\mathfrak{V}}(X, E)$, we observe that

$$(\mathfrak{Re}^{\omega}\,\mathcal{T})_{\overline{\omega}}(f) = \mathfrak{Re}\,\mathcal{T}_{\omega}(\mathfrak{Re}\,f \cdot \overline{\omega}) + i\,\mathfrak{Re}\,\mathcal{T}_{\omega}(\mathfrak{Im}\,f),$$

hence

$$\mathfrak{Re}\,((\mathfrak{Re}^{\omega}\,\mathcal{T})_{\overline{\omega}}(f)) = \mathfrak{Re}\,\mathcal{T}_{\omega}(\mathfrak{Re}\,f \cdot \overline{\omega}) = \mathfrak{Re}\,\mathcal{T}(\mathfrak{Re}^{\omega}\,f \cdot \omega).$$

Using this, we evaluate

$$\begin{aligned} \mathfrak{p}_{\mathfrak{u}}((\mathfrak{Re}^{\omega}\,\mathcal{T})_{\overline{\omega}}) &= \sup\{\mathfrak{Re}((\mathfrak{Re}^{\omega}\,\mathcal{T})_{\overline{\omega}}(f)) \mid f \in \mathfrak{u}\} \\ &= \sup\{\mathfrak{Re}\,\mathcal{T}(\mathfrak{Re}^{\omega}\,f \cdot \omega) \mid f \in \mathfrak{u}\} = \mathfrak{p}_{\mathfrak{w}}(\mathcal{T}) = \mathfrak{p}_{\mathfrak{w}|Y}(\mathcal{T}) \end{aligned}$$

by the above, and likewise $\mathfrak{p}_{\mathfrak{u}}((\mathfrak{Re}^{\omega}\,\mathcal{S})_{\overline{\omega}}) = \mathfrak{p}_{\mathfrak{w}|Y}(\mathcal{S})$. Since $\mathfrak{w}|_Y$ is $\mathfrak{C}$-superharmonic by 3.3.14(a) and $\mathfrak{U}$-bounded by the above, hence an element of $\widetilde{\mathfrak{C}}$, we infer that

$$\mathfrak{p}_{\mathfrak{u}}((\mathfrak{Re}^{\omega}\,\mathcal{S})_{\overline{\omega}}) = \mathfrak{p}_{\mathfrak{w}|Y}(\mathcal{S}) \le \mathfrak{p}_{\mathfrak{w}|Y}(\mathcal{T}) = \mathfrak{p}_{\mathfrak{u}}((\mathfrak{Re}^{\omega}\,\mathcal{T})_{\overline{\omega}}).$$

Thus $(\mathfrak{Re}^{\omega}\,\mathcal{S})_{\overline{\omega}} \prec_{\widetilde{\mathfrak{C}}} (\mathfrak{Re}^{\omega}\,\mathcal{T})_{\overline{\omega}}$. The argument for the imaginary parts is similar. For the subsequent statements, we use the set

$$\mathfrak{u} = \{h \cdot \omega \mid h \in \mathcal{C}(X),\ 0 \le h|_Y \le \wedge_{i=1}^{n} \mathfrak{Re}^{\omega} f_i|_Y\}|_Y,$$

where $f_1, \ldots, f_n \in \mathcal{H}$ such that $f = \wedge_{i=1}^{n} \mathfrak{Re}^{\omega} f_i|_Y \ge 0$. This set is $\mathfrak{U}$-bounded by our assumptions and $\mathfrak{C}$-superharmonic by Proposition 3.3.14(c). We infer that $\mathfrak{p}_{\mathfrak{u}}((\mathfrak{Re}^{\omega}\,\mathcal{S})_{\overline{\omega}}) \le \mathfrak{p}_{\mathfrak{u}}((\mathfrak{Re}^{\omega}\,\mathcal{T})_{\overline{\omega}})$ holds by our first argument, and continue to calculate that

$$\mathfrak{p}_{\mathfrak{u}}((\mathfrak{Re}^{\omega}\,\mathcal{T})_{\overline{\omega}}) = \sup\{\mathfrak{Re}^{\omega}\,\mathcal{T}(h) \mid 0 \le h|_Y \le f\} = (\mathfrak{Re}^{\omega}\,\mathcal{T})_+(f),$$

and likewise $\mathfrak{p}_{\mathfrak{u}}((\mathfrak{Re}^{\omega}\,\mathcal{S})_{\overline{\omega}}) = (\mathfrak{Re}^{\omega}\,\mathcal{S})_+(f)$. This shows $(\mathfrak{Re}^{\omega}\,\mathcal{S})_+(f) \le (\mathfrak{Re}^{\omega}\,\mathcal{T})_+(f)$ as claimed. The remaining arguments for the negative and imaginary parts are similar. □

3.3.16 Examples. (a) Let $X = \mathbb{C}$ and let $\mathfrak{V}$ be the neighborhood system of compact convergence (see Example 1.2.7(b)) generated by the neighborhoods $\mathfrak{u}_K = \mathfrak{u}_{\mathbb{B}}^{\rho_K}$, where $\mathbb{B}$ is the unit ball of $\mathbb{C}$, K is a compact subset of $X = \mathbb{C}$ and $\rho_K(x) = 1$ for $x \in K$ and $\rho_K(x) = +\infty$, else. We have $\mathcal{C}_{\mathfrak{V}}(\mathbb{C}, \mathbb{C}) = \mathcal{C}(\mathbb{C}, \mathbb{C})$ in this case. The dual of $\mathcal{C}_{\mathfrak{V}}(\mathbb{C}, \mathbb{C})$ is represented by the space of all regular complex-valued Borel measures on $\mathbb{C}$ with compact support (see Theorem 2.6.1). In the settings of this subsection, let K be a fixed compact subset of X, let $\mathfrak{K} = \{\mathfrak{u}_K\}$, $\mathcal{G} = \emptyset$, and let $\mathcal{H}$ be the subspace of all analytic functions in $\mathcal{C}(\mathbb{C}, \mathbb{C})$. We shall use the completion $\widetilde{\mathfrak{C}}$ of $\mathfrak{C}$ for the results of this subsection. The assumptions of Corollary 3.3.5, Proposition 3.3.14 and Corollary 3.3.15 are satisfied with $Y = K$, with $\omega = 1$, the constant function $x \mapsto 1$ for φ_y and $\rho_y = \rho_K$ for all $y \in K$. For $\mathcal{T} \in \mathcal{C}(\mathbb{C}, \mathbb{C})^*$, we have $\mathfrak{p}_{\mathfrak{u}_K}(\mathcal{T}) < +\infty$ if and only if the representation measure for $\mathcal{T}$ is supported by K. In particular, the Coquet boundary $\partial_{\widetilde{\mathfrak{C}}}^{\mathfrak{u}_K}$ is a subset of K. Moreover, if x is in the topological interior of K, then $x \notin \partial_{\widetilde{\mathfrak{C}}}^{\mathfrak{u}_K}$.

Indeed, there is $\varepsilon > 0$ such that the ball $B_\varepsilon(x) = \{z \in \mathbb{B} \mid |z - x| \leq \varepsilon\}$ is contained in K. Let $0 \neq \eta \in \mathbb{C}$ and consider the functional $\mathcal{T} \in \mathcal{C}(\mathbb{C}, \mathbb{C})^*$ defined by

$$\mathcal{T}(f) = (\eta/2\pi) \int_0^{2\pi} f(x + \varepsilon e^{it})\, dt.$$

Then $\mathcal{T}(f) = \mathcal{D}_x^\eta(f)$ for all $f \in \mathcal{H}$ by the Cauchy integral formula. Furthermore, since $\mathfrak{p}_{\mathfrak{u}_K}(\mathcal{D}_x^\eta) = \mathfrak{p}_{\mathfrak{u}_K}(\mathcal{T}) = |\eta|$, we infer that $\mathcal{T} \prec_{\mathfrak{C}} \mathcal{D}_x^\eta$, hence $\mathcal{T} \prec_{\widetilde{\mathfrak{C}}} \mathcal{D}_x^\eta$. Because Corollary 3.3.5 applies and $[\mathcal{T}] \neq |\mathcal{D}_x^\eta|$, we conclude that $\mathcal{T} \nsim \mathcal{D}_x^\eta$, and the point evaluation $\mathcal{D}_x^\eta$ is not minimal in the ordering $\prec_{\widetilde{\mathfrak{C}}}$. If $\mathfrak{p}_{\mathfrak{u}_K}(\mathcal{T}) < +\infty$ for $\mathcal{T} \in \mathcal{C}(\mathbb{C}, \mathbb{C})^*$, then by Corollary 3.2.9 there exists a minimal (with respect to the order $\prec_{\widetilde{\mathfrak{C}}}$) operator $\overline{\mathcal{T}} \in \mathcal{C}(\mathbb{C}, \mathbb{C})^*$, which coincides with $\mathcal{T}$ on $\mathcal{H}$, such that $\mathfrak{p}_{\mathfrak{u}_K}(\overline{\mathcal{T}}) \leq \mathfrak{p}_{\mathfrak{u}_K}(\mathcal{T})$, and according to Corollary 3.3.5 the representation measure for $\overline{\mathcal{T}}$ is supported by the topological boundary of K.

(b) Let $X = \mathbb{R}$ and consider the neighborhood system $\mathfrak{V}$ generated by the function space neighborhood $\mathfrak{v}_\rho$, which is defined by the neighborhood function $n = \rho \cdot \mathbb{B}$ on $\mathbb{R}$, where $\mathbb{B} = [-1, +1]$ is the unit ball of $\mathbb{R}$ and $\rho(x) = (1 + x^2)$. Then $\mathcal{C}_{\mathfrak{V}}(\mathbb{R}, \mathbb{C})$ consists of all continuous complex-valued functions f on $\mathbb{R}$ such that for every $\varepsilon > 0$ there is a compact subset K of $\mathbb{R}$ such that $|f(x)| \leq \varepsilon\rho(x)$ for all $x \in \mathbb{R} \setminus K$. In the settings of this subsections, let $\mathfrak{U} = \{\mathfrak{u}_{\mathbb{B}}^\rho\}$, let $\mathcal{G} + \varnothing$ and let $\mathcal{H}$ be the subspace of $\mathcal{C}_{\mathfrak{V}}(\mathbb{R}, \mathbb{C})$ generated by the functions $f_1(x) = e^{i\pi x}$ and $f_2(x) = xe^{i\pi x}$. We shall establish that

$$\partial_{\mathfrak{C}} = \partial_{\mathfrak{C}}^{\mathfrak{u}_{\mathbb{B}}^\rho} = [-1, +1].$$

Following Lemma 3.3.11, it suffices to investigate point evaluations $\mathcal{D}_y$ in order to establish this claim. Let $\mathcal{S} \prec_{\mathfrak{C}} \mathcal{D}_y$ for $y \in \mathbb{R}$. First, suppose that $|y| > 1$. We shall demonstrate that $\mathcal{D}_y \notin \Delta_{\mathfrak{C}}$. For this, we set $\mathcal{S} = \lambda\mathcal{D}_{+1} + \kappa\mathcal{D}_{-1}$, where $\lambda = -(1/2)e^{i\pi y}(y + 1)$ and $\kappa = (1/2)e^{i\pi y}(y - 1)$. It is straightforward to verify that $\mathcal{D}_y$ and $\mathcal{S}$ coincide on the functions in $\mathcal{H}$ and that

$$\mathfrak{p}_{\mathfrak{u}_{\mathbb{B}}^\rho}(\mathcal{S}) = |\lambda|\rho(1) + |\kappa|\rho(-1) = |y + 1| + |y - 1| < 1 + y^2 = \mathfrak{p}_{\mathfrak{u}_{\mathbb{B}}^\rho}(\mathcal{D}_y).$$

Thus $\mathcal{S} \prec_{\mathfrak{C}} \mathcal{D}_y$, but $\mathcal{S} \nsim \mathcal{D}_y$ since $\mathfrak{p}_{\mathfrak{u}_{\mathbb{B}}^\rho}(\mathcal{S}) \neq \mathfrak{p}_{\mathfrak{u}_{\mathbb{B}}^\rho}(\mathcal{D}_y)$. The point evaluation $\mathcal{D}_y$ is therefore not minimal and not an element of $\Delta_{\mathfrak{C}}$. We infer that $y \notin \partial_{\mathfrak{C}}$. Now suppose that $|y| \leq 1$ and consider the function $g_y \in \mathcal{H}$ defined by $g_y(x) = (2yx + (1 - y^2))e^{i\pi(x-y)}$ for all $x \in \mathbb{R}$. We realize that

$$|g_y(x)| \leq 2|y||x| + (1 - y^2) \leq (x^2 + y^2) + (1 - y^2) = \rho(x)$$

holds for all $x \in \mathbb{R}$. Hence $g_y \in \mathcal{H} \cap \mathfrak{u}_{\mathbb{B}}^\rho$. Moreover, if $x = y$, then $|g_y(x)| = \rho(x)$. Conversely, if $|g_y(x)| = \rho(x)$, then $|x| = |y|$ and indeed $x = y$ in case that $|y| < 1$. In case that $|y| = 1$, we have $|g_y(1)| = \rho(1)$ and $|g_y(-1)| = \rho(-1)$. Now we use Proposition 3.3.10 with $\mathcal{D}_y$ in place of $\mathcal{T}$ and g_y in place of g for our argument. Then $\mathfrak{p}_{\mathfrak{u}^\rho_{\mathbb{B}}}(\mathcal{D}_y) = \rho(y) = g_y(y) = \mathcal{D}_y(g)$. In case that $|y| < 1$, we have

$$Y = \{x \in X \mid |g_y(x)| = \rho(x)\} = \{y\}$$

by the above. Proposition 3.3.10 returns that $\mathcal{S}$ is supported by $\{y\}$ and that $\mathcal{S} = [\mathcal{S}]_{\overline{\omega}}$, where $\omega = \chi_Y \cdot g_y/|g_y|$, that is, $\omega(y) = 1$. We conclude that $\mathcal{S} = \lambda \mathcal{D}_y$ with some $\lambda \geq 0$, and indeed $\lambda = 1$ since $\mathfrak{p}_{\mathfrak{u}^\rho_{\mathbb{B}}}(\mathcal{S}) = \mathfrak{p}_{\mathfrak{u}^\rho_{\mathbb{B}}}(\mathcal{D}_y) \neq 0$, also by 3.3.10. Hence $\mathcal{S} = \mathcal{D}_y$ and, therefore, $\mathcal{D}_y \in \Delta_{\mathfrak{C}}$ and $y \in \partial_{\mathfrak{C}}$. In case that $|y| = 1$, say $y = +1$, we have

$$Y = \{x \in X \mid |g(x)| = \rho(x)\} = \{-1, +1\}.$$

Thus $\mathcal{S}$ is supported by Y and $\mathcal{S} = \lambda \mathcal{D}_{+1} - \kappa \mathcal{D}_{-1}$ with $\kappa, \lambda \geq 0$ because $\omega(1) = 1$ and $\omega(-1) = -1$. For the functions $f_1, f_2 \in \mathcal{H}$, we calculate

$$-1 = \mathcal{D}_{+1}(f_1) = \mathcal{S}(f_1) = -(\lambda - \kappa)$$

and

$$-1 = \mathcal{D}_{+1}(f_2) = \mathcal{S}(f_2) = -(\lambda + \kappa).$$

Thus $\lambda = 1$ and $\kappa = 0$ and, therefore, $\mathcal{S} = \mathcal{D}_{+1}$. The case that $y = -1$ is similar. This yields $[-1, +1] \subset \partial_{\mathfrak{C}}$, and our claim follows. The assumptions of Propositions 3.3.12 and 3.3.14(a) to (c) are satisfied with $Y = [-1, +1]$, the unimodular function $\omega(x) = e^{i\pi x}$ and the functions ρ for ρ_y and $\varphi_y = |g_y|$.

The case that both E and F are ordered topological vector spaces

In this setting, we assume that both E and F are ordered topological vector spaces over $\mathbb{R}$ or $\mathbb{C}$ with closed positive cones E_+ and F_+, respectively. (See Sections 1.2 and 2.6.) We abbreviate E_- for $-E_+$ and E_+^* for the positive cone of E^*, that is, the polar of E_-. We note that $a \leq b$ holds for $a, b \in E$ if and only if $\mathfrak{Re}\,\mu(a) \leq \mathfrak{Re}\,\mu(b)$ for all $\mu \in E_+^*$. Similar notation applies to F. The order relation of E transfers to $C_{\mathfrak{V}}(X, E)$, and its positive cone is the closure of the subcone of all E_+-valued functions. We intend to investigate positive operators in $\mathfrak{L}(C_{\mathfrak{V}}(X, E), F)$, that is, operators that map E_+-valued functions into F_+ (or equivalently, E_--valued functions into F_-). For this, we shall use the previously formulated sample settings in the following way. Let $\mathfrak{U}$ be family of $C(X)$ -convex neighborhoods of 0 in $C_{\mathfrak{V}}(X, E)$ with the prescribed properties and require that each $\mathfrak{u} \in \mathfrak{U}$ contains all E_--valued functions in $C_{\mathfrak{V}}(X, E)$. Let Π be a subset of F_+^* such that the weak* closure of its conic hull is all of F_+^* and such that the balanced core $\operatorname{bcore}(\Pi^\circ)$ of its polar in F is weakly compact. This construction requires that $\bigcap_{|\gamma|=1} \gamma F_+ = \{0\}$. Indeed, if $\gamma c \in F_+$ for all $|\gamma| = 1$, then $\mu(c) = 0$ for all $\mu \in \Pi$ and $c = 0$ since the linear span of Π is weak*-dense in F^*. Moreover, the first condition on Π guarantees that $c \leq d$ holds for elements $c, d \in F$ whenever $\mathfrak{Re}\,\mu(c) \leq \mathfrak{Re}\,\mu(d)$

for all $\mu \in \Pi$. We have $\mathfrak{K} = \{\Omega_{\mathfrak{u}}^{\Pi} \mid \mathfrak{u} \in \mathfrak{U}\}$. Let $\mathfrak{C}_0$ be a collection of subsets $\Omega_{\mathfrak{w}}^{\Gamma}$ of $\mathcal{C}_{\mathfrak{V}}(X,E) \otimes F^*$, such that $\Gamma \subset \Pi$ and $\mathfrak{w} \subset \mathcal{C}_{\mathfrak{V}}(X,E)$ is $C(X)$-convex and $\mathfrak{U}$-bounded. The Choquet cone $\mathfrak{C}$ is the cone of subsets of $\mathcal{C}_{\mathfrak{V}}(X,E) \otimes F^*$ generated by $\mathfrak{K}$ and $\mathfrak{C}_0$.

We note that all operators $\mathcal{T} \in \mathfrak{L}_{\mathfrak{K}}(\mathcal{C}_{\mathfrak{V}}(X,E),F)$ are weakly compact and positive in this setting. Hence the values of their representation measures are positive operators in $\mathfrak{L}(E,F)$ (see Corollary 2.6.2). Indeed, if $\mathfrak{p}_{\Omega_{\mathfrak{u}}^{\Pi}}(\mathcal{T}) < +\infty$ for $\Omega_{\mathfrak{u}}^{\Pi} \in \mathfrak{K}$, then by our assumption $\mathfrak{u}$ contains all functions $f \in \mathcal{C}_{\mathfrak{V}}(X,E_-)$, hence $\mathcal{T}(\lambda f) \in \Pi^{\circ}$ for all $\lambda \geq 0$. That is, $(\mathcal{T}(f))(\mu) \leq 0$ for all $\mu \in \Pi$ and indeed for all μ in F_+^* since all elements of F_+^* are absorbed by Π. Thus $\mathcal{T}(f) \leq 0$ as claimed.

We continue to formulate our observations for the case that both E and F are ordered topological vector spaces over $\mathbb{C}$. The real case is similar. We further specify our settings for $\mathfrak{C}_0$, hence for $\mathfrak{C}$, as we did in the sample settings for the general case. Let $\mathcal{H}$ be a subcone of $\mathcal{C}_{\mathfrak{V}}(X,E)$, and let $\mathcal{G}$ and the sets $\mathfrak{u}_V^{\rho}$ be as stated in the general settings. That is, $\mathfrak{C}_0$ consists of all sets $\Omega_{\{f\}}^{\{\mu\}}$ for $f \in \mathcal{H}$ together with all $\Omega_{\mathfrak{u}_{V\cap G}^{\rho}}^{\{\mu\}}$ for $G \in \mathcal{G}$ and all $\mu \in \Pi$.

3.3.17 Proposition. *Suppose that $\mathcal{S} \prec_{\mathfrak{C}} \mathcal{T}$ for $\mathcal{S},\mathcal{T} \in \mathfrak{L}_{\mathfrak{K}}(\mathcal{C}_{\mathfrak{V}}(X,E),F)$.*
(a) *Then $\mathcal{S}(f) \leq \mathcal{T}(f)$ for all $f \in \mathcal{H}$.*
(b) *If the conic hull of Π is F^*, then $\mathcal{S}(\mathfrak{u}_{V\cap G}^{\rho}) \subset \overline{\mathcal{T}(\mathfrak{u}_{V\cap G}^{\rho}) + F_-}$ for all $G \in \mathcal{G}$.*

Proof. Let $\mathcal{S} \prec_{\mathfrak{C}} \mathcal{T}$ for $\mathcal{S},\mathcal{T} \in \mathfrak{L}_{\mathfrak{K}}(\mathcal{C}_{\mathfrak{V}}(X,E),F)$. (a) For $f \in \mathcal{H}$, we have $\mathfrak{Re}\,\mu(\mathcal{S}(f)) \leq \mathfrak{Re}\,\mu(\mathcal{T}(f))$ for all $\mu \in \Pi$, that is, $\mathfrak{Re}\,\mu(\mathcal{S}(f)) \leq \mathfrak{Re}\,\mu(\mathcal{T}(f))$ for all $\mu \in F_+^*$ and, therefore, $\mathcal{S}(f) \leq \mathcal{T}(f)$. (b) If $\mathcal{S}(\mathfrak{u}_{V\cap G}^{\rho}) \not\subset \overline{\mathcal{T}(\mathfrak{u}_{V\cap G}^{\rho}) + F_-}$ for $G \in \mathcal{G}$, then there are $f \in \mathfrak{u}_{V\cap G}^{\rho}$ and $\nu \in F^*$ such that

$$\mathfrak{Re}\,\nu(\mathcal{S}(f)) > \sup\{\mathfrak{Re}\,\nu(\mathcal{T}(g) + c) \mid g \in \mathfrak{u}_{V\cap G}^{\rho},\ c \in F_-\}.$$

The latter implies that $\mathfrak{Re}\,\nu(c) \leq 0$ for all $c \in F_-$, that is, $\nu \in F_+^*$. Thus

$$\begin{aligned}\sup\{\mathfrak{Re}\,\nu(\mathcal{T}(g) + c) \mid g \in \mathfrak{u}_{V\cap G}^{\rho},\ c \in F_-\}& \\ &= \sup\{\mathfrak{Re}\,\nu(\mathcal{T}(g)) \mid g \in \mathfrak{u}_{V\cap G}^{\rho}\} \\ &= \mathfrak{p}_{\Omega_{\mathfrak{u}_{V\cap G}^{\rho}}^{\{\nu\}}}(\mathcal{T}).\end{aligned}$$

By our assumption, there are $\mu \in \Pi$ and $\lambda \geq 0$ such that $\nu = \lambda\mu$. Then

$$\mathfrak{p}_{\Omega_{\mathfrak{u}_{V\cap G}^{\rho}}^{\{\mu\}}}(\mathcal{S}) \leq \mathfrak{p}_{\Omega_{\mathfrak{u}_{V\cap G}^{\rho}}^{\{\mu\}}}(\mathcal{T}) \quad \text{implies that} \quad \mathfrak{p}_{\Omega_{\mathfrak{u}_{V\cap G}^{\rho}}^{\{\nu\}}}(\mathcal{S}) \leq \mathfrak{p}_{\Omega_{\mathfrak{u}_{V\cap G}^{\rho}}^{\{\nu\}}}(\mathcal{T}),$$

contradicting the above. □

3.3.18 Proposition. *Let $\varphi_1,\dots,\varphi_n$ be continuous real-valued functions on X and let $a \in E_+$ such that the functions $\varphi_i \cdot a$ are contained in $\mathcal{H}$. Then for all $\mu \in \Pi$ the set $\Omega_{\{\varphi\cdot a\}}^{\{\mu\}}$, where $\varphi = \wedge_{i=1}^{n}\varphi_i$, is $\mathfrak{C}$-superharmonic.*

Proof. Let $\mathcal{T}, \mathcal{D}_x^T \in \mathfrak{L}_{\mathfrak{K}}(\mathcal{C}_{\mathfrak{V}}(X,E),F)$ such that $\mathcal{T} \prec_{\mathfrak{C}} \mathcal{D}_x^T$ and $\mathcal{D}_x^T$ is a $\mathfrak{K}$-extremal point evaluation. Then both $\mathcal{T}$ and $\mathcal{D}_x^T$ are positive operators, hence, likewise is $T \in \mathfrak{L}(E,F)$. Let $\mu \in \Pi$. We have

$$\mathfrak{Re}\,\mu(\mathcal{T}(\varphi_i \cdot a)) \le \mathfrak{Re}\,\mu(\mathcal{D}_x^T(\varphi_i \cdot a)) = \varphi_i(x)\mathfrak{Re}\,\mu(T(a))$$

for all $i = 1,\ldots,n$. Since $\varphi \cdot a \le \varphi_i \cdot a$, both $\mathcal{T}$, and $\mathcal{D}_x^T$ are positive and since $\mu \in F_+^*$, this yields

$$\begin{aligned}\mathfrak{Re}\,\mu(\mathcal{T}(\varphi \cdot a)) &\le \bigwedge_{i=1}^{n} \mathfrak{Re}\,\mu(\mathcal{T}(\varphi_i \cdot a))\\ &\le \bigwedge_{i=1}^{n} \varphi_i(x)\mathfrak{Re}\,\mu(T(a))\\ &= \varphi(x)\mathfrak{Re}\,\mu(T(a)),\end{aligned}$$

that is, $\mathfrak{p}_{\Omega_{\{\varphi\cdot a\}}^{\{\mu\}}}(\mathcal{T}) \le \mathfrak{p}_{\Omega_{\{\varphi\cdot a\}}^{\{\mu\}}}(\mathcal{D}_x^T)$. □

The case that both *E* and *F* are topological vector lattices

In addition to the assumptions of the preceding setting, we presume that both E and F are topological vector lattices over $\mathbb{R}$ and that all elements of the subset Π of F_+^* are lattice homomorphisms. The statement of Proposition 3.3.18 then can be strengthened in the following way.

3.3.19 Proposition. *Let $f_1,\ldots,f_n \in \mathcal{H}$. Then for all $\mu \in \Pi$ the set $\Omega_{\{f\}}^{\{\mu\}}$, where $f = \wedge_{i=1}^n f_i$, is $\mathfrak{C}$-superharmonic.*

Proof. Let $\mathcal{T}, \mathcal{D}_x^T \in \mathfrak{L}_{\mathfrak{K}}(\mathcal{C}_{\mathfrak{V}}(X,E),F)$ such that $\mathcal{T} \prec_{\mathfrak{C}} \mathcal{D}_x^T$ and $\mathcal{D}_x^T$ is a $\mathfrak{K}$-extremal point evaluation. Both $\mathcal{T}$ and $\mathcal{D}_x^T$ are positive operators, so is $T \in \mathfrak{L}(E,F)$. Let $\mu \in \Pi$. We have

$$\mu(\mathcal{T}(f_i)) \le \mu(\mathcal{D}_x^T(f_i)) = \mu(T(f_i(x)))$$

for all $i = 1,\ldots,n$. Since $f \le f_i$ and μ is a lattice homomorphism, this renders

$$\mu(\mathcal{T}(f)) \le \bigwedge_{i=1}^{n} \mu(\mathcal{T}(f_i)) \le \bigwedge_{i=1}^{n} \mu(T(f_i(x))) = \varphi(x)\mathfrak{Re}\,\mu(T(a)),$$

that is, $\mathfrak{p}_{\Omega_{\{\varphi\cdot a\}}^{\{\mu\}}}(\mathcal{T}) \le \mathfrak{p}_{\Omega_{\{\varphi\cdot a\}}^{\{\mu\}}}(\mathcal{D}_x^T)$. □

List of Symbols

https://doi.org/10.1515/9783111315478-004

Bibliography

[1] E. M. Alfsen, *Compact convex sets and boundary integrals*, Ergebnisse der Mathematik und ihrer Grenzgebiete, vol. 57, Springer, Heidelberg-Berlin-New York, 1971.

[2] B. Anger and J. Lembcke, *Hahn-Banach type theorems for hypolinear functionals*, Math. Ann., 209 (1974), 127–151.

[3] R. J. Aumann, *Integrals of set-valued functions*, J. Math. Anal. Appl., 12 (1965), 1–12.

[4] R. Bagus and E. Wibowo, *The Henstock-Kurzweil integral of set-valued functions*, Int. J. Math. Anal., 8, no. 55 (2014), 2741–2755.

[5] R. G. Bartle, *A general bilinear vector integral*, Can. J. Math., 7 (1955), 298–305.

[6] R. G. Bartle, *The elements of integration*, Wiley, New York, 1966.

[7] R. G. Bartle, N. Dunford and J. Schwartz, *Weak compactness and vector measures*, Studia Math., 15 (1956), 337–352.

[8] J. Batt, *Integraldarstellungen linearer Transformationen und schwache Kompaktheit*, Math. Ann., 147 (1967), 291–304.

[9] C. J. K. Batty, *Vector-valued Choquet theory and transference of boundary measures*, Proc. Lond. Math. Soc. (3), 60 (1990), 530–548.

[10] H. Bauer, *Maß- und Integrationstheorie*, De Gruyter, Berlin-New York, 1990.

[11] E. Bishop and K. de Leeuw, *The representation of linear functionals by measures on sets of extreme points*, Ann. Inst. Fourier (Grenoble), 9 (1959), 305–331.

[12] W. M. Bogdanowicz, *Integral representation of linear continuous operators from the space of Lebesgue-Bochner-Stieltjes summable functions into any Banach space*, Proc. Natl. Acad. Sci. USA, 54 (1965), 351–354.

[13] M. Caprio, *Advances in Choquet theories*, Dissertation, Duke University, 2022.

[14] G. Choquet, *Theory of capacities*, Ann. Inst. Fourier, 5 (1955), 131–295.

[15] G. Choquet, *Existance des représentations intégrales au moyen des points extrémaux dans les cônes convexes*, C. R. Acad. Sci. Paris, 243 (1956), 699–702.

[16] G. Choquet and J. Deny, *Ensembles semi-réticulés et ensembles réticulés de fonctions continues*, J. Math. Pures Appl., 36 (1957), 179–189.

[17] G. Choquet and P. A. Meyer, *Existance et unicité des représentations intégrales dans les convexes compacts quelconques*, Ann. Inst. Fourier (Grenoble), 13 (1963), 139–154.

[18] G. Choquet, *Mesures coniques, affines et cylindriques*, Symposia Mathematica, II (1969), 145–182.

[19] D. L. Cohn, *Measure theory*, Birkhäuser, Boston-Basel-Berlin, 1980.

[20] M. De Wilde and M. T. De Wilde-Nibes, *Intégration par rapport des mesures à valeurs vectorielles*, Rev. Roum. Math. Pures Appl., 13 (1968), 1529–1537.

[21] J. Diestel and J. J. Uhl, *Vector measures*, Mathematical surveys, vol. 15, American Mathematical Society, Providence, 1977.

[22] J. Dieudonné, *Sur le théorème de Lebesgue-Nikodým*, Can. J. Math., 3 (1951), 129–139.

[23] N. Dinculeanu, *Vector measures*, International series of monographs in pure and applied mathematics, Pergamon Press, Oxford-London-New York-Syndney-Paris, 1967.

[24] J. Dugundji, *Topology*, Allyn and Bacon series in advanced mathematics, Boston, 1966.

[25] N. Dunford, *Integration and linear operations*, Trans. Am. Math. Soc., 40 (1936), 474–494.

[26] N. Dunford and J. Schwartz, *Linear operators, Parts I and II*, Wiley Classics Library Edition, John Wiley & Sons, New York-Chichester-Brisbane-Toronto-Singapore, 1988.

[27] N. Dyn, E. Farkhi and A. Mokhov, *The metric integral of set-valued functions*, Set-Valued Var. Anal., 26 (2018), 867–885.

[28] G. A. Edgar, *A noncompact Choquet theorem*, Proc. Am. Math. Soc., 49, no. 2 (1975), 354–358.

[29] W. H. Graves, *On the theory of vector measures*, American Mathematical Society, Providence, 1977.

[30] P. R. Halmos, *Introduction to Hilbert space and the theory of spectral multiplicity*, 2nd ed., Chelsea Publishing Company, New York, 1957.

https://doi.org/10.1515/9783111315478-005

[31] P. R. Halmos, *Measure theory*, Springer, Heidelberg-Berlin-New York, 1974.
[32] O. Hustad, *A norm preserving complex Choquet theorem*, Math. Scand., 29 (1971), 272–278.
[33] K. Keimel and W. Roth, *Ordered cones and approximation*, Lecture Notes in Mathematics, vol. 1517, Springer, Heidelberg-Berlin-New York, 1992.
[34] J. Lukes, I. Netuka and J. Veselý, *Choquet's theory and the Dirichlet problem*, Expo. Math., 20 (2002), 220–254.
[35] J. Lukes, J. Malý, I. Netuka and J. Spurný, *Integral representation theory: Applications to convexity, Banach spaces and Potential theory*, De Gruyter, Berlin-New York, 2010.
[36] C. W. McArthur, *On a theorem of Orlicz and Pettis*, Pac. J. Math., 22 (1967), 297–302.
[37] G. Mokobozki, *Balayage défini par un cône convexe de fonctions numériques sur un espace compact*, C. R. Acad. Sci. Paris, 254 (1962), 803–805.
[38] G. Mokobozki, *Principe de balayage, principe de domination*, Séminaire de Choquet 1, no. 1 (1962).
[39] L. Nachbin, *Elements of approximation theory*, Van Nostrand, Princeton, 1967.
[40] W. Orlicz, *Beiträge zur Theorie der Orthogonalentwicklungen II*, Stud. Math., 1 (1929), 241–255.
[41] R. R. Phelps, *Lectures on Choquet's Theorem*, Van Nostrand, New York, 1966.
[42] J. B. Prolla, *Approximation of vector-valued functions*, Mathematical Studies, vol. 25, North Holland, 1977.
[43] M. M. Rao, *Integration with vector valued measures*, Discrete Contin. Dyn. Syst., 33 (2013), 5429–5440.
[44] F. Riesz, *Sur les opérations fonctionelles linéaire*, C. R. Acad. Sci. Paris, 149 (1909), 974–977.
[45] A. P. Robertson and W. Robertson, *Topological vector spaces*, Cambridge Texts in Mathematics and Mathematical Physics, vol. 53, Cambridge at the University Press, 1964.
[46] A. P. Robertson, *On unconditional convergence in topological vector spaces*, Proc. R. Soc. Edinb. A, 68 (1969), 145–157.
[47] W. Roth, *A new concept for a Choquet ordering*, J. Lond. Math. Soc. (2), 34 (1986), 81–96.
[48] W. Roth, *Operator-valued measures and integrals for cone-valued functions*, Lecture Notes in Mathematics, vol. 1964, Springer, Heidelberg-Berlin-New York, 2009.
[49] W. Roth, *A Korovkin type theorem for weighted spaces of continuous functions*, Bull. Aust. Math. Soc., 55 (1997), 239–248.
[50] W. Roth, *Hahn-Banach type theorems for locally convex cones*, J. Aust. Math. Soc. A, 68, no. 1 (2000), 104–125.
[51] W. Roth, *Spaces of vector-valued functions and their duals*, Math. Nachr., 285, no. 8–9 (2012), 1063–1081.
[52] W. Roth, *Choquet theory for spaces of vector-valued functions on a locally compact space*, J. Convex Anal., 21, no. 4 (2014), 1141–1164.
[53] H. L. Royden, *Real Analysis*, Macmillan Publishing Company, New York/Collier Macmillan Publishing, London, 1980.
[54] W. Rudin, *Functional analysis*, McGraw-Hill Inc., New York, 1973.
[55] W. Rudin, *Real and complex analysis*, McGraw-Hill Inc., New York, 1974.
[56] H. H. Schäfer, *Topological vector spaces*, Graduate Texts in Mathematics, vol. 3, Springer, Heidelberg-Berlin-New York, 1980.
[57] F. G. Shi and C. L. Zhou, *New set-valued integral in a Banach space*, J. Funct. Spaces, 2015 (2015), 260238.
[58] M. Talagrand, *Pettis integral and measure theory*, Memoirs of the American Mathematical Society, vol. 307, 1984.
[59] J. A. Villasana, *Some Choquet theorems*, Rocky Mt. J. Math., 16, no. 2 (1986), 241–252.
[60] J. Voigt, *A course on topological vector spaces*, Compact Textbooks in Mathematics, Springer, 2020.
[61] A. Wilansky, *Functional analysis*, Blaisdell, New York, 1964.
[62] A. Wilansky, *Modern methods in topological vector spaces*, McGraw-Hill, New York, 1978.
[63] S. Willard, *General topology*, Addison-Wesley Publishing Company, Reading, Massachusetts, 1970.

Further Reading

[64] N. U. Ahmed, *Vector and operator valued measures as controls for infinite dimensional systems: Optimal control*, Discuss. Math., Differ. Incl. Control Optim., 28 (2008), 95–131.

[65] B. Anger and C. Portenier, *Radon integrals*, Birkhäuser, Boston-Basel-Berlin, 1992.

[66] R. F. Arens, *Representation of functionals by integrals*, Duke Math. J., 17 (1950), 499–506.

[67] G. Birkhoff, *Integration of functions with values in Banach spaces*, Trans. Am. Math. Soc., 38 (1935), 357–378.

[68] S. Bochner, *Integration von Funktionen deren Werte die Elemente eines Vektorraumes sind*, Fundam. Math., 20 (1933), 262–276.

[69] S. Bochner, *Absolut-additive abstrakte Mengenfunktionen*, Fundam. Math., 21 (1933), 211–213.

[70] S. Bochner and A. E. Taylor, *Linear functionals on certain spaces of abstractly-valued functions*, Ann. Math. (2), 39 (1938), 913–944.

[71] V. I. Bogachev, *Measure Theory*, vol. 2, Springer, Heidelberg-Berlin-New York, 2007.

[72] W. M. Bogdanowicz, *A generalization of the Lebesgue-Bochner-Stieltjes integral and a new approach to the theory of integration*, Proc. Natl. Acad. Sci. USA, 53 (1965), 492–498.

[73] C. Castaing and M. Valadier, *Convex analysis and measurable multifunctions*, Lecture Notes in Mathematics, vol. 580, Springer, Heidelberg-Berlin-New York, 1977.

[74] N. D. Chakraborty and S. Basu, *Integration of vector-valued functions with respect to vector measures defined on δ-rings*, Ill. J. Math., 55, no. 2 (2011), 495–508.

[75] P. J. Daniell, *A general form of integral*, Ann. Math., 19 (1917/18), 279–294.

[76] R. B. Darst, *A note on abstract integration*, Trans. Am. Math. Soc., 99 (1961), 292–297.

[77] R. B. Darst, *A note on integration of vector-valued functions*, Proc. Am. Math. Soc., 13 (1962), 858–863.

[78] N. Dinculeanu, *Sur la représentation intégrale des certaines opérations linéaires. I*, C. R. Acad. Sci. Paris, 245 (1957), 1203–1205.

[79] N. Dinculeanu and I. Klunánek, *On vector measures*, Proc. Lond. Math. Soc., 17 (1967), 505–512.

[80] R. E. Edwards, *Vector-valued measures and bounded variation in Hilbert space*, Math. Scand., 3 (1955), 90–96.

[81] R. Engelking, *General topology, Monografie Matematyczne*, Polish Scientific Publishers, Warszawa, 1977.

[82] R. K. Goodrich, *A Riesz representation theorem*, Proc. Am. Math. Soc., 24 (1970), 629–636.

[83] G. G. Gould, *Integration over vector-valued measures*, Proc. Lond. Math. Soc. (3), 15 (1965), 193–225.

[84] N. E. Gretsky and J. J. Uhl, *Bounded linear operators on Banach function spaces of vector-valued functions*, Trans. Am. Math. Soc., 167 (1972), 263–277.

[85] D. Han, D. R. Larson, B. Liu and R. Liu, *Operator-valued measures, dilations, and the theory of frames*, Mem. Am. Math. Soc., 229 (2014), 1075.

[86] J. Hoffman-Jørgensen, *Vector measures*, Math. Scand., 28 (1971), 5–32.

[87] R. B. Holmes, *Geometric functional analysis and its applications*, Graduate Texts in Mathematics, vol. 24, Springer, Heidelberg-Berlin-New York, 1975.

[88] J. L. Kelley and I. Namioka, *Linear topological spaces*, Graduate Texts in Mathematics, vol. 36, Springer, Heidelberg-Berlin-New York, 1976.

[89] D. H. Kim, *On integration of vector-valued functions for an operator-valued measure*, East Asian Math. J., 15, no. 2 (1999), 255–265.

[90] D. R. Lewis, *Integration with respect to vector measures*, Pac. J. Math., 33 (1970), 157–165.

[91] D. R. Lewis, *On integrability and summability in vector spaces*, Ill. J. Math., 16 (1972), 294–307.

[92] P. W. Lewis, *Extension of operator-valued set functions with finite semivariation*, Proc. Am. Math. Soc., 22 (1969), 563–569.

[93] D. McLaren, S. Plosker and C. Ramsey, *On operator valued measures*, Houst. J. Math., 46, no. 1 (2020), 201–226.

[94] M. Métivier, *On strong measurability of Banach-valued functions*, Proc. Am. Math. Soc., 21 (1969), 747–748.

[95] P. Meyer-Nieberg, *Banach lattices*, Springer, Heidelberg-Berlin-New York, 1991.
[96] S. Ohba, *Extensions of vector measures*, Yokohama Math. J., 21 (1973), 61–66.
[97] B. J. Pettis, *On Integration in vector spaces*, Trans. Am. Math. Soc., 44, no. 2 (1938), 277–304.
[98] W. Roth, *Bounds for linear functionals on convex sets*, Math. Nachr., 284, no. 17–18 (2011), 2272–2286.
[99] H. H. Schäfer, *Banach lattices and positive operators*, Die Grundlehren der Mathematischen Wissenschaften in Einzeldarstellungen, vol. 215, Springer, Heidelberg-Berlin-New York, 1974.
[100] B. Simon, *Convexity: An analytic viewpoint*, Cambridge Tracts in Mathematics, Cambridge University Press, Cambridge, 2011.
[101] G. F. Stefánsson, *Integration in vector spaces*, Ill. J. Math., 45 (2001), 925–938.

Index

https://doi.org/10.1515/9783111315478-006

De Gruyter Expositions in Mathematics

Volume 59
Friedrich Haslinger
The d-bar Neumann Problem and Schrödinger Operators, 2nd Edition, 2023
ISBN 978-3-11-118290-2, e-ISBN 978-3-11-118292-6

Volume 69/3
Mariusz Urbański, Mario Roy, Sara Munday
Non-Invertible Dynamical Systems. Volume 3: Analytic Endomorphisms of the Riemann Sphere, 2023
ISBN 978-3-11-076984-5, e-ISBN 978-3-11-076987-6

Volume 73
Sergei S. Akbarov
Stereotype Spaces and Algebras, 2022
ISBN 978-3-11-078086-4, e-ISBN 978-3-11-078091-8

Volume 72
Ali Baklouti
Deformation Theory of Discontinuous Groups, 2022
ISBN 978-3-11-076529-8, e-ISBN 978-3-11-076530-4

Volume 71
Victor P. Maslov, Yurievich Oleg Shvedov†
The Canonical Operator in Multiparticle Problems and Quantum Field Theory, 2022
ISBN 978-3-11-076238-9, e-ISBN 978-3-11-076270-9

Volume 70
Aron Simis, Zaqueu Ramos
Graded Algebras in Algebraic Geometry, 2022
ISBN 978-3-11-063754-0, e-ISBN 978-3-11-064069-4

Volume 69/2
Mariusz Urbański, Mario Roy, Sara Munday
Non-Invertible Dynamical Systems. Volume 1: Finer Thermodynamic Formalism – Distance Expanding Maps and Countable State Subshifts of Finite Type, Conformal GDMSs, Lasota-Yorke Maps and Fractal Geometry, 2022
ISBN 978-3-11-070061-9, e-ISBN 978-3-11-070269-9

www.ingramcontent.com/pod-product-compliance
Ingram Content Group UK Ltd.
Pitfield, Milton Keynes, MK11 3LW, UK
UKHW050953290926
15470UKWH00006B/1801

9 783111 314501